NMR in Glycoscience and Glycotechnology

New Developments in NMR

Editor-in-chief:
William S. Price, *University of Western Sydney, Australia*

Series editors:
Sharon Ashbrook, *University of St Andrews, UK*,
Bruce Balcom, *University of New Brunswick, Canada*
István Furó, *Industrial NMR Centre at KTH, Sweden*
Masatsune Kainosho, *Tokyo Metropolitan University, Japan*
Maili Liu, *Chinese Academy of Sciences, Wuhan, China*

Titles in the series:
1: Contemporary Computer-Assisted Approaches to Molecular Structure Elucidation
2: New Applications of NMR in Drug Discovery and Development
3: Advances in Biological Solid-State NMR
4: Hyperpolarized Xenon-129 Magnetic Resonance: Concepts, Production, Techniques and Applications
5: Mobile NMR and MRI: Developments and Applications
6: Gas Phase NMR
7: Magnetic Resonance Technology: Hardware and System Component Design
8: Biophysics and Biochemistry of Cartilage by NMR and MRI
9: Diffusion NMR of Confined Systems: Fluid Transport in Porous Solids and Heterogeneous Materials
10: NMR in Glycoscience and Glycotechnology

How to obtain future titles on publication:
A standing order plan is available for this series. A standing order will bring delivery of each new volume immediately on publication.

For further information please contact:
Book Sales Department, Royal Society of Chemistry, Thomas Graham House, Science Park, Milton Road, Cambridge, CB4 0WF, UK
Telephone: +44 (0)1223 420066, Fax: +44 (0)1223 420247
Email: booksales@rsc.org
Visit our website at www.rsc.org/books

NMR in Glycoscience and Glycotechnology

Edited by

Koichi Kato
National Institutes of Natural Sciences, Okazaki, Japan
Email: kkatonmr@ims.ac.jp

and

Thomas Peters
University of Luebeck, Germany
Email: thomas.peters@chemie.uni-luebeck.de

ROYAL SOCIETY OF CHEMISTRY

THE QUEEN'S AWARDS FOR ENTERPRISE: INTERNATIONAL TRADE 2013

New Developments in NMR No.10

Print ISBN: 978-1-78262-310-6
PDF eISBN: 978-1-78262-394-6
EPUB eISBN: 978-1-78801-128-0
ISSN: 2044-253X

A catalogue record for this book is available from the British Library

The Royal Society of Chemistry is a charity, registered in England and Wales, Number 207890, and a company incorporated in England by Royal Charter (Registered No. RC000524), registered office: Burlington House, Piccadilly, London W1J 0BA, UK, Telephone: +44 (0) 207 4378 6556.

For further information see our website at www.rsc.org

Printed in the United Kingdom by CPI Group (UK) Ltd, Croydon, CR0 4YY, UK

Preface

The carbohydrate chains displayed on proteins and lipids mediate a broad range of physiological and pathophysiological events on cell surfaces as well as inside cells. It is estimated that at least 80% of all proteins are glycoproteins. Glycosylation affects not only the physical but also the biological properties of proteins, *e.g.* functional protein–protein interactions or signaling. At present, the enormous complexity of glycosylation patterns precludes simple concepts that would explain the functional role of glycosylation. One of the largest challenges in this respect is the analysis of glycosylation patterns, which cannot be deduced directly from the genome. Glycosylation is orchestrated by a huge number of highly specific glycosyltransferases, and, therefore, glycan chains must be considered as secondary gene products. As a prominent example, glycosylation of humanized antibodies used in cancer therapy is of critical importance. Wrong glycosylation patterns may cause adverse immune reactions, and eventually anaphylactic shock. Therefore, the precise analysis of glycosylation patterns of humanized monoclonal antibodies used in therapy is of utmost importance. However, the complexity and heterogeneity of dynamic glycan structures often discouraged researchers from actively challenging and addressing this important issue. Recent advances in carbohydrate NMR spectroscopy alongside with rapid glyco-technological progress have delivered a fresh perspective on the potential of NMR for the analysis of glycan structures, glycan dynamics and glycan–protein interactions. Therefore, this volume should be timely and useful not only for NMR specialists but also for a broader scientific community ranging from the general field of structural biology to biochemistry and biophysics, to molecular and cellular biology, to materials science. Finally, NMR studies on glycans will have a significant impact on the development of vaccines, adjuvants, therapeutics, biomarkers and on biomass regeneration. In view of

New Developments in NMR No.10
NMR in Glycoscience and Glycotechnology
Edited by Koichi Kato and Thomas Peters

Published by the Royal Society of Chemistry, www.rsc.org

this situation, we have collected the most up-to-date NMR applications from well-known specialists in the fields of carbohydrate NMR spectroscopy.

Finally, we would like to thank Professor Masatsune Kainosho for inviting us to edit this book as part of the *New Developments in NMR* series. We also thank Dr Harriet Manning of Royal Society of Chemistry for her encouragement and assistance in putting the book together.

Thomas Peters
Koichi Kato

Contents

New Developments in NMR No.10
NMR in Glycoscience and Glycotechnology
Edited by Koichi Kato and Thomas Peters

Published by the Royal Society of Chemistry, www.rsc.org

CHAPTER 1

Intramolecular Hydrogen Bonding in Glycans in Aqueous Solution

MARCOS D. BATTISTEL[a], HUGO F. AZURMENDI[a] AND DARÓN I. FREEDBERG*[a]

[a]Laboratory of Bacterial Polysaccharides, Office of Vaccines Research and Review, Center for Biologics, Evaluation and Research, Food and Drug Administration, Silver Spring, Maryland 20903, USA
*E-mail: Daron.Freedberg@fda.hhs.gov

1.1 Introduction

Researchers in glycan structure determination spend a lot of time chasing their own tails. Because the molecular structures of glycans are more flexible than proteins and DNA, it is sometimes difficult to definitively delineate glycan "structure". A perusal of the literature will give one the impression that, for a given glycan in solution, multiple structural possibilities are generally predicted by modeling; however, it is challenging to determine which possibilities are the most reasonable due to ambiguities present in both experimental and modeling data. Thus, solution studies, especially NMR spectroscopy, should aim to provide far more definitive, quantitative parameters to improve model parameterization. Consequently, computer modeling can be more predictive than it is presently. Naturally, this will lead to a more detailed and complete understanding of the forces that dictate

New Developments in NMR No.10
NMR in Glycoscience and Glycotechnology
Edited by Koichi Kato and Thomas Peters

Published by the Royal Society of Chemistry, www.rsc.org

glycan structure and conformations. Ultimately, it may even provide a framework to understand and predict chemical reactivity of glycans. One extremely important structural parameter known to contribute to structural stability is hydrogen bonding (Hbonding) in aqueous solution and it is the focus of this chapter.

Hbonding is a vital weak inter/intra-molecular interaction, which is especially important in biological molecules. On one hand, Hbonds can be seen as a putative proton transfer; on the other, Hbonds can stabilize biological structural motifs in proteins, DNA and RNA without proton transfer.

The strength of Hbonds ranges from 1–40 kJ mol^{-1}.[1] Though strong for a non-bonded interaction, it is only a fraction of the strength of a typical covalent bond. Thus, in many biomolecules multiple Hbonds are used to increase the stability of different structural motifs in proteins and nucleic acids.

Hbonding is defined as an interaction between an electronegative atom and a hydrogen atom covalently bonded to an electronegative atom, such as fluorine, oxygen, nitrogen and sometimes carbon. The electrons in the X–H polar covalent bond are delocalized toward the former of the two electronegative atoms (Figure 1.1).

As stated previously, Hbonding is a recurrent structural feature found in proteins and nucleic acids. We were convinced that Hbonds are important in glycans too. Therefore, we sought evidence in support of Hbonding in oligo- and polysaccharides and the methods we developed may be useful in other applications as well. Hbonding in solution has only recently been directly detected in glycans using NMR.[2–7] Details of Hbonding motifs are different in proteins and nucleic acids; therefore, we reasoned that Hbonding in glycans may also differ from any other class of biomolecule.

Hbonds in glycans may differ due to flexibility, or conformational motion in biomolecules, which may cause Hbonds in glycans to be more transient than they are in proteins and nucleic acids. This in turn hampers detection of Hbonds in glycans. Furthermore, the Hbonds expected in glycans are likely to involve multiple OH groups. These are structurally different than those in proteins and nucleic acids where the NH-based Hbond is far more prominent and also dictates structure. Therefore, NMR pulse programs must be specifically tailored for glycan Hbond detection. These considerations drove us to first detect Hbonds in short helical peptides having metastable Hbonds,[8] and

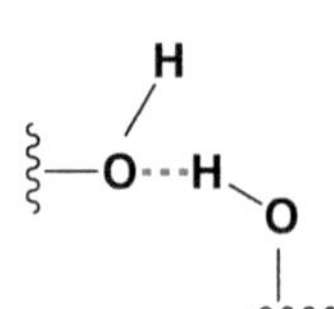

Figure 1.1 Hydrogen bond scheme of a molecular fragment bearing an OH group. The dashed red line highlights a hydrogen bond where electron density on the oxygen atom is donated to a hydrogen atom covalently bonded to a different oxygen atom. The latter oxygen atom may be a water molecule as well.

consequently metastable structure, before attempting to detect metastable Hbonds mimicking those in glycans. X-Ray data for Hbonding in oligo- and polysaccharides can be found in Rao *et al.*[9] and Jeffrey and Saenger.[1] In this chapter, we limit the discussion to Hbonding in solution.

In oligosaccharides, individual residues are linked between the anomeric carbon (C1) of one residue and virtually any other carbon of the next unit *via* an oxygen atom. Three-dimensional conformation is defined by the torsions ϕ and ψ (Figure 1.2). The anomeric signal isolation makes these resonances useful as NMR "handles". Thus, they provide valuable information for sequential assignment. However, because anomeric proton signals commonly overlap with the water resonance, glycan NMR experiments have traditionally been carried out in D_2O as opposed to H_2O. In early glycan NMR studies, presaturation was used to remove residual ^1H signals in the HOD peak. Although these procedures enable the observation of anomeric protons, information on all exchangeable protons in the molecule, such as OHs and NHs, is lost through chemical exchange with the water, whose magnetization is saturated. As we will see later, these exchangeable protons could otherwise be used to gather structural data.[10,11] More recently, water suppression in NMR spectra can be achieved by ensuring that water (H_2O, as opposed to D_2O) magnetization remains or ends up on the *z*-axis and, therefore, with the help of gradients, is not detected. Figure 1.3 shows a comparison of sucrose spectra with no water suppression, presaturation and modern water suppression.

1.2 NH Hydrogen Bonding

Hbonding from amide NH groups in proteins has been known for many years. In solution, it was usually inferred from H/D exchange in solution NMR spectroscopy or temperature-coefficients of the amide NH chemical shifts.

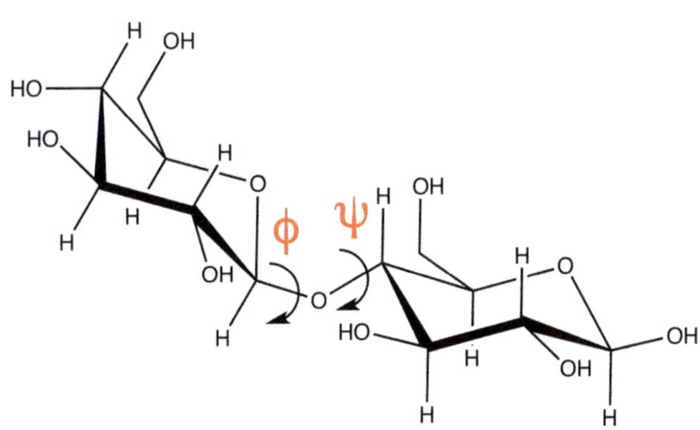

Figure 1.2 A model disaccharide highlighting that the two monosaccharide residues are linked through an oxygen atom, which on one side is bonded to an anomeric carbon of one glycan residue and the other is bonded to another carbon, in this case C4, of the other glycan.

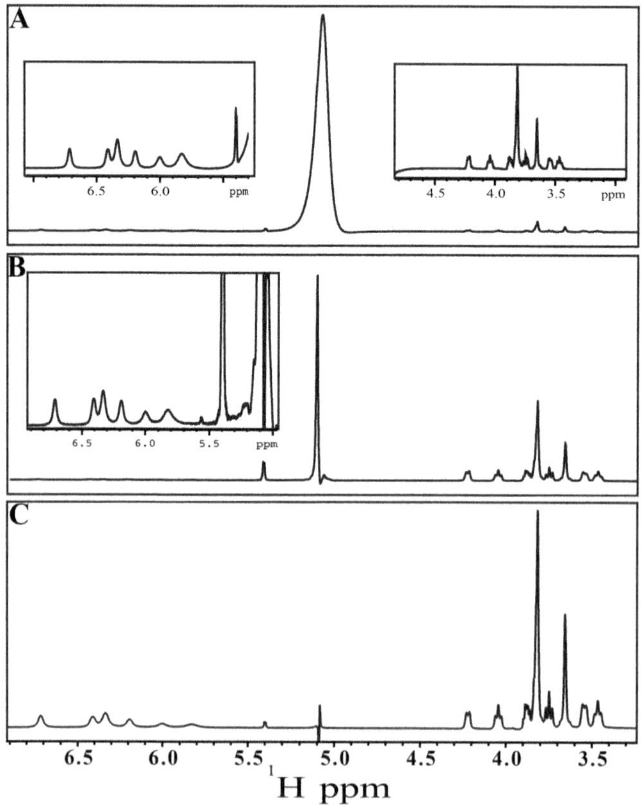

Figure 1.3 ^{1}H NMR spectra of a 300 mM sample of sucrose dissolved in H_2O (with 10% D_2O for locking purposes) taken at −10 °C. The spectra resulting from three NMR experiments with the overall goal of reducing the water resonance are shown: Panel **A.** No water suppression, demonstrating that direct observation of any ^{1}H's, especially those of the OH groups is nearly impossible in ~55 M H_2O; Panel **B.** Water presaturation, where the C–H resonances are observable and the OHs are somewhat observable, but very much attenuated relative to the C–H's; and Panel **C.** water suppression using a P3-9-19 WATERGATE scheme showing that the water resonance is completely suppressed and the OH's are observable and are on the same scale as C–H's.

In this experiment, solvent exposed NH groups readily exchange with water.[12,13] Therefore, the temperature coefficient of NH residues' chemical shift under these conditions closely mimics that of water (~0.09 ppm per °C), unless the NH group is involved in Hbonding. In contrast to solvent exposed NHs, the exchange rate of Hbonded NH groups is slowed. Although the temperature coefficient method is useful, direct evidence of Hbonds would be more desirable.

In 1999 Cordier *et al.* and Wang *et al.* reported the direct detection of Hbonds and measurement of $^{3h}J_{NC'}$ couplings in fully ^{15}N,^{13}C enriched proteins.[2,6] This allowed direct spectral assignments of Hbond donors and acceptors. The NMR spectral correlation is due to N–H···O=C orbital overlap of amide NH with the complementary carbonyl oxygen's orbitals. Similar data were reported for nucleic acids by Dingley *et al.*, though not for amide NHs.[3] Significantly, this method facilitated the measurement of coupling constants in these Hbonds, which were found to be in the range of 0.2–1.0 Hz. Later, the couplings were shown to correlate to Hbond distances from crystal structures and indicate Hbond strength.[7] This showed that Hbonding results in orbital overlap and that it could be detected in aqueous solutions of proteins and nucleic acids, even when the *J*-coupling is small.

Widmalm and co-workers predicted that the structure of the *E. coli* O142 polysaccharide (PS) should contain a potentially detectable level of Hbonding. These authors based their predictions on experimentally determined NOE's and restrained MD simulations of the pentasaccharide, which is the repeat unit of the O142 PS.[14] The Hbonds were predicted to extend from three NH groups in GalNAc-**A**, GalNAc-**B** and GalNAc-**C** rings to oxygen atoms. However, the Hbond with the highest percentage was predicted to be present in only 20% of the pentasaccharide structures by MD, while the others were predicted to be present in smaller amounts.

Inspired by the results in proteins, Norris *et al.* attempted to detect similar correlations in the *E. coli* O142 PS.[15] Unfortunately, direct (through-bond) evidence for the Hbond in this PS was difficult to demonstrate, probably due to exchange broadening combined with a low percentage of amides NH's in the Hbonded state. Nevertheless, these authors reported a measurable temperature dependence of the NH chemical shifts of GalNAc-**A**, GalNAc-**B** and GalNAc-**E** in the O142 PS. This temperature dependence differed from that of H_2O and supported the presence of Hbonding for these three amide NH groups.

Another metric used in the O142 study was the temperature-dependence of the heteronuclear NH *J*-coupling. In principle, this works like the temperature-dependence of the NH chemical shift. Similar to chemical shifts, coupling constants are averaged as a function of temperature. While the ^{1}H is localized on the ^{15}N, the $^1J_{NH}$ coupling reflects the average of the coupling of the ^{1}H to the ^{15}N while bound to ^{15}N (~92 Hz) and the value when it is not bound to ^{15}N (0 Hz). As the temperature rises and ^{1}H exchange with water becomes faster, the ^{1}H lifetime on the ^{15}N is shorter hence the averaged coupling is reduced. In an Hbonded NH group, the $^1J_{NH}$ would be expected to change little as a function of temperature because Hbonding would preclude it from exchanging with water. The $^1J_{NH}$ temperature dependence confirmed that GalNAc-**A** and GalNAc-**B** are Hbonded for at least part of the time, while GalNAc-**C** is not, allowing the authors to assign the reduced temperature coefficient displayed by the NH in GlcNAc-**E** to chemical (conformational) exchange rather than Hbonding.

The $^1J_{NH}$ coupling method is more sensitive than chemical shift temperature coefficients to the on–off rate of amide NH's and may therefore be more useful to indirectly deduce Hbonding in glycans. As stated previously, in some cases, the $^1J_{NH}$ coupling method may also help in distinguishing Hbonding from conformational motion, both of which modulate the chemical shift from NH exchange with water.

Although Hbonding in the O142 PS was not directly demonstrated, the difficulties were not trivial, as conformational motion, PS heterogeneity and potentially low Hbonded populations likely contributed to the direct detection of HN···C=O correlations in the O142 PS. Furthermore, isotopic labels did not result in the types of correlations observed by the groups of Grzesiek, Feigon, Barfield, Torchia and Bax for nucleic acids and proteins.[2,4,5]

However, in 2012, Battistel *et al.* reported the direct observation of three NH-based Hbonds in a tetramer of α2-8 *N*-acetylneuraminic (sialic) acid, SiA_4.[16] The studies began in a similar vein to the O142 PS studies, by attempting to detect indirect evidence of NH Hbonding. First, a ^{13}C,^{15}N fully labeled SiA_4 was prepared. Then, good chemical shift dispersion in the NH region was achieved at reduced temperatures (below 15 °C). Subsequently, measurements of NH/ND exchange rates using the SOLEXSY pulse sequence[17] showed that three of the four NH's exchanged with H_2O at a rate constant of approximately 1 s^{-1}, while the remaining NH exchanged with a rate constant of 3 s^{-1}. This strongly suggested that three of the four NHs are Hbonded at −10 °C.

Still, the results did not provide direct evidence for Hbonding in SiA_4. Battistel *et al.* then used an NMR pulse sequence akin to the HNCO devised by Grzesiek and co-workers,[2] but with minor changes to optimize the experiment for the chemical shifts in glycans, because, as stated earlier, they differ in structure from proteins and nucleic acids. At first, no direct correlations were observed because the ^{13}C carrier frequency had been set to the C=O region. This suggested that the Hbonded NHs, if present, were not bonded to C=O groups. In hindsight, this may not be surprising: solution Hbonding patterns in glycans are not well established and these patterns may differ from those in proteins and nucleic acids. The ^{13}C carrier was then set to multiple different values (100 ppm, 80 ppm and 60 ppm) in order to ensure coverage of the entire ^{13}C spectrum. The result was a correlation from the ^{15}N of the *N*-acetyl groups in residues I, II and III to the C8 of the same residue (Figure 1.4(A)). Although this correlation peak did not incontrovertibly dictate that Hbonds were present in residues I, II and III of SiA_4, they did strongly support their presence and the notion that the NH's are Hbonded to the glycosidic oxygen atoms (O8) in each of these units. Hbonding from the NH···O8 in SiA_4 was further confirmed by an experiment similar to the same HNCO. In this experiment a direct correlation from $NH_i \rightarrow C2_{i+1}$, where $i = 2, 3$, was detected. Thus Hbonding in α2-8 SiA_4 from the NH → O8 in residues I, II, and III was established at −10 °C.

A simple calculation of the structure with only these Hbonding restraints led to a left-handed helical structure where the helix contains two residues per turn. The helix was thus dubbed a 1_2 helix. Interestingly, the torsional angles in this structure violate the exo-anomeric effect.[18] Were the exo-anomeric effect to play a major role in structure for SiA_4, a four-residue per turn helix with a larger diameter would be expected. While the Hbonds and many of the NOE's supported both structures, the mentioned 1_2 helix and a similar helix and a 1_4 helix, the calculated NOEs from each of the computer models for the 1_2 helix and 1_4 helix showed that over 25 of the experimental NOEs are in disagreement with the 1_4 helix structural model, while no such disagreements were found for the 1_2 helix. Models of the resulting structure extended to 10 residues are shown in Figure 1.4. This helix is in agreement with research published by Yongye *et al.* who found that SiA_3 yielded NOEs and torsions that, together with calculations, predicted a subpopulation of structures that contain a 1_2 helix.[19] The left-handed helix and inter-residue angles are also consistent with crystallographic results by Schulz *et al.* where SiA_3 and SiA_5 are found in a helical conformation in the active site.[20]

Interestingly, the obtained helical solution disagrees with crystallographic results of SiA octamers[21] as well as other crystallographic results of SiA oligomers (see Figure 13 in Nagae *et al.*). It is not yet clear whether the structural features in the crystallographic studies or those found in solution are more relevant for biological function.

These results have significant implications for three-dimensional glycan structure. First, this Hbonding may help in restricting motion about the glycosidic torsions. Second, even short oligomers can adopt a three-dimensional fold, in this case, it is a helical fold reinforced by NH···O Hbonding, where the oxygen is glycosidic. Third, even if not completely rigidified at higher temperatures, this helical conformation is likely to be a conformation that is found as part of an equilibrium mixture of conformers. Finally and most importantly, the Hbonding in SiA_4 may be more general and may aid in explaining some PS structure–function relationships, especially in PS-antibody (Ab) binding.

1.3 OH Detection and Hydrogen Bonding

Amide NH's exchange relatively slowly and can therefore be used as handles in many experiments. However, not all glycans contain NH groups, and consequently Hbonding in such glycans cannot be studied as already discussed. Hbonding in glycans can also occur *via* the various OH groups in glycans. The oxygen in water has tetrahedral bond angles if the lone pairs are considered. Glycans can be likened to water on a carbon stick. Like the oxygen in water, the carbon atoms in glycans are tetrahedral. This allows the OH groups to easily fit into the water's tetrahedral Hbonding structure. Consequently, the OH groups in glycans are extremely similar to H_2O's OH groups and exchange is more facile than for many other exchangeable hydrogen atoms. In a study of these OH groups, Battistel *et al.* demonstrated that with appropriate water suppression, Hbonding in glycan OH groups could be directly observed.[11,22,23]

The key to observing these OH Hbonds is a combination of modern water suppression (p3919 WATERGATE) combined with reduced temperatures to slow down the OH exchange rate with H_2O. While OH group resonances

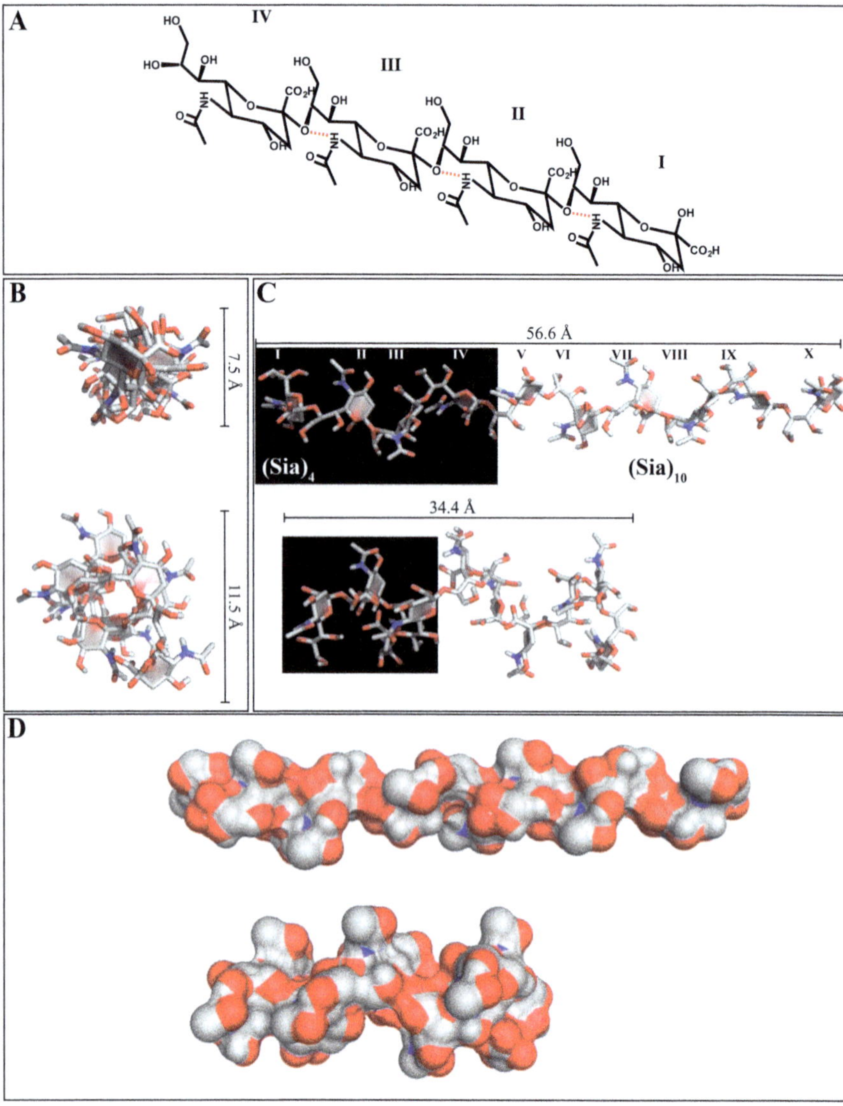

Figure 1.4 NH-based hydrogen bonds in SiA_4 (panel **A**) in a chemical structure showing the hydrogen bonds detected in the tetramer extend from the NH of a given SiA residue to a glycosidic oxygen and forms a 7-membered ring. Panel **B**. A view down the long helical axis of the two possible helices, 1_2 and 1_4. It is easy to observe the different helical diameters (7.5 Å *vs.* 11.5 Å) as well as the resulting space inside the helices. Panel **C**. Side-on view of stick models showing the two possible helical models for SiA_4, a 1_2 and 1_4 helix. The area shown against the black square highlights SiA_4 obtained from NMR experimental data with each model. The remainder is obtained using the tetramer's parameters to predict the structure of a decamer using either the 1_2 and 1_4 model. Panel **D**. Space filling representations of the 1_2 and 1_4 helical models.

could be observed at ambient temperatures, the lines were broad in aqueous solution and consequently, correlations were difficult to observe under these conditions.

Batta and Kover showed that C–H resonances could be correlated to the attached OH resonance and the water resonance could be well suppressed using a WATERGATE pulse sequence element with gradients.[24,25] Triple axis gradients provide significantly enhanced water suppression. In studies, carried out using 1.7 M sucrose at −3 °C, these authors were able to assign the OH resonances using a ^{1}H–^{1}H TOCSY sequence. Although these studies broke the barrier for assigning OHs, there are approximately 7 molecules of water for each sucrose molecule dissolved in solution at these high concentrations.[26] Ideally, one would like to detect and assign the OH groups at lower concentrations.

Battistel *et al.* showed that an HSQC–TOCSY (10 ms DIPSI-2 mixing time) improved resolution in the indirect dimension by using ^{13}C chemical shifts at 300 mM sucrose in a conventional NMR probe.[11] This alleviates spectral overlap in the crowded ^{1}H spectral regions and thus improves the ability to assign OH groups in glycan NMR spectra. Assigning the OH groups now opens the door to the ability to use the OH groups as handles in NOE experiments as well as in magnetization transfer experiments, however, at the present time, it is not easy to see how the latter could be accomplished.

Battistel *et al.* then showed that the HSQC–TOCSY with WATERGATE could be extended using a longer mixing time during the TOCSY portion (30 ms DIPSI-2) facilitating magnetization transfer across Hbonds in glycans, when present.[27] The sucrose Hbond detected by Battistel *et al.* with the HSQC–TOCSY using WATERGATE, a through-bond *J*-coupling experiment. The H1f → C2G and H2g → C1f Hbonds (Figure 1.5) were the ones that had

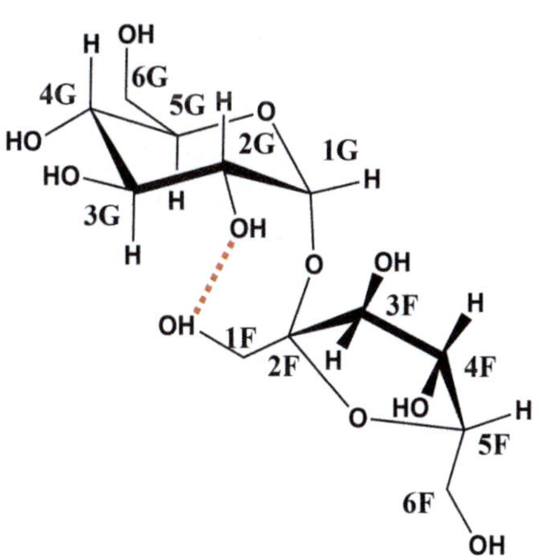

Figure 1.5 Structure depicting the H bonds detected in an aqueous solution of sucrose. Two correlations were detected: H1f → C2G (from the OH group of F1 to the carbon of G2) and H2g → C1F (from the OH group of G2 to the carbon of F1), indicating the presence of a flip-flop H bond.

gone undetected for many decades. While present in the neutron structure,[28] its existence had not yet been directly detected in solution.[10,29–32] The low signal-to-noise ratio (SNR) for this cross peak suggests one of two scenarios. In the first, line broadening of the OH groups hampers significant magnetization transfer due to relaxation. In the second, a small population of sucrose is Hbonded, leading to a low SNR for this peak. The latter is more consistent with molecular dynamics, thus Battistel *et al.* ascribed this low SNR to the low population of the Hbonded form. In both scenarios, the Hbonded form may be in rapid exchange with water and would therefore be challenging to detect.

1.3.1 Measuring OH Exchange Rate Constants for Inferring Hydrogen Bonding

At higher temperatures the OH resonances broaden and become difficult to detect and also to correlate. One can envision that reduced exchange OH rates can still imply the presence of Hbonds, even when they are not directly detectable using HSQC–TOCSY. As mentioned in an earlier section, amide NH exchange rates in disordered proteins were recently published using SOLEXSY.[17] Based on earlier studies of OH and NH exchange at reduced temperatures in mixed solvents,[10,30,33–39] we believed that this idea could be applied to glycans to measure OH/OD exchange rates in water without a co-solvent.

In a 1:1 mixture of H_2O/D_2O, only 50% of the OH groups are deuterated. This leads to two resonances in the carbon spectrum, one for a ^{13}C atom that is attached to an OD group and thus shows a deuterium isotope effect and the other that has the chemical shift of a ^{13}C atom that is attached to an OH.[40,41] Although the OH groups are exchanging with the OD groups, the exchange rate constant is slower than the frequency difference between the ^{13}C–OH and the ^{13}C–OD. The exchange rate can be measured from one-dimensional ^{13}C spectra (Figure 1.6).

A variant of the HSQC–TOCSY, called INTOXSY (Figure 1.7(A)), can be used to measure the exchange rate constants.[42] In this variant, magnetization is transferred from ^1H to ^{13}C, allowed to evolve in ^{13}C and then transferred back to ^1H, as in an HSQC. Then, the ^1H magnetization is refocused from anti-phase back to in-phase magnetization. This in-phase ^1H magnetization is then placed on the *z*-axis and a variable delay time is incorporated for H/D exchange which is followed by a 90° pulse and DIPSI-2 mixing to transfer ^1H magnetization between the C–H and the OH groups. ^{13}C atoms bearing an OH now become detectable through the **H**–C to **H**–OC coupling. ^{13}C atoms bearing an OD group are not detectable. However, as in SOLEXSY, the cross-peak intensities/volumes of ^{13}C atoms that, before the variable delay time, bore an OD but after the delay bear an OH are seen to be exponentially increasing in intensity at the end of the variable delay period. The cross-peak intensities/volumes of ^{13}C atoms that before the variable delay time bore an OH and now bear an OD are seen to be exponentially decreasing in intensity at the end

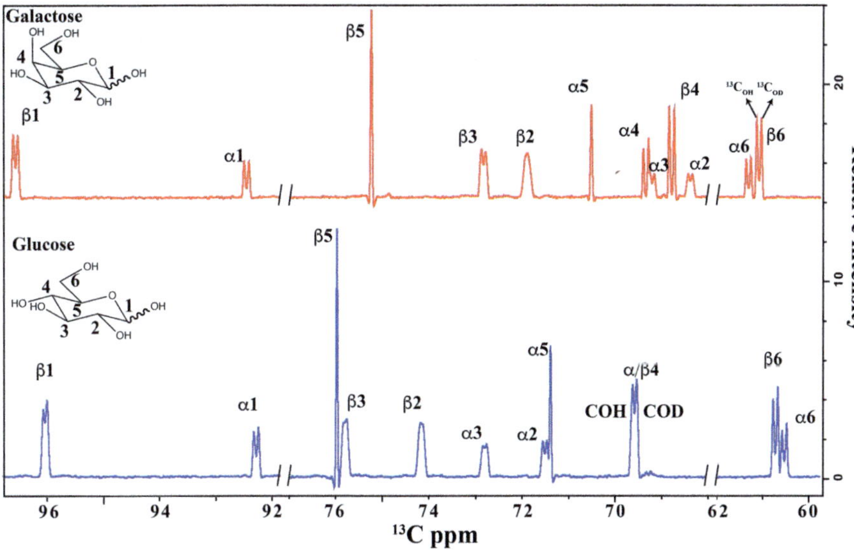

Figure 1.6 One dimensional ^{13}C NMR spectra of galactose (top) and glucose (bottom) in a solution of 1 : 1 $H_2O:D_2O$. The spectra clearly show the effect of isotopic deuterium substitution, which in many cases produces two chemical shifts, one for C–OH (less shielded) and the other for C–OD (more shielded). Using this spectrum, the ^{13}C spectrum and two additional spectra (one obtained in 100% D_2O, the other obtained in 100% H_2O) allows for the use of line shape analysis to deduce OH/OD exchange rate constants.

of the variable delay period. ^{13}C atoms that bore an OD and exchanged for an OH also contain a ^{1}H T_1 component, which can be fit to a bi-exponential function that fits both exponential rates (Figure 1.7(B)).

The INTOXSY NMR experiment described here provides data that can be used to infer Hbonding, even in cases where observation of Hbonds cannot be directly observed or where the correlation is weak. One such example is a recently published study of sialyl Lewis-X (sLe^X). Here OH groups of two glycan residues, Gal{II} and Fuc{III}, which surround a central residue, GlcNAc{II}, form Hbonds with the *N*-acetyl group, causing the *N*-acetyl to rapidly flip between the two OH groups, Fuc{III}OH2 and Gal{I}OH2. This facilitates Hbonding between the carbonyl of the *N*-acetyl group and these two OH groups. Evidence for *N*-acetyl flipping conformational motion was first found by NMR temperature dependent linewidths. However, the details were revealed by MD simulations and Hbonding was further supported by INTOXSY measurements.[42] OH/OD exchange rate constants were $<7\ s^{-1}$, confirming that these OHs are protected from exchange with solvent. OH/OD exchange rates were used in a recent intermolecular complex of Lewis-X hydroxyl groups.[41]

Presently, it is difficult to estimate the strength of intramolecular Hbonds in glycans. While it is possible to use the populations available from MD

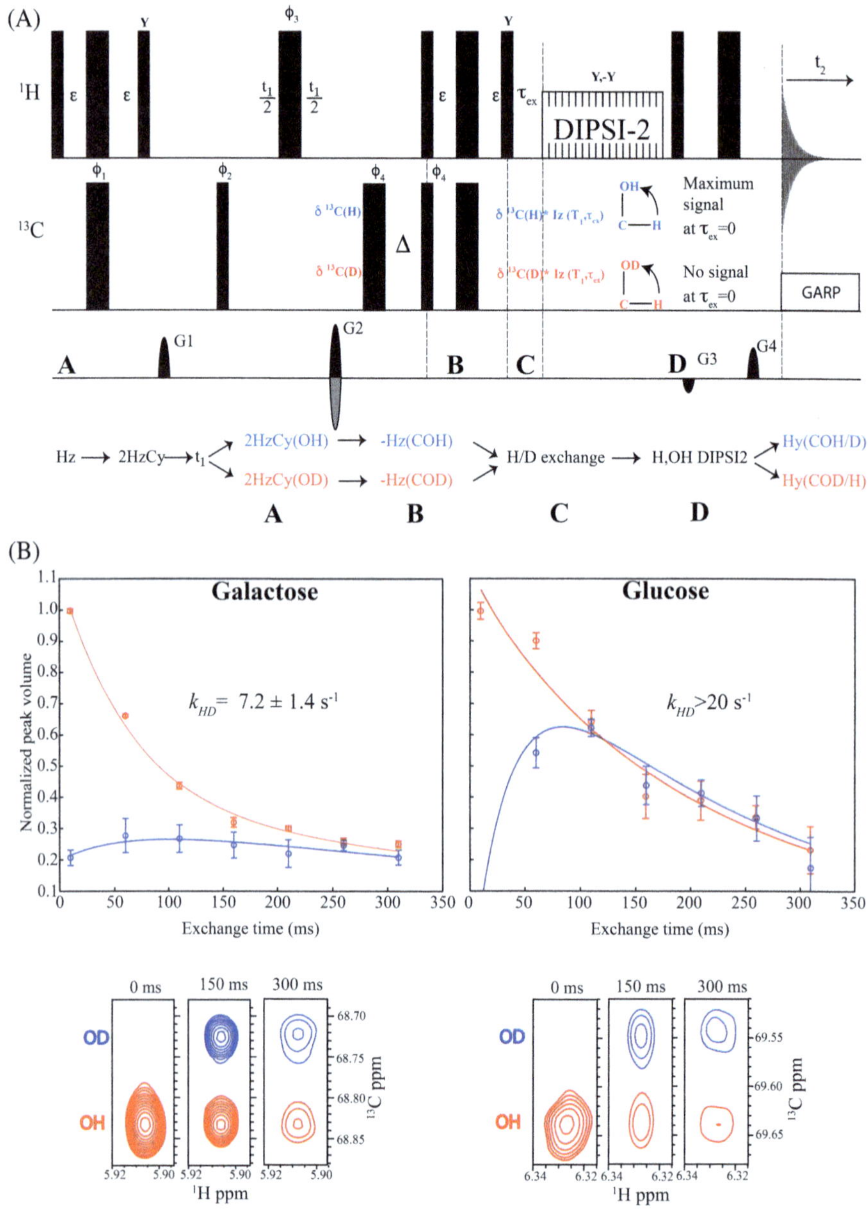

Figure 1.7 Panel **A.** INTOXSY pulse sequence highlighting the most salient features of the magnetization transfer pathway. The delay ε was set to 1.72 ms. The exchange time (Ex) was varied from 0 to 250 ms in 10, 20 or 50 ms increments. DIPSI-2 mixing time was set to 10 ms. ^{13}C atoms frequency labeled when bearing OH are depicted in blue whereas ^{13}C frequency labeled when bearing OD are in red. The pulse sequence is separated in sections: (a) HSQC; (b) Refocusing ^{1}H magnetization; (c) Exchange time and (d) ^{1}H,^{1}H TOCSY transfer. Panel **B.** Plots showing that OH/OD exchange and ^{1}H T_1 modulate isotope-discriminated peak intensity, and yield the resulting INTOXSY exchange rate profiles.

simulations, it is widely accepted that the best estimates of ground state energies are accurate to about ±1 kcal mol^{-1}. Given this level of uncertainty, it is difficult to say whether small populations of Hbonded forms predicted by MD simulations are present in the same amounts in solution.

One parameter that would help in this regard is *J*-coupling ($^{1}H–^{13}C$ and $^{1}H–^{1}H$) because the magnitude of the *J*-coupling could lend insight into stability and strength of the Hbonding interaction. O'Leary and co-workers reported some of these *J*-couplings in model compounds and found that they range from 0.15 to 0.35 in these compounds.[43,44] These small values imply that *J*-values may be challenging to measure, especially in cases of smaller populations of H-bonded glycans.

1.4 Hydrogen Bonding Involving C–H Bonds as Donors

As stated in the introduction, C–H Hbonds are controversial for a few reasons. First, it is unknown if they truly contribute to stability of biomolecules. Second, they are weaker than many Hbonds and finally, it is hard to see how they fit into the general definitions of Hbonds. Since the active H atom in the Hbond is attached to a carbon atom, it is not intuitive to predict when these C–H's will be involved in Hbonding.

Nevertheless, such C–H based Hbonds have been reported in crystal structures[1,9] and in solution.[45] Interestingly, these C–H Hbonds are from Hα in proteins, which can be said to be slightly acidic as they are in the α position to a carbonyl group. However, the carbon atom in a C–H group is not typically thought of as an electronegative group. Furthermore, the hydrogen atoms in these C–H Hbonds are usually not easily exchanged or transferred under biologically relevant solution conditions. It is with this preface that we present studies that identified and demonstrated evidence for C–H Hbonds in Lewis-X (Le^{X}) based antigens.

In 2013, Zierke *et al.* published a study of the Le^{X} trisaccharide conjugated to a protein, which predicted the existence of an Hbond between the H5 of Fuc{II} and the O5 of Gal{III}.[46] In structural studies, Gal{III} appeared to be situated directly above the H5 of Fuc{II}. This hydrogen displays an unusually deshielded chemical shift and was the basis of the authors' proposal.

To confirm the proposed Hbond, Zierke *et al.* covalently attached Le^{X} to a protein. Presumably, now that the trisaccharide is part of a larger molecule, molecular tumbling would be slowed, which should facilitate NOE measurement in the slow tumbling limit. These measurements provided structural data consistent with an Hbond, but direct NMR evidence of direct orbital overlap (*J*-coupling) was not reported in this paper. Nevertheless, the NOEs provided validation of the structural proximity of Fuc{II}H5 to Gal{III}'s pyranose oxygen (Figure 1.8).

Interestingly, Pederson and Prestegard reported that Le^{X} binds to DC-SIGN in two slightly different conformations. One where H5 of Fuc{II} is closer to the pyranose oxygen of Gal{III} and the other when H5 is closer to

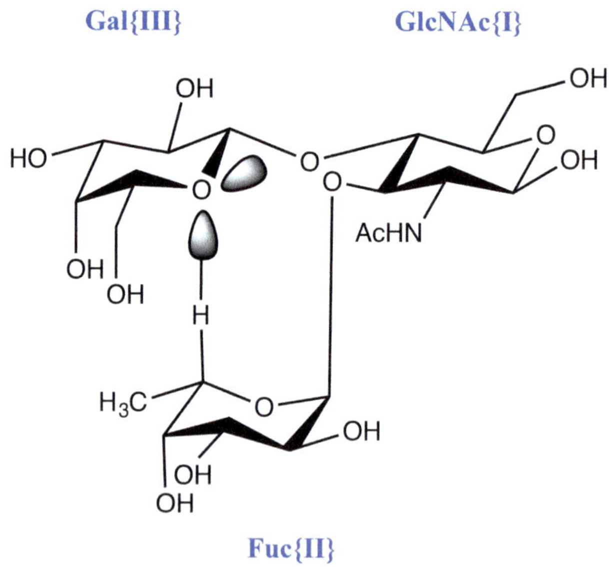

Figure 1.8 Structure of the Lewis-X trisaccharide, highlighting the H bond predicted by Zierke *et al.* from Fuc{II}H5 to Gal{III}O5. Direct NMR evidence for this H bond was observed in the analogous pentasaccharide Sialyl Lewis-X.[48]

the glycosidic oxygen between Gal{III} and GlcNAc{I};[47] both conformations are consistent with NOEs.

Recently, the existence of the CH···O Hbond in sLe^X, a Le^X analog, was shown by NMR spectroscopy.[48] In this report Battistel *et al.* showed direct through-bond correlations from Fuc{III}H5 to Gal{IV}C1 (these are the same as Fuc{II} and Gal{III}, respectively in the Le^X trisaccharide). Additional correlations from Fuc{III}H5 to GlcNAc{II}C4 and Gal{IV}H1 to Fuc{III}C5 were also observed.

Altogether, the NMR correlations in Battistel *et al.*'s paper support Zierke's original hypothesis of an Hbond from Fuc{III}H5 to Gal{IV}O5. In addition, Battistel *et al.*'s report also supports Pederson and Prestegard's NOE based hypothesis of an additional Hbond from Fuc{III}H5 to GlcNAc{II}O4.

1.5 Hydrogen Bonding Between OH and Carboxylate Groups

The cases discussed so far are representative of the typical situation in glycans where Hbonds involving OHs have relatively small interaction energies, thus representing conformations that are not as populated as others. Despite that disadvantage, Hbonds could be detected by means of through-bond NMR experiments. An unexpected reversal of situation was found when studying the structure adopted by free α(2-8) PSA oligosaccharides in solution.[49] Because historically the focus was on characterizing the, arguably more important, PSA inter-residue α-linkages, little effort was dedicated into looking at the reducing end, which is mostly found in the β-configuration (90% in the monomer, 97% in the dimer and higher oligomers). Azurmendi *et al.* report results from MD showing that, thanks to the axial orientation of the anomeric OH, the latter forms a stable Hbond with the carboxylate oxygen of the following residue (red dashed line in Figure 1.9(A)), representing over 90% of the MD population (Figure 1.9(B), bottom panel). Direct evidence was obtained using an HSQMBC.[50] Additionally, three independent lines of measurements provide the required evidence in support of both, the Hbond and the MD predicted structure for the first two residues at the reducing end of α(2-8) PSA OS: OH exchange rate values, HSQC–ROESY experiments, and residual dipolar couplings (RDCs).

OH exchange rate constants in SiA OS have been recently published.[51] The exchange rate for OH2 in the dimer is 3.4 s^{-1}, about half that of the next low value and an order of magnitude slower than the faster exchanging OHs. The trend is mostly conserved for longer OS of PSA. As mentioned previously, through-bond NMR experiments confirm the Hbond and both MD and the exchange rate indicated that HO2 was Hbonded we also performed ROESY experiments under varied conditions trying to correlate the HO2 with predicted nearby CHs through-space. Figure 1.9(B) shows specific predictions from the MD trajectory, as there is a one to one correspondence between SiA{I}HO2–SiA{II}O1 Hbond condition (bottom panel) and the generally accepted ROE observation limit of 5 Å. In other words, observation of ROESY correlations for the atoms indicated in the top three panels in Figure 1.9(B) would imply the presence of the predicted Hbond. In fact, as reported in Azurmendi *et al.*, the HSQC–ROESY experiments show correlations between all the atoms indicated in Figure 1.9(A). The same report shows that experimental RDC values measured at room temperature are well reproduced for the MD trajectory using SVD, thus supporting the notion that the Hbond is still present at relatively high temperatures as predicted by MD. Finally, the Hbond was confirmed through a long-range coupling measurement.

This Hbond could help explain the unusual behavior of the free Sialic acid oligomers, as suggested by Jennings and co-workers.[52] It also may be the cause of the shift in the $\beta:\alpha$ equilibrium ratio of the reducing ring on going from monomer (90 : 10) to dimer (97 : 3) and higher order oligomers.

1.6 Future Directions

Clearly, Hbonding in glycans is an important area of research. Because they can provide additional data, structures derived from data indicating Hbonding are likely to be more precise than those that do not utilize Hbonds. Direct detection of Hbonding can also provide new data for a more complete theory of glycan structure.

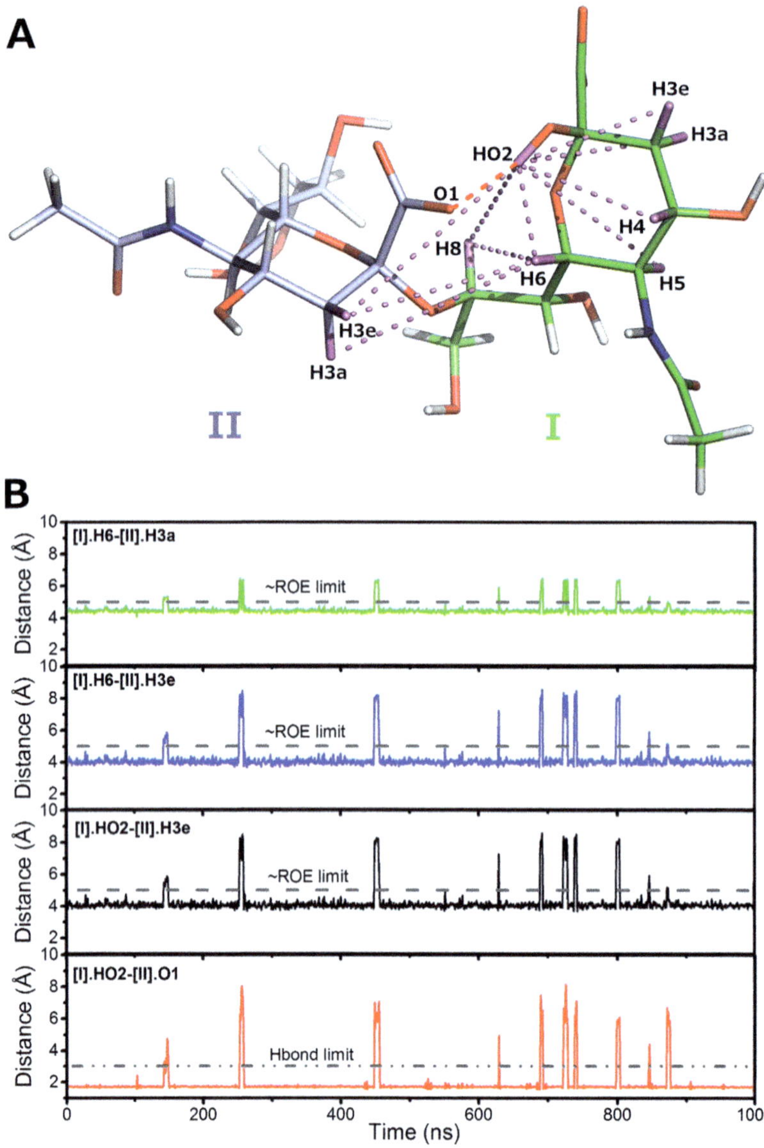

Figure 1.9 H bonding in SiA_2 between OH2 of residue 1 and the CO_2^- of residue 2. Panel **A.** Dotted lines showing ^1H–^1H internuclear cross-peaks, which indicates close internuclear distances, as obtained from ROESY experiments on this molecule in H_2O. Panel **B.** plots of internuclear distances as a function of a 1 μs trajectory showing the effects of Hbond breaking on internuclear distances. When the Hbond breaks HO2–O1′, HO2–H3e′, H6–H3e′ and H6 to H3a′ all increase beyond the distance that would be detectable by ROESY measurements. Note that the times while the H bond is broken are short compared to the overall trajectory.

The importance of *J*-coupling in establishing Hbonding should not be underestimated. While direct detection of Hbonding was not in the O142 PS, in the remaining examples given here, a finite long-range *J*-coupling between atoms in an Hbond and a remote ^{13}C was used to directly establish Hbonding in sucrose, SiA_4, SiA_2 and sLe^X.

It is likely that, until regular Hbonding patterns can be established, through bond *J*-coupling will continue to be the method of choice to establish Hbonding in glycans. Naturally, this raises the issue of the *J*-coupling values in these Hbonds, which have so far been elusive in aqueous solution. However, the magnitude of the *J*-coupling may shed light on the strength of Hbonds in glycans. These will be important targets in future studies. O'Leary and co-workers have already begun to pave the way for these studies.[43,44]

As can be seen in the studies presented, the direct detection of Hbonds can occupy an NMR spectrometer for up to one week. Future studies of Hbonding glycans may be facilitated by more sensitive detection, such as the ever-increasing sensitivity of NMR cryoprobe technology and higher magnetic field strength. More importantly, higher magnetic fields may allow the detection and study of Hbonded glycans at higher temperatures due to increased resolution (larger number of Hz in one ppm) at higher magnetic fields.

The studies presented here show that Hbonding glycan can be detected and that these data have important consequences for structure determination in glycans. It is our hope that Hbonding will continue to be a source of new structural data in glycan structure determination.

References

1. G. A. Jeffrey and W. Saenger, *Hydrogen Bonding in Biological Structures*, Springer-Verlag, Berlin Heidelberg, 1991.
2. F. Cordier and S. Grzesiek, *J. Am. Chem. Soc.*, 1999, **121**, 1601.
3. A. Dingley and S. Grzesiek, *J. Am. Chem. Soc.*, 1998, **120**, 8293.
4. A. Dingley, J. E. Masse, R. D. Peterson, M. Barfield, J. Feigon and S. Grzesiek, *J. Am. Chem. Soc.*, 1999, **121**, 6019.
5. A. Dingley and S. Grzesiek, *J. Am. Chem. Soc.*, 1998, **120**, 8293.
6. Y. X. Wang, J. Jacob, F. Cordier, P. Wingfield, S. J. Stahl, S. Lee-Huang, D. A. Torchia, S. Grzesiek and A. Bax, *J. Biomol. NMR*, 1999, **14**, 181.
7. G. Cornilescu, B. E. Ramirez, M. K. Frank, G. M. Clore, A. M. Gronenborn and A. Bax, *J. Am. Chem. Soc.*, 1999, **121**, 6275.
8. D. I. Freedberg, R. M. Venable, A. Rossi, T. E. Bull and R. W. Pastor, *J. Am. Chem. Soc.*, 2004, **126**, 10478.
9. V. S. R. Rao, P. K. Qasba, P. V. Balaji and R. Chandrasekaran, *Conformation of Carbohydrates*, Harwood Academic Publishers, Amsterdam, 1998.
10. L. Poppe and H. van Halbeek, *Nat. Struct. Biol.*, 1994, **1**, 215.
11. M. D. Battistel, R. Pendrill, G. Widmalm and D. I. Freedberg, *J. Phys. Chem. B*, 2013, **117**, 4860.
12. J. T. Davis, S. Hirani, C. Bartelett and B. R. Reid, *J. Biol. Chem.*, 1994, **269**, 3331.

13. T. Blundell and A. Almond, *Magn. Reson. Chem.*, 2007, **45**, 430.
14. C. Landersjö and G. Widmalm, *Biopolymers*, 2002, **64**, 283.
15. S. E. Norris, J. Landström, A. Weintraub, T. E. Bull, G. Widmalm and D. I. Freedberg, *Biopolymers*, 2011, **97**, 145.
16. M. D. Battistel, M. Shangold, L. B. Trinh, J. Shiloach and D. I. Freedberg, *J. Am. Chem. Soc.*, 2012, **134**, 10717.
17. V. Chevelkov, Y. Xue, D. K. Rao, J. D. Forman-Kay and N. R. Skrynnikov, *J. Biomol. NMR*, 2010, **46**, 227.
18. M. Frank, Personal Communication, 2011.
19. A. B. Yongye, J. Gonzalez-Outeirino, J. Glushka, V. Schultheis and R. Woods, *Biochemistry*, 2008, **47**, 12493.
20. E. C. Schulz, D. Schwarzer, M. Frank, K. Stummeyer, M. Muhlenhoff, A. Dickmanns, R. Gerardy-Schahn and R. Ficner, *J. Mol. Biol.*, 2010, **397**, 341.
21. M. Nagae, A. Ikeda, M. Hane, S. Hanashima, K. Kitajima, C. A. Sato and Y. Yamaguchi, *J. Biol. Chem.*, 2013, **288**, 33784.
22. C. N. Beecher, R. P. Young, D. J. Langeslay, L. J. Mueller and C. K. Larive, *J. Phys. Chem. B*, 2014, **118**, 482.
23. D. J. Langeslay, R. P. Young, S. Beni, C. N. Beecher, L. J. Mueller and C. K. Larive, *Glycobiology*, 2012, **22**, 1173.
24. M. Piotto, V. Saudek and V. Sklenar, *J. Biomol. NMR*, 1992, **2**, 661.
25. G. Batta and K. E. Kover, *Carbohydr. Res.*, 1999, **320**, 267.
26. S. B. Engelsen and S. Perez, *Carbohydr. Res.*, 1996, **292**, 21.
27. A. J. Shaka and A. Pines, *J. Magn. Reson.*, 1987, **71**, 495.
28. G. M. Brown and H. A. Levy, *Acta Crystallogr., Sect. B: Struct. Crystallogr. Cryst. Chem.*, 1973, **29**, 790.
29. K. Bock and R. U. Lemieux, *Carbohydr. Res.*, 1982, **100**, 63.
30. B. Adams and L. Lerner, *J. Am. Chem. Soc.*, 1992, **114**, 4827.
31. L. Poppe and H. van Halbeek, *J. Am. Chem. Soc.*, 1992, **114**, 1092.
32. S. Q. Sheng and H. van Halbeek, *Biochem. Biophys. Res. Commun.*, 1995, **215**, 504.
33. L. Poppe and H. van Halbeek, *J. Am. Chem. Soc.*, 1991, **113**, 363.
34. C. N. Beecher, R. P. Young, D. J. Langeslay, L. J. Mueller and C. K. Larive, *J. Phys. Chem. B*, 2014, **118**, 482.
35. G. Nestor, L. Kenne and C. Sandstrom, *Org. Biomol. Chem.*, 2010, **8**, 2795.
36. B. Hakkarainen, L. Kenne, M. Lahmann, S. Oscarson and C. Sandstrom, *Magn. Reson. Chem.*, 2007, **45**, 1076.
37. B. Bekiroglu, L. Kenne and C. Sandstrom, *Carbohydr. Res.*, 2004, **339**, 2465.
38. A. Vishnyakov, G. Widmalm, J. Kowalewski and A. Laaksonen, *J. Am. Chem. Soc.*, 1999, **121**, 5403.
39. S. Bagley, M. Odelius, A. Laaksonen, G. Widmalm, S. Liaaen-Jensen, J. Balzarini, B. Fransson, U. Ragnarsson and G. W. Francis, *Acta Chem. Scand.*, 1994, **48**, 792.
40. S. Hanashima, A. Ikeda, H. Tanaka, Y. Adachi, N. Ohno, T. Takahashi and Y. Yamaguchi, *Glycoconjugate J.*, 2014, **31**, 199.

41. S. Hanashima, K. Kato and Y. Yamaguchi, *Chem. Commun. (Camb.)*, 2011, **47**, 10800.
42. M. D. Battistel, H. F. Azurmendi and D. I. Freedberg, *J. Phys. Chem. B.*, 2017, **121**, 683.
43. M. Fierman, A. Nelson, S. I. Khan, M. Barfield and D. J. O'Leary, *Org. Lett.*, 2000, **2**, 2077.
44. N. M. Loening, C. E. Anderson, W. S. Iskenderian, C. D. Anderson, S. D. Rychnovsky, M. Barfield and D. J. O'Leary, *Org. Lett.*, 2006, **8**, 5321.
45. F. Cordier, M. Barfield and S. Grzesiek, *J. Am. Chem. Soc.*, 2003, **125**, 15750.
46. M. Zierke, M. Smiesko, S. Rabbani, T. Aeschbacher, B. Cutting, F. H. Allain, M. Schubert and B. Ernst, *J. Am. Chem. Soc.*, 2013, **135**, 13464.
47. K. Pederson, D. A. Mitchell and J. H. Prestegard, *Biochemistry*, 2014, **53**, 5700.
48. M. D. Battistel, H. F. Azurmendi, M. Frank and D. I. Freedberg, *J. Am. Chem. Soc.*, 2015, **137**, 13444.
49. H. F. Azurmendi, M. D. Battistel, J. Zarb, F. Lichaa, A. Negrete Virgen, J. Shiloach and D. I. Freedberg, 2017, submitted.
50. R. T. Williamson, A. V. Buevich, G. E. Martin and T. Parella, *J. Org. Chem.*, 2014, **79**, 3887–3894.
51. H. Shinar, M. D. Battistel, M. D. Mandler, F. Lichaa, D. I. Freedberg and G. Navon, *Carbohydr. Res.*, 2014, **389**, 165.
52. F. Michon, J.-R. Brisson and H. J. Jennings, *Biochemistry*, 1987, **26**, 8399.

CHAPTER 2

NMR Spin-Couplings in Saccharides: Relationships Between Structure, Conformation and the Magnitudes of J_{HH}, J_{CH} *and* J_{CC} *Values*

MATTHEW J. HADAD[a], WENHUI ZHANG[a], TOBY TURNEY[a], LUKE SERNAU[a], XIAOCONG WANG[c], ROBERT J. WOODS[c], ANDREW INCANDELA[a], IVANA SURJANCEV[a], AMY WANG[a], MI-KYUNG YOON[a], ATTICUS COSCIA[a], CHRISTOPHER EUELL[a], REAGEN MEREDITH[a], IAN CARMICHAEL[b] AND ANTHONY S. SERIANNI*[a]

[a]Department of Chemistry and Biochemistry, University of Notre Dame, Notre Dame, IN 46556-5670, USA; [b]The Radiation Laboratory, University of Notre Dame, Notre Dame, IN 46556-5670, USA; [c]Complex Carbohydrate Research Center, University of Georgia, Athens, GA 30602, USA
*E-mail: aseriann@nd.edu

New Developments in NMR No.10
NMR in Glycoscience and Glycotechnology
Edited by Koichi Kato and Thomas Peters

Published by the Royal Society of Chemistry, www.rsc.org

2.1 Introduction and Background

The NMR spin–spin coupling constant (spin–spin coupling, spin-coupling, scalar coupling, or *J*-coupling) plays a pivotal role in modern liquid-phase NMR spectroscopy.[1] These parameters, typically measured in Hz, have magnitudes and signs independent of the static magnetic field B_0, and report on molecular properties ranging from the *s*-character of covalent bonds[2] to molecular torsion angles from which molecular conformation in solution can be determined (the terms "*dihedral angle*" and "*torsion angle*" have different definitions, wherein the former refers to the angle between two planes connecting four different atoms that are not necessarily sequentially covalently bonded, and the latter is defined by four atoms connected by three bonds). In addition to structural applications, *J*-couplings provide functional information on molecules; for example, *J*-couplings can be used to determine the pK_a values of ionizable functional groups in solution.[3,4] Knowledge of *J*-couplings is often required for the proper design and execution of modern multi-dimensional NMR experiments, where they are used to calculate appropriate delay or spin-evolution times in some homo- and heteronuclear pulse sequences.[5–7] Cogent discussions of the theory of *J*-couplings abound in the scientific literature.[8,9]

Nuclear spin communication between atoms having non-zero spin-quantum numbers is most often a "through-bond" phenomenon, although situations exist where "through-space" effects are appreciable.[10] Focusing solely on the "through-bond" mechanism for spin-1/2 nuclei, *J*-couplings are measurable across one, two and three covalent bonds, and less often across four or five bonds (the latter are denoted long-range couplings). The number of bonds comprising the coupling pathway is indicated by a superscript, such as ^1J, ^2J and ^3J for *J*-couplings across one, two and three bonds, respectively. These pathways have generic descriptors; ^1J is denoted a *direct* coupling, ^2J is denoted a *geminal* coupling, and ^3J is denoted a *vicinal* coupling.[11] The magnitudes of *J*-couplings (*i.e.*, their absolute values) between a given pair of spin-1/2 nuclei depend in part on the number of covalent bonds between the coupled nuclei, with ^1J values considerably larger in magnitude than ^2J and ^3J values. For spin-1/2 nuclei such as ^{13}C and ^{1}H, there is no consistent relationship between ^2J and ^3J values and coupling magnitude; in some cases, magnitudes can be larger for ^2J than for ^3J.

Subscripts are added to the ^1J, ^2J and ^3J symbolisms to identify the coupled nuclei and in some cases to more fully define the coupling pathway. Thus, a vicinal *J*-coupling between ^{13}C and ^{1}H is denoted $^3J_{CH}$, but this descriptor provides no information on the identity of the two intervening nuclei. This information can be included if desired; for example, $^3J_{CCCH}$ or $^3J_{COCH}$ describe vicinal ^{13}C–^{1}H coupling pathways, with the former pathway being a C–C–C–H pathway and the latter a C–O–C–H pathway. The nature of the intervening atoms in a coupling pathway affects the magnitude of the coupling interaction, all else being equal.

In addition to magnitude, *J*-couplings possess a sign, with the general rule of thumb for ^{13}C and ^{1}H being that ^{1}J values (*e.g.*, $^{1}J_{CH}$) are positive (+) in sign, ^{2}J values (*e.g.*, $^{2}J_{CC}$) are negative (−) in sign, and ^{3}J values (*e.g.*, $^{3}J_{CH}$) are positive (+) in sign (the alternating signs are predicted by the simplified Dirac model of *J*-couplings).[1,12] As appealing as this rule may be, it is not reliable in saccharides, where it is not uncommon to encounter $^{2}J_{CH}$ and $^{2}J_{CC}$ values that have (+) signs (Scheme 2.1).

In cyclic systems, *J*-couplings may be determined by more than one through-bond pathway. For example, *J*-coupling between C1 and C4 in methyl β-D-glucopyranoside **1** occurs *via* two vicinal pathways: C1–C2–C3–C4 and C1–O5–C5–C4 (Scheme 2.2). This coupling is denoted $^{3+3}J_{C1,C4}$ to indicate that two vicinal pathways contribute to the observed coupling. Likewise, in the β-D-ribofuranosyl ring of RNA (Scheme 2.2), *J*-coupling between C1′ and C4′ occurs *via* the C1′–C2′–C3′–C4′ (vicinal) pathway and the C1′–O4′–C4′ (geminal) pathway, and is denoted $^{2+3}J_{C1',C4'}$. In these cases, it is generally expected that the experimental *J*-coupling equals the algebraic sum of the individual *J*-couplings contributed by each pathway (see Section 2.7.4).[13,14]

In mono- and oligosaccharides containing only ^{13}C and ^{1}H magnetic nuclei, the following types of *J*-couplings exist: $^{2}J_{HCH}$, $^{3}J_{HCCH}$, $^{3}J_{HCOH}$, $^{4}J_{HCCCH}$, $^{4}J_{HCOCH}$, $^{1}J_{CH}$, $^{1}J_{CC}$, $^{2}J_{CCH}$, $^{2}J_{COH}$, $^{2}J_{CCC}$, $^{2}J_{COC}$, $^{3}J_{CCCH}$, $^{3}J_{CCOH}$, $^{3}J_{COCH}$, $^{3}J_{CCCC}$, $^{3}J_{COCC}$, $^{4}J_{CCCCH}$ and $^{4}J_{COCCH}$. $^{2}J_{COH}$ and $^{3}J_{CCOH}$ values involve spin-coupling to an hydroxyl hydrogen, and their measurement in water is sometimes complicated by intrinsic exchange of the hydroxyl hydrogen with solvent. *Intra-residue J*-couplings are associated with pathways within a single residue and report on the structural characteristics of that residue. *Inter-residue J*-couplings are associated

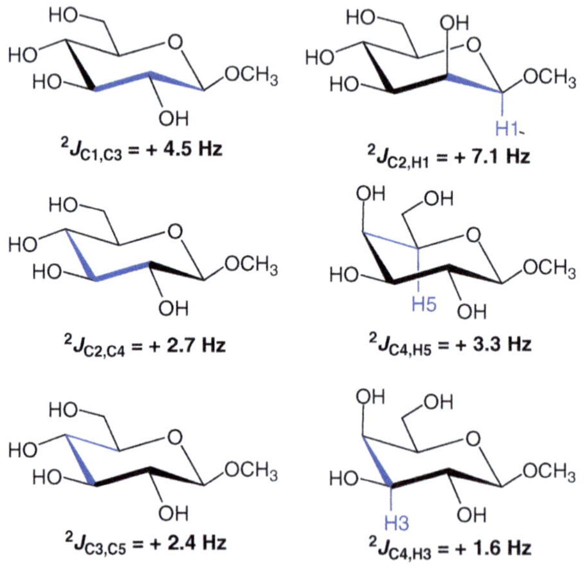

Scheme 2.1 A selection of $^{2}J_{CCH}$ and $^{2}J_{CCC}$ values in methyl β-D-aldohexopyranosides that are positive in sign.

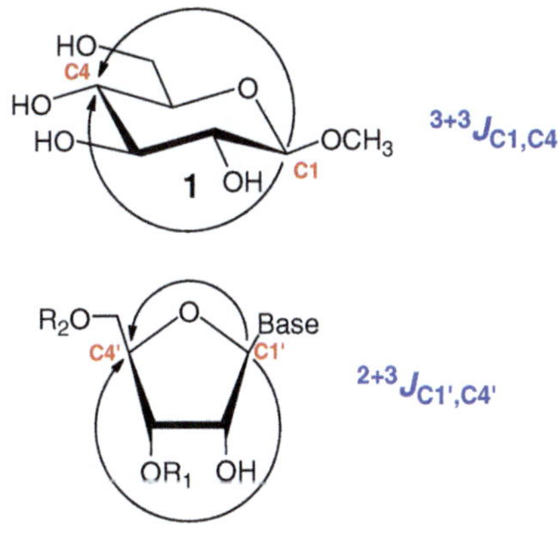

Scheme 2.2 Examples of dual-pathway *J*-couplings in aldohexopyranosyl and aldopentofuranosyl rings.

with pathways that span two contiguous residues, typically across an *O*-glycosidic linkage, and are useful in determining the conformations of these linkages. The different types of coupling pathways found in saccharides are illustrated in Figure 2.1 using the monosaccharide, methyl β-D-glucopyranoside **1** and the disaccharide, methyl β-D-galactopyranosyl-(1 → 4)-β-D-glucopyranoside (methyl β-lactoside) **2**.

Other common spin-1/2 nuclei in saccharides include ^{15}N and ^{31}P. The former are often found in *N*-acetyl side-chain substituents of aminosugars, and the latter are often associated with phosphate ester substituents. The common types of *J*-couplings involving ^{31}P and ^{15}N in saccharides are illustrated in Scheme 2.3.

Since ^{13}C and ^{15}N are low abundance isotopes of carbon (1.1%) and nitrogen (0.37%), respectively, it is often difficult to measure *J*-couplings involving these atoms without isotopic enrichment. With the advent of very high field NMRs, measurements of ^{1}H–^{15}N spin-couplings can be achieved without ^{15}N enrichment (because one of the coupling partners (^{1}H) is highly abundant).[15] In contrast, ^{13}C–^{13}C and ^{13}C–^{15}N spin-couplings commonly require enrichment of at least one of the coupled nuclei to render the measurement practical (*e.g.*, there is a 1 in ~10 000 probability of encountering ^{13}C atoms at both coupled sites in an unlabeled molecule; this probability increases to ~1 : 100 if one of the coupled carbons is enriched to ~100%). A recent review has appeared on the synthesis of isotopically labeled saccharides.[16]

2.2 Empirical Predictions of $^{2}J_{CH}$ and $^{2}J_{CC}$ Values in Saccharides

Knowledge of the signs of geminal ^{13}C–^{1}H and ^{13}C–^{13}C spin-couplings is essential for accurate interpretations of these couplings in terms of saccharide structure. This knowledge is particularly important for ^{2}J values that

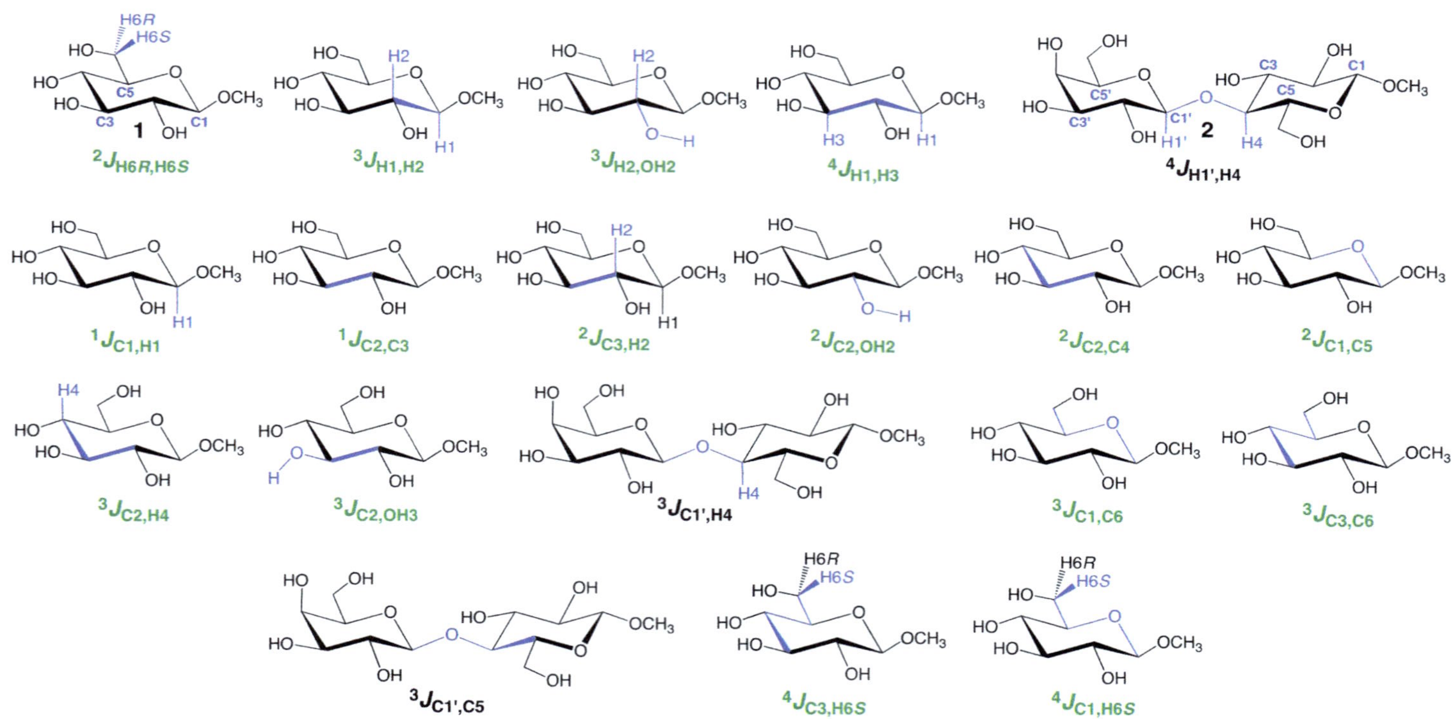

Figure 2.1 Examples of J_{HH}, J_{CH} and J_{CC} found in saccharides. *Intra-residue J*-couplings (green) are shown for the monosaccharide, methyl β-D-glucopyranoside **1**, and *inter-residue J*-couplings (black) are shown for the disaccharide, methyl β-lactoside (methyl β-D-galactopyranosyl-(1 → 4)-β-D-glucopyranoside) **2**.

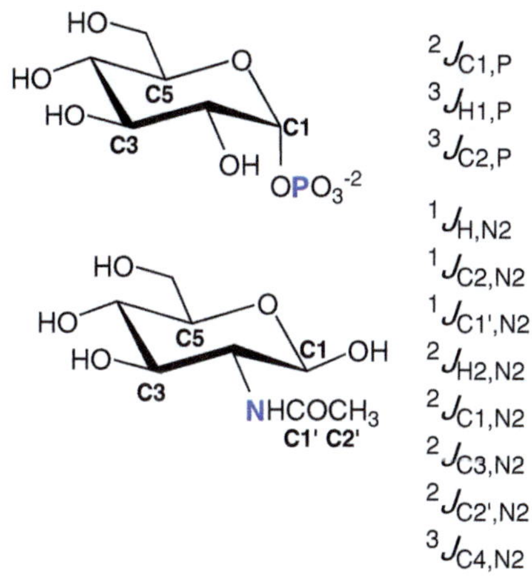

Scheme 2.3 *J*-couplings in saccharides containing ^{31}P and ^{15}N nuclei.

change sign as a function of conformation, and failure to recognize this sign change can lead to erroneous conclusions about (a) the dynamic range of the coupling (*i.e.*, its sensitivity to a given structural change) and/or (b) the structure/conformation of the element under scrutiny. *J*-Coupling signs can be assigned in three ways: (1) theoretical calculations; (2) empirical rules; and (3) experiment. Theoretical calculations using density functional theory (DFT) provide a reliable means of determining coupling signs in saccharides for $J > \sim 0.2$ Hz, and in the absence of empirical rules or experimental data, provide an attractive alternative to establishing coupling signs. The use of DFT to calculate NMR *J*-couplings in saccharides has been reviewed recently[17] and will not be discussed here. The reliability of these calculations, however, requires validation from experiment, and empirical rules evolved originally from experimental sign determinations. Two empirical rules to predict ^{2}J signs in saccharides for couplings having magnitudes >1 Hz are discussed first, followed by a discussion of experimental approaches to make sign determinations.

An empirical method to predict $^{2}J_{CCH}$ signs, known as the projection rule, was first described by Bock and Pedersen.[18] This rule is based on an inspection of a Newman projection viewed down the C–C bond of the C–C–H coupling pathway. Electronegative substituents appended to both carbons are projected onto an axis anti to the C–H bond, giving angles of 0°, ±60° or ±120°. The cosines of these angles are summed, and the resulting projection sum is correlated with the magnitude and sign of the $^{2}J_{CCH}$ value using a reference curve (Figure 2.2).[18] The method is illustrated for $^{2}J_{C2,H1}$ in methyl β-D-mannopyranoside **3** (Scheme 2.4) for which a projection sum of +2 correlates with an ~+8 Hz coupling (Figure 2.2). The experimental $^{2}J_{C2,H1}$ in **3** is +7.1 Hz (Scheme 2.1). This method does not predict coupling magnitudes quantitatively, but sign predictions are very reliable for couplings >1 Hz. The projection rule was developed

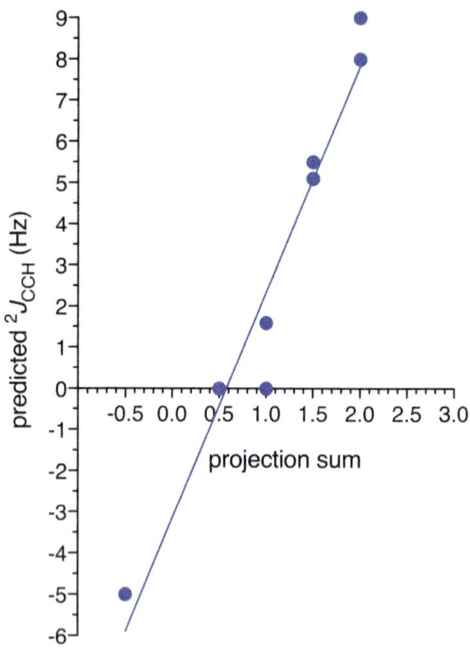

Figure 2.2 Plot correlating the projection sum with the magnitude and sign of $^2J_{CCH}$ in saccharides. A linear fit gives the equation $y = 5.49x - 3.15$. This plot was modified from that originally reported by Bock and Pedersen.[18]

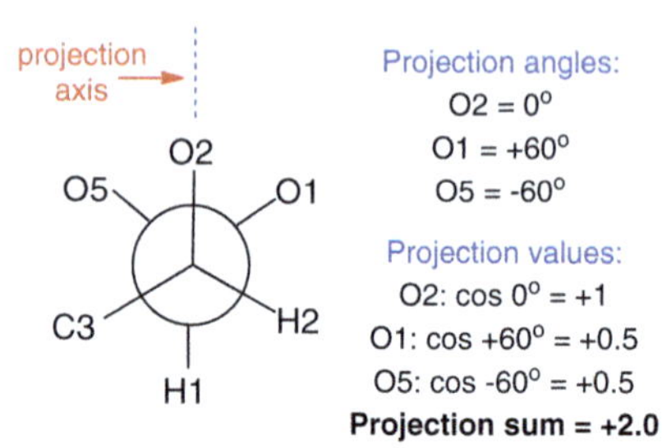

Scheme 2.4 Newman projection down the C2–C1 bond of methyl β-D-mannopyranoside 3 used to predict the sign of $^2J_{C2,H1}$ (see Scheme 2.1).

specifically for C–C fragments bearing three oxygen substituents to overcome the limitations of Perlin's rule.[19,20] However, the method appears to be generally applicable to C–C–H fragments bearing one (*e.g.*, deoxysugars) and two oxygens (Scheme 2.5), although the slopes of the latter reference plots probably differ from that shown in Figure 2.2. In addition, the method is strictly applicable to saccharides bearing only oxygen electronegative substituents. It is noteworthy that $^2J_{CCH}$ values in saccharides range from ~−5 to ~+8 Hz (Figure 2.2), giving a dynamic range of ~13 Hz.

This range exceeds that for vicinal ^{13}C–^{1}H *J*-couplings in saccharides, which are always (+) in sign and assume maximal values of 8–10 Hz (see Section 2.7.6).

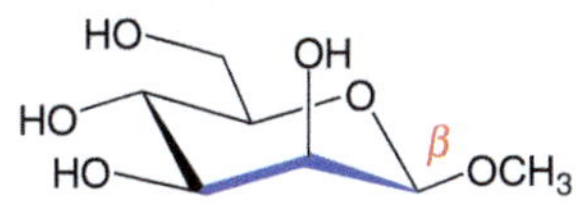

3

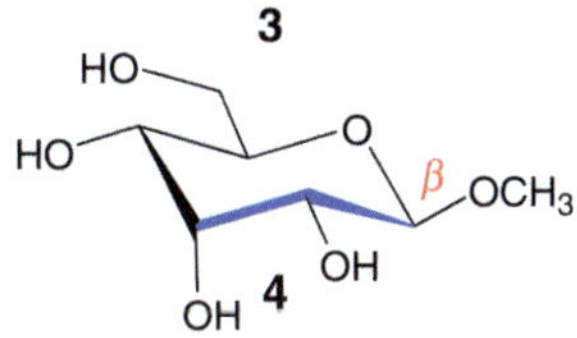

4

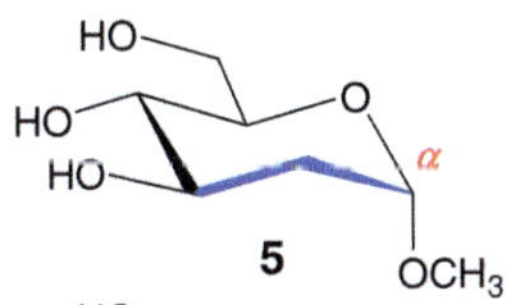

5

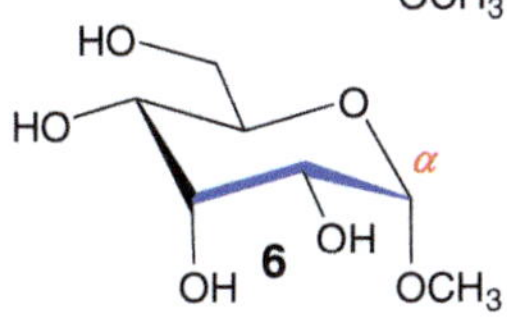

6

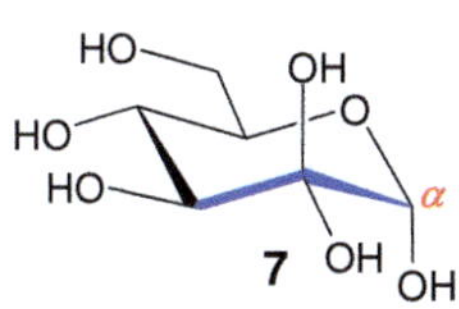

7

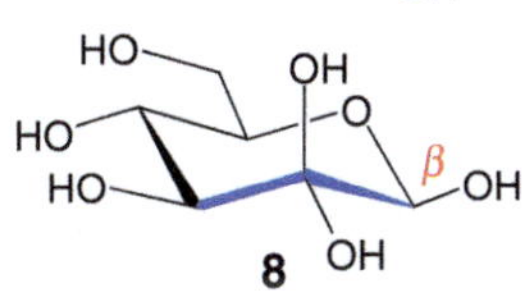

8

An empirical rule similar to the projection rule has been developed to predict $^{2}J_{CCC}$ and $^{2}J_{COC}$ signs in saccharides. This method is denoted the projection resultant (PR) method and was first described by Church *et al.*[21] The method utilizes two Newman projections to treat the two C–C bonds that comprise the C–C–C coupling pathway. The method is illustrated in Scheme 2.6 for the prediction of $^{2}J_{C1,C3}$ in methyl β-D-glucopyranoside **1** and methyl α-D-allopyranoside **6**. Projection resultant (PR) values of +2.5 and −0.5 are obtained for the β-*gluco* and α-*allo* rings, corresponding to experimental couplings of +4.6 Hz and −2.4 Hz, respectively (Table 2.1). These data, along with $^{2}J_{C1,C3}$ values calculated by DFT, are plotted in the reference graph shown in Figure 2.3. This graph resembles that originally published in 1996,[21] but contains new data for $^{2}J_{C1,C3}$ values measured in the pyranose hydrate forms of D-glucosone (α-pyranose **7**, PR = +2.5, $^{2}J_{C1,C3}$ = +2.2 Hz; β-pyranose **8**, PR = +4.0, $^{2}J_{C1,C3}$ = +7.0 Hz). These new data align well with prior results, and appreciably extend the plot in the (+) PR dimension. The plot also incorporates DFT-calculated

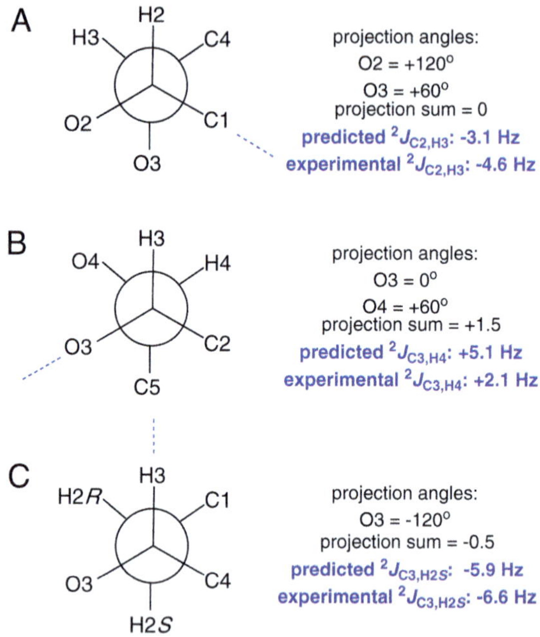

Scheme 2.5 Projection sums for, and predicted and experimental values of, ${}^2J_{C2,H3}$ (A) and ${}^2J_{C3,H4}$ (B) in methyl β-D-allopyranoside **4**, and (C) ${}^2J_{C3,H2S}$ in methyl 2-deoxy-α-D-glucopyranoside **5**. Predicted values were calculated using the equation given in Figure 2.2 (legend).

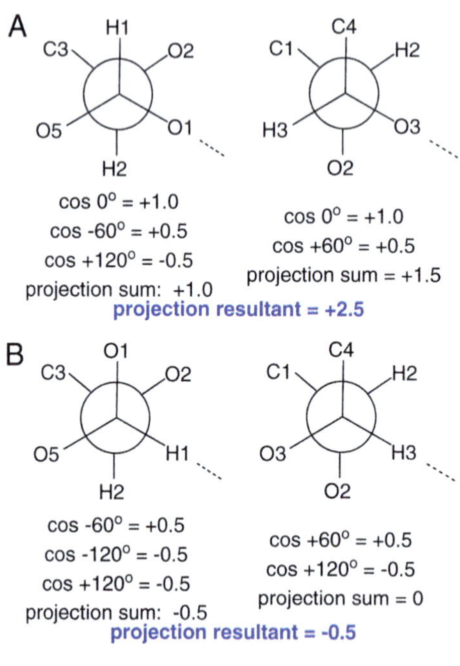

Scheme 2.6 Newman projections for the C1–C2 and C2–C3 bonds of methyl β-D-glucopyranoside **1** (A) and methyl α-D-allopyranoside **6** (B) used to predict the magnitudes and signs of ${}^2J_{C1,C3}$ in both compounds by the projection resultant method.[21]

Table 2.1 Experimental and calculated $^2J_{C1,C3}$ values in several methyl aldohexopyranosides and their 2- and 3-deoxy analogs.

Cmpd	$^2J_{C1,C3}$ (Hz) *Expt*	*Calc*
Group 1		
α-Glc	*nc*[a]	(+)0.1
α-Man	*nc*	(+)0.1
β-Allo	*nc*	(+)0.3
β-Altro	*nc*	(+)0.3
2-Deoxy-α-Glc	(−)2.3	(−)1.5
2-Deoxy-β-Glc	(−)1.8	(−)2.4
3-Deoxy-α-Glc	(−)2.0	(−)1.3
3-Deoxy-α-Man	(−)1.8	(−)1.1
Group 2		
β-Glc	(+)4.6	(+)4.5
β-Man	(+)4.0	(+)3.9
2-Deoxy-β-Allo	(−)1.8	(−)1.4
3-Deoxy-β-Glc	(+)1.0	(+)1.7
3-Deoxy-β-Man	(+)1.1	(+)1.3
Group 3		
α-Allo	(−)2.4	(−)2.0
2-Deoxy-α-Allo	(−)3.5	(−)2.9
α-Altro	*nc*	

[a]*nc* = no coupling observed, $J < 0.7$ Hz. Calculated values were obtained by DFT. Experimental signs were determined by crosspeak displacement in 2D ^{13}C–^{13}C COSY spectra of triply-^{13}C-labeled compounds.[26,27]

$^2J_{C1,C3}$ values to illustrate the high degree of agreement between $^2J_{CCC}$ measured experimentally and values predicted from theory. Linear fitting of the data in Figure 2.3 gives the equation $y = 2.14x - 2.66$ ($R^2 = 0.94$). An inspection of Figure 2.3 shows that $^2J_{C1,C3}$ values range from ~−3 Hz to ~+7 Hz, giving a dynamic range of ~10 Hz. This range exceeds that associated with vicinal ^{13}C–^{13}C *J*-couplings in saccharides, which are always (+) in sign and assume maximal values of 5–6 Hz (see Section 2.7.6).

A similar empirical treatment of $^2J_{COC}$ values shows that these geminal *J*-couplings are (−) in sign in saccharides in all circumstances examined to date.[21–23] These couplings occur commonly in aldopyranosyl rings (*i.e.*, $^2J_{C1,C5}$) and across the *O*-glycosidic linkages of oligo- and polysaccharides (*e.g.*, $^2J_{C1',C4}$ in **2**, Figure 2.1).

2.3 Experimental Determinations of Spin-Coupling Signs in Saccharides

Experimental techniques to determine spin-coupling signs depend on whether the coupling interaction is homonuclear (*e.g.*, J_{HH}, J_{CC}) or heteronuclear (*e.g.*, J_{CH}). Homonuclear coupling signs are determined using a variant of the 2D

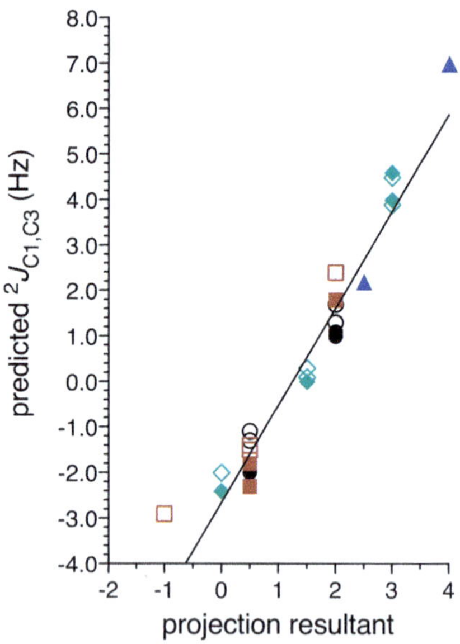

Figure 2.3 Relationship between $^2J_{C1,C3}$ and PR values in aldopyranosyl rings. Filled, experiment; open, calculated. Black, 3-deoxy; red, 2-deoxy; green, O1–O2–O3 oxy (see Table 2.1). Blue triangles are experimental data for D-glucosone hydrates (7α and 8β, respectively). A linear fit takes the form $J = 2.14\ \mathrm{PR} - 2.66$ ($R^2 = 0.94$).

COSY-90 method, COSY-45, in which the second 90° pulse in the pulse sequence is replaced by a 45° pulse.[24] The method is illustrated in Figure 2.4 for the determination of J_{HH} signs in 2,3-dibromopropionic acid **9**, whose Newman projection is shown in Scheme 2.7. In **9**, there are three hydrogens denoted H_A, H_M and H_X (three-spin system) to show that the chemical shift *difference* (in Hz) between the signals of the coupled hydrogens is >10 times the J-coupling between them, such that little or no non-first-order effects (virtual coupling) are observed in the spectrum.[25] The 1D ^{1}H spectrum of **9** appears along the diagonal, and an inspection of the *orientations* (tilt) of the off-diagonal elements (cross-peaks) allows a determination of the *relative* signs of the three J_{HH} values ($^2J_{HA,HM}$, $^3J_{HA,HX}$ and $^3J_{HM,HX}$).[24] The cross-peaks for the H_A–H_X and H_M–H_X interactions (vicinal) have the same orientations (negative slopes), and both differ from that observed for the H_A–H_M interaction (geminal) (positive slope), indicating that $^3J_{HH}$ and $^2J_{HH}$ values have opposite signs. Since it can be assumed that $^3J_{HH}$ values have positive signs in these systems, then $^2J_{HA,HM}$ must have a negative sign.

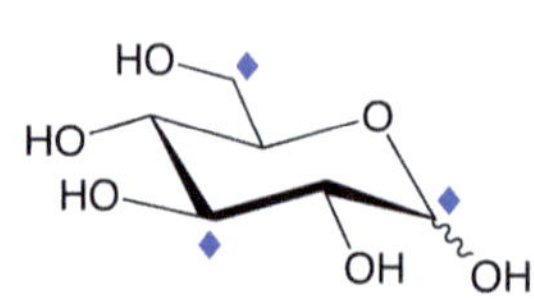

D-[1,3,6-$^{13}C_3$]glucose **10**

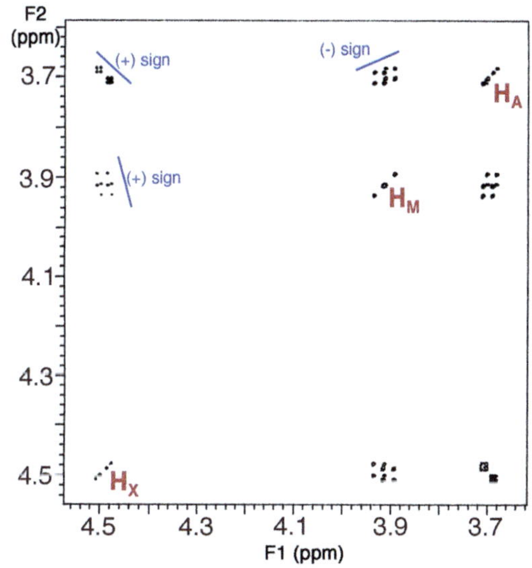

Figure 2.4 2D ^1H–^1H COSY-45 spectrum of **9** showing the 1D resonances from H_A, H_M and H_X along the diagonal (Scheme 2.7), and the detection of cros-speak "tilt" to determine the relative signs of $^2J_{HA,HX}$, $^3J_{HA,HM}$ and $^3J_{HM,HX}$ (see text).

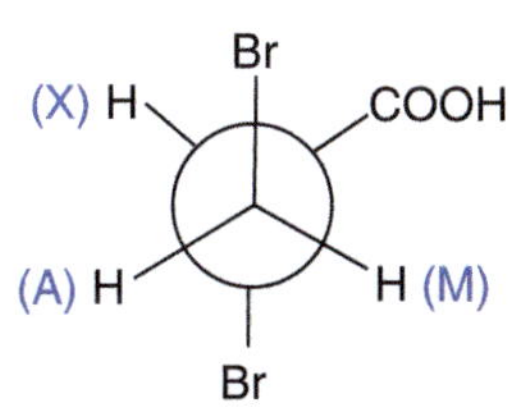

Scheme 2.7 Newman projection viewing down the C2–C3 bond of 2,3-dibromo-propionic acid **9**.

The application of COSY-45 to determine J_{HH} signs is straightforward since ^1H is a highly abundant spin-1/2 nucleus. Application of the same technique to determine the signs of J_{CC} values is not practical unless the molecule is multiply labeled with ^{13}C to allow measurements of multiple J_{CC} values in the same molecule.[26,27] For example, if D-glucose **10** is triply labeled with ^{13}C at C1, C3 and C6, then the labeled carbons form a mutually coupled, three-spin network ($^2J_{C1,C3}$, $^3J_{C1,C6}$ and $^3J_{C3,C6}$) like that found in **9**, and the ^{13}C chemical shifts of the coupled spins are distributed roughly equally over ~30 ppm (Figure 2.5), forming an AMX spin system similar to that shown in Figure 2.4. Importantly, since it can be assumed that the two $^3J_{CC}$ values have positive signs, the absolute sign of the geminal $^2J_{C1,C3}$ can be determined from the relative cross-peak orientations in the COSY-45 spectrum.

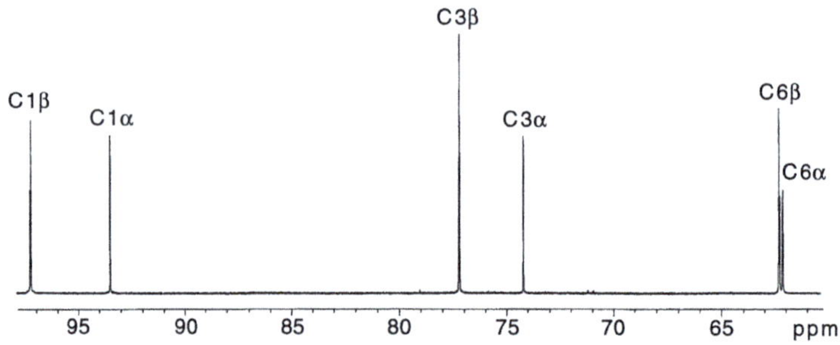

Figure 2.5 $^{13}C\{^1H\}$ NMR spectrum of **10**, showing signals from only the three labeled carbons in the α- and β-pyranose forms. Note the AMX character of the spectrum, and the signal multiplicities due to $^2J_{C1,C3}$, $^3J_{C1,C6}$ and $^3J_{C3,C6}$ (see text).

The application of COSY-45 to **10** is shown in Figure 2.6.[26] In this analysis, the sign of $^2J_{C1,C3}$ is the focus of attention. Since cross-peak tilts for all three interactions are identical (all have positive slopes), then all three J_{CC} values must have the same sign. Since $^3J_{C1,C6}$ and $^3J_{C3,C6}$ are assumed to have (+) signs, then $^2J_{C1,C3}$ must also have a (+) sign. This type of experimental data was used to assign some of the $^2J_{C1,C3}$ signs in Table 2.1.

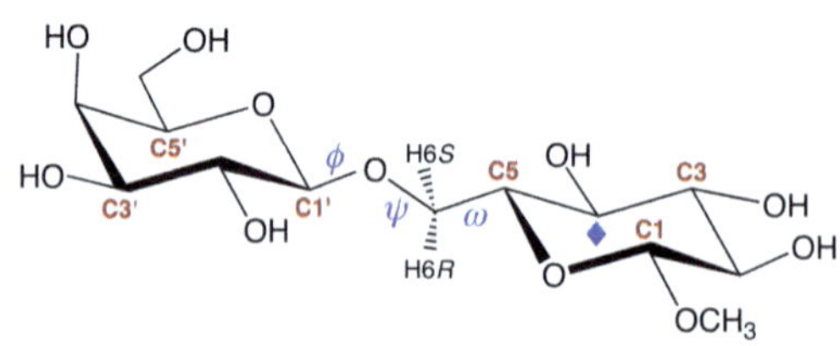

methyl β-D-galactopyranosyl-(1→6)-β-D-[4-^{13}C]glucopyranoside **11**

Sign determinations for heteronuclear *J*-couplings are made by inspection of cross-peak displacements in ^1H–^1H DQF-COSY or ^1H–^1H TOCSY 2D spectra.[28,29] Both methods give similar data, but TOCSY provides more extensive spin correlations that are particularly useful for sign determinations. This method is illustrated for the disaccharide, methyl β-D-galactopyranosyl-(1 → 6)-β-D-[4-^{13}C]glucopyranoside **11**. In **11**, the *gluco* residue is ^{13}C-labeled at C4, which allows measurements of multiple intra-residue J_{CH} values: $^1J_{C4,H4}$, $^2J_{C4,H3}$, $^2J_{C4,H5}$, $^3J_{C4,H2}$, $^3J_{C4,H6}$ and $^3J_{C4,H6'}$. Inspection of the 2D ^1H–^1H TOCSY spectrum of **11** (Figure 2.7) provides information on the signs of $^2J_{C4,H3}$ and $^2J_{C4,H5}$ through inspection of the paired TOCSY cross-peaks observed for the on-diagonal H4 signals. The H4 on-diagonal signal is split into two sets of multiplets separated by the value of $^1J_{C4,H4}$ (~+145.6 Hz). Being at an internal location in the ring, and given the relatively large $^3J_{HH}$ values in the *gluco* ring,[30] TOCSY transfer from H4 to the remaining C–H hydrogens in the ring is observed through the off-diagonal elements (cross-peaks). Inspection of the

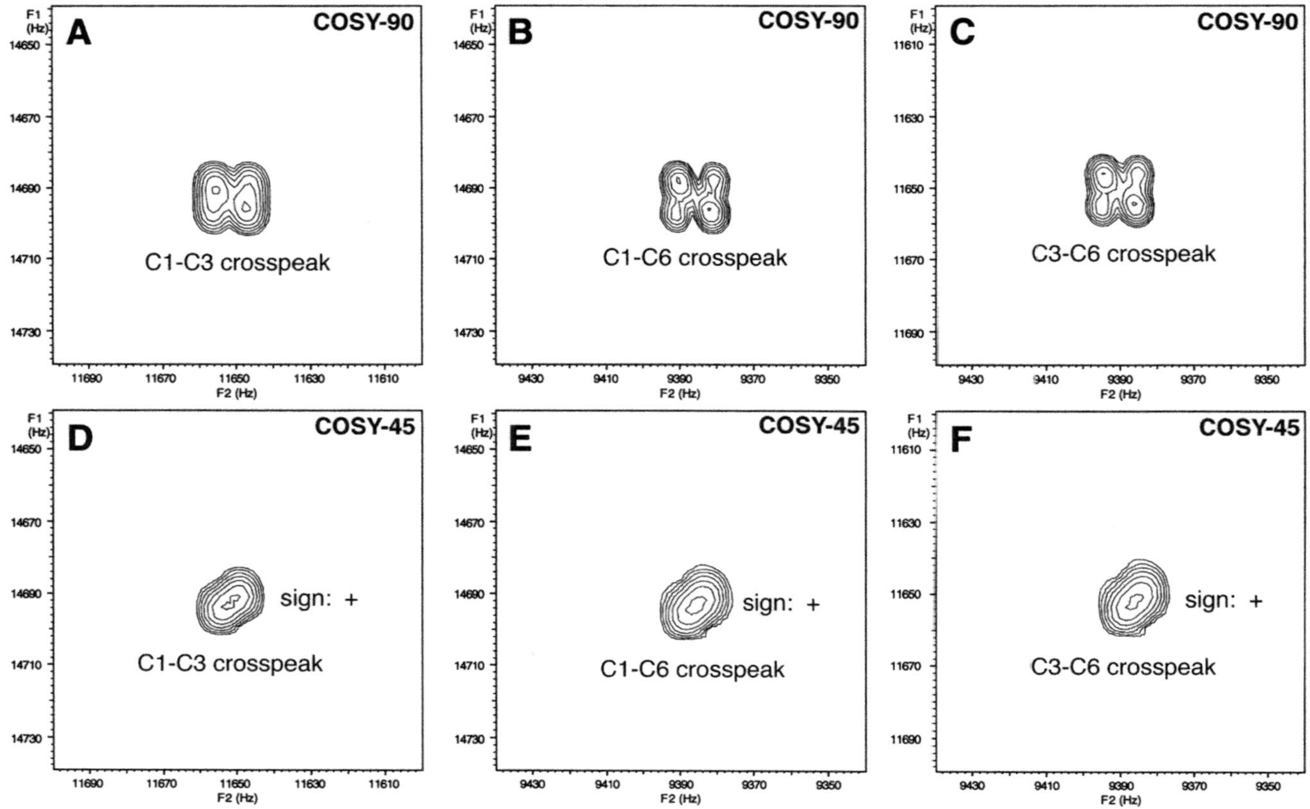

Figure 2.6 Partial ^{13}C–^{13}C COSY-90 (A–C) and COSY-45 (D–F) spectra of **10**, showing the C1–C3, C1–C6 and C3–C6 crosspeaks. Note the tilting of the COSY-45 signals, which is identical for all three interactions. Since the two $^{3}J_{CC}$ have (+) signs, the geminal $^{2}J_{C1,C3}$ must also have a (+) sign.

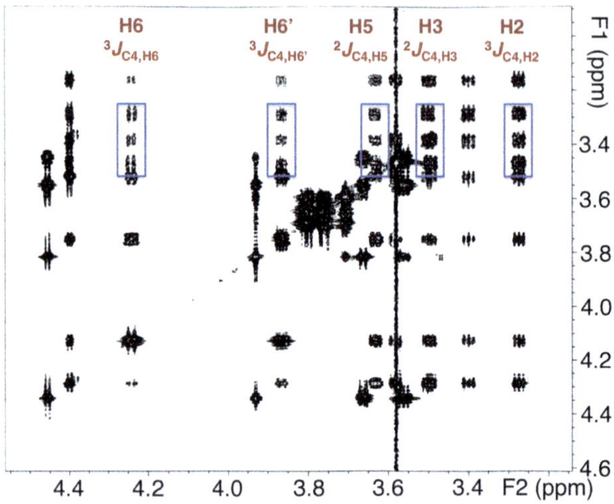

Figure 2.7 Partial ^{1}H–^{1}H 2D TOCSY spectrum of **11** (600 MHz) highlighting the correlations between H4 and the remaining hydrogens in the ^{13}C-labeled *gluco* ring. Inspection of the displacements of the paired crosspeaks (blue boxes) gives the magnitudes and signs of the indicated J_{CH} values involving the labeled carbon (see text).

relative displacements of the five paired cross-peaks gives information about relative signs, as shown in Figure 2.7 (each component of a paired cross-peak is offset along F2 by an amount equal to the ^{13}C–^{1}H coupling involving C4, and the *direction* of the offset provides information on the coupling sign). An internal reference is provided by the H4–H2, H4–H6 and H4–H6′ paired correlations, which are offset by $^3J_{C4,H2}$ (+0.8 Hz), $^3J_{C4,H6}$ (+2.7 Hz), and $^3J_{C4,H6'}$ (+0.9 Hz), respectively, in the same direction that is associated with (+) signs (vicinal $^3J_{CH}$ are assumed to be positive). A comparison of these reference offsets to those observed for the geminal couplings gives the signs of the latter: $^2J_{C4,H3} = -5.3$ Hz, and $^2J_{C4,H5} = -2.8$ Hz.

Digital resolution (Hz/point) is critical for quantitative measurements of cross-peak offsets in order to extract accurate J_{CH} values, and the F2 dimension is used most often for this purpose because it normally has the greater digital resolution after data processing. When optimized, J_{CH} values as small as 0.4 Hz or less can be measured reliably.

2.4 Second-Order Behavior in ^{1}H and ^{13}C NMR Spectra of Saccharides

The backbones of polypeptides contain hydrogens attached to the amide nitrogens and C_α carbons of each amino-acid residue, and these hydrogens are spin-coupled across three bonds ($^3J_{HNC\alpha H}$; Scheme 2.8). Intra-residue $^3J_{HNC\alpha H}$ values are used in conjunction with inter-residue ^{1}H–^{1}H nuclear

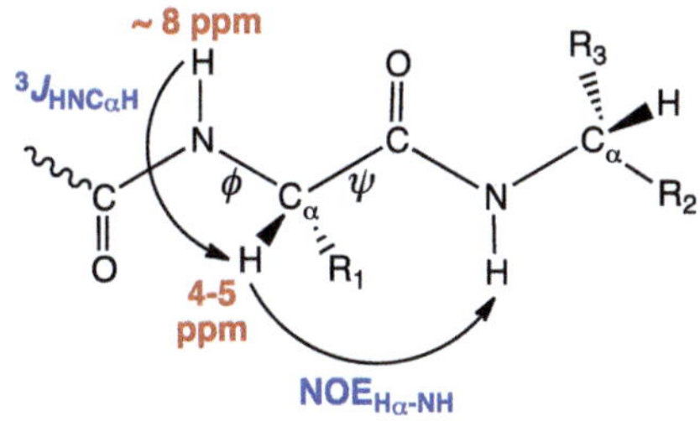

Scheme 2.8 Backbone atoms of a polypeptide, showing the chemical shifts of the NH and $C_\alpha H$ hydrogens, the intra-residue $^3J_{HNC\alpha H}$, and the inter-residue NOE connectivity between $C_\alpha H$ and NH (see www.bmrb.wisc.edu/ref_info/statful.htm).

Overhauser effects (NOEs) between C_α hydrogens and the amide hydrogens of the next residue to determine protein 3D structure.[31] This experimental approach is effective because the chemical shift difference (in Hz) between the spin-coupled N$\underline{H}$ and $C_\alpha\underline{H}$ hydrogens ($\delta_{N\underline{H}}$ ~ 8 ppm; $\delta_{C\alpha\underline{H}}$ ~ 4 ppm; ~2400 Hz difference on a 600 MHz spectrometer) is very large compared to the *J*-coupling between them (2–10 Hz depending on the value of ϕ), thus allowing accurate values of $^3J_{HNC\alpha H}$ to be extracted directly from the spectrum. In contrast, multiple ^{1}H signals arise from most monosaccharide residues (*e.g.*, seven CH hydrogens in a simple aldohexopyranosyl ring) and they occur over a narrow chemical shift range, typically 3.5–4.5 ppm. Thus, at 600 MHz, 5–7 signals are found over a span of ~600 Hz, each coupled to at least one other hydrogen in the same residue. It is therefore common to encounter situations where the chemical shifts of two mutually spin-coupled hydrogens are separated in Hz by an amount very similar to the *J*-coupling between them. This situation leads to non-first-order (strong coupling) behavior,[32] and direct use of spectral line positions does not give accurate values of the *J*-couplings. The only way to obtain the correct *J*-couplings is through spectral simulation.[33]

This effect is illustrated with two isotopomers of methyl β-D-galactopyranosyl-(1 → 6)-β-D-glucopyranoside **11**. The 600 MHz ^{1}H NMR spectrum of **11** singly labeled with ^{13}C at C1 of the Gal residue is shown in Figure 2.8. Significant overlap of the H3 and H4 signals for the Glc moiety occurs at ~3.55 ppm, and the resulting non-first-order behavior in the H2, H5, H6 and H6′ multiplets prevents the extraction of accurate *J*-couplings without spectral simulation. These distortions are eliminated when H3 in **11** is replaced with deuterium, as shown in Figure 2.9. In the deuterated disaccharide, the H2 and H4 signals appear as doublets, and the H5, H6 and H6′ signals exhibit essentially first-order splitting patterns. Closer inspection of Figures 2.8 and 2.9 reveals that each of the H6 and H6′ multiplets contain three *J*-couplings: a $^2J_{HCH}$, a $^3J_{HCCH}$ and a $^3J_{COCH}$. The $^3J_{COCH}$ values report on the C1′–O1′–C6–H6*R* and C1′–O1′–C6–H6*S* torsion angles (ψ) across the internal *O*-glycosidic linkage of **11**, as discussed below.

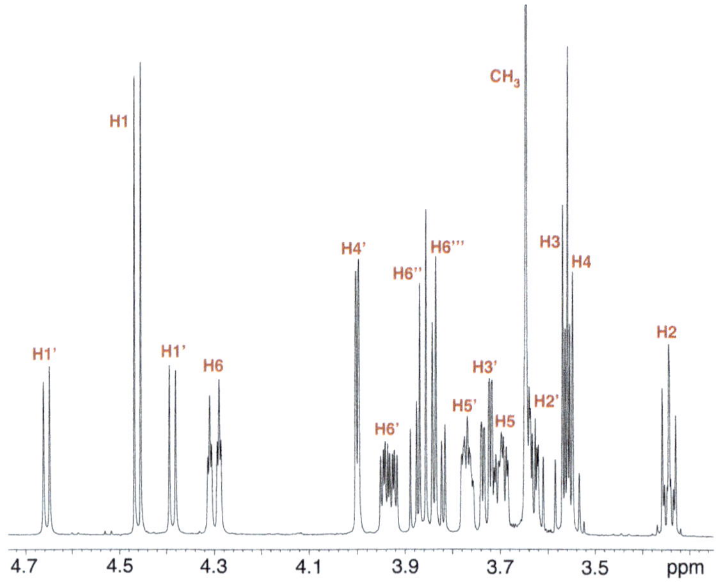

Figure 2.8 The 600 MHz ^{1}H NMR spectrum of methyl β-D-[1′-^{13}C]galactopyranosyl-(1 → 6)-β-D-glucopyranoside **11** in ^2H_2O showing signal assignments (unprimed = Glc residue). Significant overlap of the H3 and H4 signals at ~3.55 ppm causes non-first-order behavior in the H2, H5, H6 and H6′ multiplets (H6 and H6′ reside in the Glc residue; H6″ and H6‴ reside in the Gal residue).

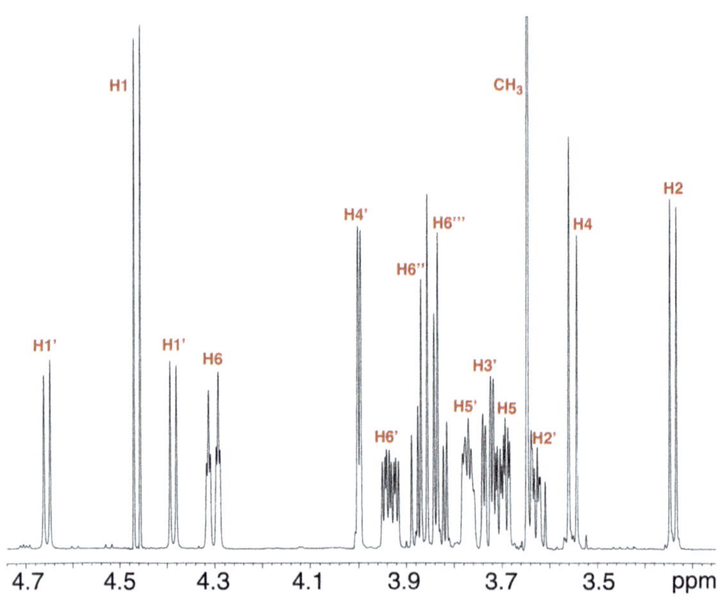

Figure 2.9 The 600 MHz ^{1}H NMR spectrum of methyl β-D-[1′-^{13}C]galactopyranosyl-(1 → 6)-β-D-[3-^{2}H]glucopyranoside **11** in ^2H_2O showing signal assignments (unprimed = Glc residue). Selective insertion of ^{2}H at C3 of the Glc residue eliminates the non-first-order behavior observed in Figure 2.8 (see text).

In addition to inserting ^{2}H as a means of eliminating non-first-order effects in ^{1}H NMR spectra of saccharides, the insertion of ^{13}C can achieve the same end.[34] The large magnitude of $^1J_{CH}$ values[30] (~140–170 Hz) splits the ^{1}H signal of the ^{13}C-bonded hydrogen and in doing so often removes the signal overlap (the effectiveness of this strategy also depends on the magnitude of B_0). For example, the 600 MHz ^{1}H NMR spectrum of methyl β-D-galactopyranosyl-(1 → 6)-β-D-[4-^{13}C]glucopyranoside **11** is essentially first order (the large splitting of the H4 signal removes its overlap with the H3 signal), thus permitting accurate determinations of J_{CH} values from 2D TOCSY cross-peak displacements as shown in Figure 2.7.

2.5 Structural Elements of Saccharides and Factors that Influence Them

The term "structure" is hierarchical in nature, and specifying the levels of saccharide structure that can be investigated profitably by NMR *J*-couplings is important to the ensuing discussion. As discussed previously,[35] NMR is not a preferred analytical method to determine saccharide primary structure (*e.g.*, identifying the types of residues present in an oligosaccharide and how these residues are linked glycosidically). This information can be more easily obtained, for example, by chemical/enzymic (*e.g.*, glycan sequencing)[36] and analytical (*e.g.*, mass spectrometry)[37] methods. The latter methods give more rapid and definitive results, and require smaller sample quantities than those required by NMR, often by orders of magnitude. One major advantage of NMR, in addition to being non-destructive to valuable samples, is that it can be used to investigate three-dimensional structure and molecular interactions in solution and in the solid-state. Defining time-dependent structural properties in solution is important to understanding the functional properties of most biomolecules, and few biomolecules are more conformationally flexible than saccharides. Assigning saccharide three-dimensional structure in solution thus involves identifying and quantifying an ensemble of conformations in chemical exchange, or more specifically, determining conformer exchange equilibria and kinetics and how they affect biological function. These types of applications represent the unique contributions that NMR makes to saccharide structure determination.

The major conformational elements of biologically important saccharides are as follows: (a) ring conformation (furanose and pyranose pseudorotation); (b) exocyclic hydroxymethyl conformation (ω); (c) exocyclic C–O bond conformation (θ); (d) exocyclic *N*-acetyl group conformation; and (e) *O*-glycosidic bond conformation (ϕ/ψ) (Scheme 2.9). Each element is discussed below to provide the structural background needed for the ensuing treatment of NMR *J*-couplings and how they are used, often in conjunction with other NMR parameters, to study these elements. It should be appreciated that conformational studies in general do not have, as

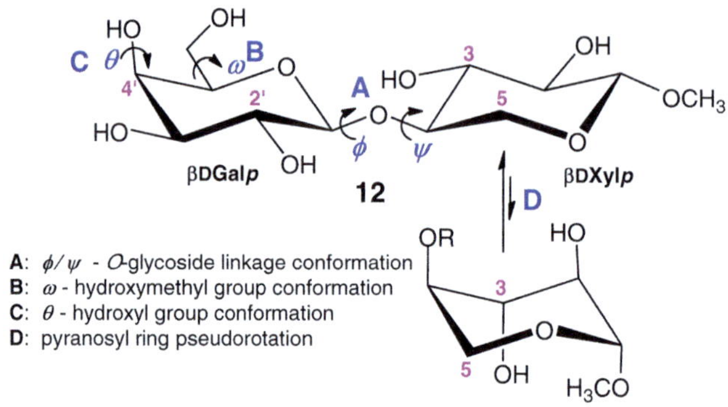

Scheme 2.9 Some conformational elements in saccharides, illustrated in the disaccharide, methyl β-D-galactopyranosyl-(1 → 4)-β-D-xylopyranoside **12**.

their singular end goal, the assignment of preferred conformational equilibria and the time-scales of conformer exchange in solution. Rather, this "*What*" knowledge is a prerequisite to answering higher-level questions, namely: (1) *Why* are particular conformational properties favored? And (2) *How* do specific structural properties influence biological properties such as binding specificity and receptor affinity? Furthermore, not only are solution conformations of interest, but also conformations bound by receptors, wherein the latter may or may not include a subpopulation of solution conformers.

2.5.1 Ring Conformation

The conformational dynamics of five- (furanose) and six-membered (pyranose) rings are described by pseudorotational itineraries (Scheme 2.10).[38,39] For furanoses, two parameters are needed to assign ring conformation: P, the phase angle, and τ_m, the puckering amplitude. Two idealized furanose conformers are denoted envelope (E) and twist (T), and there are ten of each. E forms have four contiguous coplanar ring atoms, and T forms have three contiguous coplanar ring atoms. Out-of-plane atoms are identified by either a superscript or subscript; for example, the ^3E form (also denoted C3-*endo*) of methyl 2-deoxy-β-D-*erythro*-pentofuranoside **13** orients C3 above the ring plane defined by C4–O4–C1–C2 (*i.e.*, C3 is on the same side of the ring as C5) (Scheme 2.11). Twenty furanose conformers interconvert *via* the specific pathway shown in Scheme 2.10, which does not involve the planar form (all ring atoms coplanar). The pseudorotational model assumes that the planar form is less stable than non-planar forms, but this assumption may not be valid for all ring structures and configurations. Importantly, while there are twenty idealized furanose conformers (Scheme 2.10), there are in fact an infinite number because τ_m can assume an infinite number of values for a given value of P. Computations show τ_m and P are inter-dependent.[40]

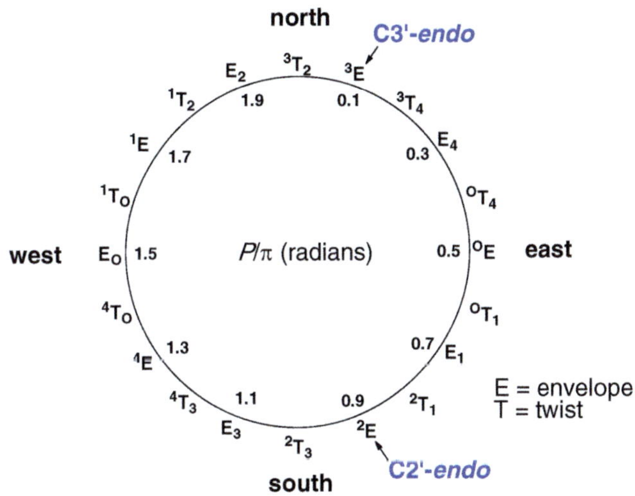

Scheme 2.10 Pseudorotational itinerary of a D-aldofuranose ring.

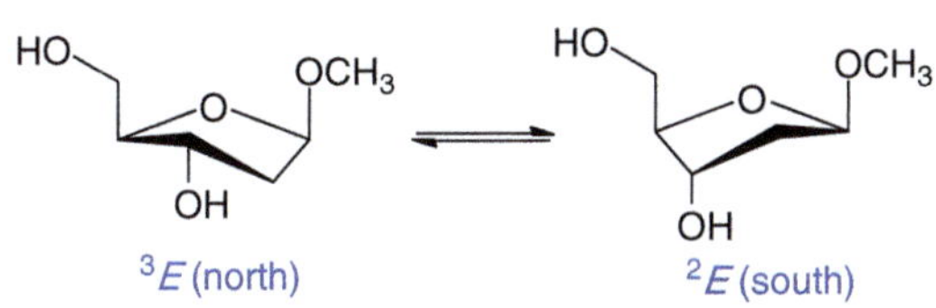

Scheme 2.11 ^3E (C3-*endo*)–^2E (C2-*endo*) exchange showing the conformation of each ring form of **13**. This exchange can occur *via* "east" or "west" forms in the pseudorotational itinerary (see Scheme 2.10), but often one of these pathways is more favored than the other.

An alternative pathway to furanose pseudorotation is ring inversion[40,41] in which two non-planar forms interconvert *via* the planar form:

$$^3E \leftrightarrows \text{planar} \leftrightarrows {}^2E$$

This mechanism is considered less preferred than pseudorotation because of the involvement of the planar form as an intermediate and the associated energetic penalty.

Pyranose ring pseudorotation requires three parameters (polar coordinates) to specify particular pyranose ring conformers on the surface of a sphere: θ, ϕ and Q, with the latter two analogous to P and τ_m in furanose pseudorotation, respectively.[39,42] The θ term specifies "slices" through this pseudorotational sphere, with each cross-section describing a different pseudorotational itinerary like that shown in Scheme 2.10 but involving different types of ring conformers such as boats (B), envelopes (E), half-chairs (H), and skew (S) (twist-boat, T) forms. Pyranose ring conformational dynamics often involves the two "idealized" forms that occupy the north ($\theta = 0°$) and south

($\theta = 180°$) poles of the itinerary, namely the $^4C_1$ and $^1C_4$ chair forms, respectively. Chair form interconversion takes the form,

$$^4C_1 \leftrightarrows \text{[non-planar intermediates]} \leftrightarrows {}^1C_4$$

where the forms in brackets may or may not contribute significantly to the population of forms present in solution. These intermediates may include boat, envelope, skew and/or half-chair forms.[39] Experimentally, it is difficult to characterize the conformational equilibria of furanose and pyranose rings rigorously, for reasons discussed in Section 2.6.

Structural factors influencing the conformational equilibria of five- and six-membered rings include: (a) stereoelectronic (*endo*-anomeric)[43] effects that influence the relative stabilities of C1–O1 bond orientation (axial *vs.* equatorial in pyranoses; *quasi*-axial *vs.* *quasi*-equatorial in furanoses); (b) *gauche* effects[43,44] that determine optimal O–C–C–O torsion angles for vicinal hydroxyl groups; (c) steric effects of axial *vs.* equatorial ring substituents; (d) effect of C2–O2 bond orientation on C1–O1 bond orientation (Δ2 effect)[45]; and (e) differential solvation effects. These and other factors conspire to determine the relative stabilities of ring conformers in five- and six-membered rings. In general, furanose rings are considered to be conformationally flexible, giving rise to conformationally heterogenous solutions. Activation barriers to conformer exchange are low (typically $<\sim 5$ kcal mol^{-1}).[46] In contrast, many pyranosyl rings are considered to be conformationally rigid, such as those having the β-*gluco* configuration, where the $^4C_1$ form is highly preferred for the D-isomers. Activation barriers are considered to be very high for *gluco* conformer exchange in solution.[47] In contrast, pyranosyl rings having the *ido* configuration are much more flexible, leading to solutions containing multiple conformations.[48] In this case, conformer exchange is rapid on the NMR time-scale, leading to NMR spectra that are averaged in a manner that reflects the relative populations of conformers.

2.5.2 Exocyclic Hydroxymethyl Conformation

This conformational element refers to an exocyclic CH_2OH fragment appended to C5 of aldohexopyranosyl rings or to C4 of aldopentofuranosyl rings. Larger carbon-based exocyclic fragments are found in the furanose forms of aldohexoses (HOH_2C–CH(OH)–; ethylene glycol), *N*-acetyl-neuraminic acid (HOH_2C–CH(OH)–CH(OH)–; glycerol), and other saccharide derivatives. The conformations of these larger fragments will not be treated here.

Focusing on the CH_2OH fragment, there are three idealized (perfectly staggered) conformers available: *gg* (*gauche–gauche*; g^-); *gt* (*gauche-trans*; g^+); and *tg* (*trans-gauche*; *t*). This nomenclature uses O5 and O6 as the reference atoms as shown in Scheme 2.12. Factors that determine the distribution of these rotamers in solution include (a) the *gauche* effect,[43,44,49] which favors the *gg* and *gt* forms (O5 and O6 *gauche*) and (b) steric interactions, especially between O6 and O4. Statistical analyses of crystallographic data

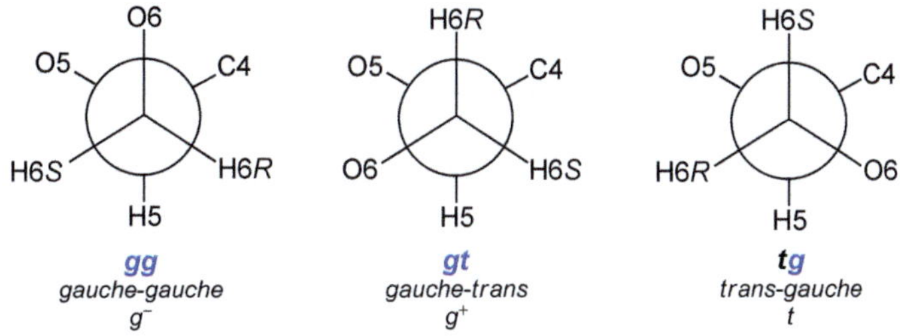

Scheme 2.12 Idealized staggered rotamers about the C5–C6 bond of aldohexopyranosyl rings.

reveal C5–C6 rotamers that are not perfectly staggered but have O5–C5–C6–O6 torsion angles within ±15° of the idealized structures shown in Scheme 2.12.[17,49]

2.5.3 Exocyclic C–O Bond Conformation

Two types of exocyclic C–O bonds are commonly found in saccharides: (1) those involving free hydroxyl groups, and (2) those involved in linkages between the monosaccharide residues of oligo- or polysaccharides. The latter bonds are discussed in Section 2.7.6. Exocyclic C–OH bonds are expected to rotate freely in most saccharides in solution, although conformational bias probably exists (see below). This rotation, however, is not to be ignored, because it exerts a significant effect on saccharide structure due to the lone-pair orbitals on oxygen. DFT calculations show that these orbitals interact with the antibonding orbitals of proximal C–H and C–C bonds, and thus their lengths change as the C–O bond rotates. This behavior renders the solution structure of a monosaccharide, and its intrinsic reactivity, distinct from its bound structure and reactivity because the rotameric distribution of any given C–O bond differs significantly in the solution (free) and bound states. In most cases, C–O bonds are "rigidified" in the bound state, wherein only one rotamer or an ensemble of closely related rotamers exist. This "rigidification" alters the intrinsic reactivity of the saccharide and must be taken into account when attempting to understand, for example, the origins of rate enhancements of enzymes that act on saccharide substrates. Molecular dynamics (MD) simulations of saccharides show that the three rotamers about a C–O bond are not perfectly staggered and are often not equally populated, and that these conformational biases originate mainly from intra- and/or intermolecular (solvent) H-bonding interactions. At present, limited experimental data are available that address C–O rotamer behavior quantitatively in solution. The effect of solvent on this behavior is also expected to be significant, largely caused by changes in the relative strengths of the intra- and intermolecular H-bonding contributions.

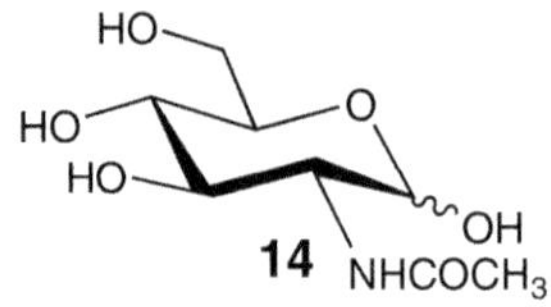

14

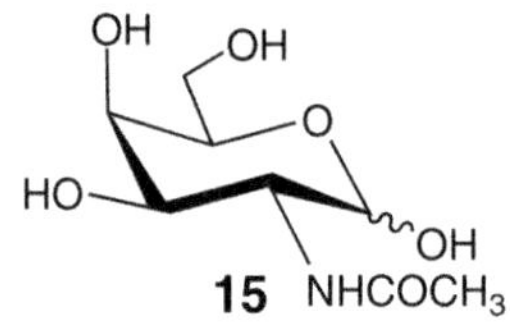

15

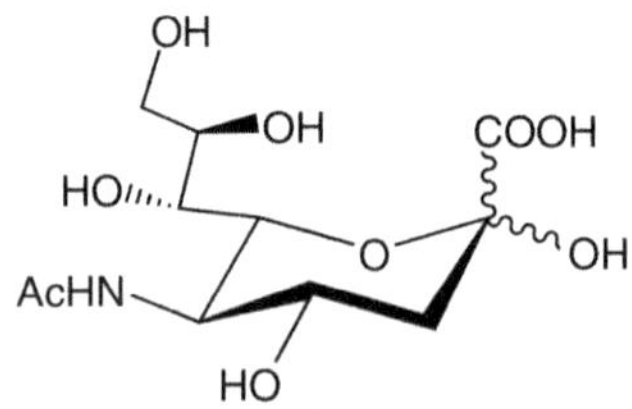

16

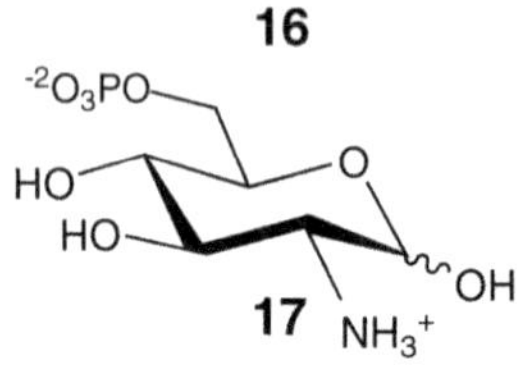

17

2.5.4 Exocyclic *N*-Acetyl Group Conformation

The *N*-acetyl substituent is a common covalent modification found in biologically important monosaccharides. Examples include *N*-acetyl-D-glucosamine (GlcNAc) **14**, *N*-acetyl-D-galactosamine (GalNAc) **15**, and *N*-acetyl-neuraminic acid (Neu5Ac) **16**. Throughout biochemistry, the *N*-acetyl group serves to reversibly mask the positive charge of free amines, such as those associated with lysine residues in proteins.[50] In the present context, 2-amino-2-deoxy-D-glucosamine 6P **17**, derived from the glycolytic metabolite D-fructose 6-phosphate (F6P), is converted enzymatically to its *N*-acetyl derivative, with acetyl-CoA serving as the source of the acetyl group.[51] Incorporation of an *N*-acetyl group alters the non-covalent bonding properties of the aminosugar, thus affecting mechanisms of substrate recognition by saccharide binding proteins and enzymes. The NH and C=O bonds serve as H-bond donors and/or acceptors, while the hydrophobic CH_3 group can penetrate hydrophobic pockets in a receptor active site.[52] Whether there are enzymes *in vivo* that control *N*-acetylation and de-*N*-acetylation of saccharides in a manner similar to that associated with histone modification remains to be determined.

From a structural standpoint, there are two degrees of freedom in the *N*-acetyl group of a 2-acetamido sugar that define its conformation

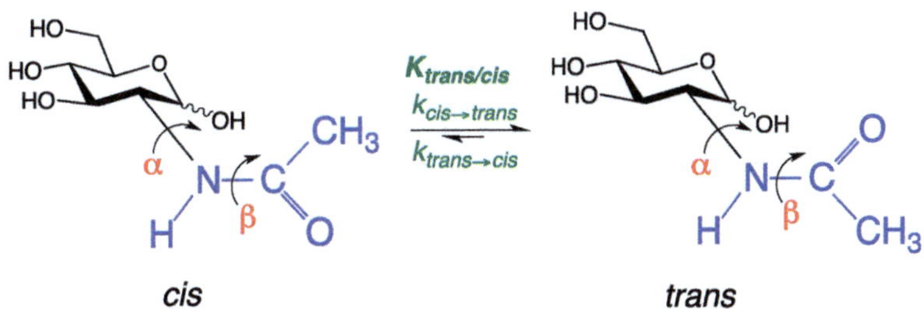

Scheme 2.13 *N*-Acetyl side-chain structure in the 2-acetamido sugar, GlcNAc **14**, showing the freely rotatable C2–N2 bond α and the conformationally restricted NH–CO (amide) bond β.

(Scheme 2.13): (1) rotation α about the C2–N2 bond; and (2) rotation β about the NH–CO (amide) bond. The α bond has, in principle, complete access to the full 360° rotational itinerary. The β (amide) bond is much more hindered in rotation (activation barrier of ~16–20 kcal mol^{-1}),[53,54] with only two planar conformational states accessible (Scheme 2.13): (1) *cis* (C2 eclipsed with CH_3), and (2) *trans* (C2 *trans* to CH_3). Both degrees of freedom affect the ability of this side-chain to H-bond either internally or to external partners. In crystal structures, the amide bond in 2-acetamido sugars is not completely planar, and is normally found in the *trans* configuration.[54]

2.5.5 *O*-Glycosidic Linkage Conformation

Monosaccharides are joined covalently by *O*-glycosidic bonds to produce oligo- and polysaccharides. These linkages can be classified into two groups: (1) those comprised of two bonds, and (2) those comprised of three bonds. Examples are shown in Scheme 2.14 for methyl α-D-mannopyranosyl-(1 → 3)-β-D-mannopyranoside **18** and methyl α-D-mannopyranosyl-(1 → 6)-α-D-mannopyranoside **19**. The two C–O bonds in **18** and **19** are denoted *phi* (ϕ) and *psi* (ψ), the former involving the anomeric carbon and the latter involving the aglycone carbon. The additional C–C bond in **19** is denoted *omega* (ω) and describes hydroxymethyl conformation (Section 2.4.2) in the context of a glycosidic linkage. Factors that affect the values of *phi* and *psi* include stereoelectronic effects (*exo*-anomeric effect; ϕ only),[55,56] and steric effects, which influence both C–O torsions. In most cases, the *exo*-anomeric effect[55–57] favors ϕ torsion angles in which the aglycone carbon is *anti* (or approximately so) to the vicinal carbon atom across the linkage (*e.g.*, C2′ *anti* to C3 in **18**; C2′ *anti* to C6 in **19**). In Scheme 2.14, the linkages involve acetal carbons (anomeric carbons) bearing hydrogens. In some cases, ketal or ketal-like carbons are involved in *O*-glycosidic linkages; a typical example is αNeu5AcβGalOCH_3 **20** in which the α-ketoacid *N*-acetyl-neuraminic acid (Neu5Ac, see **16**) is linked 2 → 6 to D-galactose. In this linkage, the anomeric carbon of Neu5Ac (C2′)

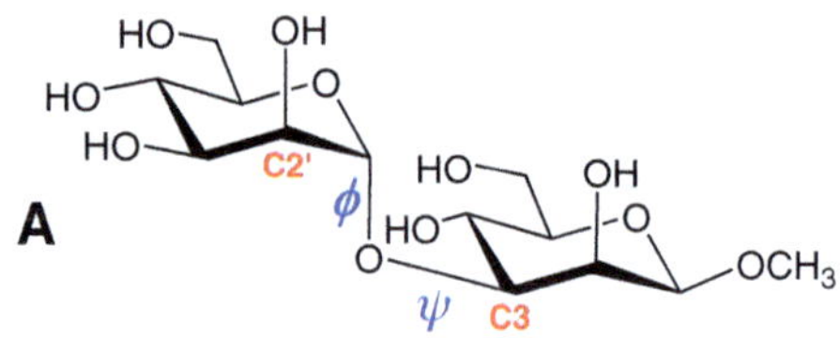

αMan13βManOCH$_3$ **18**

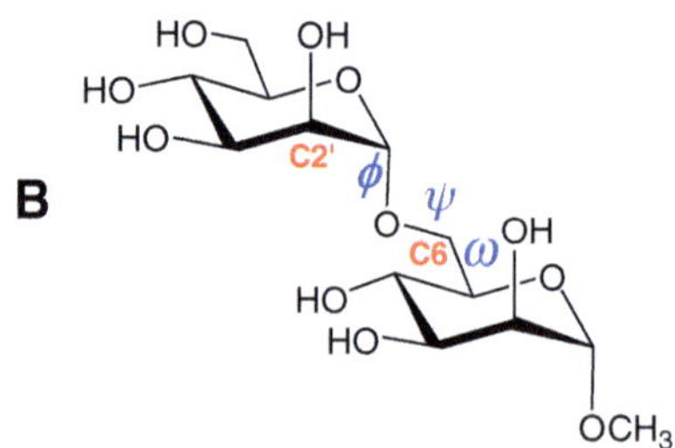

αMan16αManOCH$_3$ **19**

Scheme 2.14 Common types of *O*-glycosidic linkages in oligosaccharides, illustrated with the Man–Man disaccharides **18** and **19**. (A) Linkage comprised of C–O bonds, ϕ and ψ. (B) Linkage comprised of two C–O bonds ϕ and ψ and a C–C bond ω.

bears no hydrogen, which has implications for NMR assessments of linkage conformation (see Section 2.7.6).

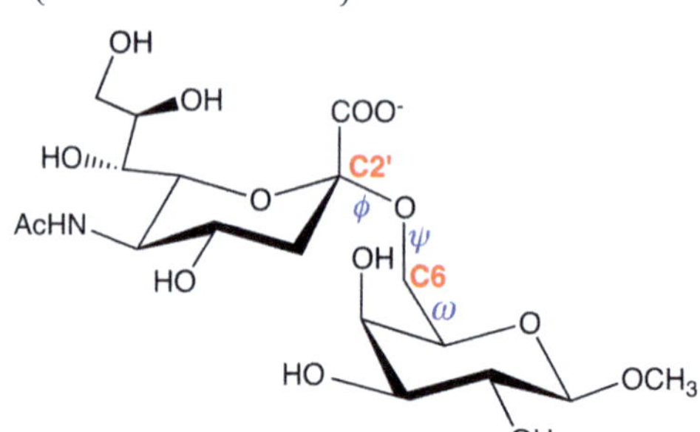

αNeu5AcβGalOCH$_3$ **20**

2.6 Assets and Limitations of *J*-Couplings to Determine Saccharide Structure

NMR *J*-couplings report on local structure because they are mainly "through-bond" parameters (as opposed to "through-space"). This feature is an asset and a limitation. *J*-couplings provide remarkable insight into molecular structure; they are exquisitely sensitive to bond lengths, bond angles and bond torsions along the coupling pathway, and to the relative disposition of electronegative substituents along the coupling pathway.

If these individual dependencies can be defined, then *J*-couplings can be used to investigate not only structural but also functional properties. For example, *J*-couplings have the potential to report on H-bond strength in solution.[58] However, because they are local reporters of molecular structure, *J*-couplings cannot be used by themselves to determine the global fold of an oligosaccharide. The latter information is more likely obtained from residual dipolar couplings (RDCs),[59] whose magnitudes are determined in part by a through-space mechanism, although in practice short (local) and long-distance information should be used collectively to assign global fold.

A notable limitation of *J*-couplings is that they can be difficult to interpret when the molecule is conformationally flexible and when this motion is very rapid relative to the frequency of the NMR observation (fast exchange limit), as found for many saccharide exchange events (see below). Under these conditions, *J*-couplings average linearly, meaning that if two conformers C_a and C_b contribute equally to the population, and *J* values of 5 Hz and 10 Hz are observed in C_a and C_b, then the experimentally observed *J* will be 0.5 (5 Hz) + 0.5 (10 Hz) = 7.5 Hz. This behavior contrasts with that of NOEs, rotating frame NOEs (ROEs), and RDCs which average non-linearly, behavior that evolves from their r^{-3} or r^{-6} dependencies on internuclear distance, where conformers containing shorter distances contribute more to the averaged experimental distance than dictated by their actual populations in solution.[60] Linear averaging of *J*-couplings is an attractive property when the identities of the individual conformers undergoing exchange, and the *J*-couplings associated with them, are known. However, when this is not the case, which is not uncommon, deconvoluting the experimentally observed *J*-coupling into populations of conformers in solution can prove challenging.

Concurrent with *J*-coupling exchange averaging is chemical shift averaging. In general, the uncertainty in the resonance frequency of an NMR signal is related to the lifetime τ by the relationship,

$$\Delta\nu = h/(2\pi\tau) \tag{2.1}$$

which shows that the resonance frequency ν (in Hz) can be measured precisely when the lifetime is long (large $\tau \rightarrow$ small $\Delta\nu \rightarrow$ more precise line position).[61] As lifetime decreases, $\Delta\nu$ increases, an effect known as lifetime broadening. In practical terms, if two sites with signals at ν_a and ν_b are undergoing chemical exchange, and the lifetime of the two states is long relative to the difference $|\nu_a - \nu_b|$, then two separate signals are observed. On the other hand, if τ is small relative to the separation, then line-broadening of the two signals is observed (exchange broadening), and both signals coalesce at τ_{coal}, the coalescence lifetime. The latter can be calculated[61] from

$$\tau_{coal} = (\sqrt{2}\pi\Delta\nu)^{-1} = k^{-1} \tag{2.2}$$

Thus, for a spin-coupled nucleus undergoing rapid chemical exchange between two conformational states with chemical shifts of ν_a and ν_b, with both states equally populated, a single resonance will be observed at $(\nu_a + \nu_b)/2$, that is, the chemical shifts will be averaged linearly. In many saccharide exchange processes at room temperature and at typical static magnetic fields B_0 (*e.g.*, 600 MHz), a single average resonance is observed (shift averaging) on which is superimposed the linear averaging of the *J*-coupling. This type of *J*-coupling averaging is distinct from averaging caused by the effects of spin relaxation, where very rapid nuclear relaxation effectively decouples the spin interaction and the line-splitting is lost. That is, if we view $\Delta\nu$ in eqn (2.2) as a *J*-value, and if the lifetime of the nucleus in a particular environment is short compared to the reciprocal of the scalar coupling, then *J*-coupling will not be observed, and only a single coalesced line accrues.

It should be evident from eqn (2.2) that τ_{coal} for a given exchange process will be field-dependent, because the separation, $\Delta\nu$, in Hz varies linearly with B_0. Thus, whether an exchange process is fast or slow on the NMR time-scale under a given set of solution conditions depends on the frequency of the NMR observation.

Estimates of the time-scales of common structural motions in saccharides are as follows: (1) pyranosyl ring pseudorotation, $>10^{-6}$ s; (2) furanosyl ring pseudorotation, $<10^{-9}$ s; (3) hydroxymethyl group rotation, $<10^{-8}$ s; (4) C–O (hydroxyl group) bond rotation, $<10^{-9}$ s; (5) C–N bond rotation (aminosugars), $>10^{-6}$ s; (6) *cis–trans* isomerization of amide bonds in *N*-acetyl side-chains, very slow; (7) C–O bond rotation (*O*-glycosidic linkages); $>10^{-6}$–10^{-9} s. These time-scales are estimated in some cases from molecular dynamics (MD) simulations and are rough approximations. Furanosyl ring pseudorotation and C–O bond rotation appear to occur on sub-nanosecond timescales. Pyranosyl ring pseudorotation, C–N bond rotation, and C–O bond rotation occur more slowly (microseconds or greater), and context appears to affect linkage time-scales appreciably, with some more dynamic than others. The slowest process is *cis–trans* amide bond exchange, which occurs on a time-scale slow enough to allow individual signals for the *cis* and *trans* forms to be observed by NMR (*i.e.*, slow exchange).[54] These behaviors are shown on the time-scale plot in Figure 2.10 with those associated with other biological processes for reference. With the exception of amide *cis–trans* isomerization, all other dynamic processes in saccharides occur rapidly enough to give time-averaged chemical shifts and *J*-couplings in NMR spectra measured at room temperature.

NMR *J*-couplings are very abundant in saccharides comprised of only carbon, oxygen and hydrogen atoms. For example, the β-D-glucopyranosyl ring **21** contains fifty J_{HH}, J_{CH} and J_{CC} values involving the carbon-bound hydrogens (Scheme 2.15), some of which report on ring conformation, and others on the ω (C5–C6) and θ (C6–O6) torsion angles in the hydroxymethyl fragment. Sixteen *J*-couplings are also available if the hydroxyl hydrogens are involved as coupled nuclei; these couplings include $^2J_{C2,O2H}$, $^3J_{H2,O2H}$, $^3J_{C1,O2H}$, $^3J_{C3,O2H}$, $^2J_{C3,O3H}$, $^3J_{H3,O3H}$, $^3J_{C2,O3H}$, $^3J_{C4,O3H}$, $^2J_{C4,O4H}$, $^3J_{H4,O4H}$, $^3J_{C3,O4H}$, $^3J_{C5,O4H}$, $^2J_{C6,O6H}$, $^3J_{H6R,O6H}$, $^3J_{H6S,O6H}$ and $^3J_{C5,O6H}$. This abundance of parameters produces

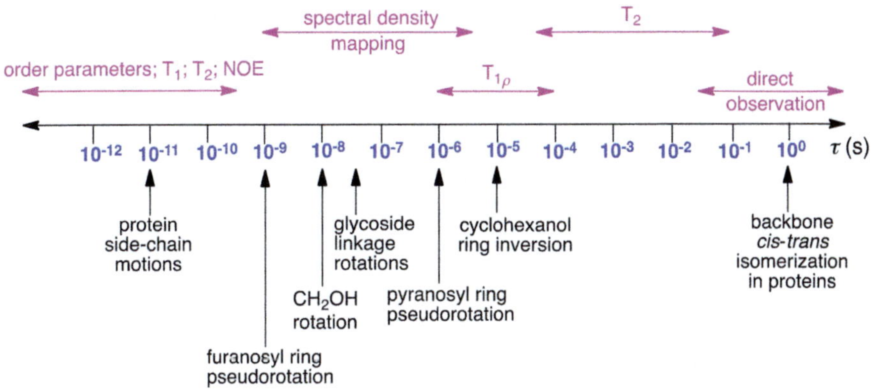

Figure 2.10 Approximate time-scales for different molecular motions in saccharides and other biomolecules, and NMR techniques appropriate for their detection and measurement.

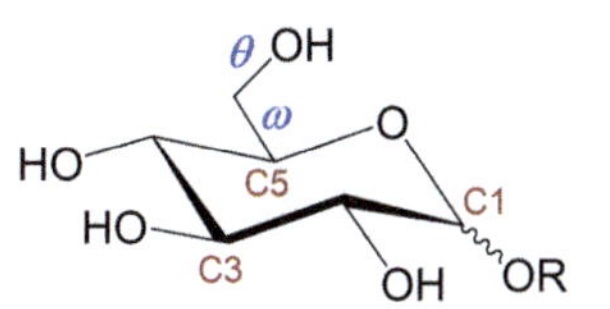

α,β-D-glucopyranosyl ring **21**

J_{HH}: $^3J_{H1,H2}$, $^3J_{H2,H3}$, $^3J_{H3,H4}$, $^3J_{H4,H5}$, $^3J_{H5,H6R}$, $^3J_{H5,H6S}$, $^2J_{H6R,H6S}$ — **Total = 7**

$^1J_{CH}$: $^1J_{C1,H1}$, $^1J_{C2,H2}$, $^1J_{C3,H3}$, $^1J_{C4,H4}$, $^1J_{C5,H5}$, $^1J_{C6,H6R}$, $^1J_{C6,H6S}$ — **Total = 7**

$^2J_{CH}$: $^2J_{C1,H2}$, $^2J_{C2,H1}$, $^2J_{C2,H3}$, $^2J_{C3,H2}$, $^2J_{C3,H4}$, $^2J_{C4,H3}$, $^2J_{C4,H5}$, $^2J_{C5,H4}$, $^2J_{C5,H6R}$, $^2J_{C5,H6S}$, $^2J_{C6,H5}$ — **Total = 11**

$^3J_{CH}$: $^3J_{C1,H3}$, $^3J_{C1,H5}$, $^3J_{C2,H4}$, $^3J_{C3,H1}$, $^3J_{C3,H5}$, $^3J_{C4,H2}$, $^3J_{C4,H6R}$, $^3J_{C4,H6S}$, $^3J_{C5,H1}$, $^3J_{C5,H3}$, $^3J_{C6,H4}$ — **Total = 11**

$^1J_{CC}$: $^1J_{C1,C2}$, $^1J_{C2,C3}$, $^1J_{C3,C4}$, $^1J_{C4,C5}$, $^1J_{C5,C6}$ — **Total = 5**

$^2J_{CC}$: $^2J_{C1,C3}$, $^2J_{C1,C5}$, $^2J_{C2,C4}$, $^2J_{C3,C5}$, $^2J_{C4,C6}$ — **Total = 5**

$^3J_{CC}$: $^3J_{C1,C6}$, $^3J_{C3,C6}$ — **Total = 2**

$^{3+3}J_{CC}$: $^{3+3}J_{C1,C4}$, $^{3+3}J_{C2,C5}$ — **Total = 2**

Scheme 2.15 J_{HH}, J_{CH} and J_{CC} values in an α,β-D-glucopyranosyl ring **21**. *J*-Couplings to the hydroxyl hydrogens are not listed. Values in black relate to ring conformation; values in green relate to conformations about ω and/or θ.

a level of *redundancy* in structure determination that is unmatched by other NMR parameters. This redundancy, when used, provides a higher level of confidence in conformational assignments provided that the relationships between conformation of the structural element and the magnitudes/signs of the pertinent *J*-couplings differ (*i.e.*, non-overlapping correlations). Redundancy also applies to studies of furanose ring conformations, *N*-acetyl side-chains, and C–O bond conformations for free hydroxyl groups and *O*-glycosidic linkages, as discussed in Section 2.7.

2.7 Core Relationships Between *J*-Coupling and Saccharide Structure

Structural interpretations of *J*-couplings in saccharides depend on the availability of quantitative relationships between their magnitudes and signs and one or more structural elements such as a bond length, angle or torsion. Tacit recognition of the importance of oxygen and nitrogen lone-pair orbitals on saccharide structure, and indirectly on saccharide *J*-couplings, is vital to deriving these relationships. Some of these relationships are discussed here.

2.7.1 High Abundance of Oxygen and Nitrogen Lone-Pair Orbitals in Saccharides

While it is natural to view saccharide structure solely in terms of the types and locations of covalent bonds, it is equally, if not more important to pay attention to the lone-pair orbitals in these structures (Scheme 2.16). Saccharides possess these orbitals, mainly residing on oxygen and nitrogen substituents, in greater abundance and density than any other class of biomolecule. This feature contributes to the well-known stereoelectronic properties of saccharides involving the anomeric carbon, such as the *endo*-anomeric effect[55,56,62] (dictates the relative stabilities of the axial and equatorial orientations of anomeric C–O bonds in furanosyl and pyranosyl rings) and the *exo*-anomeric effect[55,57] (dictates the relative stabilities of rotamers involving anomeric C–O bonds, especially in *O*-glycosidic linkages). However, the impact of these lone-pairs transcends their roles involving anomeric carbons. Quantum

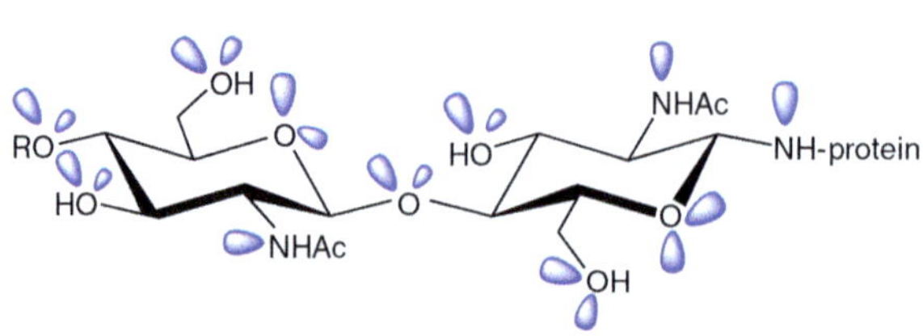

Scheme 2.16 Illustration of the high lone-pair density in the βGlcNAc14βGlcNAc disaccharide that connects *N*-glycans to human glycoproteins. Structural effects of the indicated nitrogen lone-pairs will be mitigated by delocalization into the amide bonds.

chemical calculations show that the rotation of exocyclic C–O bonds exerts a significant effect on saccharide covalent structure, meaning that bond lengths and angles are significantly affected by C–O bond conformation.[17] This effect is caused by changes in the relative dispositions of lone-pair orbitals with respect to proximal covalent bonds (bonding and anti-bonding orbitals) as a C–O bond is rotated. Indeed, the *intrinsic* reactivities of saccharides free in solution and bound to a receptor differ in part because of differences in the distribution of C–O rotamers in both states. Because of the strong correlation between C–O bond conformation and saccharide structure, notably bond lengths and angles, it is not surprising that C–O bond conformation exerts a major effect on NMR *J*-couplings, and efforts to treat *J*-values quantitatively must come to terms with these contributions. However, evaluating C–O bond conformation in saccharides experimentally and computationally is challenging, as discussed below. In the long term, however, such evaluations will be vital to progress in the field of structural glycobiology.

To illustrate the conformational behaviors of exocyclic C–O bonds and their effects on saccharide structure, density functional theory (DFT) and aqueous molecular dynamics (MD) simulations were applied to a model monosaccharide, methyl β-D-glucopyranoside **1** (Scheme 2.17) (for details on how these calculations are performed, see ref. 17). In both calculations, three C–O bond torsions were fixed, one (C1–O1) was allowed to optimize from its initial value, and one (C2–O2) was rotated in 15° increments through 360° (DFT) or allowed to rotate freely (MD). Rotamer definitions for the C2–O2 bond in **1** are shown in Scheme 2.18. The results of these calculations are shown in Figure 2.11. Both methods suggest that two of the rotamers (θ = +60° and 180°) are favored, although the two methods give different relative abundances, while the third (θ = 300°) is considerably less populated.

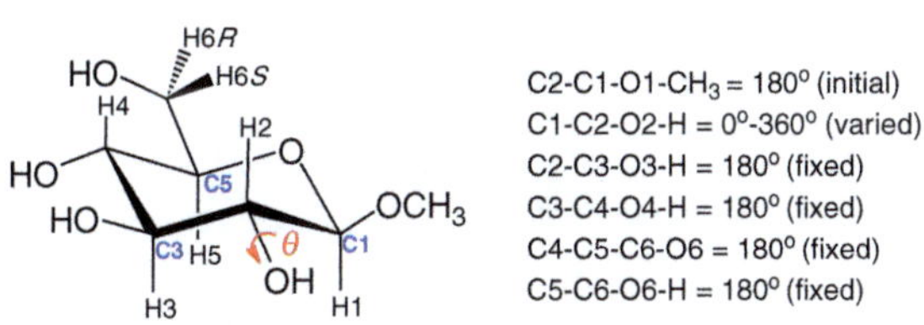

Scheme 2.17 Methyl β-D-glucopyranoside (**1**) DFT and MD model.

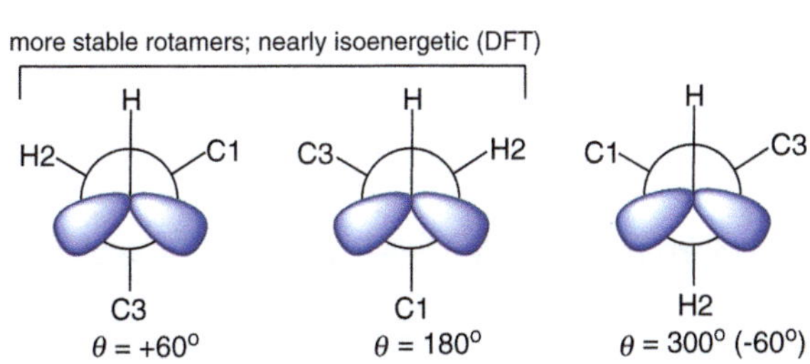

Scheme 2.18 Definitions of the C2–O2 rotamers in **1** (Scheme 2.17).

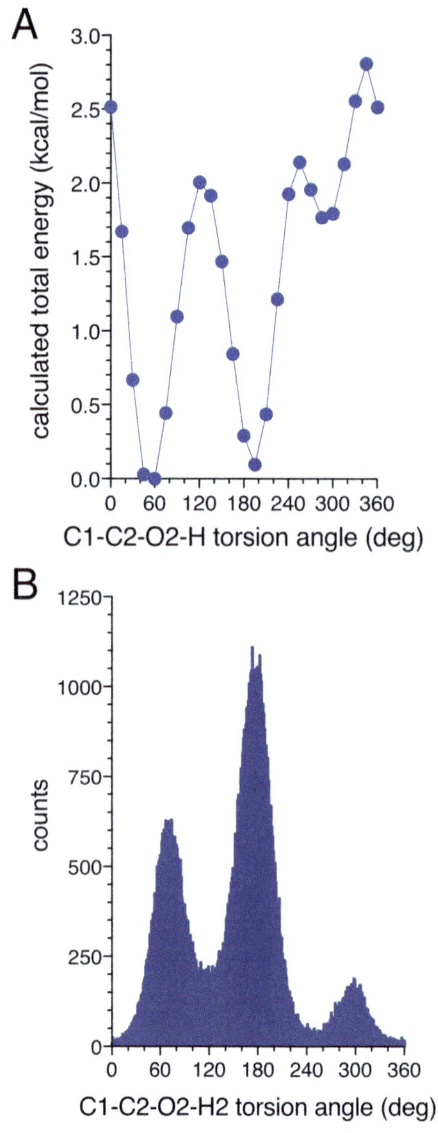

Figure 2.11 (A) Effect of rotating the C2–O2 bond on the total energy of **1** (Scheme 2.17) determined from DFT calculations. (B) Distribution of C2–O2 bond rotamers in **1** predicted by aqueous MD simulation (1 μs simulation).

The three rotamers are close to ideal (*i.e.*, perfectly staggered), behavior much different than observed for C–O bonds involved in *O*-glycosidic linkages.[17] A question that arises from these findings is why the $\theta = 300°$ rotamer is destabilized. In this rotamer, both lone-pair orbitals on O2 experience equivalent destabilizing interactions with a lone-pair orbital on either O1 or O3. In the remaining two rotamers, one of these interactions is replaced by an attractive H-bonding interaction between O2H (donor) and a lone-pair orbital (acceptor) on either O1 or O3. Contributions to the energetics arising from 1,3-diaxial interactions involving H1 and H3 and either a lone-pair or O2–H bond on C2 are neglected in this treatment. What is clear, however, is that the populations of a given C–O rotamer will be heavily influenced by interactions of its lone-pair orbitals and hydrogen with vicinal hydroxyl groups. In the present case, the adjacent C–O bond torsions were fixed to simplify the interpretation of the data, but in reality potent synergy between adjacent C–O rotamers could produce more complex behaviors, possibly leading to non-ideal rotamers, while H-bonding to solvent might also influence C–O rotamer behavior.

Examples of lone-pair (LP) interactions that influence covalent bond lengths in saccharides are summarized in Scheme 2.19, illustrated for C–H bonds only. Vicinal LP interactions lengthen C–H bonds, and 1,3- and 1,4-interactions shorten them (see below). Similar effects are observed for C–C bonds. These bond length effects are important because C–C and C–H bond lengths influence J_{HH}, J_{CH} and J_{CC} values significantly. A lack of knowledge of C–O bond conformations in solution thus introduces uncertainties in the structural interpretations of *J*-couplings. In the following discussion, the model system shown in Scheme 2.17, in which the C2–O2 bond in methyl β-D-glucopyranoside **1** is rotated and its effect on calculated structures and *J*-couplings determined by DFT, will be used to illustrate the behaviors of specific types of J_{CH} and J_{CC} values in saccharides.

A. One-Bond ^{13}C–^{1}H and ^{13}C–^{13}C Spin-Couplings. Direct *J*-couplings depend strongly on bond length, with shorter bonds (greater *s*-character) corresponding to larger *J*-values.[2,63] Because of the strong dependence on bond length, lone-pair effects on $^1J_{CH}$ and $^1J_{CC}$ values are significant in saccharides. For example, vicinal lone-pair orbitals (see Scheme 2.19) lengthen both C–H and C–C (not shown) bonds, and this effect is often reflected in smaller $^1J_{CH}$ (and $^1J_{CC}$) values. It thus becomes apparent why structural interpretations of experimental $^1J_{CH}$ and $^1J_{CC}$ values in saccharides depend heavily on knowledge of the behaviors of exocyclic C–O torsions in solution (see above).

The effects of C2–O2 bond conformation in **1** on $r_{C1,H1}$, $r_{C2,H2}$, and $r_{C3,H3}$ are shown in Figure 2.12. The vicinal lone-pair effect on C–H bond length

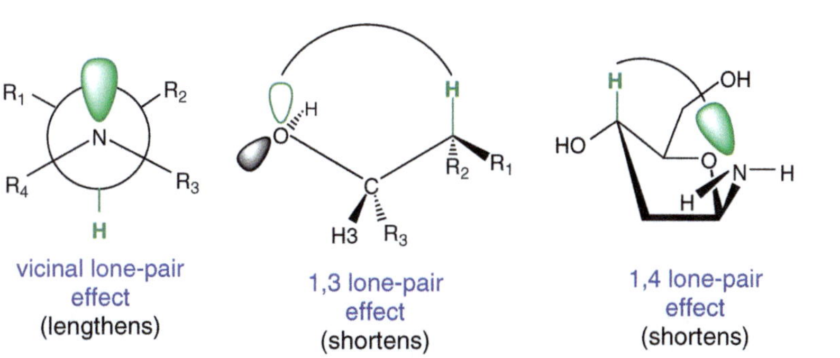

Scheme 2.19 Effect of oxygen and nitrogen lone-pair orbitals on saccharide C–H bond lengths.

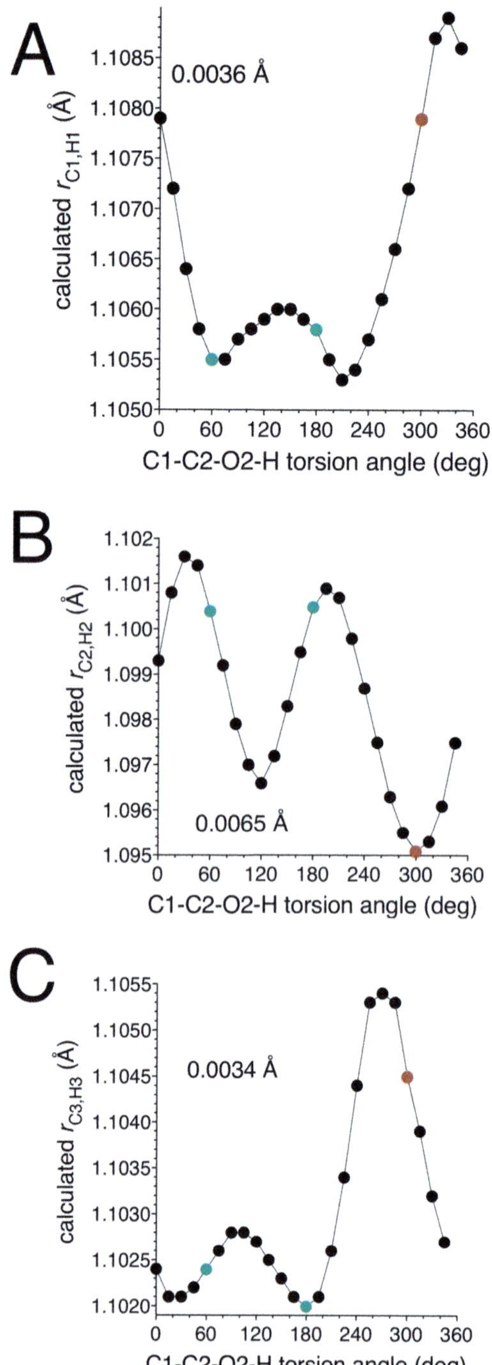

Figure 2.12 Effect of C2–O2 bond rotation in **1** on (A) $r_{C1,H1}$, (B) $r_{C2,H2}$, and (C) $r_{C3,H3}$. Values inserted into each plot show the overall change in bond length (in Å) as the C2–O2 bond is rotated through 360°. Points shown in green and red are those found in the three perfectly staggered C2–O2 rotamers (Scheme 2.18).

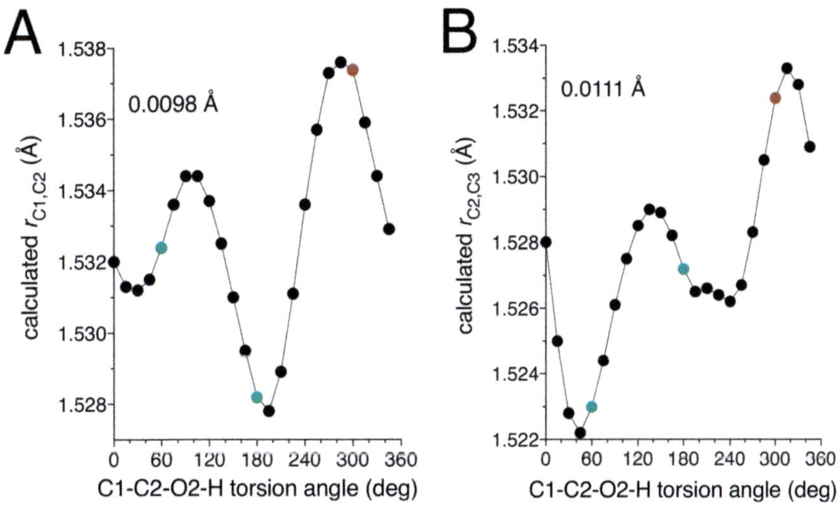

Figure 2.13 Effect of C2–O2 bond rotation in **1** on (A) $r_{C1,C2}$ and (B) $r_{C2,C3}$. Values inserted into each plot show the overall change in bond length (in Å) as the C2–O2 bond is rotated through 360°. Points shown in green and red are those found in the three perfectly staggered C2–O2 rotamers (Scheme 2.18).

is illustrated in Figure 2.12B, where $r_{C2,H2}$ is large at $\theta = +60°$ and 180°, and considerably smaller at $\theta = 300°$ (Scheme 2.18). The overall change in bond length is ~0.007 Å. 1,3-Lone-pair effects on C–H bond length are illustrated in Figure 2.12A and C. Both $r_{C1,H1}$ and $r_{C3,H3}$ are smaller at $\theta = +60°$ and 180° than at $\theta = 300°$, indicating that 1,3-interactions truncate C–H bond lengths. In both cases, the overall change in bond length is ~0.003 Å, that is, ~1/2 that found for the vicinal effect. Ring structure precludes the use of **1** to illustrate 1,4-lone-pair effects on C–H bond lengths caused by C2–O2 bond rotation.

Vicinal lone-pair effects on C–C bond lengths are shown in Figure 2.13. In the staggered C2–O2 rotamers, $r_{C1,C2}$ and $r_{C2,C3}$ are large when $\theta = +60°$ and 180°, respectively, in which an O2 lone-pair is *anti* to the C–C bond. Both bonds shorten at $\theta = 180°$ and 60°, respectively, in which the O2–H bond is *anti* to the C–C bond. Interestingly, $r_{C1,C2}$ and $r_{C2,C3}$ are greatest at $\theta = 300°$, that is, when O2 lone pairs are *anti* to both C–C bonds. In the latter case, the vicinal lone-pair effect appears additive; bond elongation for contiguous C–C bonds in saccharide rings is enhanced when exocyclic hydroxyl oxygen lone-pairs are *anti* to both bonds. The latter behavior has important implications for the geminal *J*-coupling, $^2J_{C1,C3}$ (see Section 2.7.2).

1,3-Lone-pair effects are also observed for C–O bonds, as shown in Figure 2.14A and B. The C1–O1 bond is short at $\theta = 180°$ and 300°, and elongated at $\theta = 60°$. Likewise, the C3–O3 bond is short at $\theta = 60°$ and 300°, and elongated at $\theta = 180°$. The effect of C2–O2 bond rotation on $r_{C2,O2}$ in **1** is shown in Figure 2.14C; bond lengths are minimal in the three staggered rotamers

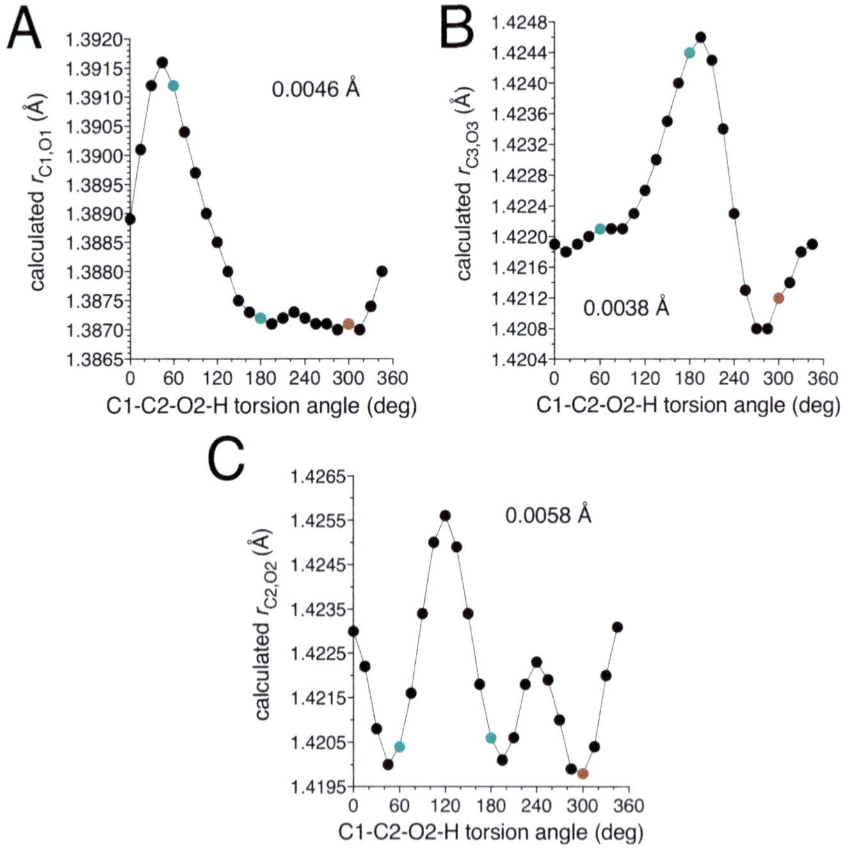

Figure 2.14 Effect of C2–O2 bond rotation in **1** on (A) $r_{C1,O1}$, (B) $r_{C3,O3}$ and (C) $r_{C2,O2}$. Values inserted into each plot show the overall change in bond length (in Å) as the C2–O2 bond is rotated through 360°. Points shown in green and red are those found in the three perfectly staggered C2–O2 rotamers (Scheme 2.18).

($\theta = 60°$, 180° and 300°), and are elongated in the three eclipsed rotamers ($\theta = 0°$, 120° and 240°), the latter effect likely caused by steric compression. Elongation is greatest at $\theta = 120°$ where the two O2 lone-pairs eclipse the C1–C2 and C2–C3 bonds.

Plots shown in Figures 2.12 and 2.13 can be related to corresponding plots of calculated $^1J_{CH}$ and $^1J_{CC}$ values as a function of C2–O2 bond rotation in **1** (Figure 2.15). A consistent relationship emerges between r_{CH} and $^1J_{CH}$; corresponding plots in Figures 2.12 and 2.15 are almost mirror images about the *x*-axis, and in general, shorter C–H bonds produce larger $^1J_{CH}$. This relationship, however, is not consistently observed for $^1J_{C2,H2}$. The relationship holds in the staggered forms, but factors in addition to bond length apparently influence $^1J_{C2,H2}$ in other C2–O2 rotamers. Note that in the region of $\theta = 60–120°$, both $r_{C2,H2}$ and $^1J_{C2,H2}$ are decreasing. In the

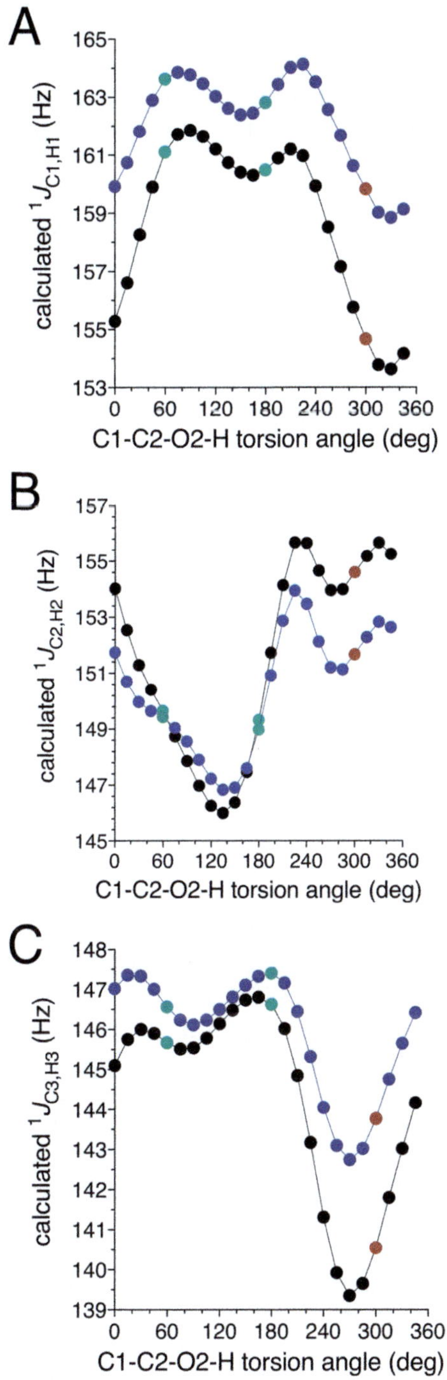

Figure 2.15 Effect of C2–O2 bond rotation in **1** on (A) $^1J_{C1,H1}$, (B) $^1J_{C2,H2}$, and (C) $^1J_{C3,H3}$. Points shown in green and red are those found in the three perfectly staggered C2–O2 rotamers (Scheme 2.18). Black points, no solvent; blue points, with solvent. All geometry optimizations were conducted with solvent.

eclipsed C2–O2 rotamers, $r_{C2,H2}$ (0°) > $r_{C2,H2}$ (240°) > $r_{C2,H2}$ (120°), but $^1J_{C2,H2}$ (240°) > $^1J_{C2,H2}$ (0°) > $^1J_{C2,H2}$ (120°). The steric compression effects observed for $r_{C2,O2}$ (Figure 2.14C) in eclipsed rotamers probably contribute to the aberrant behavior of $^1J_{C2,H2}$.

An important feature of the plots in Figure 2.15 is the effect of solvent on calculated $^1J_{CH}$ values (for details on including solvent in DFT calculations, see ref. 17). Including solvent water (modeled as a surrounding polarizable dielectric continuum, *i.e.*, without allowance for explicit H-bonds) almost uniformly increases the calculated $^1J_{CH}$ value at a given C2–O2 torsion angle and truncates the dynamic range. This behavior mimics experimental observations (see Figure 2.16), where transitioning to a less polar solvent (*e.g.*, water to tetrahydrofuran (THF)) results in smaller experimental $^1J_{CH}$ (we assume that unsolvated (vacuum) DFT data more closely correspond to experimental data obtained in less polar solvents). It should be noted that the effect of including solvent water in DFT calculations of NMR *J*-couplings is most significant for $^1J_{CH}$ and $^1J_{CC}$; the effect is much smaller (and within the errors of the calculations) for geminal and vicinal ^{13}C–^1H and ^{13}C–^{13}C *J*-couplings (see Sections 2.7.2 and 2.7.3).

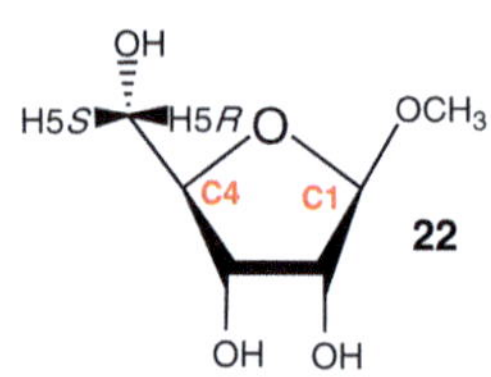

22

Related bond length effects are observed on ^1J values in other saccharide ring systems. For example, $^1J_{C4,H4}$ values in methyl β-D-ribofuranoside **22** depend on ring conformation. In the E_4 conformer ($P/\pi = 0.3$) (see Scheme 2.10), the C4–H4 bond is *anti* to one of the lone-pair orbitals on O4 (ring oxygen), whereas in ^4E ($P/\pi = 1.3$) neither lone-pair orbital on O4 is *anti* to the C4–H4 bond. On this basis, $^1J_{C4,H4}$ is predicted to be larger in ^4E than in E_4 in which the bond is elongated, and this behavior is confirmed by DFT calculations (Figure 2.17).

Superimposed on lone-pair orbital effects on ^1J magnitude are bond orientation effects. In general, a given C–H bond is shorter when equatorial (or *quasi*-equatorial) than when the same bond is axial (or *quasi*-axial), although exceptions exist. Thus, $^1J_{C1,H1}$ is smaller in a β-D-aldopyranosyl ring ($^4C_1$ form) than in the corresponding α-D-aldopyranosyl ring ($^4C_1$ form), the former orienting the C1–H1 bond axial and the latter equatorial.[64–66] The difference is typically ~10 Hz, and the effect in this case is accentuated by vicinal lone-pair effects arising from the ring oxygen that further lengthens the C1–H1 bond in the β-anomer in which one of the O5 lone-pairs is *anti* to the axial C1–H1 bond. In a similar vein, in an aldofuranosyl ring such as **22**, $r_{C2,H2}$ is *quasi*-axial and longer in the ^2E conformer than in E_2, where it is *quasi*-equatorial; in this case ring pseudorotation affects $r_{C2,H2}$ appreciably as its orientation changes from *quasi*-axial to *quasi*-equatorial.[67] Similar behavior is observed for C–C bonds. For example, $r_{C4,C5}$ in **22** is shorter in E_4 in which the bond is

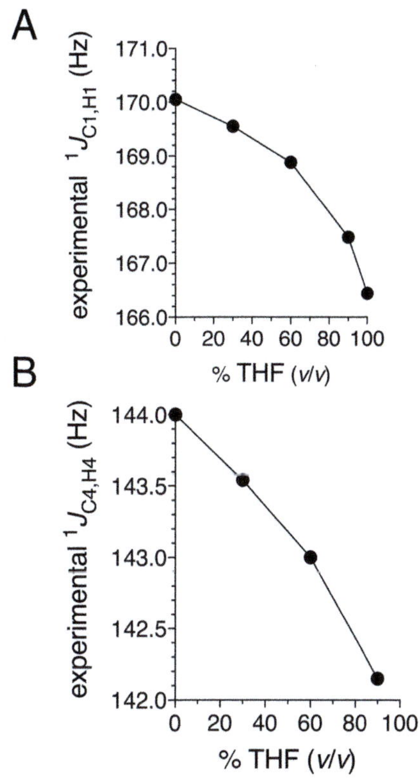

Figure 2.16 Effect of solvent on $^1J_{CH}$ values in methyl α-D-glucopyranoside **23**. (A) $^1J_{C1,H1}$. (B) $^1J_{C4,H4}$. Solvent was varied from 100% ^2H_2O to mixed THF-^2H_2O solvents (v/v) up to ~100% THF. Data were collected at 25 °C using **23** selectively labeled at either C1 or C4 with ^{13}C. Concentrations of glycoside were ~100 mM in solvents containing ^2H_2O and ~6 mM in pure THF. $^1J_{CH}$ values (±0.1 Hz) were measured from ^{1}H NMR spectra.

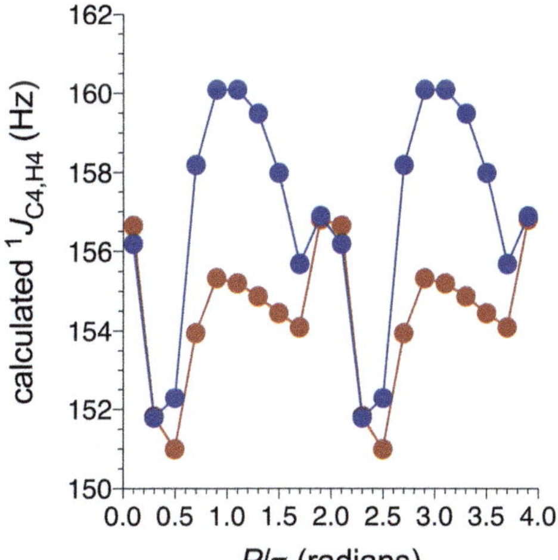

Figure 2.17 Effect of furanose ring conformation on $^1J_{C4,H4}$ in **22** determined by DFT. The two curves pertain to two different combinations of C2–O2 and C3–O3 bond conformations used in the calculations. In both cases, $^1J_{C4,H4}$ is smallest at conformations near E_4 (P/π = 0.3) in which the C4–H4 bond is *anti* to an O4 lone-pair orbital.

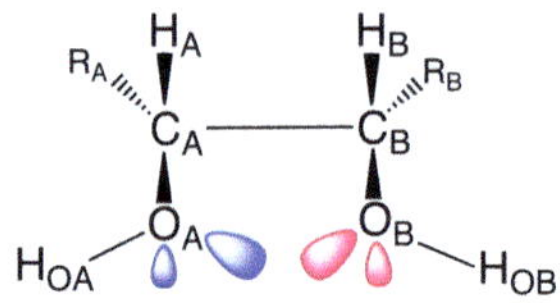

Scheme 2.20 Model ethylene diol fragment illustrating vicinal and 1,3-lone pair effects on proximal C–H and C–C bond lengths in saccharides.

quasi-equatorial than in ^4E in which the bond is *quasi*-axial.[68] This behavior renders $^1J_{C4,C5}$ larger in E_4 than in ^4E.[68]

In light of this discussion, we can now inspect a saccharide vicinal diol fragment (Scheme 2.20) that contains C–H, C–O and C–C bonds. Rotating the C_A–O_A and C_B–O_B bonds modulates C–H and C–C bond lengths involving C_A and C_B *via* vicinal lone-pair effects (Scheme 2.19), resulting in changes in $^1J_{CA,HA}$, $^1J_{CB,HB}$ and $^1J_{CA,CB}$. Additionally, 1,3-lone-pair effects are also operating (Scheme 2.19). For example, rotating the C_A–O_A bond modulates 1,3-lone-pair interactions with the C_B–H_B bond. These vicinal lone-pair and 1,3-lone pair effects oppose one another, one lengthening the bond, the other shortening it (Scheme 2.19). Interestingly, rotating the C_A–C_B bond while keeping both C–O bond conformations fixed also affects $^1J_{CA,CB}$, with that conformation having O_A and O_B eclipsed (*cis*) giving a smaller coupling than when both oxygen atoms are *trans*.[69] It appears, however, that C_A–C_B bond rotation affects $^1J_{CA,CB}$ less than do rotations about the individual C_A–O_A and C_B–O_B bonds.[69]

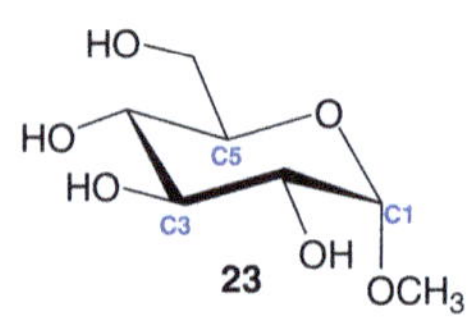

23

Thus far we have focused on intramolecular factors that affect r_{CH} and r_{CC} and thus $^1J_{CH}$ and $^1J_{CC}$ in saccharides. Intermolecular effects are also present, arising mainly from solvent interactions. Solvent effects on $^1J_{CH}$ values[70] are greater than those for $^1J_{CC}$ values, presumably because one of the coupled atoms (H) is terminal and potentially more exposed to solvent interactions. An example of this dependence is shown in Figure 2.16. In general, decreasing solvent polarity reduces $^1J_{CH}$ values. In methyl α-D-glucopyranoside **23**, $^1J_{C1,H1}$ = 170.1 Hz in ^2H_2O and 166.4 Hz in THF. Likewise, $^1J_{C4,H4}$ = 144.0 Hz in ^2H_2O and 142.1 Hz in 90% THF/10% ^2H_2O. These limited data suggest that the solvent effect is site-independent, and at least two factors contribute to it. In ^2H_2O, exocyclic OH groups are likely involved in H-bonding with solvent that will (1) affect C–O rotamer populations (and thus lone-pair interactions with the C–H bonds) and (2) affect r_{CH} directly *via* the H-bond itself (see below). In a less polar solvent like THF, vicinal OH groups have a greater tendency to H-bond with one-another, resulting in altered C–O bond conformations, altered vicinal lone-pair interactions with the C–H bonds, altered r_{CH}, and altered $^1J_{CH}$. Hydrogen bonding to THF is also restricted to the solvent

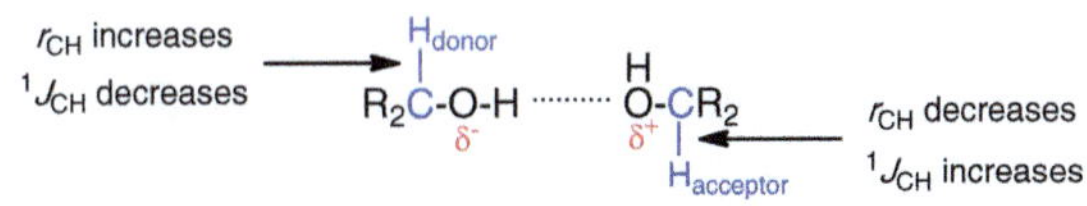

Scheme 2.21 Effect of hydrogen bonding on C–H bond lengths and corresponding $^1J_{CH}$ values in alcohols.

serving only as a H-bond acceptor, which will affect the direction of change in r_{CH} and in $^1J_{CH}$.

Hydrogen bonding of exocyclic hydroxyl groups with solvent water also affects C–H bond lengths, and this effect has been investigated as a means of measuring the strengths of H-bonds in solution.[58] Whether an H-bond lengthens or shortens a C–H bond in a H–C–OH fragment depends on whether the OH group is serving as a donor or as an acceptor (Scheme 2.21). If it serves as a donor, r_{CH} increases, and the corresponding $^1J_{CH}$ decreases. If the OH group serves as an acceptor, then r_{CH} decreases and the corresponding $^1J_{CH}$ increases. The extent to which the $^1J_{CH}$ values change depends on the strength of the H-bonding interaction. A quantitative relationship between H-bond strength and $^1J_{CH}$ has been proposed when the OH group of a secondary alcohol serves as a donor; $^1J_{CH}$ decreases by ~0.2 Hz per ~1 kJ mol^{-1} increase in H-bond enthalpy.[58b] The decrease in $^1J_{CH}$ is attributed to an increased overlap of the H-bonding σ orbital with the antibonding σ^* orbitals of the vicinal C–H bonds. It is interesting to speculate that, since $^1J_{CH}$ values tend to increase in water relative to THF (Figure 2.17), perhaps the hydroxyl groups of saccharides are more likely on average to serve as acceptors in H-bonding to solvent water than donors. From a statistical viewpoint, this seems reasonable in that the saccharide has only one opportunity to serve as a donor, whereas water has two.

The above treatment underscores the fact that multiple intra- and intermolecular factors interact in saccharides to influence r_{CH} and r_{CC} values. Structural interpretations of $^1J_{CH}$ and $^1J_{CC}$ values depend on identifying and specifying their contributions to the observed J-coupling. To date, it has been difficult to interpret ^1J values quantitatively in saccharides despite suggestions from theory that they could serve as useful structural constraints.

2.7.2 Two-Bond (Geminal) ^{13}C–^{1}H and ^{13}C–^{13}C Spin-Couplings

The magnitudes and signs of $^2J_{CH}$ and $^2J_{CC}$ in saccharides vary widely, and both configuration and conformation influence these J-couplings. For $^2J_{CH}$, experimental values range from ~+7 Hz to ~−6 Hz,[30,71] and experimental $^2J_{CC}$ values range from ~+5 Hz to ~−3 Hz.[21,23] Here we focus on the effects of C–O bond conformation on ^2J values.

For $^2J_{CH}$, two pathways pertain: C–C–H and C–O–H. In general, $^2J_{COH}$ values are similar for all hydroxyl groups in saccharides, are negative in sign,[72] and are unaffected by C–O bond rotation (*e.g.*, rotation of the C2–O2 bond in **1** (Scheme 2.17) gives calculated $^2J_{C2,OH2}$ values of −2.5 ± 0.2 Hz). Experimental

$^2J_{C2,OH2}$ values are typically −2 to −4 Hz in close agreement with the DFT calculations.[72] Other than possible effects of the C–O–H bond angle or H-bonding on $^2J_{COH}$, there does not appear to be much useful structural information encoded in $^2J_{COH}$ values (*e.g.*, rotation of the C2–O2 bond in **1** gives C2–O2–H bond angles of 107.7° ± 1.0°). Their measurement in aqueous solution is not straightforward given the propensity of hydroxyl protons to exchange with the water protons.[73]

For C–C–H pathways, two oxygen substitution effects pertain. For example, the C_A–C_B–H_B coupling pathway in Scheme 2.20 contains one OH substituent (O_AH_{OA}) appended to the coupled carbon and a second (O_BH_{OB}) is appended to the carbon (C_B) bearing the coupled hydrogen (H_B). Rotations about the two C–O bonds do not exert equivalent effects on $^2J_{CA,HB}$. The relative magnitudes of these effects can be seen by examining the effects of C2–O2 bond rotation in **1** (Scheme 2.17) on $^2J_{C2,H3}$ and $^2J_{C3,H2}$ (Figure 2.18). $^2J_{C3,H2}$ is more affected by the rotation than is $^2J_{C2,H3}$. $^2J_{C3,H2}$ values are similar (~−4.6 Hz) at $\theta = 60°$ and 300°, and shift to a less negative value (~−3 Hz) at $\theta = 180°$. An O2 lone-pair orbital is *anti* to the C2–H2 bond at $\theta = 60°$, and *anti* to the C2–C3 bond at $\theta = 300°$. At $\theta = 180°$, both bonds are *anti* to an O2 lone-pair, and the additive effect of lengthening both bonds presumably contributes to the less negative coupling. $^2J_{C2,H3}$ is slightly more negative at $\theta = 60°$ (~−4.6 Hz) than at $\theta = 180°$ and 300° (~−4.2 Hz). In this case the C2–C3 and C3–H3 bonds are subject to vicinal and 1,3-lone-pair effects, respectively, that are apparently competitive (1,3 at $\theta = 60°$, vicinal and 1,3 at $\theta = 180°$, and vicinal at $\theta = 300°$).

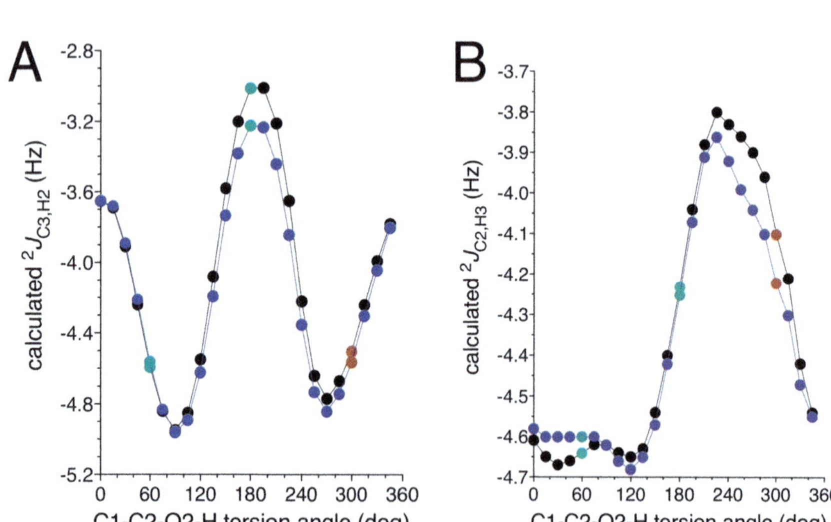

Figure 2.18 Effect of C2–O2 bond rotation in **1** on calculated (A) $^2J_{C3,H2}$ and (B) $^2J_{C2,H3}$. Black points, no solvent; blue points, with solvent. The overall change in $^2J_{CCH}$ in (A) (~2 Hz) is larger than that in (B) (~1 Hz). Both calculated $^2J_{CCH}$ are ~−4 Hz, which is consistent with a projection sum of 0 (see Figure 2.2).

For $^2J_{CC}$, two pathways are of interest in saccharides, C–C–C and C–O–C. The latter pathway is associated with experimental values of ~−2 Hz.[21,23] This pathway commonly involves the anomeric (C1) carbon and C5 in aldopyranosyl rings. The anomeric carbon bears an OH group, and configuration at that carbon affects the $^2J_{C1,C5}$ value. For example, $^2J_{C1,C5}$ in methyl α-D-glucopyranoside **23** is ~−2 Hz, whereas $^2J_{C1,C5}$ in methyl β-D-glucopyranoside **1** is ~0 Hz.[23] Rotation of the terminal C–O bond does not influence $^2J_{C1,C5}$ values appreciably.

$^2J_{CCC}$ Values for coupling pathways in which all three carbons bear a single hydroxyl group are heavily affected by configuration at the terminal coupled carbons, and less so by configuration at the central carbon (Table 2.1). Conformation about the C–O bond involving the central carbon is a major determinant of the J_{CCC}, while conformation about the C–O bonds involving the terminal coupled carbons exerts a minor influence. These dependencies are illustrated in Figure 2.19 in which the effects of rotation about the C2–O2 bond in **1** (Scheme 2.17) on $^2J_{C1,C3}$ and $^2J_{C2,C4}$ are interrogated. Calculated $^2J_{C1,C3}$ in **1** have positive signs as expected (Table 2.1; Figure 2.3) and show a bicyclic dependence on C2–O2 bond rotation (Figure 2.19A). Calculated J-couplings at θ = +60° and 180° are similar (~+4.4 Hz), whereas that at θ = 300° is much more positive (~+7.0 Hz). In the former group, one O2 lone-pair orbital is *anti* to either the C1–C2 or C2–C3 bond, and in the latter, both bonds are *anti* to an O2 lone-pair orbital, which elongates both bonds (Figure 2.13) and shifts the J-coupling to a more positive value. The overall change in $^2J_{C1,C3}$ is ~3 Hz, which contrasts with the smaller overall change

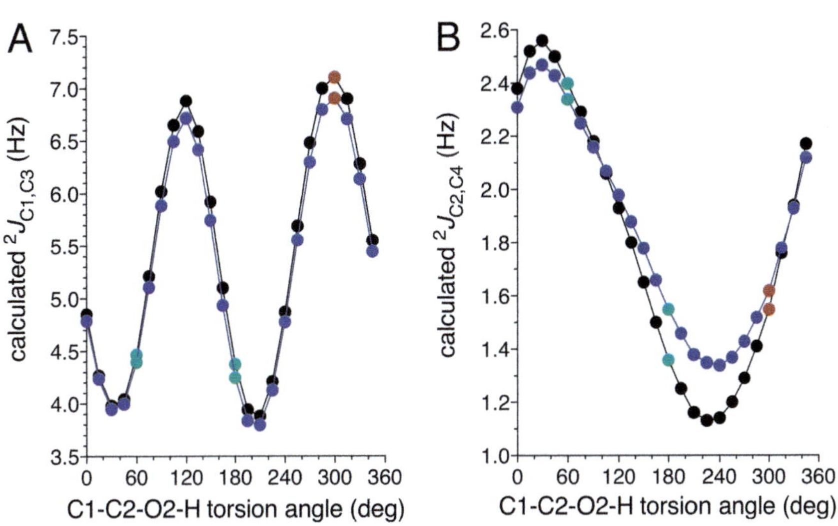

Figure 2.19 Effect of C2–O2 bond rotation in **1** on (A) $^2J_{C1,C3}$ and (B) $^2J_{C2,C4}$. Black points, no solvent; blue points, with solvent. Points shown in green and red are those found in the three perfectly staggered C2–O2 rotamers (Scheme 2.18).

of ~1.2 Hz observed for $^{2}J_{C2,C4}$ (Figure 2.19B). Calculated $^{2}J_{C2,C4}$ values are similar (~+1.5 Hz) at θ = 180° and 300°, whereas that at θ = +60° is more positive (~+2.4 Hz). Note that in the former group, one of the O2 lone-pair orbitals is *anti* to the C2–C3 bond, but when θ = +60°, the O2–H bond is *anti* to the C2–C3 bond. This observation suggests that orienting a lone-pair in the coupling plane renders the $^{2}J_{CCC}$ value slightly more negative (or less positive).

Rotations of exocyclic C–O bonds exert modest effects on remote intra-ring C–C–C and C–O–C coupling pathways in saccharides. These effects can be seen in the plots shown in Figure 2.20, where the effects of C2–O2 bond rotation in **1** on $^{2}J_{C3,C5}$ and $^{2}J_{C1,C5}$ are shown. While the effects are small, they point to a role that H-bonding plays in affecting *J*-couplings. Note the similar $^{2}J_{C3,C5}$ values for θ = +60° and 300°, and the shift to a more positive coupling at θ = 180°. The latter conformation allows an H-bonding interaction between O2H and a lone-pair orbital on O3. A similar effect is

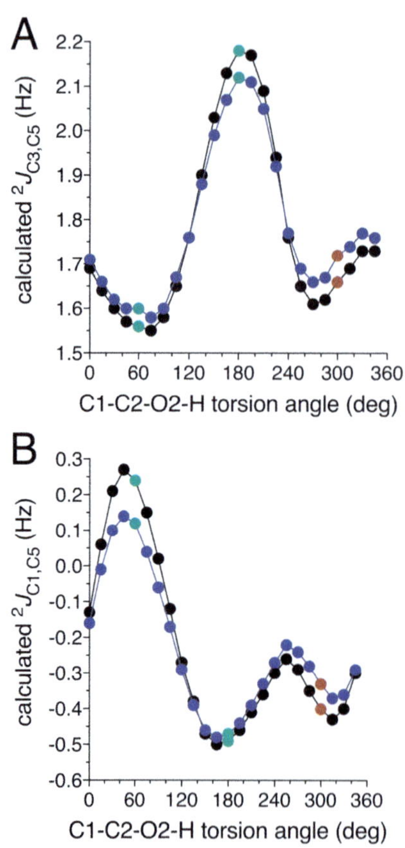

Figure 2.20 Effect of C2–O2 bond rotation in **1** on (A) $^{2}J_{C3,C5}$ and (B) $^{2}J_{C1,C5}$. Black points, no solvent; blue points, with solvent. Points shown in green and red are those found in the three perfectly staggered C2–O2 rotamers (Scheme 2.18).

observed for $^{2}J_{C1,C5}$. In this case, the slightly negative and similar values observed at $\theta = 180°$ and $300°$ are shifted to a slightly more positive value at $\theta = +60°$, the latter allowing H-bonding between O2H and a lone-pair orbital on O1. This weak H-bonding behavior may also be reflected in changes in $r_{C1,H1}$ and $r_{C3,H3}$ shown in Figure 2.12. In both cases, the two C2–O2 rotamers associated with the shorter C–H bonds (caused by 1,3-lone-pair interactions) produce slightly different C–H bond lengths. The slightly shorter C–H bond is associated with the C2–O2 rotamer that allows H-bonding between O2H and the OH group appended to the same carbon as the C–H bond. For example, $r_{C3,H3}$ is slightly shorter at $\theta = 180°$ than at $\theta = +60°$; the former allows the H-bond. Likewise, $r_{C1,H1}$ is slightly shorter at $\theta = +60°$ than at $\theta = 180°$; the former allows the H-bond. These data suggest that when the oxygen in an HO–C–H fragment serves as an H-bond acceptor, the C–H bond in the fragment shortens. This behavior mimics that shown in Scheme 2.21, even though the patterns shown therein were derived from studies of intermolecular H-bonding systems.

Unlike observations made for $^{1}J_{CH}$ (Figure 2.15), calculated $^{2}J_{CH}$ and $^{2}J_{CC}$ are not much affected by the inclusion of solvent in the calculations (see Figures 2.18–2.20).

2.7.3 Three-Bond (Vicinal) ^{13}C–^{1}H and ^{13}C–^{13}C Spin-Couplings

Vicinal *J*-couplings depend primarily on the torsion angle between the coupled nuclei.[11,74] There are four different pathways in saccharides classified according to the types of atoms comprising the pathway (in-pathway atoms): (1) C–C–C–H, (2) C–O–C–H, (3) C–C–C–C and (4) C–O–C–C. These four scaffolds can be decorated with terminal hydroxyl groups (*i.e.*, those appended to the coupled carbon or carbons) and with internal hydroxyl groups (*i.e.*, those appended to non-terminal in-pathway carbons). These different pathways are illustrated in the monosaccharide, methyl β-D-glucopyranoside **1**, in Scheme 2.22.

In general, vicinal coupling pathways display a sinusoidal dependence of the ^{3}J on the torsion angle θ subtended by the coupled atoms, and equations for these pathways take the general form:

$$^{3}J_{CX} = A + B\cos\theta + C\cos 2\theta + D\sin\theta + E\sin 2\theta \tag{2.3}$$

where $^{3}J_{CX}$ is the experimentally observed *J*-coupling, and θ is the torsion angle in degrees. What distinguishes one ^{3}J from another are the values of the coefficients A–E. Quantitative relationships between ^{3}J and θ are known as Karplus relationships,[11,74] which typically show maxima at $\theta = 0°$ and $180°$, and a minimum at $\theta = 90°$. Curve amplitudes (difference between the smallest and largest *J*-value) and displacements (phase-shifting of the curve along θ) are strong functions of the intervening in-pathway atoms (which affect amplitude) and/or the location of hydroxyl substituents (which affects the displacement).

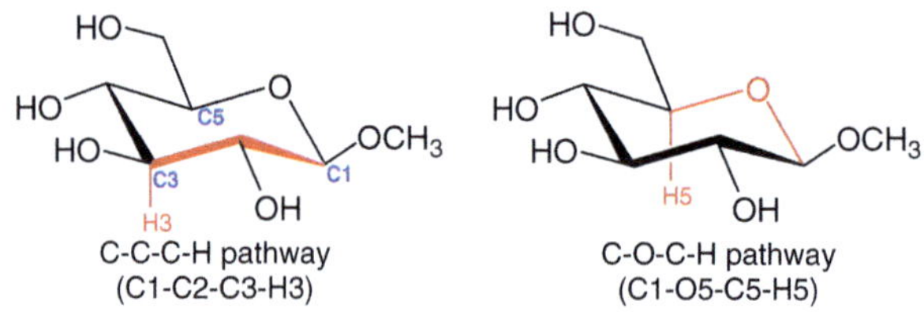

Scheme 2.22 Different three-bond C–H and C–C coupling pathways found in methyl β-D-glucopyranoside **1**.

In aldopyranosyl rings such as **1**, the vicinal C–H and C–C coupling pathways are often associated with relatively fixed torsion angles because many of these rings are conformationally rigid (often in chair conformers) or at least highly constrained to a few closely related conformers. Thus, only two torsion angles for these pathways are sampled: ~±60° and ~180° (the angles are not ideal because the rings are not perfect chairs). In **1**, coupling pathways involving atoms found exclusively within the ring are constrained to ~±60° torsions; these include C1–C2–C3–H3, C2–C3–C4–H4, C3–C2–C1–H1, C3–C4–C5–H5, C4–C3–C2–H2, C1–O5–C5–H5, and C5–O5–C1–H1. In **1**, experimental values for these *J*-couplings and the associated torsion angles obtained by X-ray crystallography, are given in Table 2.2. Vicinal C–H pathways also involve the exocyclic hydroxymethyl group atoms; these pathways include C4–C5–C6–H6*R* (1.2 Hz in **1**) C4–C5–C6–H6*S* (2.4 Hz in **1**) and C6–C5–C4–H4 (3.6 Hz in **1**).[75] The latter pathway has a torsion angle fixed by the ring (~−51° in the X-ray structure), but the other two can, in principle, sample torsion angles ranging from 0–360° since the C5–C6 bond samples at least three staggered conformers in solution (Scheme 2.12).

The situation is different for $^3J_{CC}$. Intra-ring $^3J_{CC}$ values exist in aldohexopyranosyl rings but are dual pathway *J*-couplings (Scheme 2.2) whose analysis is more complex (see below). Two single pathway $^3J_{CC}$ involve the exocyclic C6 as a coupled nucleus: $^3J_{C1,C6}$ and $^3J_{C3,C6}$.[23,76,77] The former is a C–O–C–C pathway and the latter a C–C–C–C pathway, and the torsion angles associated with both are constrained by the pyranosyl ring. These torsion angles are −175.2° for $^3J_{C1,C6}$ and −171.7° for $^3J_{C3,C6}$ in the X-ray structure of **1**.

Some of the characteristic features of C–O–C–C coupling pathways in saccharides are illustrated through DFT calculations on simple model structures **24–26** and the notations shown in Scheme 2.23. A_1, A_2 and A_3, and A_1', A_2' and A_3', denote substituents attached to the terminal (coupled) carbons, C_α and C_γ, respectively. S_1 and S_2 denote substituents attached to

Table 2.2 Experimental vicinal intra-ring ^{13}C–^{1}H spin-couplings in **1**.

	Experimental value[a] (Hz)	C–H torsion angle[b] (deg)
$^{3}J_{C1,H3}$	1.2	+65.7
$^{3}J_{C2,H4}$	~0	−68.9
$^{3}J_{C3,H1}$	1.1	−58.7
$^{3}J_{C3,H5}$	2.2	+65.9
$^{3}J_{C4,H2}$	1.1	+64.4
$^{3}J_{C1,H5}$	2.3	−56.2
$^{3}J_{C5,H1}$	1.1	+52.2

[a]Experimental values were taken from ref. 30 and 75. Values are accurate to ±0.1 Hz and were measured at ~25 °C in $^{2}H_2O$ solvent.
[b]Torsion angles were obtained from an X-ray structure (hemihydrate) of **1** (unpublished).

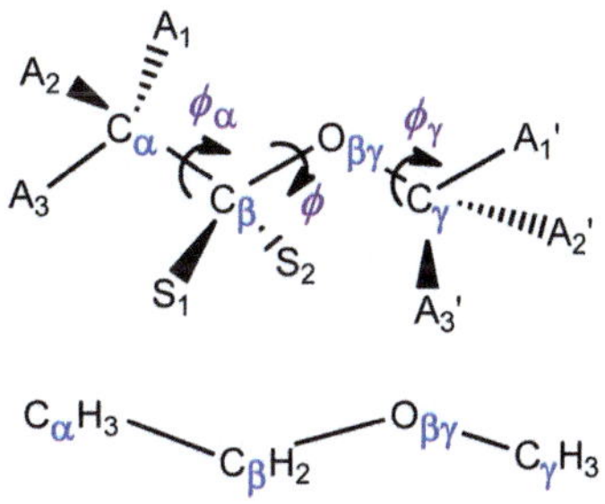

methyoxyethane **24**

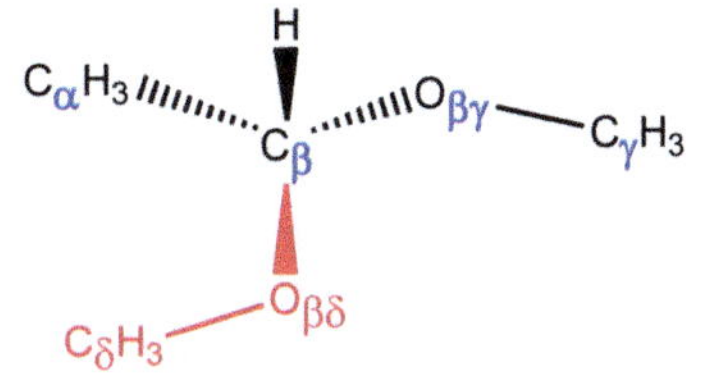

acetaldehyde dimethyl acetal **25**

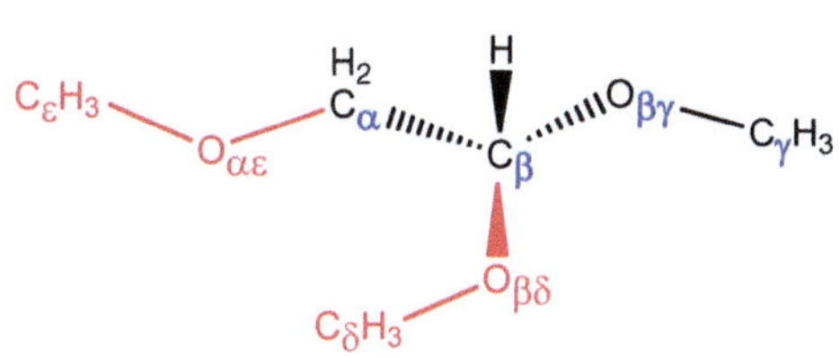

glycolaldehyde dimethyl acetal **26**

Scheme 2.23 Acyclic model structures used to study the behavior of J_{CH} and J_{CC} values in saccharides.

the internal carbon, C_β. Torsion angles defining rotations about the C_α–C_β and C_β–$O_{\beta\gamma}$ bonds along the coupling pathway are denoted ϕ_α and ϕ, respectively. In **25**, the C_α–C_β–$O_{\beta\delta}$–C_δ torsion angle was fixed at 180° in the calculations. In **26**, both the C_α–C_β-$O_{\beta\delta}$–C_δ and C_β–C_α–$O_{\alpha\varepsilon}$–C_ε torsion angles were fixed at 180°.

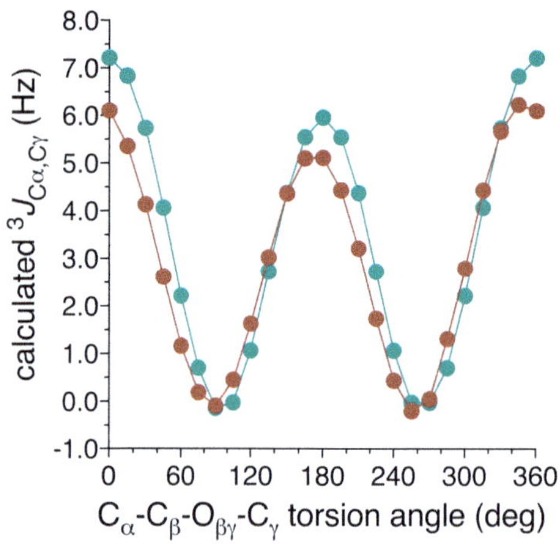

Figure 2.21 Dependence of calculated $^3J_{CCOC}$ values in **24** and **25** on the C_α–C_β–*O*–C_γ torsion angle. Green points, $^3J_{CCOC}$ in **24**; green line, eqn (2.4). Red points, $^3J_{CCOC}$ in **25**; red line, eqn (2.5).

Calculated $^3J_{CCOC}$ values in **24** (Figure 2.21) show a simple dependency on the C_α–C_β–$O_{\beta\gamma}$–C_γ torsion angle ϕ described by eqn (2.4),

$$^3J_{CCOC} = 3.44 + 0.78\cos\phi + 3.51\cos 2\phi \qquad (2.4)$$
$$\text{RMSE} = 0.14 \text{ Hz}$$

where the RMSE is the root mean square error calculated from data fitting. Eqn (2.4) describes $^3J_{CCOC}$ behavior in a minimally substituted C–O–C–C coupling pathway, and serves as a reference state to evaluate the effects of terminal and internal electronegative substituents on $^3J_{CCOC}$. Calculated $^3J_{COCC}$ in **24** range from 0–~7 Hz, with the coupling at ϕ = 0° (~7.2 Hz) larger than the coupling at ϕ = 180° (~6.0 Hz). All of the calculated $^3J_{CCOC}$ values are positive in sign.

Calculated $^3J_{CCOC}$ values in **25** show a dependence on the C_α–C_β–$O_{\beta\gamma}$–C_γ torsion angle similar to that found for **24**, but the curve is shifted slightly to the left (Figure 2.21) and curve amplitude is reduced by ~1 Hz. Data for **25** were fit to an extended equation according to Pachler,[78] giving eqn (2.5):

$$^3J_{CCOC} = 2.81 + 0.78\cos\phi + 2.68\cos 2\phi - 0.079\sin\phi - 1.06\sin 2\phi \qquad (2.5)$$
$$\text{RMSE} = 0.18 \text{ Hz}$$

The small phase shift is caused by the presence of an internal electronegative substituent (OCH_3) in **25** that is absent in **24**.[79] This substitution generally truncates the amplitude of $^3J_{CCOC}$ Karplus curves, as shown in Figure 2.21. Calculated $^3J_{CCOC}$ values in **24** at C_α–C_β–$O_{\beta\gamma}$–C_γ torsion angles of +60° and −60° are identical (2.3 Hz) due to molecular symmetry, but in **25** the value at +60° (1.0 Hz) is smaller than that at −60° (2.9 Hz). Inspection of the latter two rotamers (Scheme 2.24) shows that $O_{\beta\delta}$ is *anti* to C_γ at +60°, whereas at −60°, it is *gauche*. The former arrangement is known to truncate $^3J_{HH}$

values[80,81] and appears to affect $^3J_{CCOC}$ values similarly. This "asymmetry" also contributes to the observed phase shifting of the curve for **25** (Figure 2.21).

Model structure **26** can be used to evaluate the effect of terminal electronegative substituents, and the combined effects of internal and terminal electronegative substitution, on $^3J_{CCOC}$ values.

Calculated $^3J_{CCOC}$ values in **26** are shown as a function of ϕ and ϕ_α in Figure 2.22. $^3J_{CCOC}$ Values depend heavily on ϕ, and 2D slices at the three perfectly staggered values of ϕ_α (Figure 2.23) show dependencies similar to those observed for **25**.

Calculated $^3J_{CCOC}$ values in **26** also depend on ϕ_α, as shown by 2D slices through the hypersurface in Figure 2.22 at fixed values of ϕ (Figure 2.24). Previous experimental work showed[23,82] that $^3J_{CCOC}$ values increase by ~+0.7 Hz when a terminal oxygen substituent is "in-plane" in a C–C–O–C coupling pathway in which the terminal coupled carbons are anti-periplanar. The present results confirm this increase; $^3J_{CCOC}$ increases by ~+1.2 Hz when only

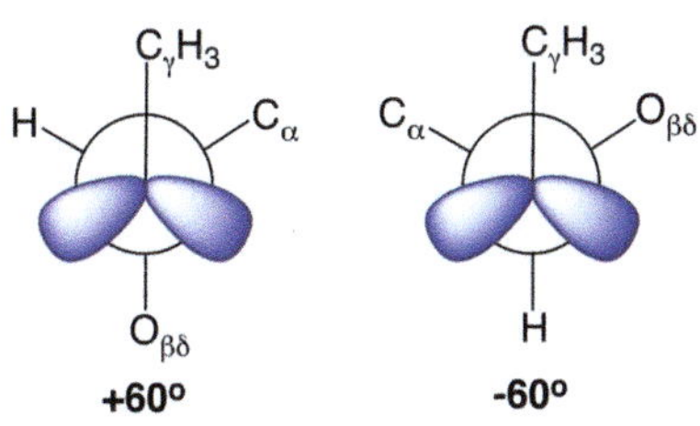

Scheme 2.24 Newman projections for two different conformations of the C_β–$O_{\beta\gamma}$ bond in **25**.

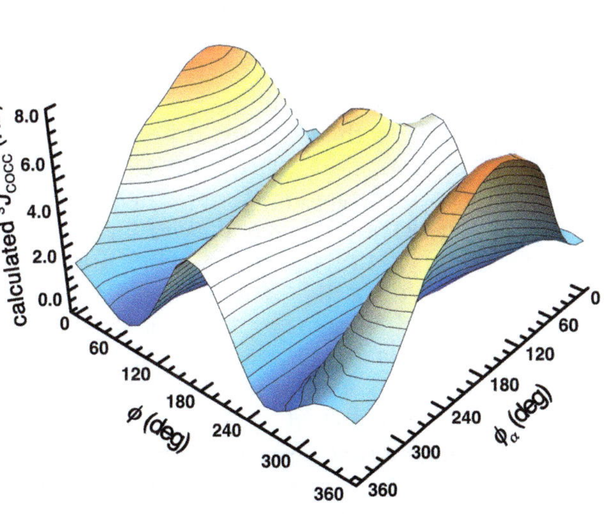

Figure 2.22 3D Hypersurface showing calculated $^3J_{CCOC}$ values in **26** as a function of torsion angles ϕ and ϕ_α.

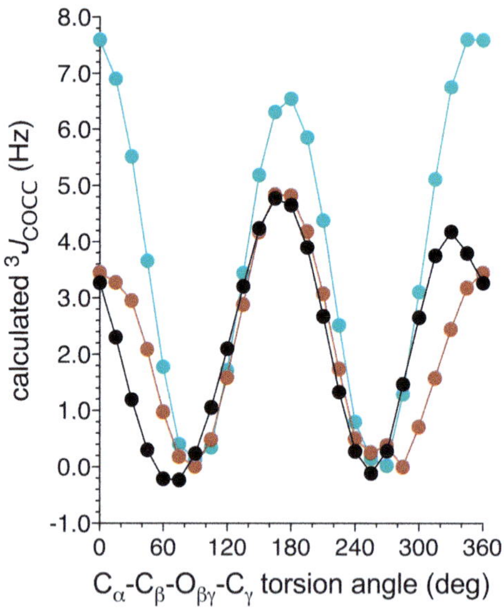

Figure 2.23 2D Slices of data in Figure 2.22 showing the dependencies of $^3J_{COCC}$ on ϕ at the three perfectly staggered values of ϕ_α. Green points, ϕ_α = *trans*; red points, $\phi_\alpha = g^+$; black points, $\phi_\alpha = g^-$.

perfectly staggered ϕ_α rotamers are considered. This increase is observed in Figure 2.23 at C_α–C_β–$O_{\beta\gamma}$–C_γ = 180° when ϕ_α = *trans* compared to when ϕ_α = +60° (g^+) and −60° (g^-). Likewise, inspection of the data in Figure 2.24 for ϕ = 180° shows an overall change in $^3J_{CCOC}$ of ~1.5 Hz (staggered structures only).

The effects of terminal electronegative substituents on $^3J_{COCC}$ are revealed in experimental $^3J_{C1,C6}$ values in anomeric pairs of aldohexopyranoses ($^4C_1$ conformers) (Table 2.3). In all cases, $^3J_{C1,C6}$ values are smaller in α-anomers than in β-anomers by ~0.8 Hz. Small differences in C1–O5–C5–C6 torsion angles may also contribute to the observed differences. Superimposed on the effect of C1 configuration is the effect of C5–C6 bond conformation on $^3J_{C1,C6}$ values. This effect is illustrated in Figure 2.25 in which $^3J_{C1,C6}$ values in model structures **27** and **28** were calculated as a function of O5–C5–C6–O6 torsion angles. These data confirm the O1 "in-plane" effect; calculated $^3J_{C1,C6}$ values are uniformly larger in **28** than in **27** by ~0.9 Hz. In addition, in both structures, $^3J_{C1,C6}$ is larger when the O5–C5–C6–O6 torsion angle is 180° (*tg* rotamer; O6 lies in the C1–O5–C5–C6 coupling plane) compared to torsion angles of +60° (*gt* rotamer) and −60° (*gg* rotamer) (Scheme 2.12).

The behavior of $^3J_{C1,C6}$ is mimicked by $^3J_{C3,C6}$, although in the latter case configuration at the intervening C4 must also be considered.[77] An inspection of the dependence of $^3J_{C3,C6}$ in **28** and **29** shows maximal coupling at O5–C5–C6–O6 torsion angles near +60° (Figure 2.26), that is, in the *gt* conformer (Scheme 2.12) in which O6 lies in the C3–C4–C5–C6 coupling plane.

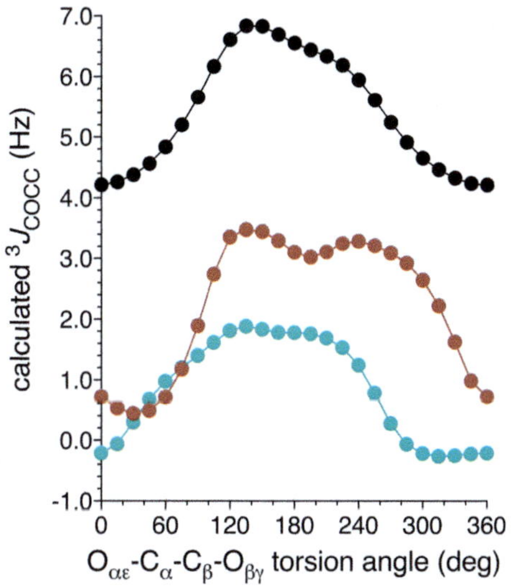

Figure 2.24 2D Slices of data in Figure 2.22 showing the dependencies of $^3J_{COCC}$ on ϕ_α at the three perfectly staggered values of ϕ. Green points, ϕ = 60°; black points, ϕ = 180°; red points, ϕ = 300°.

Table 2.3 Experimental $^3J_{C1,C6}$ values in Anomeric Aldopyranoses.[a]

Compound	$^3J_{C1,C6}$ (Hz)	J_β–J_α (Hz)
α-D-galactopyranose	3.6	0.9
β-D-galactopyranose	4.5	
α-D-glucopyranose	3.3	0.9
β-D-glucopyranose	4.2	
α-D-mannopyranose	3.3	0.6
β-D-mannopyranose	3.9	
α-D-talopyranose	3.4	0.9
β-D-talopyranose	4.3	

[a]Data were taken from ref. 76 and 77. All compounds highly favor the $^4C_1$ ring conformation.

This behavior is independent of configuration at C3, but the latter does affect absolute values, with an equatorial O3 (which lies in the C3–C4–C5–C6 coupling plane) increasing $^3J_{C3,C6}$ values by 1–1.5 Hz. These DFT predictions correlate well with experimental $^3J_{C3,C6}$ values observed in β-D-glucopyranose (4.4 Hz) and β-D-allopyranose (3.0 Hz).[77]

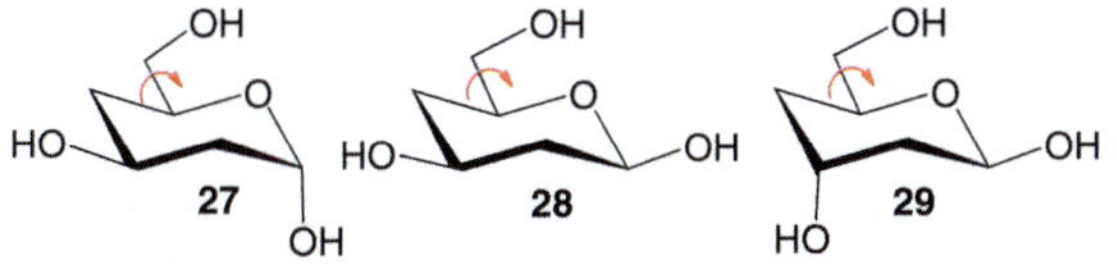

We now return to the calculated $^3J_{CCOC}$ values in **26** shown in Figure 2.22. A replot of the data (Figure 2.27) shows the scatter at discrete values of the

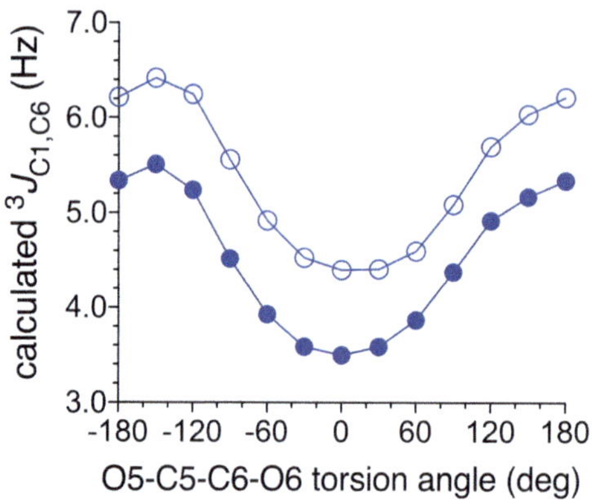

Figure 2.25 Effect of C5–C6 bond rotation on calculated $^3J_{C1,C6}$ values in **27** (closed symbols) and **28** (open symbols).

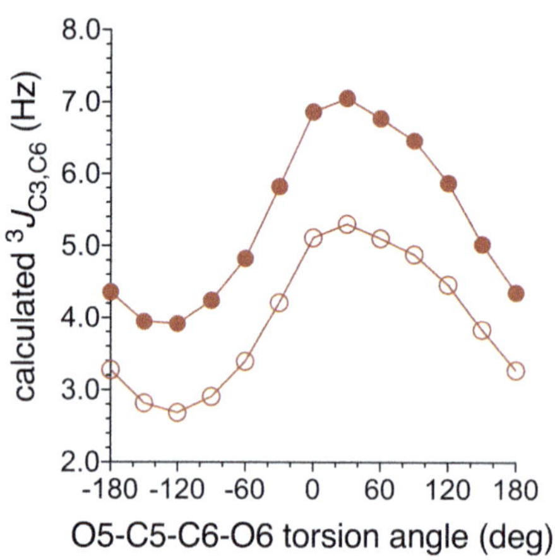

Figure 2.26 Effect of C5–C6 bond rotation on calculated $^3J_{C3,C6}$ values in **28** (closed symbols) and **29** (open symbols).

C_α–C_β–$O_{\beta\gamma}$–C_γ torsion angle ϕ caused by the effect of ϕ_α. The scatter displays a systematic dependence on ϕ, and is much greater at ϕ values near 0° than at ϕ values near 180°. When ϕ is 0°, rotating ϕ_α causes the calculated $^3J_{COCC}$ to vary from 1.3 Hz ($\phi_\alpha = 0°$) to 7.5 Hz ($\phi_\alpha = 180°$). When $\phi_\alpha = 0°$, C_γ and $O_{\alpha\varepsilon}$ are in close contact, presumably causing structural perturbations along the C_α–C_β–$O_{\beta\gamma}$–C_γ coupling pathway that reduce the $^3J_{CCOC}$ substantially. Through-space contributions have not yet been eliminated as potential contributors to the latter data scatter.

The above discussion has focused exclusively on vicinal ^{13}C–^1H and ^{13}C–^{13}C *J*-couplings within pyranosyl rings and acyclic model fragments. An important structural element of saccharides that can be investigated by the same *J*-couplings involves coupling pathways across *O*-glycosidic linkages. This application is discussed in Section 2.6.6.

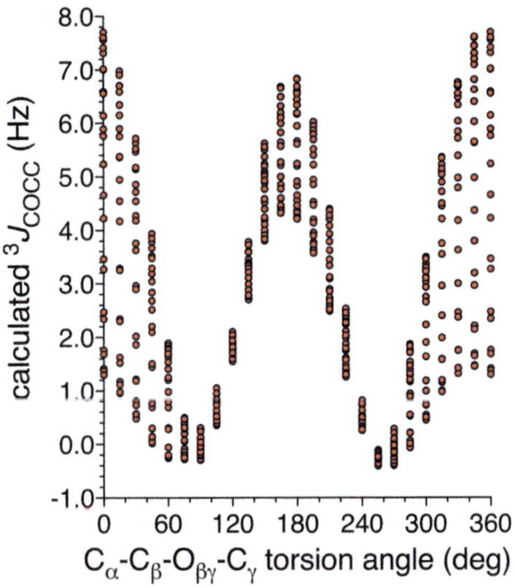

Figure 2.27 Data in Figure 2.22 displayed in two-dimensions as a function of ϕ. The dispersion observed at a given ϕ value is caused by the effect of ϕ_α on the calculated $^3J_{COCC}$ value.

2.7.4 Dual-Pathway ^{13}C–^{13}C Spin-Couplings in Aldopyranosyl and Aldofuranosyl Rings

Aldopyranosyl rings contain dual-pathway ^{13}C–^{13}C spin-couplings such as $^{3+3}J_{C1,C4}$ shown in Scheme 2.2. For $^{3+3}J_{C1,C4}$, the two pathways involve a $^3J_{COCC}$ and a $^3J_{CCCC}$. A second dual-pathway coupling also exists, $^{3+3}J_{C2,C5}$. It is generally believed that experimentally observed dual-pathway *J*-couplings are the algebraic sums of the two individual *J*-couplings arising from the two pathways.[13,14] Both pathways correspond to vicinal couplings for torsion angles of ±60°, and both constituent *J*-couplings are expected to be positive in sign. Thus, $^{3+3}J_{CC}$ values are expected to be positive in sign when observed.

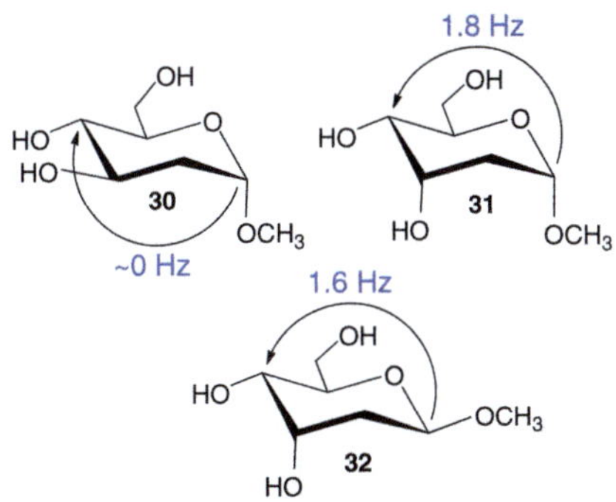

$^{3+3}J_{C1,C4}$ is very small or zero in α- and β-D-glucopyranoses and β-D-mannopyranose, but ~0.8 Hz in α-D-mannopyranose.[23] Experimental $^{3+3}J_{C1,C4}$ in methyl 2-deoxy-α-D-*arabino*-hexopyranoside (methyl 2-deoxy-α-D-glucopyranoside)

30 is also small, but $^{3+3}J_{C1,C4}$ values in methyl 2-deoxy-α-D-*ribo*-hexopyranoside (methyl 2-deoxy-α-D-allopyranoside) **31** and methyl 2-deoxy-β-D-*ribo*-hexopyranoside (methyl 2-deoxy-β-D-allopyranoside) **32** are 1.8 Hz and 1.6 Hz, respectively.[23] Thus, the conversion of ring OH substituents from equatorial to axial orientations along the "front" pathway increases $^{3+3}J_{C1,C4}$ values. This observation leads to the prediction that $^{3+3}J_{C1,C4}$ values should be relatively large in β-D-altropyranose, and an experimental coupling of 1.3 Hz is observed.[23]

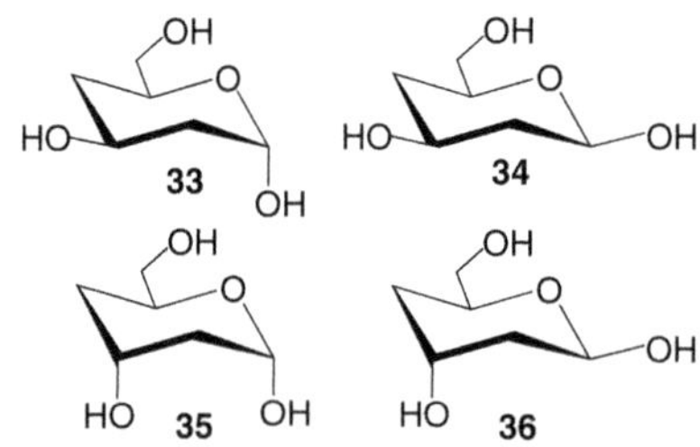

A semi-quantitative structural rationale for the above behaviors has been developed[23] based on the observation that electronegative substituents *anti* to coupled carbons in *gauche* C–O–C–C, and presumably *gauche* C–C–C–C, coupling pathways reduce $^{3}J_{CC}$ values (see discussion above; Scheme 2.24). Similar effects occur along the C1–C2–C3–C4 coupling pathway of aldopyranosyl rings. When O2 and O3 are equatorial, they are *anti* to C4 and C1, respectively, and this configuration reduces the $^{3}J_{CCCC}$ along the front pathway to a small value. When one (*e.g.*, in α-D-mannopyranose) or both (*e.g.*, in β-D-altropyranose) of these oxygens is/are axial, this reduction is mitigated and a larger coupling is observed. The effect appears additive, with β-D-altropyranose yielding a $^{3+3}J_{C1,C4}$ value (1.3 Hz)~2-fold greater than observed in α-D-mannopyranose (~0.8 Hz). Removing an equatorial substituent (deoxygenation at C2; **30–31**) produces a similar mitigating effect.

Configuration at the internal carbons of the front pathway appears to influence $^{3+3}J_{C1,C4}$ values more than do the orientations of terminal electronegative substituents on the coupled carbons. This conclusion presumes that the contribution made by the rear ($^{3}J_{COCC}$) pathway is constant at ~0 Hz and is essentially unaffected by configuration at C2 and C3.

Validation of the above treatment was obtained by calculating $^{3+3}J_{C1,C4}$ values in structures **33–36** as a function of exocyclic CH_2OH conformation (Figure 2.28). The DFT-calculated couplings were +0.5–1.0 Hz for structures **33–34** and ~+2 Hz for structures **35–36**. The effect of terminal carbon configuration (anomeric configuration in this case) is much smaller than the effect of configuration at the internal C3 carbon. Structures containing an axial O3 (**35–36**) gave larger couplings by ~1 Hz compared to those containing an equatorial O3 (**33–34**). These results affirm the importance of internal electronegative substituent effects on $^{3}J_{CCCC}$ values for both single-[77] and dual[23] coupling pathways.

$^{3+3}J_{C2,C5}$ values are small or zero in α- and β-D-glucopyranoses and α- and β-D-mannopyranoses.[23] By analogy to $^{3+3}J_{C1,C4}$, configuration at C3 and C4

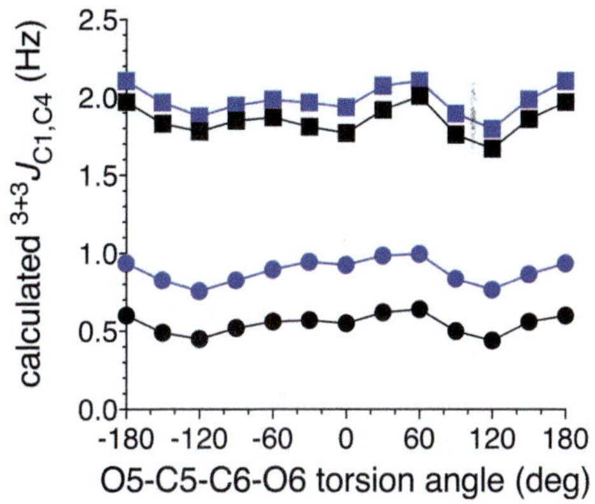

Figure 2.28 Effect of exocyclic CH_2OH conformation on calculated $^{3+3}J_{C1,C4}$ values in **33** (blue circles), **34** (black circles), **35** (blue squares), and **36** (black squares).

affects the contribution from the C2–C3–C4–C5 pathway, and configuration at C1 affects the contribution from the C2–C1–O5–C5 pathway. In the *galacto* and *talo* configurations, $^{3+3}J_{C2,C5}$ is ~1.5 Hz in α-anomers, and virtually zero in the corresponding β-anomers.[23] In these configurations, O4 is axial, and this factor presumably enhances the C2–C3–C4–C5 contribution by ~1.5 Hz. This "axial" enhancement is mitigated by changing O1 from an axial to an equatorial orientation. This interpretation leads to the (yet unconfirmed) prediction of a non-zero coupling in β-D-gulopyranose **37**, since the two axial contributions within the C2–C3–C4–C5 pathway (O3 and O4) are expected to override the single equatorial contribution from the C2–C1–O5–C5 pathway (O1). Furthermore, experimental $^{3+3}J_{C2,C5}$ in 3-deoxy-α-D-*ribo*-hexopyranose (3-deoxy-α-D-glucopyranose) **38** is 0.8 Hz, whereas no coupling is observed in the corresponding β-pyranose **39**.[23] These results confirm the correlation between axial oxygen substituents and $^{3+3}J_{CC}$ enhancement in aldopyranosyl rings.

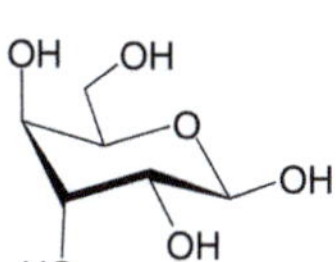
37

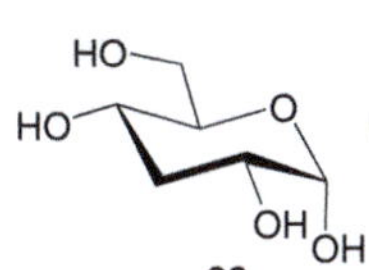
38

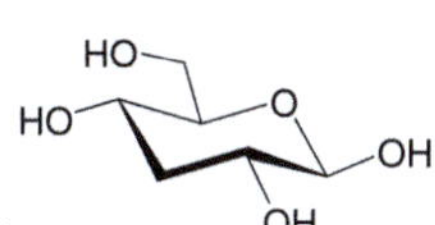
39

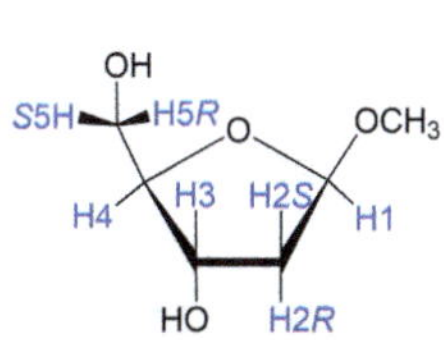

methyl 2-deoxy-β-D-ribofuranoside **40**

Fixed and initial torsions in DFT calculations:
C2-C1-O1-CH_3 = 180° (initial)
C2-C3-O3-H = 180° (fixed)
C3-C4-C5-O5 = 180° (fixed)
C4-C5-O5-H = 180° (fixed)

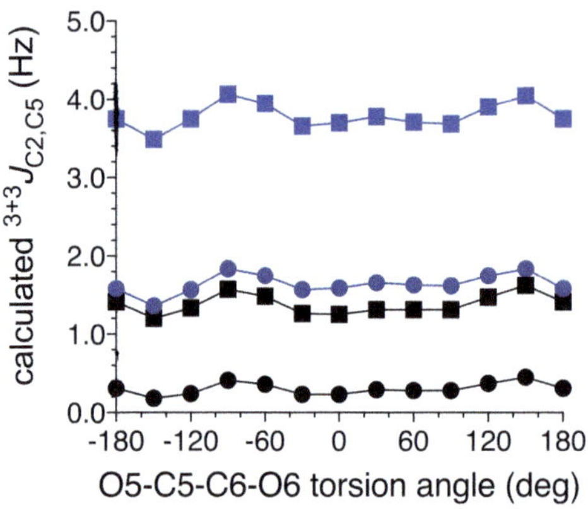

Figure 2.29 Effect of exocyclic CH_2OH conformation on calculated $^{3+3}J_{C2,C5}$ values in **33** (blue circles), **34** (black circles), **35** (blue squares), and **36** (black squares).

The above interpretations were tested[23] by computing $^{3+3}J_{C2,C5}$ values in **33–36** as a function of exocyclic CH_2OH conformation (Figure 2.29). Three groups of couplings were observed: ~0 Hz in **34**; ~+1.5 Hz in **33** and **36**; and ~+4 Hz in **35**. In **34**, both component pathways gave couplings of ~0 Hz, which is consistent with both pathways containing an equatorial C–O bond (O1 and O3 are *anti* to C5). Each component pathway elicits a +1.5–2.0 Hz coupling in **35**, given that each contains an axial C–O bond (O1 and O3 are *gauche* to C5). In **33** and **36**, one component pathway contains an axial and the other an equatorial C–O bond, giving a coupling of intermediate magnitude.

Similar analyses can be applied to aldofuranosyl rings, which also contain dual-pathway couplings (Scheme 2.2). In these ring systems, ring conformation plays a central role in dictating the observed couplings, thus complicating the analysis. Unlike aldopyranosyl rings, one of the constituent pathways involves a geminal $^2J_{CC}$. For example, in methyl 2-deoxy-β-D-ribofuranoside (methyl 2-deoxy-β-D-*erythro*-pentofuranoside) **40**, $^{2+3}J_{C1,C3}$ involves $^2J_{C1,C2,C3}$ and $^3J_{C3,C4,O4,C1}$; $^{2+3}J_{C1,C4}$ involves $^2J_{C1,O4,C4}$ and $^3J_{C1,C2,C3,C4}$; and $^{2+3}J_{C2,C4}$ involves $^2J_{C2,C3,C4}$ and $^3J_{C2,C1,O4,C4}$. DFT calculations of $^{2+3}J_{C1,C3}$ and $^{2+3}J_{C1,C4}$ in **40** as a function of ring conformation are shown in Figure 2.30. For $^{2+3}J_{C1,C4}$, the geometries of the "rear" C1–O4–C4 geminal pathways are largely unchanged in the $^3T_2$ (north) and $^2T_3$ (south) ring conformers (Scheme 2.10), but the "front" C1–C2–C3–C4 vicinal pathways differ, with the C3–O3 bond *quasi*-equatorial in $^3T_2$ (O3 *anti* to C1) and *quasi*-axial in $^2T_3$ (O3 *gauche* to C1). Calculated $^{2+3}J_{C1,C4}$ values are ~+1.9 Hz in $^2T_3$ and ~+0.2 Hz in $^3T_2$ (Figure 2.30); the smaller coupling is observed along the "front" vicinal pathway where O3 is *anti* to C1, as expected from studies of $^{3+3}J_{CC}$ in aldopyranosyl rings.[23] A similar approach can be taken for $^{2+3}J_{C1,C3}$. Calculated $^{2+3}J_{C1,C3}$ are +2.1 Hz in E_2 (P/π

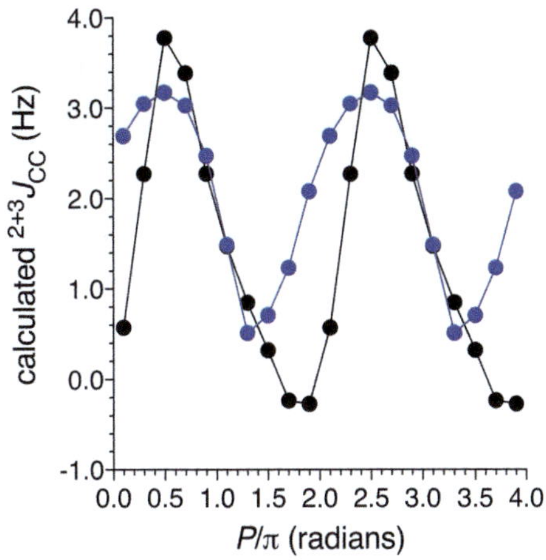

Figure 2.30 Dependence of $^{2+3}J_{C1,C3}$ (blue symbols) and $^{2+3}J_{C1,C4}$ (black symbols) in **40** on ring conformation, determined by DFT.

= 1.9) and +2.5 Hz in ^{2}E (P/π = 0.9) (Figure 2.30). Electronegative substituent effects on the C1–C2–C3 pathway are small (the C1–O1 and C3–O3 bonds are *quasi*-axial/*quasi*-equatorial in both conformers) while the C1–O4–C4–C3 torsion angle is unchanged at ~0° in both conformers. Additional experimental and theoretical work remains to be done to fully understand intra-ring $^{2+3}J_{CC}$ values in furanosyl ring systems.

2.7.5 Four-Bond ^{1}H–^{1}H and ^{13}C–^{1}H Spin-Couplings in Aldopyranosyl Rings

Long-range *J*-couplings are defined as those associated with coupling pathways containing more than three bonds between the coupled atoms. Within typical aldopyranosyl rings, these couplings include $^{4}J_{HCCCH}$, $^{4}J_{CCCCH}$ and $^{4}J_{COCCH}$. The small magnitudes of intra-ring $^{4}J_{HH}$ render them measurable only in high-resolution NMR spectra, and incorrect assignments can occur if care is not taken during spectral analysis. Despite these limitations, intra-ring $^{4}J_{HH}$ values provide useful confirmatory information about ring configuration and conformation when detected and when their signs are known.

Intra-ring four-bond ^{1}H–^{1}H pathways yield $^{4}J_{HH}$ values that depend on pathway structure, as shown from DFT calculations on model structures **1**, **23**, **41** and **42** in Scheme 2.25 (*gt* rotamers only): +0.9 ± 0.2 Hz for H_{ax}–C–X–C–H_{ax} pathways; ~+0.1 Hz for H_{eq}–C–X–C–H_{eq} pathways; and −1.0 ± 0.1 Hz for H_{eq}–C–X–C–H_{ax} pathways, where X can be H–C–OH or O. When X = H–C–OH, configuration at the carbon does not appear to affect $^{4}J_{HH}$ values appreciably.

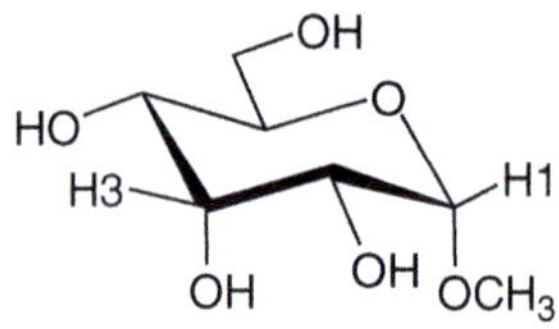

methyl α-D-allopyranoside **41**

$^4J_{H1,H3}$ = +0.1 Hz

$^4J_{H1,H5}$ = -1.2 Hz

$^4J_{H2,H4}$ = +0.9 Hz

$^4J_{H3,H5}$ = -1.0 Hz

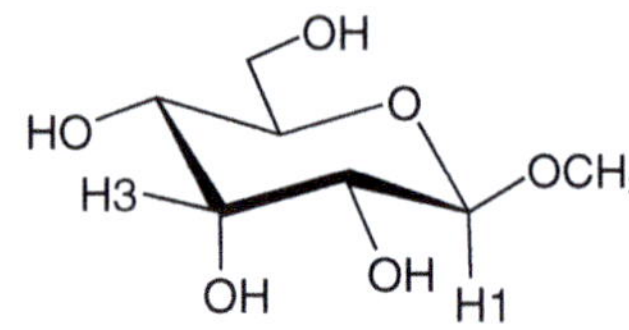

methyl β-D-allopyranoside **42**

$^4J_{H1,H3}$ = -0.9 Hz

$^4J_{H1,H5}$ = +1.2 Hz

$^4J_{H2,H4}$ = +1.0 Hz

$^4J_{H3,H5}$ = -1.0 Hz

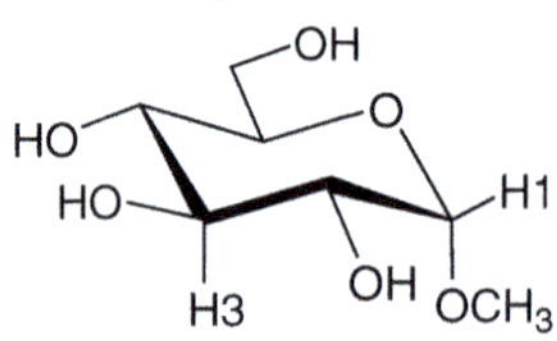

methyl α-D-glucopyranoside **23**

$^4J_{H1,H3}$ = -0.9 Hz

$^4J_{H1,H5}$ = -1.2 Hz

$^4J_{H2,H4}$ = +0.7 Hz

$^4J_{H3,H5}$ = +0.7 Hz

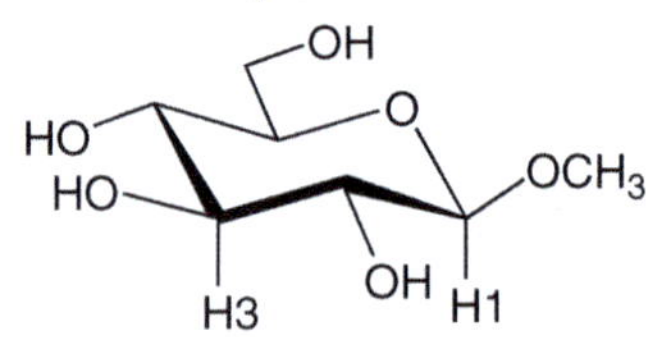

methyl β-D-glucopyranoside **1**

$^4J_{H1,H3}$ = +0.7 Hz

$^4J_{H1,H5}$ = +1.1 Hz

$^4J_{H2,H4}$ = +0.7 Hz

$^4J_{H3,H5}$ = +0.7 Hz

Scheme 2.25 Calculated $^4J_{HH}$ in several aldohexopyranosyl rings.

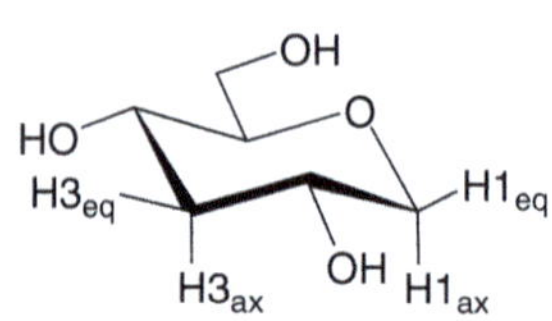

43

$^4J_{H1ax,H3ax}$ = +0.4 Hz

$^4J_{H1ax,H3eq}$ = -0.8 Hz

$^4J_{H1eq,H3ax}$ = -0.7 Hz

$^4J_{H1eq,H3eq}$ = +1.5 Hz

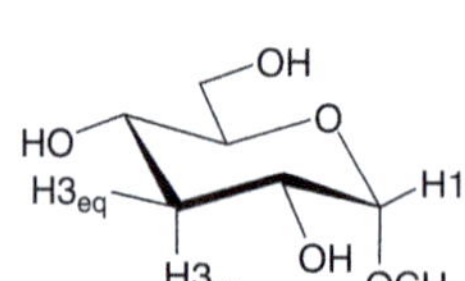

44

$^4J_{H1,H3ax}$ = -1.0 Hz

$^4J_{H1,H3eq}$ = +0.3 Hz

$^4J_{H1,H5}$ = -1.2 Hz

$^4J_{H2,H4}$ = +0.6 Hz

$^4J_{H3ax,H5}$ = +0.6 Hz

$^4J_{H3eq,H5}$ = -0.7 Hz

The findings shown in Scheme 2.25 call into question the common assumption that planar W-shaped (zig-zag) four-bond coupling pathways in pyranosyl rings (*e.g.*, ${}^{4}J_{H1,H3}$ in **41**) produce the largest ${}^{4}J_{HH}$ values. The DFT data show that this pathway structure gives the smallest absolute coupling. To investigate the question further, DFT calculations were conducted on 1,3-dideoxy model structure **43** to determine whether terminal electronegative substituents affect ${}^{4}J_{HH}$ behavior. The results of these calculations are as follows: ${}^{4}J_{H1eq,H3eq}$ = +1.5 Hz; ${}^{4}J_{H1ax,H3eq}$ = −0.8 Hz; ${}^{4}J_{H1eq,H3ax}$ = −0.7 Hz; ${}^{4}J_{H1ax,H3ax}$ = +0.4 Hz (see structure **43**). Thus, for intra-ring H–C–C–C–H pathways devoid of terminal OH substituents, the largest ${}^{4}J_{HH}$ (absolute value) is observed for W-shaped pathways in which both of the coupled hydrogens are equatorial (H_{eq}–H_{eq}) as expected from earlier work by Barfield and coworkers.[83] However, when these pathways bear hydroxyl substituents appended to the carbons bearing the coupled hydrogens, the associated ${}^{4}J_{HH}$ values are nearly zero, while ${}^{4}J_{HH}$ values for the H_{ax}–H_{ax} and H_{ax}–H_{eq} pathways are larger and of comparative magnitude but opposite sign. Additional tests of these relationships were conducted using methyl 3-deoxy-α-D-glucopyranoside **44** and **45** as model compounds. In **44**, coupling between $H1_{eq}$ and $H3_{eq}$ is +0.3 Hz, whereas coupling between $H1_{eq}$ and $H3_{ax}$ is −1.0 Hz. In **45**, coupling between $H1_{eq}$ and $H3_{eq}$ is +0.3 Hz, whereas coupling between $H1_{ax}$ and $H3_{eq}$ is −1.0 Hz. These results show that when both carbons bearing the coupled hydrogens do not bear an OH group, the W-shaped H_{eq}–C–C–C–H_{eq} pathway yields large and positive ${}^{4}J_{HH}$ values. If either carbon, or both carbons, is (are) hydroxylated, the associated ${}^{4}J_{Heq,Heq}$ values are small.

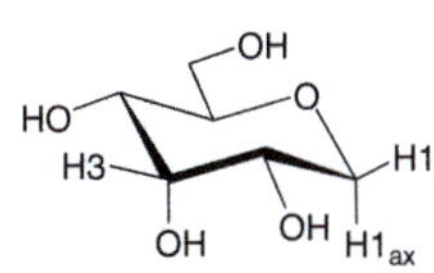

45

${}^{4}J_{H1ax,H3}$ = -1.0 Hz
${}^{4}J_{H1eq,H3}$ = +0.3 Hz
${}^{4}J_{H1ax,H5}$ = +1.1 Hz
${}^{4}J_{H1eq,H5}$ = -0.9 Hz
${}^{4}J_{H2,H4}$ = +0.9 Hz
${}^{4}J_{H3,H5}$ = -1.0 Hz

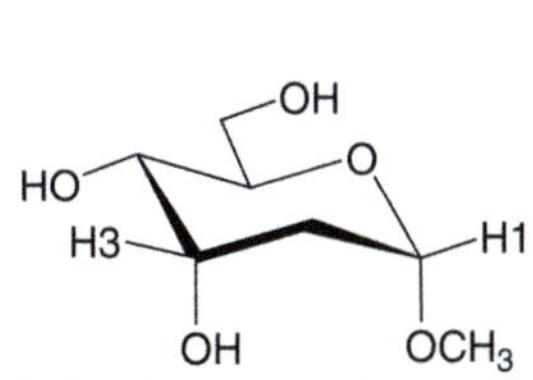

methyl 2-deoxy-α-D-allopyranoside **31**

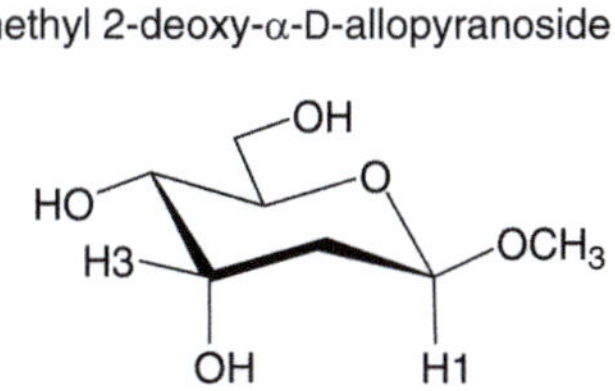

methyl 2-deoxy-β-D-allopyranoside **32**

Contributions of C–O bond conformation to intra-ring $^4J_{HH}$ values were evaluated in computations on methyl 2-deoxy-α-D-allopyranoside **31** and methyl 2-deoxy-β-D-allopyranoside **32** in which the three perfectly staggered rotamers about the C1–O1 and C3–O3 bonds were investigated. Calculated $^4J_{HH}$ values are shown in Tables 2.4 and 2.5. These calculations show that C1–O1 and C3–O3 bond conformations exert a relatively small effect on $^4J_{H1,H3}$ (overall change <0.2 Hz) when the coupled hydrogens are axial-equatorial (structure **32**) (Table 2.4). When the coupled hydrogens are both equatorial (W-shaped arrangement; structure **31**) (Table 2.5), however, C–O rotational effects are significant (overall change ~1.4 Hz), taking into account that the $^4J_{H1,H3}$ value changes sign depending on the relative conformations of the C1–O1 and C3–O3 bonds.

Table 2.4 Calculated $^4J_{HCCCH}$ values in methyl 2-deoxy-β-D-allopyranoside **32** as a function of C1–O1 and C3–O3 bond torsions.

C1–O1/ C3–O3 bond torsions[a]	$^4J_{H1,H3}$	$^4J_{H3,H5}$	$^4J_{H1,H5}$	$^4J_{H2ax,H4}$	$^4J_{H2eq,H4}$
60°/60°	−0.97	**−1.04**	+1.11	+1.04	−0.35
180°/60°	−1.07	**−1.03**	+1.14	+1.04	−0.39
300°/60°	**−1.09**	−1.03	+1.16	+1.09	**−0.32**
60°/180°	−0.94	−1.03	+1.11	**+0.85**	−0.57
180°/180°	−1.06	−1.03	+1.14	+0.85	**−0.61**
300°/180°	−1.07	−1.03	**+1.17**	+0.90	−0.55
60°/300°	**−0.92**	−1.04	**+0.92**	+1.12	−0.36
180°/300°	−1.05	−1.03	+0.96	+1.13	−0.40
300°/300°	−1.06	−1.03	+0.98	**+1.18**	−0.34
Range[b]	0.17	0.01	0.25	0.33	0.29

[a]C1–O1 = C2–C1–O1–OCH_3 torsion angle; C3–O3 = C2–C3–O3–H torsion angle.
[b]Range = maximum dispersion of calculated values (difference between values shown in bold).

Table 2.5 Calculated $^4J_{HCCCH}$ values in methyl 2-deoxy-α-D-allopyranoside **31** as a function of C1–O1 and C3–O3 bond torsions.

C1–O1/ C3–O3 bond torsions[a]	$^4J_{H1,H3}$	$^4J_{H3,H5}$	$^4J_{H1,H5}$	$^4J_{H2ax,H4}$	$^4J_{H2eq,H4}$
60°/60°	−0.46	**−0.94**	−1.08	+0.88	−0.23
180°/60°	+0.07	−1.00	−1.27	+0.93	−0.33
300°/60°	+0.43	**−1.02**	**−1.28**	+0.94	−0.37
60°/180°	−0.62	−0.94	−1.08	**+0.71**	−0.50
180°/180°	−0.22	−0.98	−1.26	+0.73	**−0.59**
300°/180°	+0.13	−1.01	−1.28	+0.77	−0.58
60°/300°	**−0.64**	−0.98	**−0.97**	+1.02	**−0.12**
180°/300°	+0.43	−1.02	−1.24	+1.03	−0.33
300°/300°	**+0.79**	−1.02	−1.26	**+1.08**	−0.32
Range[b]	1.43	0.08	0.31	0.37	0.47

[a]C1–O1 = C2–C1–O1–OCH_3 torsion angle; C3–O3 = C2–C3–O3–H torsion angle.
[b]Range = maximum dispersion of calculated values (difference between values shown in bold).

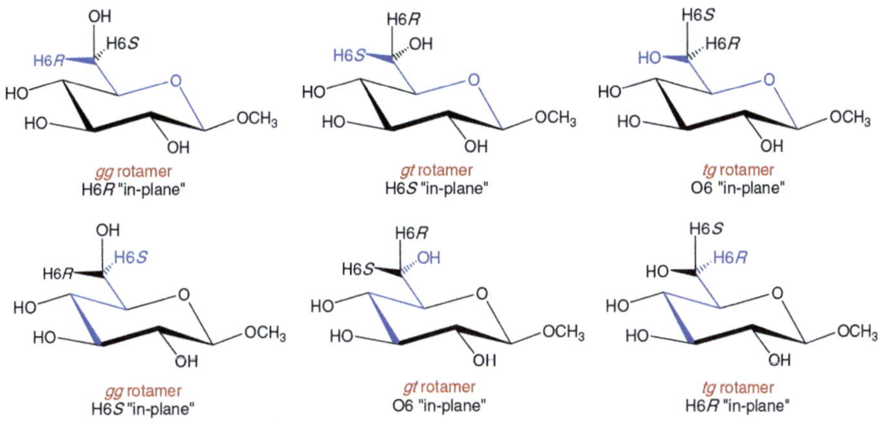

Scheme 2.26 Four-bond ^{13}C–^{1}H coupling pathways in methyl β-D-glucopyranoside **1** involving H6*R* and H6*S* and either C1 or C3.

While the results of DFT calculations of intra-ring $^4J_{HH}$ values in saccharides reveal internally consistent trends that may prove useful for structure studies, it must be kept in mind that the calculated values are small and the errors in absolute values and trends could be significant. Additional experimental measurements remain to be made to validate the $^4J_{HH}$ behaviors predicted by DFT.

Perhaps less appreciated are the $^4J_{CCCCH}$ and $^4J_{COCCH}$ values involving the exocyclic hydroxymethyl hydrogens of aldohexopyranosyl rings (Scheme 2.26).[84] When either H6*R* or H6*S* in the exocyclic hydroxymethyl fragment lies in the C1–O5–C5–C6 or C3–C4–C5–C6 coupling plane, a measurable $^4J_{CH}$ is observed (~1–2 Hz). For example, in the *gg* rotamer of **1**, H6*R* is "in-plane" with C1, leading to a measurable $^4J_{C1,H6R}$ (Scheme 2.26). Since H6*S* is "out-of-plane" in the *gg* rotamer, only a very small or zero $^4J_{C1,H6S}$ is observed. In contrast, H6*S* is "in-plane" with C3 in the *gg* rotamer, leading to a measureable $^4J_{C3,H6S}$, whereas $^4J_{C3,H6R}$ is small or zero.

A quantitative assessment of this behavior is obtained from DFT calculations on model structure **46** in which the C5–C6 bond is rotated through 360° to determine how hydroxymethyl conformation influences $^4J_{C1,H6R/H6S}$ and $^4J_{C3,H6R/H6S}$ values (Figure 2.31). These results show that (a) $^4J_{CH}$ values vary from ~−0.5 Hz to ~+1.5 Hz, and (b) the largest (most positive) couplings are observed for "in-plane" (W-shaped, zig-zag) pathway geometries. Thus, $^4J_{C1,H6S}$ is ~+1.5 Hz in *gt*, and ~−0.4 Hz in *gg* and *tg*, while $^4J_{C1,H6R}$ is ~+1.0 Hz in *gg* and ~−0.3 Hz in *gt* and *tg* (Figure 2.31A). Similar behaviors are observed for $^4J_{C3,H6R}$ and $^4J_{C3,H6S}$ (Figure 2.31B). It should be noted that $^4J_{CH}$ values for in-plane H6*R* or H6*S* in the *gg* conformer are smaller relative to those for other in-plane orientations for reasons that are unclear. It is noteworthy that the C3–C4–C5–C6–H6*R*/*S* coupling pathway appears to behave similarly to the C6–C5–C4–C3–H3 pathway (both are C–C–C–C–H pathways), as indicated by the calculated *J*-couplings in Scheme 2.27; in both cases, the

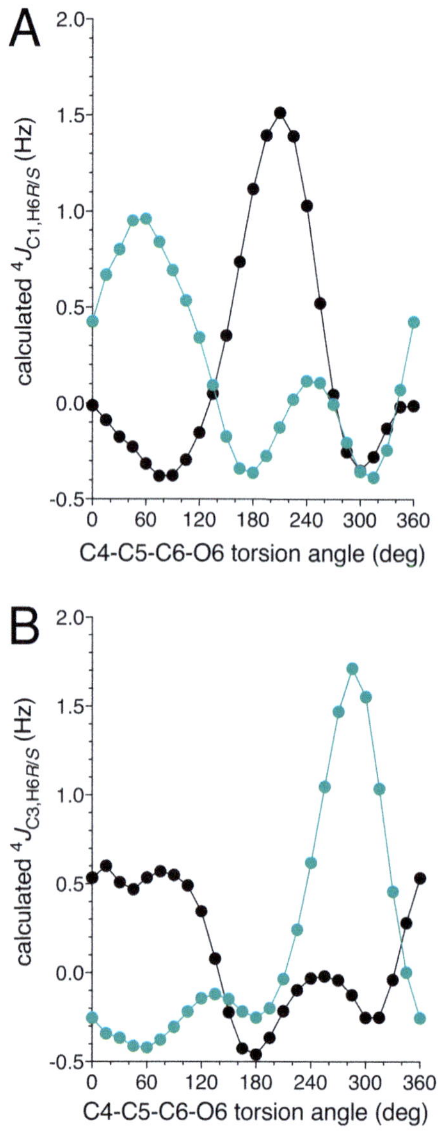

Figure 2.31 Effect of C5–C6 bond rotation in **46** on calculated $^4J_{C1,H6R}$ and $^4J_{C1,H6S}$ values (A), and on calculated $^4J_{C3,H6R}$ and $^4J_{C3,H6S}$ values (B). Green, H6*R*; black, H6*S*.

in-plane pathway yields a $^4J_{CH}$ of ~+1.2 Hz, and the out-of-plane pathway yields a $^4J_{CH}$ of ~−0.3 Hz. This correlation is not observed for the analogous C1–O5–C5–C6–H6*R*/*S* and C6–C5–O5–C1–H1 pathways. In the latter pathway, both "in-plane" and "out-of-plane" geometries give the same $^4J_{CH}$ (~−0.3 Hz). This behavior is not fully understood, but is not unexpected given that the structures of the two pathways differ, one being a C–O–C–C–H pathway and the other a C–C–O–C–H pathway.

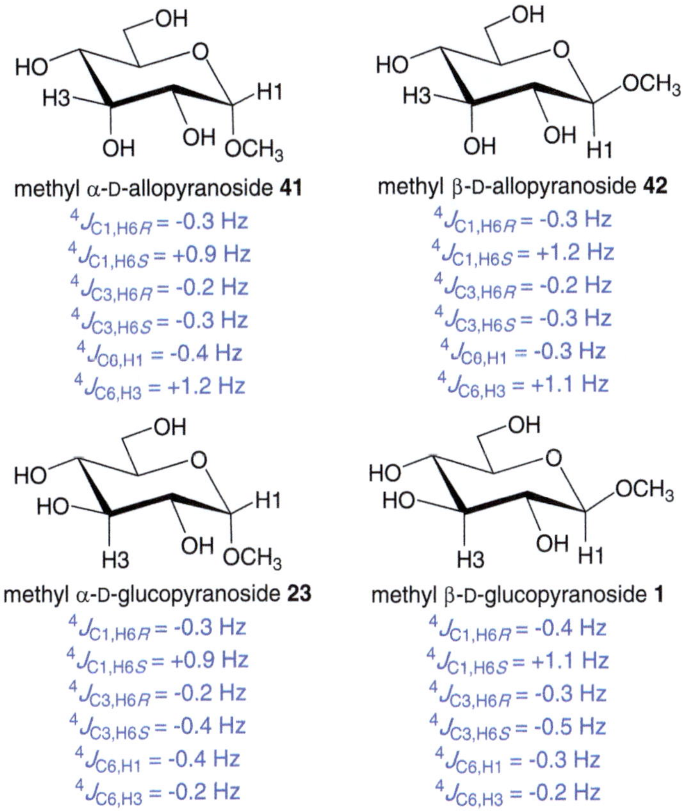

Scheme 2.27 Calculated $^4J_{CH}$ values in several aldohexopyranosyl rings (*gt* rotamers).

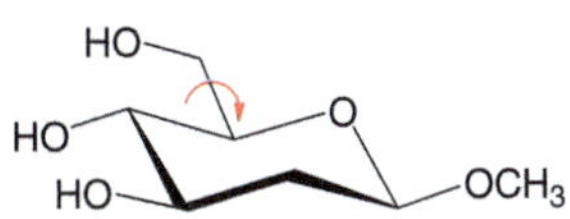

46

In addition to the H–C–O–C–H and H–C–C–C–H pathways that exist within monosaccharide residues, a H–C–O–C–H pathway exists across *O*-glycosidic linkages of oligosaccharides (Figure 2.1). The *trans-O*-glycoside $^4J_{HH}$ values associated with the latter pathways have been measured experimentally in an effort to apply them as constraints in determinations of linkage conformation in solution.[85] Their behavior is shown in Figure 2.32, where *trans-O*-glycoside $^4J_{H1',H4}$ values were calculated by DFT as a function of the *phi* (ϕ) (H1′–C1′–O1–C4) and *psi* (ψ) (C1′–O1–C4–H4) torsion angles across the glycosidic linkage of methyl β-D-galactopyranosyl-(1 → 4)-β-D-glucopyranoside **2** (Scheme 2.28). Calculated $^4J_{H1',H4}$ values vary from ~−1.5 Hz to ~+1.5 Hz. Importantly, these data can be compared to those for the intra-ring H–C–O–C–H pathway pertinent to $^4J_{H1,H5}$ (Scheme 2.25) in order to test for internal consistency between these related pathways (*e.g.*, the intra-ring

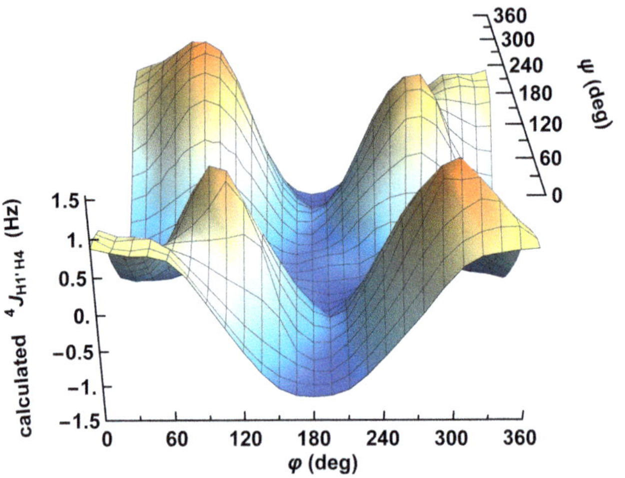

Figure 2.32 Effect of the *phi* (H1′-C1′-O1′-C4) and *psi* (C1′-O1′-C4-H4) torsion angles in **2** (Scheme 2.28) on calculated *trans-O*-glycoside $^4J_{H1',H4}$ values determined by DFT.

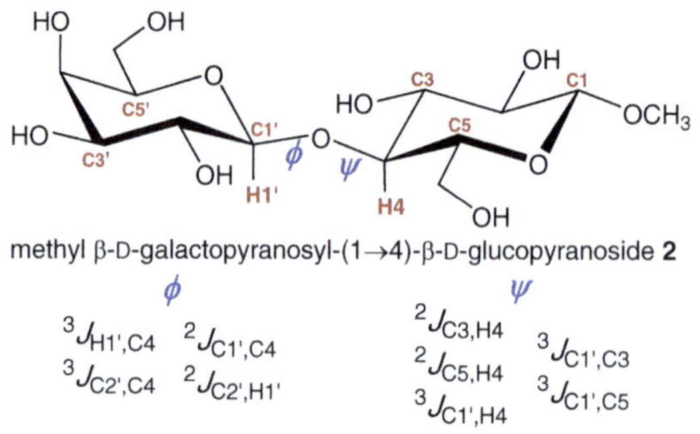

Scheme 2.28 Spin-couplings across a β-(1 → 4)-linkage.

H1–C1–O5–C5–H5 pathways in **1** and **42** and the inter-ring H1′–C1′–O1′–C4–H4 pathway in **2**). To simplify this comparison, a 2D slice through the hypersurface in Figure 2.32 at *phi* = +60° is pertinent because this torsion angle is highly favored by **2** in solution based on stereoelectronic considerations (*exo*-anomeric effect)[55–57] and NMR data.[86] This 2D slice is shown in Figure 2.33. Focusing only on the perfectly staggered values of *psi*, $^4J_{H1',H4}$ correlates with *phi*/*psi* pairs as follows: +60°/+60° = +0.4 Hz; +60°/−60° = +1.4 Hz; +60°/180° = −0.9 Hz. Two of these *phi*/*psi* combinations mimic the two intra-ring H1–C1–O5–C5–H5 pathways shown in Scheme 2.25: the β-*allo* **42**/β-*gluco* **1** structures are mimicked by the +60°/−60° pair, and the α-*allo* **41**/α-*gluco* **23** structures are mimicked by the +60°/180° pair.

The former set have similar $^4J_{HH}$ (~+1.2 Hz in the monosaccharides, +1.4 Hz in the linkage), as do the latter (−1.2 Hz in the monosaccharides, −0.9 Hz

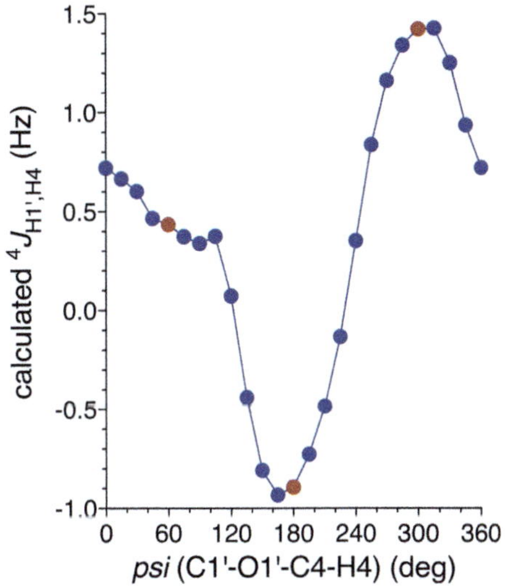

Figure 2.33 The effect of the *psi* (C1′–O1′–C4–H4) torsion angle in **2** on calculated *trans-O*-glycoside $^4J_{H1',H4}$ values at *phi* (H1′–C1′–O1′–C4) = 60°, determined by DFT. Points in red correlate with perfectly staggered values of *psi* (60°, 180° and 300°).

in the linkage), demonstrating that the intra- and inter-ring pathways behave the same with respect to the magnitudes and signs of $^4J_{HH}$ values. The third linkage pathway for the +60°/+60° pair has no mimic in Scheme 2.25; it is associated with a $^4J_{HH}$ value (+0.4 Hz) that is intermediate between those for the 60°/180° and 60°/−60° pairs.

2.7.6 Spin-Couplings Across *O*-Glycosidic Linkages in Oligosaccharides

An important application of spin-couplings in saccharide structure analysis involves those that are sensitive to *O*-glycosidic linkage conformation. These linkages commonly appear in two forms (see Section 2.4.5): those comprised of two bonds (Scheme 2.28), and those comprised of three bonds (Scheme 2.29). Nine (9) spin-couplings are useful conformational probes for two-bond linkages such as that found in methyl β-D-galactopyranosyl-(1→4)-β-D-glucopyranoside **2** (Scheme 2.28). Four are sensitive to *phi* (ϕ) and five are sensitive to *psi* (ψ). The general behaviors of these *J*-couplings were discussed above using the simplified model structures **24–26** (Scheme 2.23). The general structural dependencies of the two vicinal pathways ($^3J_{H1',C4}$, $^3J_{C2',C4}$) and the geminal $^2J_{C2',H1'}$ sensitive to *phi* were discussed in Sections 2.6.3 and 2.6.2, respectively. The geminal $^2J_{C1',C4}$ mimics $^2J_{C1,C5}$ values in aldopyranosyl rings (Section 2.6.2) and appears to show a stronger dependence on *phi* than on *psi*;[87] at present it appears to be the least robust *J*-coupling for *phi* analysis.

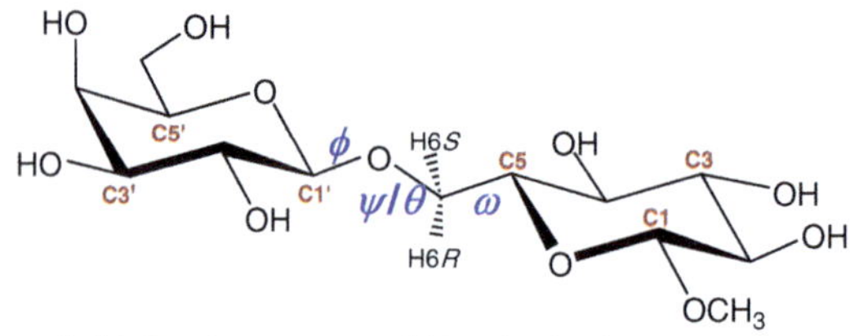

methyl β-D-galactopyranosyl-(1→6)-β-D-glucopyranoside **11**

ϕ	ψ/θ	ω
$^2J_{C1',C6}$	$^2J_{C5,H6R}$	$^2J_{H6R,H6S}$
$^2J_{C2',H1'}$	$^2J_{C5,H6S}$	$^2J_{C5,H6R}$
$^3J_{H1',C6}$	$^2J_{C6,H5}$	$^2J_{C5,H6S}$
$^3J_{C2',C6}$	$^3J_{C1',H6R}$	$^3J_{H5,H6R}$
	$^3J_{C1',H6S}$	$^3J_{H5,H6S}$
	$^3J_{C1',C5}$	$^3J_{C4,H6R}$
		$^3J_{C4,H6S}$

Scheme 2.29 Spin-couplings across a β-(1 → 6)-linkage.

The geminal $^2J_{C2',H1'}$ is a useful probe of *phi* because, as discussed in Section 2.6.2, it depends more heavily on rotation about the C1′–O1′ bond (*phi*) than on rotation about the C2′–O2′ bond. For *psi*, three vicinal *J*-couplings pertain: $^3J_{C1',C3}$, $^3J_{C1',C5}$ and $^3J_{H1',C4}$. Two $^2J_{CCH}$ values, $^2J_{C3,H4}$ and $^2J_{C5,H4}$, are also sensitive to *psi*; rotation about the C–O bond involving the carbon bearing the coupled hydrogen (*psi*) affects $^2J_{CCH}$ more than C–O bond rotation involving the coupled carbon (Section 2.6.2).

For linkages involving three bonds, two are C–O bonds and the third is a C–C bond. This situation is illustrated in Scheme 2.29 in methyl β-D-galactopyranosyl-(1 → 6)-β-D-glucopyranoside **11**. In this case, a total of 17 *J*-couplings are available: (a) four sensitive to *phi*; (b) six sensitive to *psi*; and (c) seven sensitive to ω. The ω torsion angle pertains to hydroxymethyl group rotation (Section 2.4.2) and is investigated by the same *J*-couplings used for the assignment of exocyclic hydroxymethyl group conformation in monosaccharides. To extract maximal information on three-bond linkages, stereochemical assignments of the H6*R* and H6*S* NMR signals are necessary. In addition, as shown in Scheme 2.29, the *psi* torsion angle is also denoted θ because conformation about this C–O bond can also be assigned from the C5–C6 side of the linkage. The latter is possible because $^2J_{C5,H6R}$, $^2J_{C5,H6S}$ and $^2J_{C6,H5}$ depend on both ω and θ, and parameterized equations can be derived containing both torsion angles.[75] These *J*-couplings can be used to obtain correlated information for ω and θ in these linkages, as discussed in prior work.[75]

While it is clear that multiple *J*-coupling constraints exist across *O*-glycosidic linkages, less clear is whether generalized equations can be derived to treat them regardless of structural context. For example, can the same

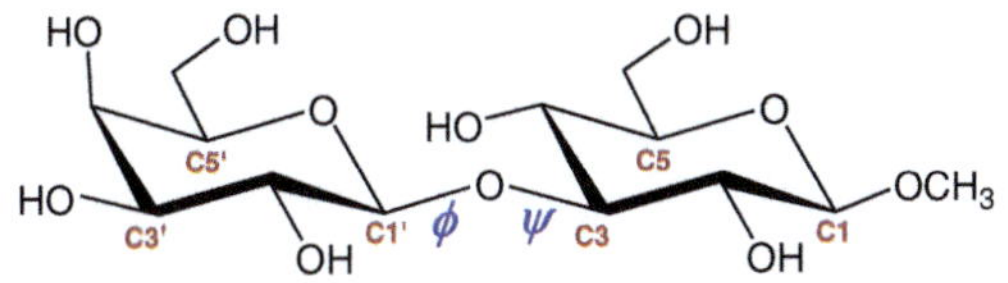

methyl β-D-galactopyranosyl-(1→3)-β-D-glucopyranoside **47**

ϕ: $^3J_{H1',C3}$, $^2J_{C1',C3}$, $^3J_{C2',C3}$, $^2J_{C2',H1'}$

ψ: $^2J_{C2,H3}$, $^2J_{C4,H3}$, $^3J_{C1',H3}$, $^3J_{C1',C2}$, $^3J_{C1',C4}$

Scheme 2.30 Spin-couplings across a β-(1 → 3)-linkage.

equations be used to treat *J*-couplings across the linkages in Schemes 2.28 and 2.30, the former a β-(1 → 4) linkage and the latter a β-(1 → 3) linkage? Likewise, what effect does anomeric configuration have on these *J*-couplings and their parameterization? Answers to these questions are still under investigation in this laboratory. However, to demonstrate that these *J*-couplings can be parameterized, ten (10) equations pertinent to the linkage found in methyl β-D-galactopyranosyl-(1 → 4)-β-D-xylopyranoside **12** (Scheme 2.9) follow (in these equations, *phi* (ϕ) is defined as H1′–C1′–O1′–C4 and *psi* (ψ) is defined as C1′–O1′–C4–H4):

J-Couplings in **12** Sensitive to *phi* (ϕ):

$$^2J_{C1',C4} = -2.32 + 0.52\cos(\phi) - 0.73\sin(\phi) - 0.07\cos(2\phi) - 0.30\sin(2\phi)$$

rms 0.4 Hz

$$^3J_{H1',C4} = 3.65 - 1.54\cos(\phi) - 0.27\sin(\phi) + 3.51\cos(2\phi) + 0.91\sin(2\phi)$$

rms 0.5 Hz

$$^3J_{C2',C4} = 1.79 + 0.05\cos(\phi) - 0.11\sin(\phi) - 0.30\cos(2\phi) + 1.89\sin(2\phi)$$

rms 0.4 Hz

$$^2J_{C2',H1'} = -4.25 - 0.23\cos(\phi) - 0.07\sin(\phi) + 0.66\cos(2\phi) - 0.81\sin(2\phi)$$

rms 0.3 Hz

J-Couplings in **12** Sensitive to *psi* (ψ):

$$^3J_{C1',H4} = 3.79 - 1.77\cos(\psi) + 0.15\sin(\psi) + 4.00\cos(2\psi) + 0.12\sin(2\psi)$$

rms 0.6 Hz

$$^{3}J_{C1',C3} = 1.78 + 0.07\cos(\psi) + 0.48\sin(\psi) - 0.75\cos(2\psi) + 1.80\sin(2\psi)$$

rms 0.4 Hz

$$^{3}J_{C1',C5} = 2.45 + 0.66\cos(\psi) - 0.30\sin(\psi) - 1.53\cos(2\psi) - 2.55\sin(2\psi)$$

rms 0.9 Hz

$$^{2}J_{C3,H4} = -3.96 + 0.58\cos(\psi) - 0.32\sin(\psi) - 0.52\cos(2\psi) - 0.53\sin(2\psi)$$

rms 0.3 Hz

$$^{2}J_{C5,H4} = -3.09 + 0.88\cos(\psi) + 0.35\sin(\psi) - 0.71\cos(2\psi) + 1.26\sin(2\psi)$$

rms 0.3 Hz

$$^{2}J_{C3,C5} = 2.47 - 0.24\cos(\psi) - 0.13\sin(\psi) + 1.29\cos(2\psi) - 0.18\sin(2\psi)$$

rms 0.2 Hz

The general behaviors of $^3J_{COCC}$ values across *O*-glycosidic linkages were discussed in Section 2.6.3 using model structures **24–26** (Scheme 2.23). $^2J_{CCH}$, $^2J_{COC}$ and $^3J_{COCH}$ values that are sensitive to the *phi* (ϕ) or *psi* (ψ) *O*-glycosidic torsion angles in **12** can also be evaluated through DFT studies of model structures **25** and **26**. In **25**, calculated $^2J_{C\alpha,C\beta,H}$ values vary from +1.7 Hz to +4.5 Hz when the C_α–C_β–$O_{\beta\gamma}$–C_γ torsion angle is rotated through 360° (Figure 2.34A). For the three perfectly staggered rotamers, calculated couplings are: +60°, +4.5 Hz; −60°, +2.1 Hz; 180°, +2.5 Hz. The most positive $^2J_{C\alpha,C\beta,H}$ is observed in the +60° rotamer in which both of the coupled atoms (C_α and H) are *anti* to a lone-pair orbital on $O_{\beta\gamma}$ (Scheme 2.31). In the two remaining staggered rotamers, only one of the coupled atoms is *anti* to an $O_{\beta\gamma}$ lone-pair orbital, which reduces the $^2J_{CCH}$ to less positive values. A large positive $^2J_{C\alpha,C\beta,H}$ is also observed in the conformer having the C_α–C_β and C_β–H bonds eclipsed with a lone-pair orbital on $O_{\beta\gamma}$ (C_α–C_β–$O_{\beta\gamma}$–C_γ torsion angle of 240°). Projection sums (Figure 2.2) calculated for the three staggered rotamers are identical (+1.0; Scheme 2.31) since configuration at C_β is constant in these rotamers, and predicted $^2J_{CCH}$ values are ∼+2 Hz (Figure 2.2). The conformational effect on $^2J_{CCH}$ is therefore apparent, in this case confirming the effect of C–O bond rotation on $^2J_{CCH}$ values when the carbon of the C–O bond being rotated bears the coupled hydrogen (see Section 2.6.2).

$^2J_{C\beta,O\beta\gamma,C\gamma}$ values in **25** vary from −0.5 to −3.3 Hz when the C_α–C_β–$O_{\beta\gamma}$–C_γ torsion angle is rotated through 360° (Figure 2.34B). For the three perfectly staggered rotamers, calculated couplings are: +60°, −0.6 Hz; −60°, −2.5 Hz; 180°, −3.0 Hz. The C_α–C_β–$O_{\beta\gamma}$–C_γ fragment in **25** mimics the C2′–C1′–O1′–C4 fragment in the β-(1→4)-linked disaccharide **2** shown in Scheme 2.28, and thus rotating the C_β–$O_{\beta\gamma}$ bond in **25** mimics rotation about *phi* (ϕ) in **2**.

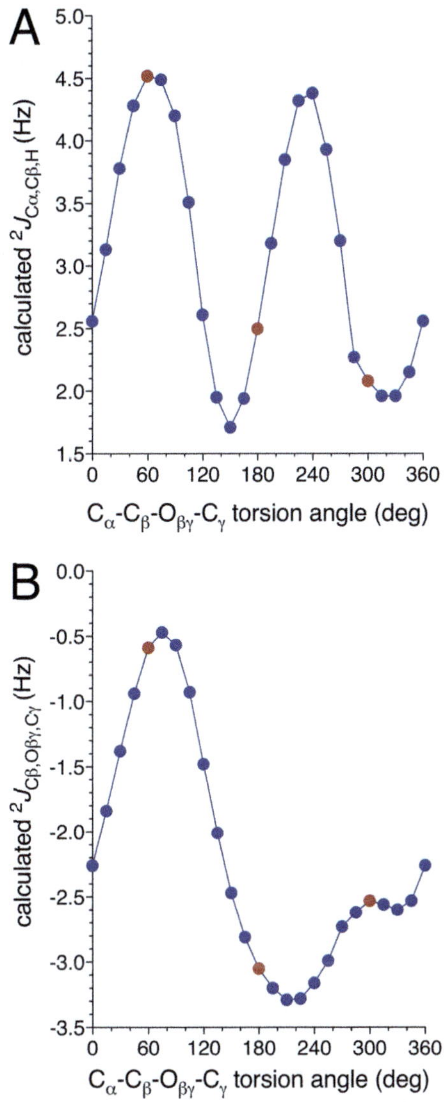

Figure 2.34 Effect of the C_α–C_β–$O_{\beta\gamma}$–C_γ torsion angle in **25** on calculated values of (A) $^2J_{C\alpha,C\beta,H}$ and (B) $^2J_{C\beta,O\beta\gamma,C\gamma}$. Data points shown in red correspond to the three perfectly staggered rotamers about the C_β–$O_{\beta\gamma}$ bond.

In most cases, stereoelectronic factors favor a C_α–C_β–$O_{\beta\gamma}$–C_γ torsion angle of 150–170°,[55–57] suggesting that *trans-O*-glycoside $^2J_{COC}$ values should be −2 to −3 Hz (see below). The behavior of $^2J_{C\beta,O\beta\gamma,C\gamma}$ in **25** for the three perfectly staggered C_α–C_β–$O_{\beta\gamma}$–C_γ rotamers is predicted by a modified C–O–C projection resultant[21] described in Section B (Figure 2.3). As discussed previously,[21] the three staggered C_α–C_β–$O_{\beta\gamma}$–C_γ rotamers (torsion angle equivalent to ϕ in an *O*-glycosidic linkage) are modeled by analogous $^2J_{C1,C5}$ values in aldopyranosyl

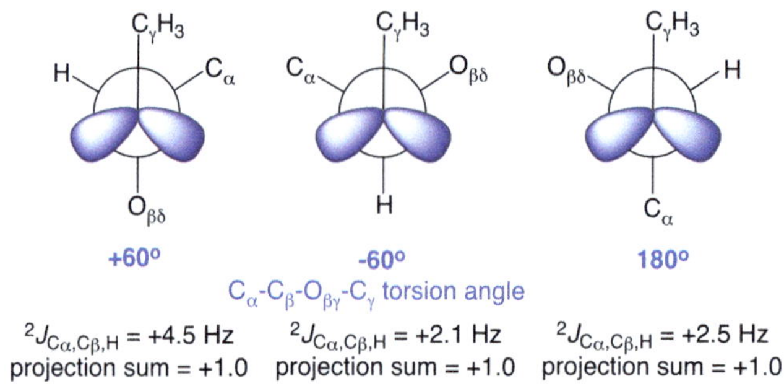

Scheme 2.31

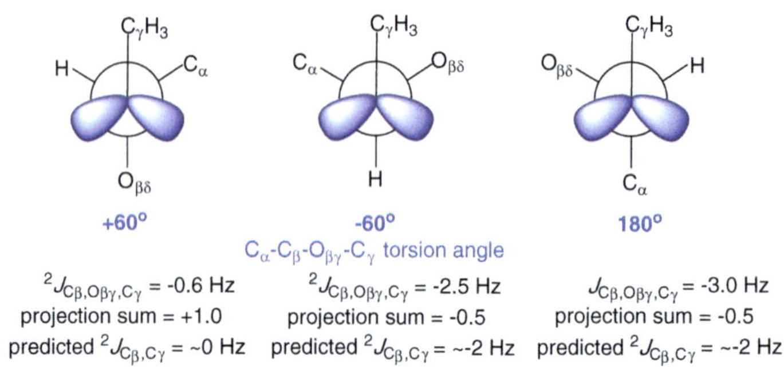

Scheme 2.32

rings, giving the predicted $^2J_{C1,C5}$ shown in Scheme 2.32. Note that the staggered rotamers about ψ give no projection sums because these fragments bear no oxygen substituents, and thus only one projection sum (that for ϕ) pertains in this treatment.[21] The treatment predicts an essentially zero value of $^2J_{C\beta,C\gamma}$ for the +60° rotamer, and equal and more negative $^2J_{C\beta,O\beta\gamma,C\gamma}$ values for the 180° and −60° rotamers. This empirical prediction reproduces the DFT results (Figure 2.34B) well, though not entirely quantitatively.

The effects of terminal electronegative substituents on $^2J_{C\alpha,C\beta,H}$ and $^2J_{C\beta,C\gamma}$ can be appraised using model structure **26**, which contains an OCH_3 group appended to C_α. Data in Figure 2.35 show the effect of the C_α–C_β–$O_{\beta\gamma}$–C_γ torsion angle on calculated $^2J_{C\alpha,C\beta,H}$ values at different $O_{\alpha\varepsilon}$–C_α–C_β–$O_{\beta\gamma}$ torsion angles. Calculated $^2J_{C\alpha,C\beta,H}$ values range from ~−2 Hz to ~+9 Hz; rotating the C_α–C_β bond changes the relative orientation of $O_{\alpha\varepsilon}$ and $O_{\beta\delta}$, which in turn affects the projection sum for the C_α–C_β–H pathway. Projection sums at $O_{\alpha\varepsilon}$–C_α–C_β–$O_{\beta\gamma}$ torsion angles of 60°, 180° and 300° are +2, +0.5 and +0.5, respectively, translating into predicted $^2J_{C\alpha,C\beta,H}$ values of ~8 Hz, ~0 Hz, and ~0 Hz, respectively, based on the graph shown in Figure 2.2. At a given $O_{\alpha\varepsilon}$–C_α–C_β–$O_{\beta\gamma}$ torsion angle, the behavior of $^2J_{C\alpha,C\beta,H}$ as the C_α–C_β–$O_{\beta\gamma}$–C_γ torsion angle is rotated through 360° mimics that found in **25** (*i.e.*, similar overall sinusoidal behavior and dynamic range but different absolute

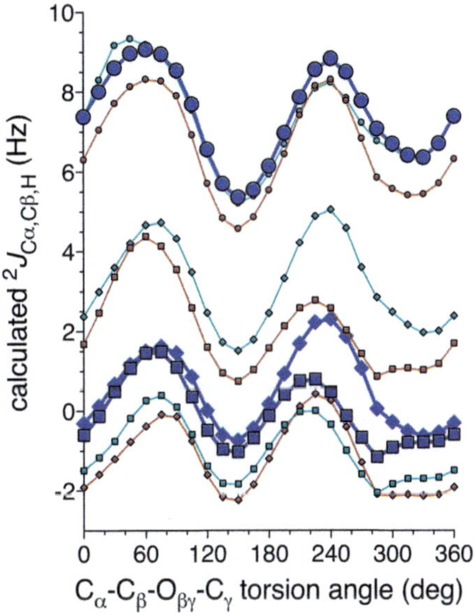

Figure 2.35 Effect of the C_α–C_β–$O_{\beta\gamma}$–C_γ torsion angle in **26** on calculated values of $^2J_{C\alpha,C\beta,H}$ at different values of the $O_{\alpha\varepsilon}$–C_α–C_β–$O_{\beta\gamma}$ torsion angle. Filled large circles, +60°; filled large triangles, 180°; filled large squares, 300°. Smaller symbols denote −30° (green) and +30° (red) increments of the $O_{\alpha\varepsilon}$–C_α–C_β–$O_{\beta\gamma}$ torsion that bracket data for the perfectly staggered rotamers.

J-couplings). The three-dimensional hypersurface showing the effects of the C_α–C_β–$O_{\beta\gamma}$–C_γ and $O_{\alpha\varepsilon}$–C_α–C_β–$O_{\beta\gamma}$ torsion angles on calculated $^2J_{C\alpha,C\beta,H}$ values in **26** is shown in Figure 2.36.

The effect of the C_α–C_β–$O_{\beta\gamma}$–C_γ torsion angle on calculated $^2J_{C\beta,O\beta\gamma,C\gamma}$ values in **26** at different $O_{\alpha\varepsilon}$–C_α–C_β–$O_{\beta\gamma}$ torsion angles is shown in Figure 2.37. Rotation about the C_α–C_β bond exerts a small effect on the calculated $^2J_{C\beta,O\beta\gamma,C\gamma}$ and the general shape and amplitude of the curves resemble that shown in Figure 2.34B for **25**, with *J*-couplings ranging from ~−0.5 Hz to ~−3.6 Hz. The full hypersurface showing the effects of the C_α–C_β–$O_{\beta\gamma}$–C_γ and $O_{\alpha\varepsilon}$–C_α–C_β–$O_{\beta\gamma}$ torsion angles on calculated $^2J_{C\beta,O\beta\gamma,C\gamma}$ values in **26** is shown in Figure 2.38.

Calculated $^3J_{C\gamma,O\beta\gamma,C\beta,H}$ values in model structures **25** and **26** (Scheme 2.23) provide baseline information on *trans-O*-glycoside $^3J_{COCH}$ values, specifically, their magnitudes, how they depend on the C–O–C–H torsion angle, and how they are affected by terminal electronegative substituents. This information is shown in Figure 2.39. The data for **25** (shown in red) show that $^3J_{C\gamma,O\beta\gamma,C\beta,H}$ values vary from 0–8 Hz, with the value for the eclipsed C–O–C–H rotamer (C_α–C_β–$O_{\beta\gamma}$–C_γ = 120°) slightly smaller than that for the trans rotamer (C_α–C_β–$O_{\beta\gamma}$–C_γ = 300°). Superimposed on this plot are analogous data for **26** (data in blue). The overall shape of the latter curve mimics that for **25**, although average curve amplitude for the trans rotamer is slightly greater than that found for **25**, suggesting a small in-plane effect from $O_{\alpha\varepsilon}$. When the $O_{\alpha\varepsilon}$–C_α–C_β–$O_{\beta\gamma}$

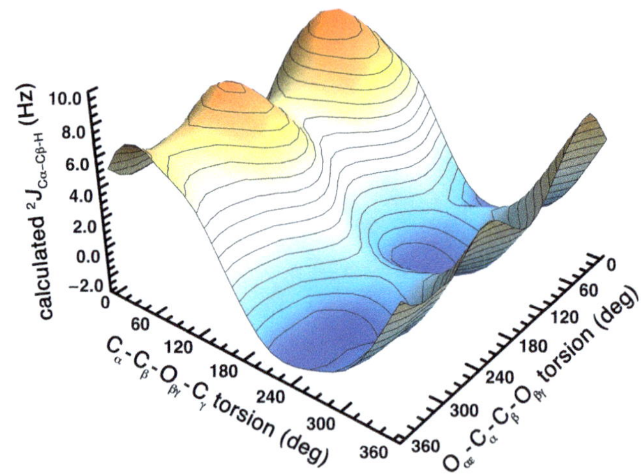

Figure 2.36 Hypersurface showing the dependence of $^2J_{C\alpha,C\beta,H}$ in **26** on the C_α–C_β–$O_{\beta\gamma}$–C_γ and $O_{\alpha\varepsilon}$–C_α–C_β–$O_{\beta\gamma}$ torsion angles determined by DFT.

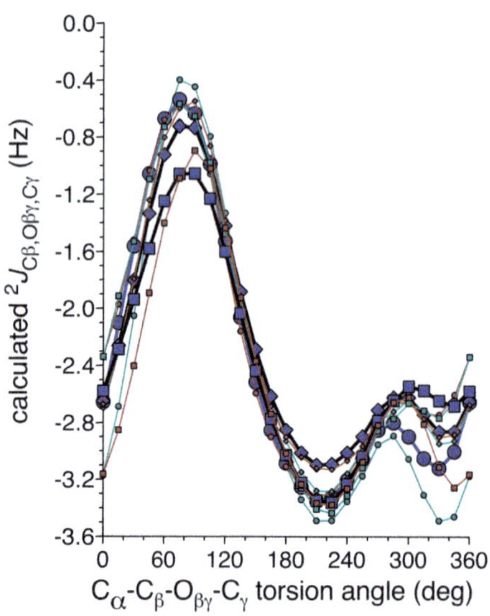

Figure 2.37 Effect of the C_α–C_β–$O_{\beta\gamma}$–C_γ torsion angle in **26** on calculated values of $^2J_{C\beta,O\beta\gamma,C\gamma}$ at different values of the $O_{\alpha\varepsilon}$–C_α–C_β–$O_{\beta\gamma}$ torsion angle. Filled large circles, +60°; filled large triangles, 180°; filled large squares, 300°. Smaller symbols denote −30° (green) and +30° (red) increments of the $O_{\alpha\varepsilon}$–C_α–C_β–$O_{\beta\gamma}$ torsion that bracket data for the perfectly staggered rotamers.

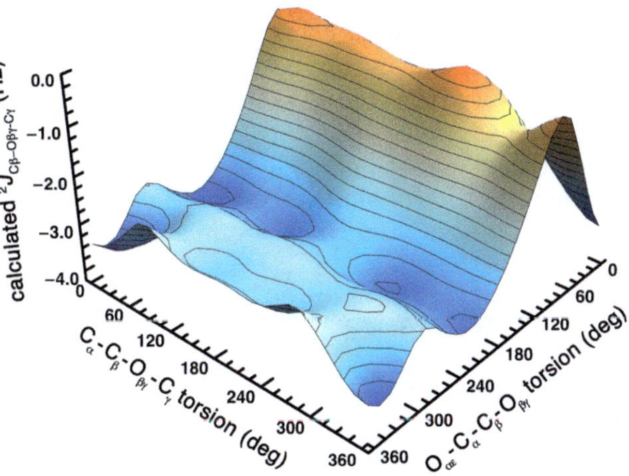

Figure 2.38 Hypersurface showing the dependence of $^2J_{C\beta,O\beta\gamma,C\gamma}$, in **26** on the C_α–C_β–$O_{\beta\gamma}$–C_γ and $O_{\alpha\varepsilon}$–C_α–C_β–$O_{\beta\gamma}$ torsion angles determined by DFT.

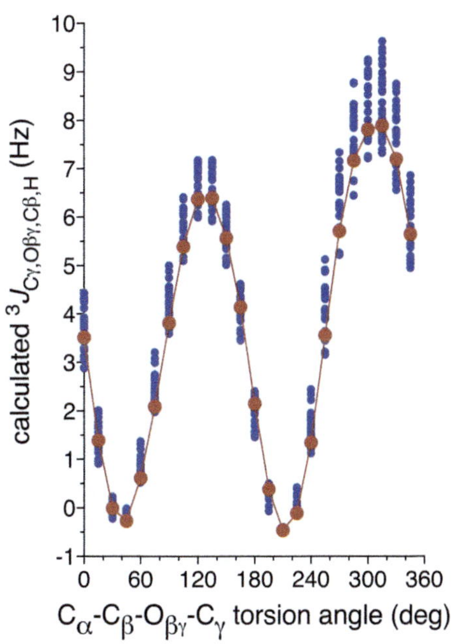

Figure 2.39 Dependence of calculated $^3J_{C\gamma,O\beta\gamma,C\beta,H}$ values in **25** (red) and **26** (blue) on the C_α–C_β–$O_{\beta\gamma}$–C_γ torsion angle. Data scatter at discrete values of the C_α–C_β–$O_{\beta\gamma}$–C_γ torsion angle in **26** is caused by the effect of rotating the $O_{\alpha\varepsilon}$–C_α–C_β–$O_{\beta\gamma}$ torsion angle (see Scheme 2.23). Figure 2.40 shows the same data for **26** as a 3D hypersurface.

torsion angle is +60°, the calculated $^3J_{COCH}$ is +9.0 Hz, whereas the value is ~7.6 Hz in the other two staggered rotamers. When the $O_{\alpha\varepsilon}$–C_α–C_β–$O_{\beta\gamma}$ torsion angle is +60°, $O_{\alpha\varepsilon}$ is *anti* to the coupled hydrogen on C_β.

Close inspection of the data in Figure 2.39 shows that calculated $^3J_{C\gamma,O\beta\gamma,C\beta,H}$ values differ for the two *gauche* C–O–C–H torsion angles, with the C–O–C–H torsion of −60° (C_α–C_β–$O_{\beta\gamma}$–C_γ = +60°) giving a smaller value than that observed for the C–O–C–H torsion angle of +60° (C_α–C_β–$O_{\beta\gamma}$–C_γ = 180°). In the former conformer, the coupled carbon (C_γ) is *anti* to $O_{\beta\delta}$, which truncates the *gauche* $^3J_{COCH}$. This latter effect is similar to that observed for $^3J_{C\alpha,C\gamma}$ (Scheme 2.24).

The range of $^3J_{C\gamma,O\beta\gamma,C\beta,H}$ values observed in **26** at discrete C_α–C_β–$O_{\beta\gamma}$–C_γ torsion angles (Figure 2.39) is caused by rotating the C_α–C_β bond (see Scheme 2.23). As observed for $^3J_{C\alpha,C\gamma}$ in **26** (Figure 2.27), the data scatter is not uniform in width, but varies with the C_α–C_β–$O_{\beta\gamma}$–C_γ torsion angle, although the relative variability (width of scatter across the plot) is smaller than found for $^3J_{C\alpha,C\gamma}$. Nevertheless, both $^3J_{COCH}$ and $^3J_{COCC}$ values in **26** are affected by the orientation of $O_{\alpha\varepsilon}$ relative to the coupling pathway. The full 3D hypersurface showing the dependence of $^3J_{C\gamma,O\beta\gamma,C\beta,H}$ on both the C_α–C_β–$O_{\beta\gamma}$–C_γ and $O_{\alpha\varepsilon}$–C_α–C_β–$O_{\beta\gamma}$ torsion angles in **26** is shown in Figure 2.40.

An attractive feature of *J*-couplings across *O*-glycosidic linkages is that more than one *J*-coupling is sensitive to each C–O or C–C bond torsion angle comprising these linkages. This redundancy improves the assignment of C–O (and C–C) bond conformation provided that (a) accurate equations relating each *J*-coupling to a given C–O and/or C–C bond torsion angle are available, and (b) a mathematical algorithm is available to fit the set of

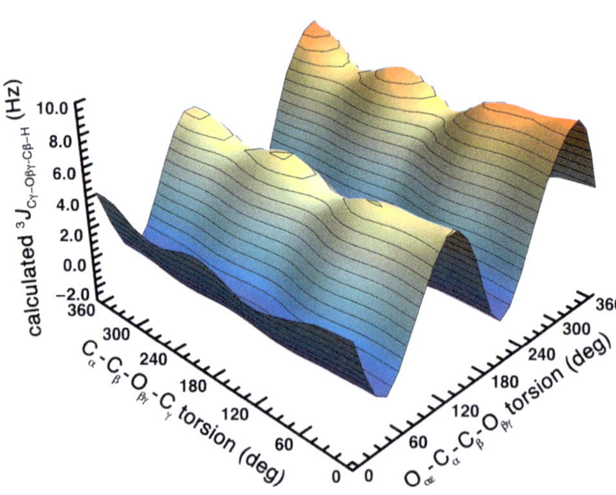

Figure 2.40 Three-dimensional plot showing the dependence of calculated $^3J_{C\gamma,O\beta\gamma,C\beta,H}$ values in **26** on the C_α–C_β–$O_{\beta\gamma}$–C_γ and $O_{\alpha\varepsilon}$–C_α–C_β–$O_{\beta\gamma}$ torsion angles (see Scheme 2.23). Figure 2.39 shows the same data presented as a 2D plot.

redundant experimental *J*-couplings to a conformational model. The above discussion focused on the use of DFT to meet requirement (a), and while these parameterizations are not devoid of error caused mainly by incomplete treatments of electronegative substituent and/or lone-pair effects on calculated *J*-couplings, the approach nevertheless provides reliable equations in most instances (*e.g.*, see parameterized equations for **12**).

While studies of redundant *O*-glycosidic *J*-couplings in oligosaccharides are ongoing and beyond the scope of this review, the value of these constraints is briefly considered here for the following three β-(1 → 4)-linked disaccharides (Scheme 2.33): methyl β-D-galactopyranosyl-(1 → 4)-β-D-glucopyranoside **2**, methyl β-D-galactopyranosyl-(1 → 4)-β-D-xylopyranoside **12**, and methyl β-D-galactopyranosyl-(1 → 4)-β-D-allopyranoside **48**. These disaccharides contain structural differences in the vicinity of their internal *O*-glycosidic linkages. If **2** is considered the "reference state", then **12** lacks the exocyclic hydroxymethyl group appended to C5 of **2**, and **48** contains an axial O3 instead of the equatorial O3 found in **2** (Scheme 2.33). The effects of these structural changes on experimental *trans*-glycosidic *J*-couplings are shown in Table 2.6. Note that the ensembles of *J*-couplings sensitive to ϕ are similar in each disaccharide, whereas the ensembles of

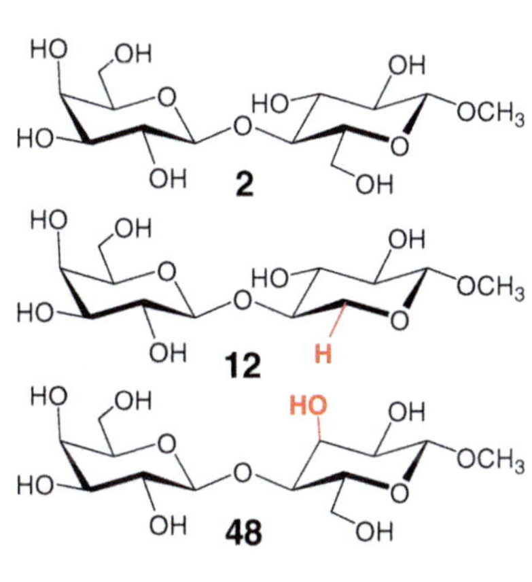

Scheme 2.33 Three β-(1→4)-linked disaccharides containing different covalent structure in the vicinity of their internal *O*-glycosidic linkages.

Table 2.6 *Trans*-glycoside NMR *J*-couplings[a] in β-(1 → 4)-linked disaccharides **2**, **12** and **48**.

	Phi (ϕ)			*Psi* (ψ)		
Disaccharide	$^2J_{C1',C4}$	$^3J_{C4,H1'}$	$^3J_{C2',C4}$	$^3J_{C1',H4}$	$^3J_{C1',C3}$	$^3J_{C1',C5}$
βGal(1 → 4)βGlcOCH$_3$ **2**	−2.0	3.9	3.1	5.0	*br*	2.0
βGal(1 → 4)βXylOCH$_3$ **12**	−1.9	4.0 ± 0.2	2.9	4.4	1.8	0.8
βGal(1 → 4)βAllOCH$_3$ **48**	−1.8		3.1	4.7	*br*	3.0

[a]In Hz ± 0.1 Hz; ~25 °C; in ^2H_2O; *br* denotes broadened signal (J < 0.7 Hz).

J-couplings sensitive to *ψ* differ. Qualitatively, these results strongly suggest that the conformational preferences about *ϕ* are essentially identical in these disaccharides, but those about *ψ* differ. This conclusion can be made with confidence because the respective parameterized equations for the *J*-couplings in these disaccharides are identical (*i.e.*, the differences in structure do not affect the equations). This being the case, visual inspection of *trans*-glycosidic *J*-coupling ensembles provides a rapid and reliable means of classifying glycosidic linkages according to preferred conformation. These data also show that, at least within this type of linkage, the *ψ* torsion angle is more prone to change in response to changes in covalent structure than the *ϕ* torsion angle, a behavior likely caused by the added stereoelectronic constraints imposed on *ϕ*.

While differences in experimental *J*-coupling ensembles in **2**, **12** and **48** (Table 2.6) suggest different preferred conformations about *ψ*, these behaviors cannot be quantified without a more rigorous treatment of the data. This analysis is possible using the *MA'AT* program,[88] which generates a model for the rotamer distribution about *ψ* using DFT-parameterized equations, experimental *J*-couplings and circular statistics. For compounds **2**, **12** and **48**, the mean positions of models generated by this method are −8°, −21° and +16°, respectively. These torsion angles thus appear to obey a single-state model, and the *MA'AT* results are in close agreement with the behaviors of *ψ* predicted by aqueous MD simulations of **2**, **12** and **48** (mean positions of −3°, −20° and +12°, respectively). These findings bode well for the use of *O*-glycoside *J*-coupling ensembles as experimental constraints with which to assign linkage conformations of oligosaccharides in solution.

2.8 Measurement of NMR Spin-Couplings in Solids

Historically, vicinal (three-bond) NMR *J*-couplings (*e.g.*, $^3J_{HH}$) have been interpreted quantitatively through the use of Karplus curves,[11,74] the latter produced by measuring experimental ^3J values in compounds containing fixed (conformationally constrained) and known values of the molecular torsion angle that largely determines the magnitude of the ^3J. This approach has limitations as discussed previously.[17] With the advent of DFT methods to calculate NMR spin-couplings, the need for constrained compounds to construct Karplus curves has been reduced considerably, but the usefulness of this computational approach hinges on the reliability of the DFT results. In solution, molecular torsion angles are typically conformationally averaged, and experimental *J*-couplings associated with them are likewise time-averaged. On the other hand, in the solid state, this averaging is reduced or eliminated. *J*-Couplings measured in a crystalline solid of known structure (determined by x-ray crystallography) can provide a direct connection between *J*-coupling magnitude and the molecular torsion angle found in the crystal.

Recent work has shown that NMR *J*-couplings between heavy spin-1/2 atoms can be measured quantitatively in the solid state.[89] This method has recently been applied to measure $^2J_{C1,C3}$ in crystalline methyl β-D-[1,3-$^{13}C_2$]glucopyranoside **1**. As shown in Table 2.1, $^2J_{C1,C3}$ in **1** is +4.6 Hz in aqueous solution. However, $^2J_{C1,C3}$ was recently measured in crystalline **1** was found to be 6.0 Hz (absolute value). The discrepancy can be explained by considering the effect of the C1–C2–O2–H torsion angle on $^2J_{C1,C3}$ (see Figure 2.19A). In solution, this torsion angle is time-averaged, but in the crystalline solid, it is +89° as determined from x-ray crystallographic analysis of the same sample used for the solid-state NMR measurements. DFT calculations on **1** as a function of the C1–C2–O2–H torsion angle show $^2J_{C1,C3}$ to be +6.1 Hz when the C1–C2–O2–H torsion angle is equal to +89° (Figure 2.19A). While studies of this type are just beginning in this laboratory, current results confirm that, counter to conventional wisdom, NMR *J*-couplings can be measured in crystalline solids, and this approach should add an important new dimension to structural studies of saccharides, either free or in the bound state, based on *J*-coupling constraints.

2.9 Concluding Remarks

The primary aims of this chapter were two-fold, namely, to (1) provide a summary of more than two decades of experimental and theoretical studies of NMR *J*-couplings in saccharides involving ^{1}H and ^{13}C, and (2) test and/or extend prior findings and conclusions through new experimental measurements and theoretical calculations. This treatment also provides a brief appraisal of new and evolving efforts in the field, for example, those involving the measurement and analysis of redundant *O*-glycosidic *J*-couplings in oligosaccharides, and the exploitation of solid-state NMR techniques to measure *J*-couplings in fixed saccharide conformations. In the coming years, research findings in these evolving areas will likely demand separate reviews. Importantly, however, this future work stands on the solid foundation provided by the extensive fundamental studies discussed herein.

While NMR *J*-couplings have long been used in saccharide structure determination, as exemplified by the early seminal work of Lemieux and coworkers,[90] much remains to be discovered about these parameters and how they are influenced by saccharide structural properties. Two factors are central to understanding the correlations between saccharide structure and *J*-couplings, namely, the high abundance of electronegative atoms such as oxygen in these molecules and the resulting high density of lone-pair orbitals residing on these atoms. These factors have been discussed at length in this chapter, and their critical roles should now be evident. Arguably they complicate structural interpretations of *J*-couplings, but they also hold enormous potential in understanding subtle structural behaviors and features that resist inquiry by other means and which may ultimately prove important in explaining the chemical and biochemical properties of these important biomolecules. Particularly interesting is the observation that some *J*-couplings display systematic dependencies on more than one structural parameter, which can be captured faithfully in mathematical form.[75] This feature leads to the potential use of *J*-couplings to interrogate correlated conformational behaviors *via* experiment, which is an opportunity not offered at present by other experimental methods, thus leaving molecular dynamics simulations as the sole tool to investigate these properties.

The high abundance of *J*-couplings in saccharides leads to a high degree of information redundancy. While prior and present studies often ignore this redundancy and tend towards using one or a few *J*-couplings to define a conformational feature, much useful information is left untapped which, if used, would significantly increase the confidence of conformational assignments and/or measurably improve investigations of conformational flexibility. This unused information has been overlooked mainly because quantitative correlations between *J*-couplings and saccharide structure have been lacking. Research over the past two decades, described herein, has focused on addressing this deficiency. The ongoing development of programs like *MA'AT* and the increasing ease of measuring accurate carbon-based *J*-couplings in ^{13}C-labeled molecules, lower the barriers to more extensive and effective use of redundant *J*-couplings, and offer a means to promote a deeper understanding of saccharide structure, especially for those saccharides that are conformationally mobile, and to illuminate their biological functions.

While *J*-couplings, being largely through-bond NMR parameters, provide a unique window through which to view carbohydrate structure, it must be appreciated that they report on local (short-range) structure only, and cannot be relied upon on their own to assign global structure in larger, more complex molecules. Other experimental NMR parameters, such as NOEs, ROEs, RDCs and paramagnetic relaxation enhancements (PREs),[91] must be brought to bear to provide longer-range structural constraints. Structural conclusions drawn from the analysis of these longer-range constraints must, however, be internally consistent with those drawn from the analysis of local structure constraints such as *J*-couplings. Indeed, this synergy should prove valuable in future studies of larger saccharides. With this goal in mind, it will be critical to validate structural predictions based on RDCs with those determined by *J*-couplings in those molecules in which both parameters can be measured simultaneously. Only then can sufficient confidence be placed in structural interpretations of RDCs in molecules whose sizes preclude the measurement of spin-couplings.

Despite the important effects of C–O bond torsions of hydroxyl groups on *J*-couplings in saccharides that have been revealed in this chapter, a thorough understanding of the behaviors of these torsions in solution remains elusive. Perhaps the greatest impediment to securing a fully quantitative understanding of *J*-couplings is the current lack of experimental data that report on these torsions in solution. In the past, this information could only be gleaned from *J*-coupling studies involving the solvent-exchangeable hydroxyl hydrogen. These couplings remain valuable but difficult to measure, but we now realize that other *J*-couplings can report on these behaviors, for example, $^2J_{CCH}$ and $^2J_{CCC}$, that do not involve the OH hydrogen as a coupled atom. This new opportunity opens the door to improved studies of these C–O torsions, which not only play a role in determining *J*-couplings due to lone-pair effects, but also play major roles in solvent interactions and in interactions with receptors.

Acknowledgements

This work was supported by a grant from the National Science Foundation to A.S. (CHE 1402744) and by a grant from the National Institutes of Health to Omicron Biochemicals, Inc. (SBIR HHSN261201500020C). A.S. is the president and CEO of Omicron Biochemicals, Inc. The Notre Dame Radiation Laboratory (I.C.) is supported by the U. S. Department of Energy Office of Science, Office of Basic Energy Science under Award Number DE-FC02-04ER15533. This is document number NDRL 5132.

References

1. (a) I. Ando and G. A. Webb, *Theory of NMR Parameters*, Academic Press, 1983, pp. 83–113; (b) H. Günther, *NMR Spectroscopy–Basic Principles, Concepts and Applications in Chemistry*, Wiley-VCH, 3rd edn, 2013, pp. 108–133.
2. T. J. Church, I. Carmichael and A. S. Serianni, *J. Am. Chem. Soc.*, 1997, **119**, 8946–8964.
3. T. Klepach, I. Carmichael and A. S. Serianni, *J. Am. Chem. Soc.*, 2008, **130**, 11892–11900.
4. W. Zhang, X. Hu, I. Carmichael and A. S. Serianni, *J. Org. Chem.*, 2012, **77**, 9521–9534.
5. R. R. Ernst, G. Bodenhausen and A. Wokaun, *Principles of Nuclear Magnetic Resonance in One and Two Dimensions*, Oxford University Press, 1987.
6. M. H. Levitt, *Spin Dynamics – Basics of Nuclear Magnetic Resonance*, Wiley, New York, 2001.
7. S. Berger and S. Braun, *200 and More NMR Experiments – A Practical Course*, Wiley-VCH, 2004.
8. N. F. Ramsey, *Phys. Rev.*, 1953, **91**, 303–307.
9. G. A. Aucar, *Concepts Magn. Reson.*, 2008, **32A**, 88–116.
10. (a) A. Saika and H. S. Gutowski, *J. Am. Chem. Soc.*, 1956, **78**, 4818–4819; (b) J.-C. Hierso, *Chem. Rev.*, 2014, **114**, 4838–4867.
11. M. Karplus, *J. Chem. Phys.*, 1959, **30**, 11–15.
12. P. Pyykkö, *Chem. Phys.*, 1977, **22**, 289–296.
13. J. L. Marshall, L. G. Faehl, R. Kattner and P. E. Hansen, *Org. Magn. Reson.*, 1979, **12**, 169–173.
14. J. L. Marshall, *Carbon-Carbon and Carbon-Proton NMR Couplings: Applications to Organic Stereochemistry and Conformational Analysis, Methods in Stereochemical Analysis*, Verlag Chemie International, Deerfield Beach, 1983.

15. G. E. Martin and C. E. Hadden, *J. Nat. Prod.*, 2000, **63**, 543–585.
16. W. Zhang, S. Zhao and A. S. Serianni, in *Methods Enzymol.*, ed. Z. Kelman, Academic Press, 2015, vol. 565, pp. 423–458.
17. T. Klepach, H. Zhao, X. Hu, W. Zhang, R. Stenutz, M. J. Hadad, I. Carmichael and A. S. Serianni, in *Glycoinformatics, Methods in Molecular Biology*, ed. T. Lütteke and M. Frank, Springer, New York, 2015, vol. 1273, pp. 289–331.
18. K. Bock and C. Pedersen, *Acta Chem. Scand., Ser. B*, 1977, **31**, 354–358.
19. J. A. Schwarcz, N. Cyr and A. S. Perlin, *Can. J. Chem.*, 1975, **53**, 1872–1875.
20. J. Schwarcz and A. Perlin, *Can. J. Chem.*, 1972, **50**, 3667–3676.
21. T. Church, I. Carmichael and A. S. Serianni, *Carbohydr. Res.*, 1996, **280**, 177–186.
22. S. Zhao, G. Bondo, J. Zajicek and A. S. Serianni, *Carbohydr. Res.*, 1998, **309**, 145–152.
23. B. Bose-Basu, T. Klepach, G. Bondo, P. B. Bondo, W. Zhang, I. Carmichael and A. S. Serianni, *J. Org. Chem.*, 2007, **72**, 7511–7522.
24. A. Bax, *Two-Dimensional Nuclear Magnetic Resonance in Liquids*, Delft University Press, 1982, pp. 82–84.
25. A. Bax and R. Freeman, *J. Magn. Reson. (1969)*, 1981, **44**, 542–561.
26. A. S. Serianni, P. B. Bondo and J. Zajicek, *J. Magn. Reson., Ser. B*, 1996, **112**, 69–74.
27. A. S. Serianni, P. B. Bondo and J. Zajicek, *Carbohydr. Res.*, 1998, **309**, 145–152.
28. A. S. Serianni and C. A. Podlasek, *Carbohydr. Res.*, 1994, **259**, 277–282.
29. J. Wu and A. S. Serianni, *Carbohydr. Res.*, 1992, **226**, 209–218.
30. C. A. Podlasek, J. Wu, W. A. Stripe, P. B. Bondo and A. S. Serianni, *J. Am. Chem. Soc.*, 1995, **117**, 8635–8644.
31. A. Bax and S. Grzesiek, *NMR of Proteins*, Springer, 1993, pp. 33–52.
32. A. Abragam, *The Principles of Nuclear Magnetism*, Oxford University Press, 1961.
33. A. M. Castillo, L. Patiny and J. Wist, *J. Magn. Reson.*, 2011, **209**, 123–130.
34. A. S. Serianni and R. Barker, *Can. J. Chem.*, 1979, **57**, 3160–3167.
35. A. S. Serianni, in *Glycoconjugates – Composition, Structure and Function*, ed. H. J. Allen and E. C. Kisailus, Marcel Dekker, Inc., New York, 1992.
36. J. U. Baenziger and D. Fiete, *J. Biol. Chem.*, 1979, **254**, 789–795.
37. J. Zaia, *Mass Spectrom. Rev.*, 2004, **23**, 161–227.
38. C. Altona and M. Sundaralingam, *J. Am. Chem. Soc.*, 1972, **94**, 8205–8212.
39. D. Cremer and J. Pople, *J. Am. Chem. Soc.*, 1975, **97**, 1354–1358.
40. E. Westhof and M. Sundaralingam, *J. Am. Chem. Soc.*, 1980, **102**, 1493–1500.
41. E. Westhof and M. Sundaralingam, *J. Am. Chem. Soc.*, 1983, **105**, 970–976.
42. S. Makeneni, B. L. Foley and R. J. Woods, *J. Chem. Inf. Model.*, 2014, **54**, 2744–2750.
43. (a) E. Juaristi and G. Cuevas, *Tetrahedron*, 1992, **48**, 5019–5087; (b) E. Juaristi and G. Cuevas, *The Anomeric Effect*, CRC Press, Boca Raton, 1995.
44. N. C. Craig, A. Chen, K. H. Suh, S. Klee, G. C. Mellau, B. P. Winnewisser and M. Winnewisser, *J. Am. Chem. Soc.*, 1997, **119**, 4789–4790.

45. J.-H. Lii, K.-S. Chen and N. L. Allinger, *J. Comput. Chem.*, 2003, **24**, 1504–1513.
46. (a) S. C. Harvey and M. Prabhakaran, *J. Am. Chem. Soc.*, 1986, **108**, 6128–6136; (b) X. Wang and R. J. Woods, *J. Biomol. NMR*, 2016, **64**, 291–305.
47. A. R. Ionescu, A. Bérces, M. Z. Zgierski, D. M. Whitfield and T. Nukada, *J. Phys. Chem. A*, 2005, **109**, 8096–8105.
48. (a) B. M. Satelle, S. U. Hansen, J. Gardiner and A. Almond, *J. Am. Chem. Soc.*, 2010, **132**, 13132–13134; (b) B. M. Satelle, B. Bose-Basu, M. Tessier, R. J. Woods, A. S. Serianni and A. Almond, *J. Phys. Chem. B*, 2012, **116**, 6380–6386.
49. R. Stenutz, I. Carmichael, G. Widmalm and A. S. Serianni, *J. Org. Chem.*, 2002, **67**, 949–958.
50. G. Legube and D. Trouche, *EMBO Rep.*, 2003, **4**, 944–947.
51. (a) D. H. Brown, *Biochim. Biophys. Acta*, 1955, **16**, 429–431; (b) R. Hurtado-Guerrero, O. G. Raimi, J. Min, H. Zeng, L. Vallius, S. Shepherd, A. F. M. Ibrahim, H. Wu, A. N. Plotnikov and D. M. F. van Aalten, *Biochem. J.*, 2008, **415**, 217–223.
52. A. P. May, R. C. Robinson, M. Vinson, P. R. Crocker and E. Y. Jones, *Mol. Cell*, 1998, **1**, 719–728.
53. G. Scherer, M. L. Kramer, M. Schutkowski, U. Reimer and G. Fischer, *J. Am. Chem. Soc.*, 1998, **120**, 5568–5574.
54. X. Hu, W. Zhang, I. Carmichael and A. S. Serianni, *J. Am. Chem. Soc.*, 2010, **132**, 4641–4652.
55. R. U. Lemieux, *Pure Appl. Chem.*, 1971, **25**, 527–548.
56. J.-P. Praly and R. U. Lemieux, *Can. J. Chem.*, 1987, **65**, 213–223.
57. F. Cloran, I. Carmichael and A. S. Serianni, *J. Am. Chem. Soc.*, 1999, **121**, 9843–9851.
58. (a) N. C. Maiti, Y. Zhu, I. Carmichael, A. S. Serianni and V. E. Anderson, *J. Org. Chem.*, 2006, **71**, 2878–2880; (b) W. Zhang, H. Zhao, I. Carmichael and A. S. Serianni, *Carbohydr. Res.*, 2009, **344**, 1582–1587.
59. (a) M. Martin-Pastor and C. A. Bush, *J. Biomol. NMR*, 2001, **19**, 125–139; (b) S. Ganguly, J. Xia, C. Margulis, L. Stanwyck and C. A. Bush, *Biopolymers*, 2010, **95**, 39–50.
60. H. C. Siebert, R. Kaptein and J. F. G. Vliegenthart, in *Lectins and Glycobiology*, ed. H.-J. Gabius and S. Gabius, Springer-Verlag, Berlin, 1993, pp. 105–116.
61. R. G. Bryant, *J. Chem. Educ.*, 1983, **11**, 933–935.
62. P. P. Graczyk and M. Mikolajczyk, *Top. Stereochem.*, 1994, **21**, 159–349.
63. T. Klepach, W. Zhang, I. Carmichael and A. S. Serianni, *J. Org. Chem.*, 2008, **73**, 4376–4387.
64. K. Bock, I. Lundt and C. Pedersen, *Tetrahedron Lett.*, 1973, **14**, 1037–1040.
65. K. Bock and C. Pedersen, *Acta Chem. Scand.*, 1975, **B29**, 258–264.
66. K. Bock and C. Pedersen, *J. Chem. Soc., Perkin Trans. 2*, 1974, 293–297.
67. A. S. Serianni, J. Wu and I. Carmichael, *J. Am. Chem. Soc.*, 1995, **117**, 8645–8650.

68. F. Cloran, I. Carmichael and A. S. Serianni, *J. Am. Chem. Soc.*, 2001, **123**, 4781–4791.
69. I. Carmichael, D. M. Chipman, C. A. Podlasek and A. S. Serianni, *J. Am. Chem. Soc.*, 1993, **115**, 10863–10870.
70. (a) S. Watanabe, I. Ando and Y. Sakamoto, *J. Mol. Struct.*, 1983, **96**, 219–224; (b) M. Pecul and Sadlej, *J. Chem. Phys.*, 2000, **255**, 137–148.
71. T. E. Walker, R. E. London, T. W. Whaley, R. Barker and N. A. Matwiyoff, *J. Am. Chem. Soc.*, 1976, **98**, 5807–5813.
72. H. Zhao, Q. Pan, W. Zhang, I. Carmichael and A. S. Serianni, *J. Org. Chem.*, 2007, **72**, 7071–7082.
73. B. Adams and L. Lerner, *J. Am. Chem. Soc.*, 1992, **114**, 4827–4829.
74. M. Karplus, *J. Am. Chem. Soc.*, 1963, **85**, 2870–2871.
75. C. Thibaudeau, R. Stenutz, B. Hertz, T. Klepach, S. Zhao, Q. Wu, I. Carmichael and A. S. Serianni, *J. Am. Chem. Soc.*, 2004, **126**, 15668–15685.
76. M. J. King-Morris and A. S. Serianni, *J. Am. Chem. Soc.*, 1987, **109**, 3501–3508.
77. J. Wu, P. B. Bondo, T. Vuorinen and A. S. Serianni, *J. Am. Chem. Soc.*, 1992, **114**, 3499–3505.
78. K. G. R. Pachler, *Tetrahedron Lett.*, 1970, **22**, 1955–1958.
79. H. Zhao, I. Carmichael and A. S. Serianni, *J. Org. Chem.*, 2008, **73**, 3255–3257.
80. H. Günther, *NMR Spectroscopy*, Wiley, New York, 1995, pp. 119–120.
81. K. G. R. Pachler, *Tetrahedron*, 1971, **27**, 187–199.
82. B. Bose, S. Zhao, R. Stenutz, F. Cloran, P. B. Bondo, G. Bondo, B. Hertz, I. Carmichael and A. S. Serianni, *J. Am. Chem. Soc.*, 1998, **120**, 11158–11173.
83. (a) M. Barfield and B. Chakrabarti, *Chem. Rev.*, 1969, **69**, 757–778; (b) M. Barfield, A. M. Dean, C. Fallick, R. Spear, S. Sternhell and P. Westerman, *J. Am. Chem. Soc.*, 1975, **97**, 1482–1492.
84. Q. Pan, T. Klepach, I. Carmichael, M. Reed and A. S. Serianni, *J. Org. Chem.*, 2005, **70**, 7542–7549.
85. (a) A. Otter and D. R. Bundle, *J. Magn. Reson., Ser. B*, 1995, **109**, 194–201; (b) S. Sheng, R. Cherniak and H. van Halbeek, *Anal. Biochem.*, 1998, **256**, 63–66.
86. M. L. Hayes, A. S. Serianni and R. Barker, *Carbohydr. Res.*, 1982, **100**, 87–101.
87. F. Cloran, I. Carmichael and A. S. Serianni, *J. Am. Chem. Soc.*, 2000, **122**, 396–397.
88. L. Sernau, T. Turney and A. S. Serianni, *MA'AT Program, Version 1.0*, available for download at: www.nd.edu/~serlab03.
89. P. Thureau, G. Mollica, F. Ziarelli and S. Viel, *J. Magn. Reson.*, 2013, **231**, 90–94.
90. R. Lemieux and S. Koto, *Tetrahedron*, 1974, **30**, 1933–1944.
91. G. M. Clore and J. Iwahara, *Chem. Rev.*, 2009, **109**, 4108–4139.

CHAPTER 3

Insights into Carbohydrate Recognition by 3D Structure Determination of Protein–Carbohydrate Complexes Using NMR

MARIO SCHUBERT[a]

[a]Department of Molecular Biology, University of Salzburg, 5020, Salzburg, Austria
*E-mail: mario.schubert@sbg.ac.at

3.1 Introduction

The recognition of carbohydrates by proteins is crucial to many interactions occurring at cell surfaces with roles in recruiting and activating immune cells,[1,2] deciding about cancer progression and metastasis,[3,4] the attachment of pathogens to host cells[5,6] and manipulating host immune pathways.[7] Glycans function as a kind of bar code decorating the cell surface or specific proteins to provide information about the type and status of the cell, which can be read by carbohydrate-recognising proteins and receptors. In addition, glycosylation at asparagines (*N*-glycosylation) is part of a control mechanism that monitors proper protein folding in the endoplasmic reticulum of eukaryotes.[8,9] The *N*-glycans change their composition during folding, and

New Developments in NMR No.10
NMR in Glycoscience and Glycotechnology
Edited by Koichi Kato and Thomas Peters

Published by the Royal Society of Chemistry, www.rsc.org

receptors that recognise these changes decide whether or not the protein is exported to the Golgi, goes to another cycle of folding or is degraded.

In general, a very important function of glycans is to provide a densely packed code—the glycocode—consisting of glycoepitopes that are read by carbohydrate-recognising protein receptors. There are estimates of 3000–7000 functional glycoepitopes in mammals,[10] of which only a fraction have been characterised. To understand the glycocode it is crucial to understand the molecular basis of their recognition by the protein receptors.

Three-dimensional structures of proteins in complex with a carbohydrate catch the proteins in action and provide atomic details of how specificity and affinity is achieved. Structure determination of proteins and protein-complexes is mainly achieved by X-ray crystallography and NMR spectroscopy, whereby crystallography contributes approximately 90% and NMR spectroscopy roughly 10% of all protein structures deposited in the Protein Data Bank (PDB). Determining 3D structures of protein–carbohydrate complexes is challenging for both methods as discussed below.

Carbohydrate-binding protein domains are often extracellular and typically contain disulfide bonds, making it difficult to obtain homogeneous and properly folded proteins in sufficient amounts, however, it is even more challenging to crystallise or isotopically label them. Oligosaccharides are often flexible and can prevent crystallisation and therefore structure determination by X-ray diffraction. NMR spectroscopy is not currently perceived as a method for 3D structure determination of protein–carbohydrate interactions, therefore this review aims to demonstrate the competitive advantages that should be considered to help unravel the enormous challenges of cracking the glycocode and its many crucial functions.

There are a number of excellent reviews on the contribution of NMR spectroscopy to the understanding of protein–carbohydrate interactions that focus on ligand detected methods, mapping the binding site(s) on the protein and the use of docking to build models.[11–15] Here we focus on three-dimensional structures of protein–carbohydrate complexes determined with NMR spectroscopy.

Whereas the number of protein–DNA and protein–RNA complex structures determined with NMR spectroscopy is steadily increasing, only a few protein–carbohydrate complex structures have so far been determined (Figure 3.1). There are many reasons that protein–carbohydrate structures lag behind including technological limitations on the NMR side and obtaining sufficient amounts of active carbohydrate-binding proteins and carbohydrate ligands. Often the glycoepitopes that are recognised are oligosaccharides that are only accessible by sophisticated chemical synthesis.

In the case of protein–DNA and protein–RNA complex structures NMR spectroscopy is actually quite competitive contributing 15% and 35%, respectively, to the complex structures with a size of <40 kDa. These contributions are much higher than the average contribution of NMR to protein structures of ~10% to the PDB. I hypothesise that, if NMR spectroscopy would be applied with the same vigor and persistency to protein–carbohydrate

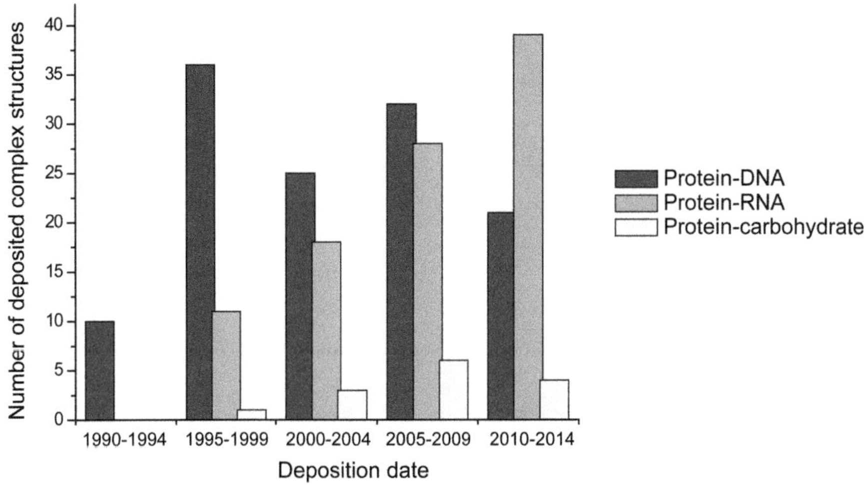

Figure 3.1 Statistics of three-dimensional NMR structures of protein–DNA, protein–RNA and protein–carbohydrate complexes deposited in the Protein Data Bank with a size of <40 kDa correlated with the year of deposition.

interactions, NMR could make a similarly strong contribution. This review serves to illustrate the huge potential of applying NMR spectroscopy to explore the world of protein–carbohydrate recognition. The protein-carbohydrate complex structures that have been determined by NMR spectroscopy can serve as a proof of principle confirming that all the methodology is available for the 3D structure determination of protein–carbohydrate complexes. Advantages, problems and pitfalls of this method are discussed and illustrated by examples.

3.2 Overview of Protein–Carbohydrate Structures Determined by NMR

As already mentioned, 14 biomolecular NMR structures that contain a carbohydrate ligand bound to a protein have been deposited in the Protein Data Bank (PDB). These structures are summarised in Table 3.1. The affinities, expressed as dissociation constants (K_d values), range from nanomolar to millimolar, covering the typical range of affinities generally found for protein–carbohydrate interactions. I chose two examples for illustrating the methodology: the fungal lectin *Coprinopsis cinerea* lectin 2 (CCL2) interacting with the trisaccharide GlcNAcβ1,4[Fucα1,3]GlcNAc with a K_d = 1 μM to represent a tight interacting protein–carbohydrate complex[16] and the complex between rhamnose binding lectin (RBL) domain of Latrophilin-1 and L-rhamnose with a K_d = 1.8 mM,[17] representing a weak interaction. A representative structure of both ensembles is shown in Figure 3.2.

Table 3.1 Protein–carbohydrate complex structures determined by NMR spectroscopy and deposited in the PDB database.

Year of deposition	PDB code	Protein name	K_d	Exchange regime	Number of intermol NOEs	Reference
1997	1ACZ	Glucoamylase[a]	6 μM and 28 μM[o]	Fast	22	55
2001	1IIY	Cyanovirin-N[b]	139 nM	Slow	37	56
2004	1T0W	Hev32[c]	130 μM	Fast	9	57
	1WCO	Nisin[d]	50 nM[p]	—[p]	36	58
2005	1ZNT	AcAMP2F18Pff/Y20Pff[e]	2 mM	Fast	13	59
2006	2ERM	FGF-1[f]	n.d.[q,r]	Fast[r]	6	44
2008	2K46	Malectin[g]	26 μM	Intermediate/slow[r]	31	60
	2JXA	Latrophilin[h]	1.8 mM	Fast	16	17
2010	2K8R	HFGF[i]	16 μM	Fast	6	61
	2KR2	Malectin[j]	50 μM	Intermediate/slow[r]	30	62
2011	2YHH	Microvirin[k]	48 μM	Slow	26	37
2012	2LIQ	CCL2[l]	1 μM	Slow	82	16
	2LL4	TgMIC4-A5[m]	110 μM	Intermediate	7	63
2013	2LVZ	ECP[n]	20 μM	Fast	(8)[r,s]	64

[a]Complex with β-cyclodextrin.
[b]Complex with Manα1,2Manα.
[c]Complex with GlcNAcβ1,4 GlcNAcβ1,4 GlcNAc.
[d]Complex with lipid II containing GlcNAcα1,4MurNAc.
[e]Complex with GlcNAcβ1,4 GlcNAcβ1,4 GlcNAc.
[f]Complex with GlcNα1,4IdoAβ1,4GlcNα1,4IdoAβ1,4GlcNα1,4IdoAβ-iPr.
[g]Complex with Glcα1,3Glc.
[h]Complex with Rha.
[i]Complex with inositol hexaphosphate.
[j]Complex with Glcα1,4Glcα.
[k]Complex with Manα1,2Man.
[l]Complex with Galβ1,4[Fucα1,3]GlcNAcβ.
[m]Complex with Galβ1,3GlcNAc.
[n]Complex with *O*-iPr-GlcNS6Sα1,4IdoA2Sα1,4GlcNS6Sα.
[o]K_d values for the two binding sites from Williamson *et al.*[65]
[p]K_d from Wiedemann *et al.*,[66] the solubility of the complex in aqueous solution was so low that the NMR structure was determined based on measurements in DMSO.
[q]n.d.: not determined.
[r]personal communication with the authors.
[s]not used for the calculations, just for verification.

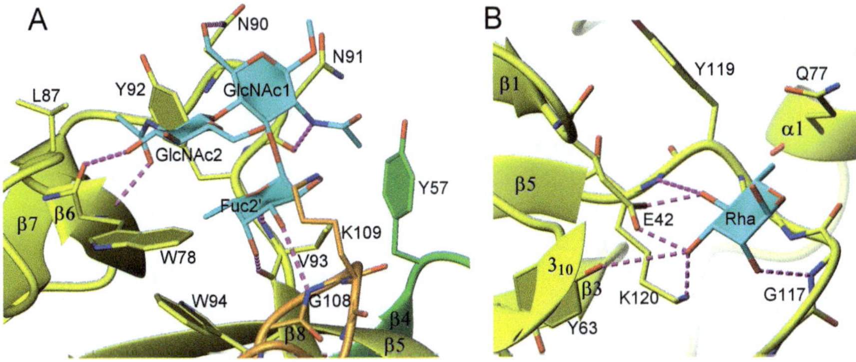

Figure 3.2 Two examples of protein–carbohydrate structures. (A) The tightly binding Coprinopsis cinerea lectin 2 (CCL2) in complex with the trisaccharide GlcNAcβ1,4[Fucα1,3]GlcNAc and (B) the complex of rhamnose binding lectin (RBL) domain of Latrophilin-1 and L-rhamnose.

3.3 Three-Dimensional Structure Determination by NMR Spectroscopy

3.3.1 The Importance of Intermolecular NOEs

The most reliable NMR method to determine 3D structures of protein–carbohydrate complexes is based on intermolecular nuclear Overhauser effects (NOEs), because an NOE directly establishes the proximity of a protein proton with a ligand proton. If NOEs are observed for example between a methyl group of the protein and several carbohydrate protons, it is a very immediate way to prove such proximity in solution. Such a direct measurement in solution is a huge advantage of NMR spectroscopy. In contrast, deducing such interactions from X-ray crystal structures is more indirect and can be biased, because (a) ligand interactions in the crystal might differ from that in solution, *e.g.* by occluding the binding site due to crystal packing[18] or when the oligosaccharide is contacting and thus bridging several protein molecules within the crystal[19,20] (b) the conditions under which crystals are formed are often quite different than those relevant for the protein–carbohydrate interactions in their natural environment and (c) positioning small saccharides into the electron density, especially for low resolution data, is error prone as illustrated by the high number of inconsistencies or errors in carbohydrate moieties of protein crystal structures.[21] Due to the direct contact information contained in intermolecular NOEs measured in solution, NMR structures determined based on them represent much more reliable structures than those obtained from theoretical calculations without experimental restraints.

The number of intermolecular NOEs critically governs the quality of a protein–carbohydrate complex structure. A handful of NOEs might be sufficient to localise the ligand at the right binding site, but it might be insufficient to properly orient the ligand. In contrast, a high number of intermolecular NOEs improves the precision and the accuracy of a complex and, thus, reveals much more atomic interactions. In the case of the CCL2–trisaccharide complex[16] 82 intermolecular NOEs could be unambiguously assigned which are schematically illustrated in Figure 3.3(A). This number is surprisingly large for a bound trisaccharide that crucially contributed to a high quality complex structure, in which a large number of individual interactions are observed as shown schematically in Figure 3.3(B). In the case of the low affinity complex between Latrophilin-1 and L-rhamnose[17] the detection of 16 intermolecular NOEs was a sufficiently large number for orienting and defining the monosaccharide (Figure 3.3(C)). These intermolecular NOEs together with 12 docking restraints based on chemical shift changes resulted in a precise structure revealing many atomic contacts (Figure 3.3(D)).

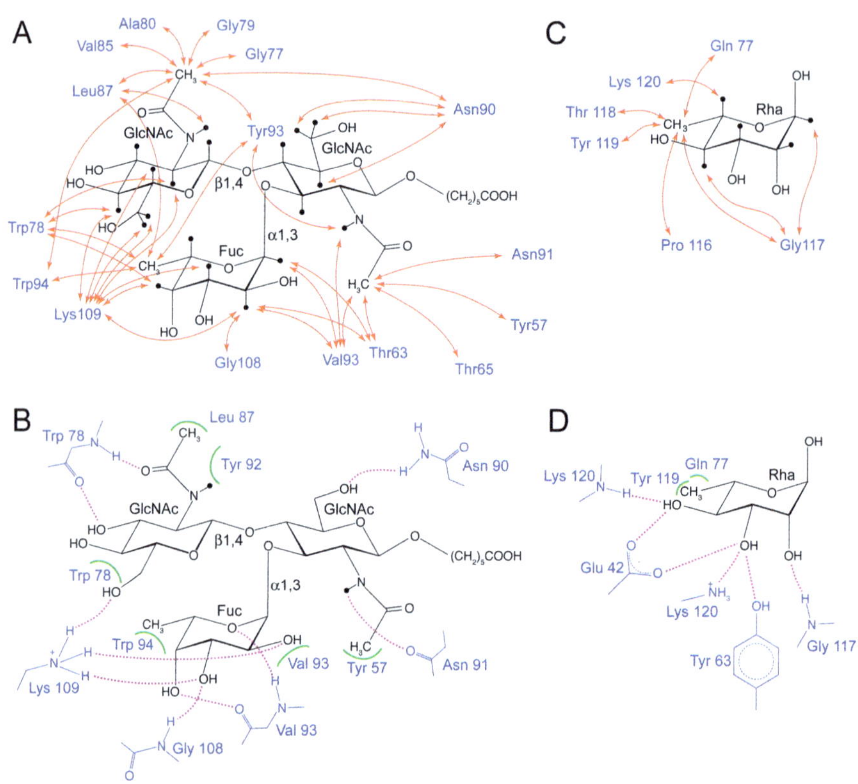

Figure 3.3 Intermolecular NOEs and extracted details of protein–carbohydrate complex structures illustrated with two complexes: (A, B) the complex between the CCL2 and the trisaccharide GlcNAcβ1,4[Fucα1,3]GlcNAc and (C, D) the complex between the Latrophilin-1 and L-rhamnose. In (A) and (C) the observed intermolecular NOEs are shown schematically. Whereas the NOEs are shown for each proton of the ligand, the involved protons on the protein side are combined for each residue. In (B) and (D) the key interactions involved in ligand recognition are displayed schematically.

3.3.2 NMR Pulse Sequences to Detect Intermolecular NOEs

Choosing the most sensitive NMR technique is a critical factor for obtaining many intermolecular NOEs with high quality. To achieve this, the protein needs to be isotope labelled with ^{13}C and ^{15}N close to 100% so that protons of the protein can be distinguished from protons of the ligand by ^{13}C/^{15}N-editing and ^{13}C/^{15}N-filter elements within an NMR pulse sequence. Numerous suitable versions of 2D and 3D ^{13}C-filtered and ^{13}C-edited ^{1}H,^{1}H NOESY experiments are described in excellent reviews.[22–24]

Two-dimensional (2D) ^{1}H–^{1}H NOESY spectra that select protons attached to ^{13}C of labelled protein in one dimension and a ^{13}C-filter to select protons attached to ^{12}C (unlabelled ligand) in the other dimension ensures the exclusive observation of intermolecular NOEs. Either the protein resonances are selected in the indirect dimension which is the case in an F1-edited F2-filtered NOESY or in the direct dimension as in the F1-filtered F2-edited NOESY.[25] The detection of the ligand resonances in the direct dimension has the advantage that the resolution is best for the unlabelled ligand. The resonances of the labelled component can be further resolved by introducing a third dimension to record ^{13}C chemical shift correlations.

About an order of magnitude in sensitivity can be gained if the measurements are carried out in D_2O, although then all NOEs involving exchangeable protons disappear. The most sensitive 3D experiment with the best resolution is an F1 ^{13}C-edited, F3 ^{13}C-filtered HSQC–NOESY described in Dominguez *et al.*[24] It combines the high resolution for the often highly degenerate ^{1}H signals of the unlabelled carbohydrate with the high sensitivity that is obtained for the signals around the water resonance due to the measurement in D_2O. High magnetic fields help to resolve chemical shift degeneracies of the carbohydrate signals together with increasing the sensitivity and the NOE build-up. Although the two indirect dimensions have lower resolution, the combination of ^{1}H with ^{13}C typically allows an unambiguous assignment of an NOE to a protein CH, CH_2 or CH_3 group. Any unnecessary elements of the pulse sequence that reduce signal intensity such as a WATERGATE (WATER-suppression by GrAdient-Tailored Excitation) element[26,27] should be avoided for measurements in D_2O.[24] For detecting intermolecular NOEs of exchangeable protons in aqueous solution, ^{13}C/^{15}N-filter elements either in the direct or indirect dimension can be used. A 3D F1 ^{13}C-edited F3 ^{13}C/^{15}N-filtered HSQC–NOESY with WATERGATE is available as the Bruker standard pulse sequence hsqcgpnoewgx33d. In addition a 3D F1 ^{13}C/^{15}N-filtered F3 ^{15}N-edited NOESY–HSQC with WATERGATE is provided as noesyhsqcf3gpwgx13d.

Critical for all these NOESY experiments is a properly chosen mixing time to avoid spin diffusion. Typical values range in between 80 and 150 ms for complex sizes of <20 kDa. For larger complexes 80–120 ms are appropriate.

3.3.3 Isotope Labelling of Oligosaccharides

In case of chemical shift degeneracies of the oligosaccharide some intermolecular NOEs cannot unambiguously assigned using only one ^{1}H dimension. Uniform isotope labelling of the carbohydrate with ^{13}C and ^{15}N can resolve such degeneracies by obtaining a better dispersion in two-dimensional ^{1}H–^{13}C and ^{1}H–^{15}N correlations or even three-dimensional correlations.[28,29] The use of uniformly ^{13}C/^{15}N-labelled oligosaccharides in complex with unlabelled protein together with 3D ^{13}C/^{15}N-filtered and ^{13}C/^{15}N-edited NOESY experiments as described above allow ambiguities in assigning intermolecular NOEs to be resolved by offering two dimensions (^{1}H and ^{13}C or ^{15}N) for the identification of the involved ligand protons. Although uniform labelling of oligosaccharides with ^{13}C and ^{15}N isotopes is very difficult, it was successfully achieved either by chemical synthesis,[30] chemoenzymatic synthesis[31] or by bacterial synthesis.[32] The obtained ^{13}C/^{15}N-isotope labelled carbohydrates include an heparan sulfate analogue,[33] hyaluronan oligomers of different length,[32] α(2,8) polysialic acid,[34] lipopolysaccharide,[35] GM3, Lewis^{x} and sialyl Lewis^{x}.[31] Although, isotope labelled oligosaccharides have to the best of my knowledge not been used so far for detecting intermolecular NOEs, they will be very important for the NMR structure determination of protein interactions with larger glycans in the future.

3.3.4 Line Broadening at the Recognition Interface

Resonances that experience large chemical shift changes upon binding are prone to line broadening, both on the protein and the ligand side, if the exchange rate falls in the same range as resonance frequency deviations upon binding. Especially protons that are involved in proton–π interactions at the protein–carbohydrate interface experience large ring current effects. Line broadening of such resonances is especially unfavourable, because aromatic rings are frequently found at carbohydrate-binding sites and such ring protons could provide intermolecular NOEs that are central to define the 3D structure of the complex. Strategies to avoid this line broadening are to increase or decrease temperature, increasing the fraction of bound ligand, or using an excess of ligand as discussed in the following sections.

The exchange regime of protein resonances can be different compared to ligand resonances, because it depends not only on the kinetics but also on the individual frequency differences between free and bound state (in Hz). Therefore, it is also dependent on the type of experiment. In the case of the CCL2–ligand complex, ^{1}H and ^{15}N signals of the protein displayed slow chemical exchange behaviour, in a 1:1 complex only signals of the bound protein are seen (Figure 3.4(A)). ^{1}H signals of the ligand also showed slow chemical exchange behaviour (Figure 3.4(B)), but ^{1}H–^{13}C cross-peaks of the ligand were never detectable in a natural abundance ^{13}C-HSQC spectrum, likely due to intermediate exchange of the ^{13}C resonances on the NMR time scale.

3.3.5 Binding Equilibria and Their Consequences on Detecting Intermolecular NOEs

To minimise line-broadening effects at the protein–carbohydrate interface, one has to take into account the fraction of bound protein as well as the

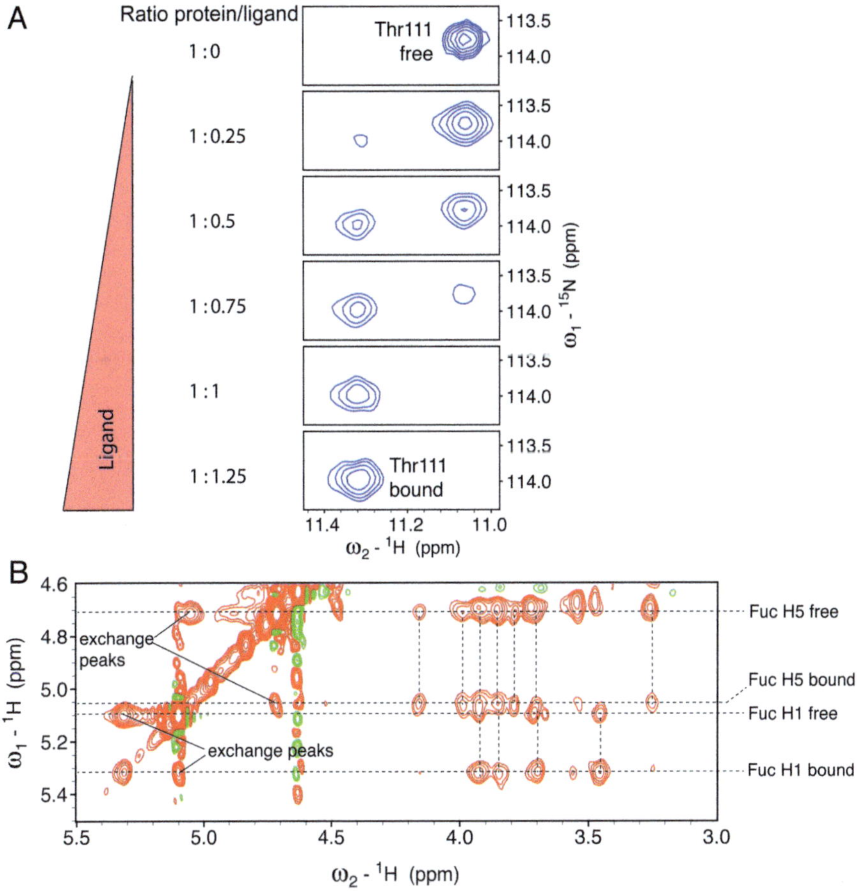

Figure 3.4 Slow exchange behaviour of protein and ligand resonances in the complex of CCL2 and fucosylated chitobiose. (A) Region of a ^{15}N-HSQC spectrum of CCL2 in the presence and absence of the ligand fucosylated chitobiose. The behaviour of the ^{1}H–^{15}N correlation signal of Thr111 with increasing amounts of ligand is shown. (B) Chemical exchange of the ligand fucosylated chitobiose in complex with CCL2 (at a protein–ligand ratio of 1:2.7) illustrated with a 2D F1 ^{13}C-filtered F2 ^{13}C-filtered ^{1}H–^{1}H NOESY recorded at 700 MHz and 310 K with a mixing time of 150 ms. Exchange peaks between free and bound resonances are visible and cross-peak patterns are similar between free and bound resonances.

fraction of bound ligand. For a single binding site interaction the fraction bound can be calculated as follows (see eqn (3.1)).[36]

$$f_b = \frac{(P_0 + L_0 + K_d) - \sqrt{(P_0 + L_0 + K_d)^2 - 4P_0L_0}}{2P_0} \tag{3.1}$$

here f_b denotes the fraction bound, P_0 the total protein concentration, L_0 the total ligand concentration and K_d the dissociation constant.

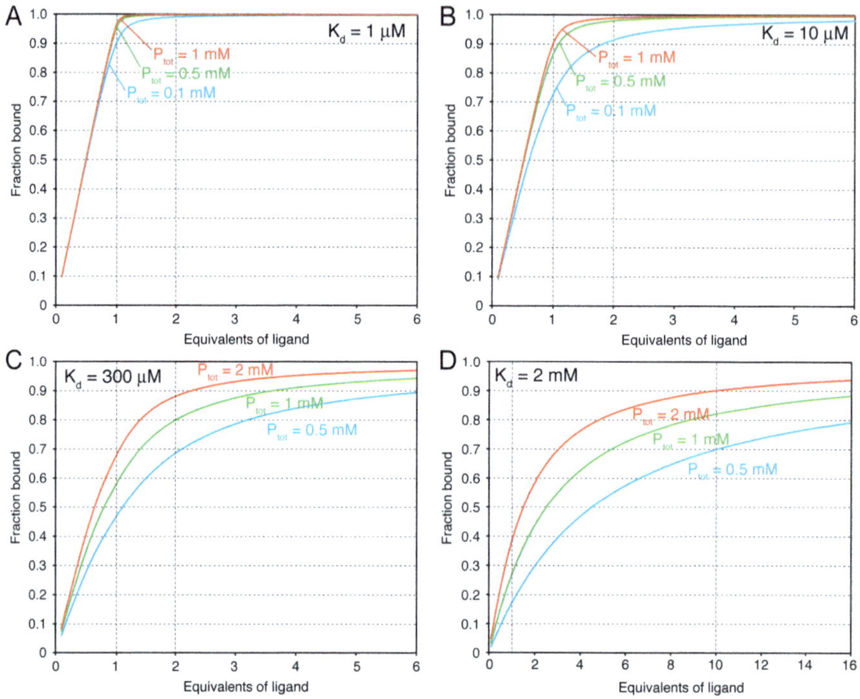

Figure 3.5 Fraction bound for a single binding-site model, calculated with eqn (3.1). Graphs for four dissociation constants are shown with some typical protein concentrations: (A) $K_d = 1$ µM, (B) $K_d = 10$ µM, (C) $K_d = 300$ µM and (D) $K_d = 2$ mM.

How the bound fraction depends on the ligand–protein ratio is illustrated for some typical cases in Figure 3.5. In case of a rather high affinity binder with a $K_d = 1$ µM over 95% saturation is achieved with a 1 : 1 complex at concentrations >0.5 mM, while only a small excess of ligand pushes the bound fraction to 100% (Figure 3.5(A)). For a ligand that binds ten times weaker ($K_d = 10$ µM) the situation changes slightly: only 90% of the protein is bound in a 1:1 complex at a concentration of 1 mM (Figure 3.5(B)). To achieve saturation an excess of ligand is required. Many protein–carbohydrate interactions display a weak to moderate affinity, which results in a different scenario. An interaction with a $K_d = 300$ µM is illustrated in Figure 3.5(C). Even at high protein concentrations of 2 mM only 70% of the protein is bound at a 1 : 1 ratio. To reach 90% saturation an excess of ligand with a ratio of 1 : 2.2 (for 2 mM protein) or 1 : 3.6 (for 1 mM protein) is required. Substantial higher ratios are required for even weaker interactions, for example $K_d = 2$ mM as shown in Figure 3.5(D). In this case a ratio of 1 : 10 is necessary to achieve 90% saturation at a protein concentration of 2 mM.

To avoid signal loss due to chemical exchange phenomena the binding sites should be saturated. As a rule of thumb a ligand concentration of 10 × K_d is required to reach the plateau of saturation. However, for the detection of intermolecular NOEs both components are required in sufficient concentrations and exchange broadening on the ligand side has to be prevented as well. There are two approaches to minimise line broadening caused by the exchange in intermolecular NOE experiments: the first approach uses high concentrations of both the protein and the ligand at a ratio of 1 : 1 to push both components close to saturation, the second approach uses the ligand in excess to have the protein fully saturated and the ligand mainly free. For this second approach the ligand has to exchange between free and bound state during the NOESY experiment so that intermolecular NOEs from the bound state can be picked up on the free ligand resonances.

3.3.6 Strategies to Obtain Good Intermolecular NOEs for High Affinity Complexes

The most common approach uses high concentrations of both the protein and the ligand at a 1 : 1 ratio to achieve high saturation for both components. This strategy is illustrated with the fungal lectin *Coprinopsis cinerea* lectin 2 (CCL2) interacting with the trisaccharide GlcNAcβ1,4[Fucα1,3]GlcNAc with a K_d = 1 μM. Unusual for this affinity range, the exchange rate is slow ($k_{ex} < 80\ s^{-1}$) so that protein amide signals of both the free and the bound protein are visible during the titration with increasing amounts of the ligand (Figure 3.4(A)). In order to observe intermolecular NOEs a concentration of ~1 mM and a 1 : 1 protein–ligand ratio was used, which resulted in a saturation of 97% on the protein side and 97% on the ligand side according to eqn (3.1). Sharp intermolecular signals were observed. In practice the ratio was verified with the disappearance of free protein signals. An excess of ligand pushed the protein resonances towards 100% of bound state leading to the simultaneous observation of free and bound ligand resonances (Figure 3.6(A)). Interestingly, both free and bound state resonances pick up intermolecular NOEs. Although the exchange regime was favourable to observe ^{1}H, ^{13}C and ^{15}N resonances on the protein side, only ^{1}H ligand signals could be observed for the bound form of the ligand. Any attempt to detect ^{1}H–^{13}C correlations at ^{13}C natural abundance did not show any bound state signals, only signals of the free ligand were detected when a slight excess of ligand was used.

The strategy with high concentration at a 1 : 1 ratio also works for fast exchanging resonances. The concentration of the protein–carbohydrate complex should be well above 10 × K_d, or even better 100 × K_d to achieve high saturation of both the protein and the ligand. The exact 1 : 1 ratio is critical, a small excess of ligand results in a significant population of free ligand, and an excess of protein reduces the fraction bound.

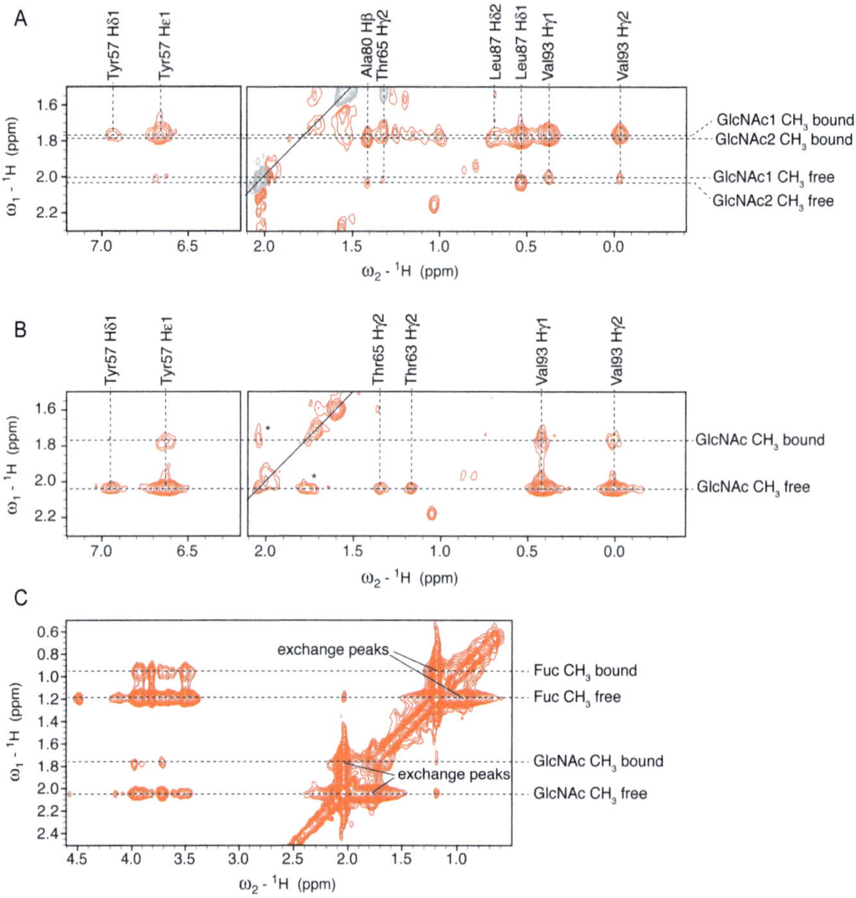

Figure 3.6 Intermolecular NOEs picked up on free and bound ligand resonances in slow exchange. (A) 2D F1 ^{13}C-filtered F2 ^{13}C-edited ^{1}H–^{1}H NOESY measured of the complex between CCL2 (1 mM) and fucosylated chitobiose (2.7 mM) at 500 MHz and 310 K with a mixing time of 150 ms. (B) 2D F1 ^{13}C-filtered F2 ^{13}C-edited ^{1}H–^{1}H NOESY measured of the interaction between CCL2 (1 mM) and Lewis^{x} (2.7 mM) measured at 500 MHz and 310 K with a mixing time of 150 ms. (C) 2D F1 ^{13}C-filtered F2 ^{13}C-filtered ^{1}H–^{1}H NOESY of the same complex as in panel B recorded at 500 MHz and 310 K with a mixing time of 150 ms.

3.3.7 Low Affinity – A Problem for Protein–Carbohydrate Structure Determination?

Counterintuitively, low to medium affinity protein–carbohydrate interactions (K_d = 10 μM–10 mM) do not prevent NMR structure determination. However, slightly different strategies have to be applied. Basically, a compromise has to be found, to achieve sufficient concentrations of both components and to minimise the line broadening of signals for both the protein and the ligand.

Typically, weak interactions are characterised by fast exchange kinetics, where resonance frequencies and line shapes of both the protein and the ligand depend very much on the grade of saturation. However, slow or intermediate exchange regimes can also occur for low affinity interactions. In any case, resonance assignments of the protein are normally achieved in the fully saturated state with an excess of ligand. Under fast exchange regimes, signal positions depend on the binding equilibrium so that differences in chemical shifts might occur between the assigned protein resonances of the fully saturated state and the observed peak positions.

For moderate affinities (K_d = 10–500 μM) the concentrations required to saturate both components in a 1 : 1 ratio are already difficult to reach. The stability and solubility of the sample set the limits in that context. Nevertheless, successful examples using high concentrations together with a 1 : 1 ratio have been reported for protein–carbohydrate and protein–RNA complexes, *e.g.* the microvirin–Manα1,2Man complex[37] with a 1.5 mM complex concentration (K_d = 48 μM) and the SRp20–RNA complex[38] with 1 mM complex concentration (K_d = 20 μM). Let us consider a protein–ligand interaction with a K_d of 300 μM, for which the calculated bound fraction is shown in Figure 3.5(C). A 1 : 1 mixture results in only 70% saturation even at the high concentration of 2 mM (for both the protein and the ligand). For a fast exchange regime the line widths of both components might be still sufficiently sharp to observe intermolecular NOEs, whereas an intermediate exchange regime would likely prevent this observation. Surprisingly, even for affinities in the micromolar range, slow exchange behaviour can occur, for example in the case of the interaction between CCL2 and Lewis^x (Galβ1,4[Fucα1,3]GlcNAcβ). Lewis^x lacks an acetamido group compared to the preferred ligand fucosylated chitobiose (GlcNAcβ1,4[Fucα1,3]GlcNAcβ; K_d = 1 μM), resulting in an almost 500-fold decrease in affinity (K_d = 456 μM). Despite this low affinity the ligand resonances are in slow exchange as shown in a 2D F1 ^{13}C-filterd F2 ^{13}C-edited NOESY (Figure 3.6(B) and (C)). The intermolecular NOE patterns are very similar to the complex with the preferred ligand fucosylated chitobiose, indicating a very similar ligand recognition and orientation. Interestingly, the intermolecular cross-peaks picked up at the resonances of the free Lewis^x ligand have much higher intensities than the corresponding cross-peaks picked up at the bound state resonances. This finding is contrary to the observation made with the preferred ligand (Figure 3.6(A)) and gives a hint how to study such interactions. The resonances of the free and bound form exchange during the NOE mixing time as indicated by strong exchange peaks in a 2D F1 ^{13}C-filterd F2 ^{13}C-edited ^{1}H–^{1}H NOESY (Figure 3.6(C)), resulting in very similar NOE patterns picked up on the free and the bound state resonances (Figure 3.6(B) and (C)).

Independent of the exchange regime, the best strategy for the structure determination of weak interactions is using an excess of ligand, which results in a complete saturation of the protein. For example, an excess of ligand resulting in <10% bound ligand has the advantage that the chemical shifts of the ligand will be very close to resonances of the free form, so that an assignment of the ligand in the bound form is not required to interpret intermolecular NOEs. The protein concentration is very important in that context, whereas a ligand excess of 1 : 10 is sufficient to reach 90% saturated protein with an $K_d = 2$ mM at a protein concentration of 2 mM, a much higher excess is required to reach the same saturation at a protein concentration of 1 mM (Figure 3.5(D)).

An extreme of this scenario is used for the transfer NOE experiment, in which a small ligand is in large excess, whereas the protein is present in very small amounts and, thus, completely saturated. In this way, the ligand resonances correspond to the free ligand ones and the small bound fraction still dominates the NOE enhancement that reports the conformation of the bound ligand. This scenario is, however, not suited to detect intermolecular NOEs since the protein concentration must be high enough to detect protein signals.

3.3.8 Influence of the Field Strength and Other Parameters on Intermolecular NOEs

The field strength has a marked influence on the exchange behaviour of the NMR resonances and thus on the observation of intermolecular NOEs. Complexes displaying intermediate to slow exchange behaviour benefit from higher field strengths, where the exchange regime is pushed towards slow exchange. With the other advantages of higher resolution, better NOE build-up and sensitivity, significant improvements are expected from 28.2 Tesla magnets (1.2 GHz systems). Such magnets raise the likelihood to observe intermolecular NOEs that could not be observed otherwise. However, complexes displaying intermediate to fast exchange might not profit from higher fields due to increased line broadening. Then a compromise regarding the magnetic field has to be experimentally found.

Beside the field strength, protein and ligand concentrations, there are many parameters that can be changed to improve the quality of intermolecular NOEs: temperature, pH, ionic strength and the type of buffer system. The affinity of charged interactions can be increased by lower salt concentrations due to a larger contribution of Coulomb attractions to the binding. A decrease in ionic strength might help for complexes with sialic acid-containing glycans or sulfated glycans that are negatively charged and for which electrostatic interactions are expected. In that case lower ionic strength might tip the balance to achieve sufficiently sharp lines on both the protein and the ligand side in order to obtain good intermolecular NOEs, similar to observations made with protein–RNA complexes.[38]

3.3.9 Other Complications and Solutions

Line broadening can also have other reasons than the exchange between free and bound states, for example slow ring flipping of Tyr and Phe side chains. Typically these side chains are flipping their aromatic rings very fast leading to average chemical shifts for both the Hδ and the Hε hydrogens. However, if sterically hindered by their environment, these flipping rates can slow down dramatically so that in extreme cases two resonances for Hδ and two for Hε can be observed.[39] When the flipping rates get close to the chemical shift differences $\Delta\delta = \delta(\mathrm{H}\delta1) - \delta(\mathrm{H}\delta2)$ or $\Delta\delta = \delta(\mathrm{H}\varepsilon1) - \delta(\mathrm{H}\varepsilon2)$ this leads to such severe line broadening that the signals disappear. This effect is important since aromatic residues are overrepresented at the binding sites for carbohydrate interactions and the ring flipping of Tyr and Phe side chains is sometimes slowed down due to the ligand. In the case of the CCL2–trisaccharide complex, the Hδ and Hε resonances of Tyr92 are broadened and not observable in most spectra. The aromatic ring of Tyr92 is stacking to the *N*-acetyl group of a GlcNAc that is slowing down the flipping rate. The Hδ resonance of Tyr92 was only observable in the complex in a 3D F1 ^{13}C-edited F3 ^{13}C-filtered HSQC–NOESY albeit with a broad line shape.

Sliding of the ligand within the binding side is another complication that might occur when repetitive units follow each other in a linear oligosaccharide. This behaviour increases the apparent affinity but hampers studying the interaction with NMR spectroscopy due to exchange between different registers of binding. This phenomenon is well-described for example by the protein–RNA interaction of hnRNP C with single stranded poly-U containing RNA oligonucleotides.[40] In that case, the exchange rate between different binding registers was faster than the applied mixing times in the NOESY experiments, so that resonances of one particular uracil picked up NOEs from different register positions within the binding site. Structure calculations with all NOE derived distance restraints resulted in a dilemma, because the restraints could not be fulfilled by one structure with one register. However, after disentangling the contributions of the different registers with the help of an alternate ^{13}C-labelled RNA sample, a 3D structure determination was achieved.[40] The recognition of glucosaminoglycans consisting of repetitive (disaccharide) units displays striking similarities to such a scenario: the binding sites for glucosaminoglycans are probably even more plastic because some seem to accommodate ligands in different orientations, thereby mainly recognising SO_4^- groups, no matter which underlying structure they are attached to, as long as the SO_4^- groups occur on similar sites with similar distances.[41–43] Nevertheless, NMR spectroscopy was successfully used to provide a three-dimensional picture of the human FGF-1 in complex with a hexasaccharide heparin analogue.[44]

3.3.10 Promising Technological Developments Applicable to Protein–Carbohydrate Complexes

Despite the advantages of using NOE-based 3D structure determination, it requires a lot of effort in terms of spectrometer time and data analysis, whereby the data complexity increases exponentially with protein size. Additionally, although NOE-derived NMR structures can reach good local accuracy, the entire structures are usually less accurate than corresponding crystal structures due to the lack of long-range structural restraints. Long-range restraints obtained by residual dipolar couplings (RDCs) and pseudo contact shifts (PCS) can significantly improve the accuracy and speed of convergence.[45–47] In particular PCS are promising for protein complexes,[48,49] they provide unique inter-molecular long-range restraints that help to orient and position the binding partner. Long-range restraints of up to 40 Å were obtained, providing very valuable restraints.[45] PCS require a metal binding tag on either the protein or the ligand and are easily measured by the chemical shift differences $\Delta\delta^{PCS}$ observed between a sample with a diamagnetic and a paramagnetic metal ion. Using different lanthanoid ions with the same chelator tag results in independent datasets providing a large amount of long-range restraints. The use of PCS for the structure determination of protein complexes evolved rapidly, which is excellently reviewed for protein–protein complexes by Hass and Ubbink.[48] This methodology should also be applicable to protein–carbohydrate complexes and promises exciting molecular details in the near future. Oligosaccharides with lanthanide binding tags have already been reported.[50–52] So far they were mainly used to elucidate the structures of the oligosaccharides, but their application to structure determination of protein–carbohydrate complexes seems feasible. The first PCS- and field-induced RDC-based structural models of a protein–carbohydrate complex were reported for Galectin-3 in complex with lactose,[53,54] although they were not deposited in the PDB.

3.4 Judging the Quality of Protein–Carbohydrate Complex Structures Determined by NMR

How reliable are NMR structures of protein–carbohydrate complexes? Figure 3.7 shows four examples of protein–carbohydrate complex structures obtained by NMR spectroscopy displaying different precision at the binding site. It seems logical that the more restraints are used, the better the precision and accuracy. However, the two examples in Figure 3.6 with most intermolecular NOEs, namely the CCL2 complex (82 intermol. NOEs) and Malectin complex (31 intermol. NOEs) display not the best precision with RMSDs (all heavy atoms) of 1.17 Å and 1.77 Å, respectively. However, for the complexes of Latrophilin and TgMIC4–A5 with RMDSs of 0.61 Å and 0.54 Å, respectively, a docking algorithm was used that applied chemical shift perturbation-derived ambiguous interaction restraints in addition to intermolecular NOEs. Also the number of restraints has to be considered in relation to the ligand size and the protein size, *e.g.* smaller ligands are easier defined. In addition, an excellent precision might be misleading, because few incorrect restraints can over-restrain the ensemble leading to an artificially high precision but worse accuracy.

Judging the accuracy of a protein–carbohydrate complex structure is more difficult. In general more specific interactions such as hydrogen bonds and H–π interactions are visible in high quality structures. Several observables can be used to verify the correctness of a structure, favourably unbiased

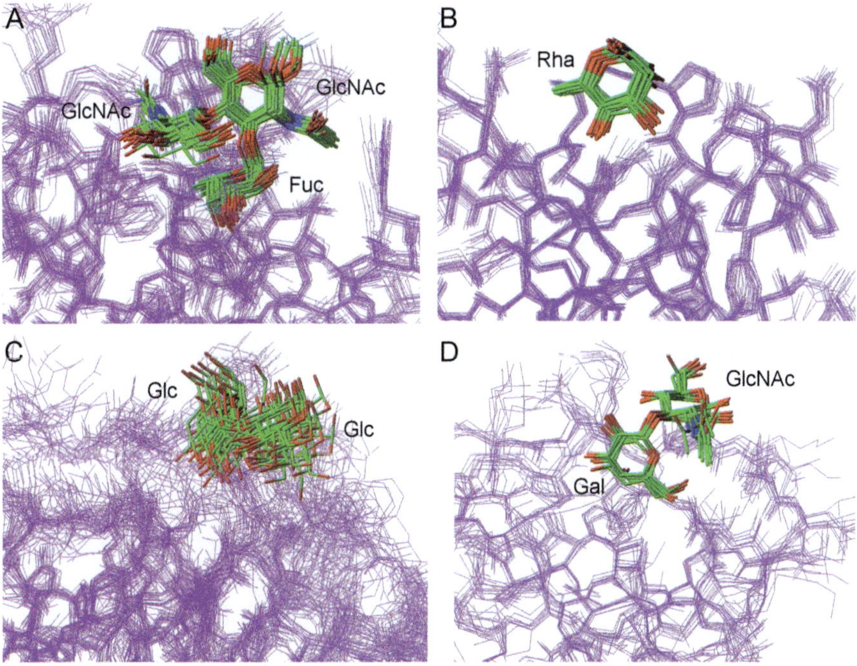

Figure 3.7 Precision of various protein–carbohydrate complexes. (A) Complex between CCL2 and GlcNAcβ1,4[Fucα1,3]GlcNAc (PDB: 2LIQ). (B) Interaction between rhamnose binding lectin (RBL) domain of Latrophilin-1 and L-rhamnose (PDB: 2JXA). (C) Malectin interacting with Glcα1,3Glc (PDB: 2K46). (D) The Apple-5 domain of *Toxoplasma gondii* microneme protein (TgMIC4 A5) in complex with lacto-*N*-biose (PDB: 2LL4).

observables that were not used for the structure determination. Chemical shift deviation during titration, both on the ligand and the protein side, is one such indicator. The structure of the CCL2–trisaccharide complex revealed three intermolecular N–H···O hydrogen bonds. Interestingly, all three showed large chemical shift deviations during the NMR titration, of which Trp78 and Gly108 represent the largest deviations of all with $\Delta\delta(^1H)$ of 1.66 and 1.21 ppm, and $\Delta\delta(^{15}N)$ of 3.4 and 2.7 ppm, respectively (Figure 3.8). This observation confirms the formation of N–H···O hydrogen bonds upon carbohydrate binding. On the ligand side, the H^N of the acetamido group of GlcNAc1 experiences a downfield shift of 0.29 ppm, confirming its participation in an N–H···O hydrogen bond to the carbonyl of Asn91 as well. Methyl–π interactions also lead to significant chemical shift deviations upon complex formation. The methyl group of GlcNAc1 that is forming a methyl–π interaction with Tyr68 is experiencing an upfield shift of 0.24 ppm. Similar observations were made for the low affinity complex between Latrophilin-1 and L-rhamnose, where the two amides forming direct N–H···O hydrogen bonds to the monosaccharide experience the largest chemical shift deviations among all N–H resonances.[17]

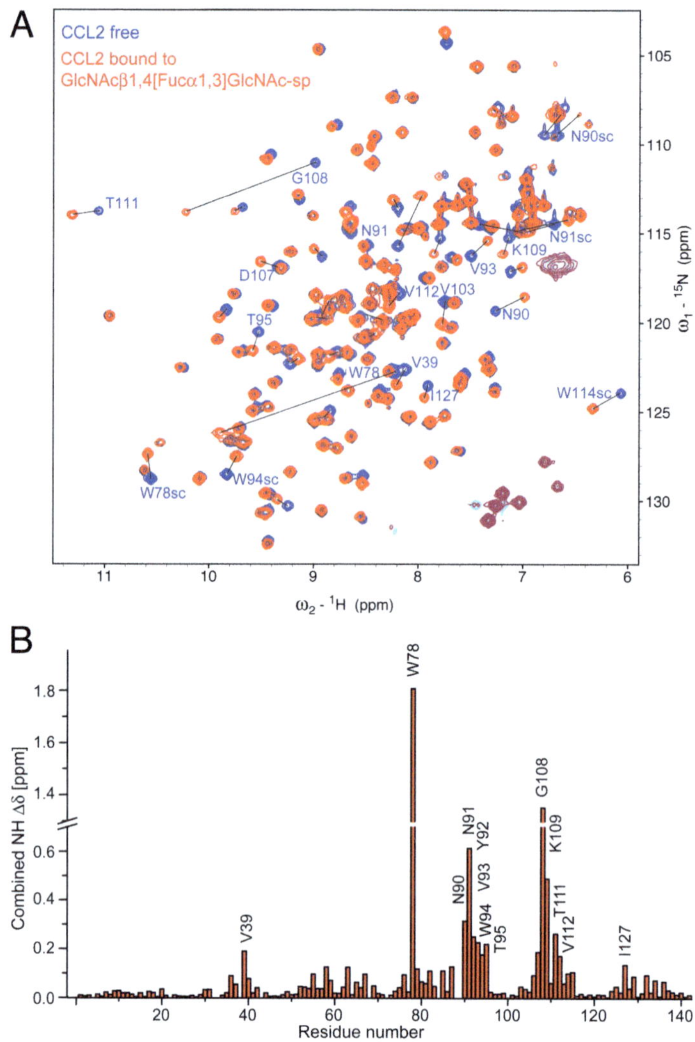

Figure 3.8 Chemical shift deviations verify the accuracy of a complex structure. (A) Overlay of two ^{15}N-HSQC spectra of CCL2 in the presence (red) and absence (blue) of the ligand fucosylated chitobiose. The NH group of each protein residue is represented by one signal whose spectral position is sensitive to the local environment. A change in the proximity of an NH, *e.g.* due to ligand binding, is picked up by this sensitive reporter and a deviation of the signal position is observed. The most prominent deviations are labelled and the corresponding signals connected by a line. (B) Plot of the combined ^{1}H and ^{15}N chemical shift deviations between free and bound CCL2. (Reproduced with permission from ref. 16, Copyright: © 2012 Schubert *et al.* This is an open-access article distributed under the terms of the Creative Commons Attribution License.)

In conclusion, chemical shift changes and expected ring current effects can be used to judge the accuracy and quality of a protein–carbohydrate complex structure. Observables that were not used for the structure determination enable an unbiased verification, similar to the separate set of reflections in X-ray crystallography for calculating R_{free}.

3.5 Conclusions

In summary, NMR spectroscopy is a very powerful method to obtain 3D structures of protein–carbohydrate complexes. Especially in the case of weak carbohydrate interactions, NMR spectroscopy offers significant advantages over X-ray crystallography due to the difficulty in obtaining high quality crystals and blurred or lacking electron density. The development of higher magnetic fields promises a further advance for the study of protein–carbohydrate interactions not only due to better sensitivity, resolution and more favourable NOE build-up, but also by providing another way to counteract detrimental exchange broadening.

With the continuous finding of new functional glycoepitopes and their involvements of protein–carbohydrate recognition interactions in almost all biological functions, the elucidation of the 'glycocode' is an area expected to grow very fast in the near future. Still underestimated in many fields of biology and medicine there is increasing evidence that weak to moderate protein–carbohydrate interactions are involved in central elements of immune function, for example self and non-self recognition, down or up-regulating immune responses. Exploring the molecular basis of such interactions that are so central to immunology and medicine will have a tremendous impact. NMR spectroscopy can make a significant contribution by providing three-dimensional structures of the protein–carbohydrate complexes.

Acknowledgements

The author thanks Wolfgang Kress (ETH Zürich, now ESBATech AG) for sharing his expertise in calculating protein–ligand equilibria, Ioannis Vakonakis (University of Oxford), Nico van Nuland (Brussels), Marta Bruix (CSIC, Madrid), Jesús Jiménez-Barbero (CIC BioGUNE, Bilbao), Claudia Muhle-Goll (Karlsruhe Institute of Technology), Michael P. Williamson (University of Sheffield) for answering my questions about details concerning their complex structures and Cameron Mackereth (IECB, Bordeaux), Peter Göttig (University of Salzburg), Bernd Simon (EMBL, Heidelberg), Sandra Scanu (TU Munich) and Florian Zauner (University of Salzburg) for valuable comments on the manuscript.

References

1. J. D. Marth and P. K. Grewal, *Nat. Rev. Immunol.*, 2008, **8**, 874–887.
2. M. Sperandio, C. A. Gleissner and K. Ley, *Immunol. Rev.*, 2009, **230**, 97–113.
3. K. S. Lau and J. W. Dennis, *Glycobiology*, 2008, **18**, 750–760.

4. Y. Y. Zhao, M. Takahashi, J. G. Gu, E. Miyoshi, A. Matsumoto, S. Kitazume and N. Taniguchi, *Cancer Sci.*, 2008, **99**, 1304–1310.
5. F. Lehmann, E. Tiralongo and J. Tiralongo, *Cell. Mol. Life Sci.*, 2006, **63**, 1331–1354.
6. G. R. Vasta, *Nat. Rev. Microbiol.*, 2009, **7**, 424–438.
7. B. Khatua, S. Roy and C. Mandal, *Indian J. Med. Res.*, 2013, **138**, 648–662.
8. A. Tannous, G. B. Pisoni, D. N. Hebert and M. Molinari, *Semin. Cell Dev. Biol.*, 2015, **41**, 79–89.
9. Y. Kamiya, T. Satoh and K. Kato, *Biochim. Biophys. Acta, Gen. Subj.*, 2012, **1820**, 1327–1337.
10. R. D. Cummings, *Mol. BioSyst.*, 2009, **5**, 1087–1104.
11. A. Arda, A. Canales, F. J. Canada and J. Jimenez-Barbero, in *Carbohydrates in Drug Design and Discovery*, ed. J. Jimenez-Barbero, F. J. Canada and M.-S. Sonsoles, The Royal Society of Chemistry, 2015, pp. 1–20.
12. V. Roldos, F. J. Canada and J. Jimenez-Barbero, *ChemBioChem*, 2011, **12**, 990–1005.
13. C. A. Bewley and S. Shahzad-ul-Hussan, *Biopolymers*, 2013, **99**, 796–806.
14. M. D. Fernandez-Alonso, D. Diaz, M. A. Berbis, F. Marcelo, J. Canada and J. Jimenez-Barbero, *Curr. Protein Pept. Sci.*, 2012, **13**, 816–830.
15. A. Clery, M. Schubert and F. H. T. Allain, *Chimia*, 2012, **66**, 741–746.
16. M. Schubert, S. Bleuler-Martinez, A. Butschi, M. A. Walti, P. Egloff, K. Stutz, S. Yan, I. B. H. Wilson, M. O. Hengartner, M. Aebi, F. H. T. Allain and M. Kunzler, *PLoS Pathog.*, 2012, **8**, e1002706.
17. I. Vakonakis, T. Langenhan, S. Promel, A. Russ and I. D. Cambell, *Structure*, 2008, **16**, 944–953.
18. I. Botos, B. R. O'Keefe, S. R. Shenoy, L. K. Cartner, D. M. Ratner, P. H. Seeberger, M. R. Boyd and A. Wlodawer, *J. Biol. Chem.*, 2002, **277**, 34336–34342.
19. H. Attrill, H. Takazawa, S. Witt, S. Kelm, R. Isecke, R. Brossmer, T. Ando, H. Ishida, M. Kiso, P. R. Crocker and D. M. F. van Aalten, *Biochem. J.*, 2006, **397**, 271–278.
20. C. Fotinou, P. Emsley, I. Black, H. Ando, H. Ishida, M. Kiso, K. A. Sinha, N. F. Fairweather and N. W. Isaacs, *J. Biol. Chem.*, 2001, **276**, 32274–32281.
21. T. Lutteke, *Acta Crystallogr., Sect. D: Biol. Crystallogr.*, 2009, **65**, 156–168.
22. A. L. Breeze, *Prog. Nucl. Magn. Reson. Spectrosc.*, 2000, **36**, 323–372.
23. I. M. Robertson, L. Spyracopoulos and B. D. Sykes, *Nato Science Peace S*, 2009, pp. 101–119.
24. C. Dominguez, M. Schubert, O. Duss, S. Ravindranathan and F. H. T. Allain, *Prog. Nucl. Magn. Reson. Spectrosc.*, 2011, **58**, 1–61.
25. R. D. Peterson, C. A. Theimer, H. H. Wu and J. Feigon, *J. Biomol. NMR*, 2004, **28**, 59–67.
26. M. Piotto, V. Saudek and V. Sklenar, *J. Biomol. NMR*, 1992, **2**, 661–665.
27. V. Sklenar, M. Piotto, R. Leppik and V. Saudek, *J. Magn. Reson., Ser. A*, 1993, **102**, 241–245.
28. S. A. Colebrooke, C. D. Blundell, P. L. DeAngelis, I. D. Campbell and A. Almond, *Magn. Reson. Chem.*, 2005, **43**, 805–815.

29. R. Harris, T. J. Rutherford, M. J. Milton and S. W. Homans, *J. Biomol. NMR*, 1997, **9**, 47–54.
30. D. Live, L. A. P. Silks and J. Schmidt, *Methods Enzymol.*, 2001, **338**, 305–319.
31. M. A. Probert, M. J. Milton, R. Harris, S. Schenkman, J. M. Brown, S. W. Homans and R. A. Field, *Tetrahedron Lett.*, 1997, **38**, 5861–5864.
32. C. D. Blundell, P. L. DeAngelis, A. J. Day and A. Almond, *Glycobiology*, 2004, **14**, 999–1009.
33. C. Laguri, N. Sapay, J. P. Simorre, B. Brutscher, A. Imberty, P. Gans and H. Lortat-Jacob, *J. Am. Chem. Soc.*, 2011, **133**, 9642–9645.
34. H. F. Azurmendi, J. Vionnet, L. Wrightson, L. B. Trinh, J. Shiloach and D. I. Freedberg, *Proc. Natl. Acad. Sci. U. S. A.*, 2007, **104**, 11557–11561.
35. W. Wang, H. J. Sass, U. Zahringer and S. Grzesiek, *Angew. Chem., Int. Ed.*, 2008, **47**, 9870–9874.
36. M. P. Williamson, *Prog. Nucl. Magn. Reson. Spectrosc.*, 2013, **73**, 1–16.
37. S. Shahzad-ul-Hussan, E. Gustchina, R. Ghirlando, G. M. Clore and C. A. Bewley, *J. Biol. Chem.*, 2011, **286**, 20788–20796.
38. Y. Hargous, G. M. Hautbergue, A. M. Tintaru, L. Skrisovska, A. P. Golovanov, J. Stevenin, L. Y. Lian, S. A. Wilson and F. H. T. Allain, *EMBO J.*, 2006, **25**, 5126–5137.
39. G. Wagner, D. Bruhwiler and K. Wuthrich, *J. Mol. Biol.*, 1987, **196**, 227–231.
40. Z. Cienikova, F. F. Damberger, J. Hall, F. H. T. Allain and C. Maris, *J. Am. Chem. Soc.*, 2014, **136**, 14536–14544.
41. A. D. DiGabriele, I. Lax, D. I. Chen, C. M. Svahn, M. Jaye, J. Schlessinger and W. A. Hendrickson, *Nature*, 1998, **393**, 812–817.
42. S. Faham, R. E. Hileman, J. R. Fromm, R. J. Linhardt and D. C. Rees, *Science*, 1996, **271**, 1116–1120.
43. J. Schlessinger, A. N. Plotnikov, O. A. Ibrahimi, A. V. Eliseenkova, B. K. Yeh, A. Yayon, R. J. Linhardt and M. Mohammadi, *Mol. Cell*, 2000, **6**, 743–750.
44. A. Canales, R. Lozano, B. Lopez-Mendez, J. Angulo, R. Ojeda, P. M. Nieto, M. Martin-Lomas, G. Gimenez-Gallego and J. Jimenez-Barbero, *FEBS J.*, 2006, **273**, 4716–4727.
45. M. Allegrozzi, I. Bertini, M. B. L. Janik, Y. M. Lee, G. H. Lin and C. Luchinat, *J. Am. Chem. Soc.*, 2000, **122**, 4154–4161.
46. I. Bertini, C. Luchinat and G. Parigi, *Concepts Magn. Reson.*, 2002, **14**, 259–286.
47. K. Chen and N. Tjandra, *Top. Curr. Chem.*, 2012, **326**, 47–67.
48. M. A. S. Hass and M. Ubbink, *Curr. Opin. Struct. Biol.*, 2014, **24**, 45–53.
49. D. Shishmarev and G. Otting, *J. Biomol. NMR*, 2013, **56**, 203–216.
50. M. Erdelyi, E. d'Auvergne, A. Navarro-Vazquez, A. Leonov and C. Griesinger, *Chem.–Eur. J.*, 2011, **17**, 9368–9376.
51. A. Canales, A. Mallagaray, J. Perez-Castells, I. Boos, C. Unverzagt, S. Andre, H. J. Gabius, F. J. Canada and J. Jimenez-Barbero, *Angew. Chem., Int. Ed.*, 2013, **52**, 13789–13793.
52. T. Yamaguchi, Y. Sakae, Y. Zhang, S. Yamamoto, Y. Okamoto and K. Kato, *Angew. Chem., Int. Ed.*, 2014, **53**, 10941–10944.

53. T. D. Zhuang, H. S. Lee, B. Imperiali and J. H. Prestegard, *Protein Sci.*, 2008, **17**, 1220–1231.
54. A. Canales, A. Mallagaray, M. A. Berbis, A. Navarro-Vazquez, G. Dominguez, F. J. Canada, S. Andre, H. J. Gabius, J. Perez-Castells and J. Jimenez-Barbero, *J. Am. Chem. Soc.*, 2014, **136**, 8011–8017.
55. K. Sorimachi, M. F. LeGalCoeffet, G. Williamson, D. B. Archer and M. P. Williamson, *Structure*, 1997, **5**, 647–661.
56. C. A. Bewley, *Structure*, 2001, **9**, 931–940.
57. N. Aboitiz, M. Vila-Perello, P. Groves, J. L. Asensio, D. Andreu, F. J. Canada and J. Jimenez-Barbero, *ChemBioChem*, 2004, **5**, 1245–1255.
58. S. T. D. Hsu, E. Breukink, E. Tischenko, M. A. G. Lutters, B. de Kruijff, R. Kaptein, A. M. J. J. Bonvin and N. A. J. van Nuland, *Nat. Struct. Mol. Biol.*, 2004, **11**, 963–967.
59. M. I. Chavez, C. Andreu, P. Vidal, N. Aboitiz, F. Freire, P. Groves, J. L. Asensio, G. Asensio, M. Muraki, F. J. Canada and J. Jimenez-Barbero, *Chem.–Eur. J.*, 2005, **11**, 7060–7074.
60. T. Schallus, C. Jaeckh, K. Feher, A. S. Palma, Y. Liu, J. C. Simpson, M. Mackeen, G. Stier, T. J. Gibson, T. Feizi, T. Pieler and C. Muhle-Goll, *Mol. Biol. Cell*, 2008, **19**, 3404–3414.
61. S. M. Kumar, H. M. Wang, S. K. Mohan, R. H. Chou and C. Yu, *Biochemistry*, 2010, **49**, 10756–10764.
62. T. Schallus, K. Feher, U. Sternberg, V. Rybin and C. Muhle-Goll, *Glycobiology*, 2010, **20**, 1010–1020.
63. J. Marchant, B. Cowper, Y. Liu, L. Lai, C. Pinzan, J. B. Marq, N. Friedrich, K. Sawmynaden, L. Liew, W. G. Chai, R. A. Childs, S. Saouros, P. Simpson, M. C. R. Barreira, T. Feizi, D. Soldati-Favre and S. Matthews, *J. Biol. Chem.*, 2012, **287**, 16720–16733.
64. M. F. Garcia-Mayoral, A. Canales, D. Diaz, J. Lopez-Prados, M. Moussaoui, J. L. de Paz, J. Angulo, P. M. Nieto, J. Jimenez-Barbero, E. Boix and M. Bruix, *ACS Chem. Biol.*, 2013, **8**, 144–151.
65. M. P. Williamson, M. F. LeGalCoeffet, K. Sorimachi, C. S. M. Furniss, D. B. Archer and G. Williamson, *Biochemistry*, 1997, **36**, 7535–7539.
66. I. Wiedemann, E. Breukink, C. van Kraaij, O. P. Kuipers, G. Bierbaum, B. de Kruijff and H. G. Sahl, *J. Biol. Chem.*, 2001, **276**, 1772–1779.

CHAPTER 4

Paramagnetic, RDC and RCSA Constraints in the Structural Analysis of Glycans

JAMES H. PRESTEGARD*[a] AND ALEXANDER ELETSKY[a]

[a]Complex Carbohydrate Research Center, University of Georgia, 315 Riverbend Rd, Athens, Georgia, 30602, USA
*E-mail: jpresteg@ccrc.uga.edu

4.1 Introduction

Glycans, or carbohydrates, are among the most diverse oligomeric structures in nature, being composed of a dozen or more sugar residues, usually in the form of five-membered (furanose) or six-membered (pyranose) rings, joined in, not just linear structures, but branched structures as well. The residues are connected by glycosidic bonds between an anomeric carbon of one residue (a keto or aldehyde carbon before ring closure) and one of the non-anomeric oxygens on the adjacent residue. The residues making up glycans do not differ primarily due to differences in functional groups (most are just carbons carrying hydroxyl groups), but due to differences in chirality at each carbon center. The linkages between residues can also be either α or β depending on how the chirality of the anomeric carbon relates to the chirality of the last chiral center in a residue. Hence the properties of glycans are intimately related to three-dimensional (3D) structure with successive residues pointing in different directions depending on the chirality of the anomeric carbon, which site

New Developments in NMR No.10
NMR in Glycoscience and Glycotechnology
Edited by Koichi Kato and Thomas Peters

Published by the Royal Society of Chemistry, www.rsc.org

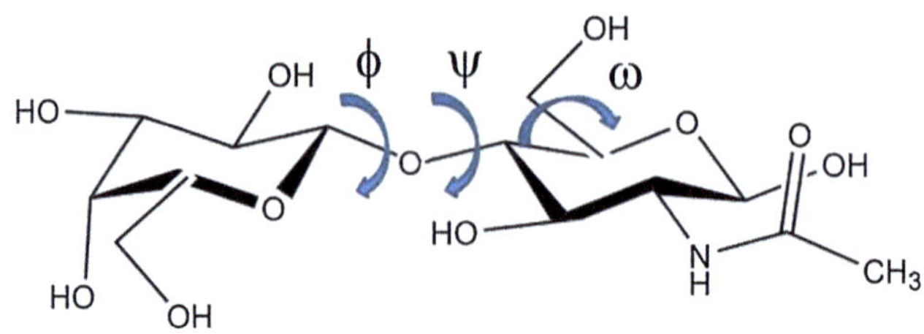

Figure 4.1 Principal structural variables in a glycan. The molecule illustrated is Galβ1-4GlcNAc. ϕ and ψ are torsion angles about the glycosidic bond. ω, the torsion angle about an exocyclic hydroxymethylene would come into play if linked to another sugar.

in the next sugar is linked, and the torsion angles of the glycosidic bonds (ϕ, ψ and sometimes ω, see Figure 4.1). Differences in structure are recognized by proteins to elicit various biological responses. Characterizing glycan structure is therefore of utmost importance in understanding function and building structural mimics that can alter physiological response and combat disease. NMR provides an excellent way of characterizing these structures, whether this is at the level of residue and linkage identification or at the level of a 3D structure that might be recognized by a protein. This chapter will focus primarily on the latter.

Characterization of 3D structures of larger glycans is not without its challenges. The lack of functional group diversity often leads to poorly resolved proton NMR spectra; most resonances occur between 3.5 and 5.0 ppm and scalar coupling of vicinal pairs can produce second order effects. Also, procedures for isotopic labeling of glycans are not as well developed as they are for proteins (although significant progress has been made). However, simplifying spectra by adding a second ^{13}C dimension can be done at natural abundance when adequate material is available to overcome sensitivity limitations. Even when resolution is adequate, traditional sources of structural information such as the nuclear Overhauser effect (NOE) give fewer structurally useful data than in proteins (typically 1–2 contacts across a glycosidic bond and perhaps an occasional long-range residue contact). This compares to the 5–10 inter-residue contacts per amino acid often seen in proteins.

To make matters worse, glycans are frequently dynamic structures, sampling a range of angles about ϕ, ψ and particularly ω. One also wants information on glycan structures, not just in solution, but when bound to a protein receptor. In some ways this reduces the need for modeling glycan dynamics, because a particular conformer is often selected in the binding process, but it adds complexity in that one would like to know where on a protein surface a glycan is bound and what parts are involved in contact with the protein. Some NOE related experiments (STD and trNOE) help address these issues (see earlier chapters). However, the lack of NOE contacts, multiple conformation sampling and the need to characterize protein-bound forms, still puts additional sources of structural information at a premium.

Many potential sources are known, and some glycan applications of residual dipolar couplings (RDCs), residual chemical shift anisotropies (RCSAs), and paramagnetic effects such as pseudo-contact shifts (PCSs) and paramagnetic relaxation enhancements (PREs) have been reviewed previously.[1–4] In this chapter we attempt to present an integrated presentation of these effects and their application.

4.2 NMR Methodology

4.2.1 RDC and RCSA

4.2.1.1 Theory for RDC and RCSA

Residual dipolar couplings (RDCs) and residual chemical shift anisotropies (RCSAs) gain their utility from an ability to report on the average orientation of bond vectors relative to a magnetic field in which they are partially oriented.[5,6] RDCs are most commonly applied in proteins, where knowing something about the orientation of each N–H vector in the polypeptide backbone of a ^{15}N-labled protein is clearly a powerful constraint on protein structure.[7] The coupled nuclei are usually spin 1/2 and, when one is a proton, indirect detection through the proton provides a substantial sensitivity advantage. For glycans ^{13}C–^{1}H bonds are often monitored. However, if magnetic moments are large and inter-nuclear distances are short, nuclear pairs do not need to be directly bonded; ^{1}H–^{1}H RDCs of vicinal pairs are also a viable source of information for glycans.

The geometric dependence of dipolar couplings displays a simple $1/r^3$ and $(1 - 3\cos^2\theta')$ dependence, where r is the inter-nuclear distance and θ' is the angle of the vector with respect to the magnetic field. This interaction averages to zero under the isotropic sampling of molecular orientations. Therefore, their effects are not normally seen in solution NMR spectra, and hence, the need for partial orientation. Partial orientation introduces five additional variables that describe the extent of orientation, the anisotropy of orientation, and the axis directions for an orientation tensor relative to a molecular frame. There are two commonly used expressions for residual dipolar coupling (D_{ij}) between a pair of spin 1/2 magnetic nuclei i and j, one in terms of an alignment tensor (eqn (4.1)) and one in terms of an order tensor (eqn (4.2)). They differ in that eqn (4.1) assumes one knows the direction of the alignment frame, or will find this frame in some iterative optimization process. Eqn (4.2) uses a molecular frame, as given in crystal structures or by common modeling software.

$$D_{ij} = \frac{D_{aij}}{r^3}\left[\left(3\cos^2\theta - 1\right) + \frac{3}{2}R\sin^2\theta\cos(2\phi)\right] \tag{4.1}$$

$$D_{ij} = \frac{D_{\max ij}}{r^3}\sum_{k,l} S_{kl}\cos(\theta_k)\cos(\theta_l) \tag{4.2}$$

In eqn (4.1) the angles θ and ϕ do not relate to the magnetic field, but are now the polar coordinates representing the orientation of the inter-nuclear vector in the principal alignment frame, D_{aij} is a measure of the degree of alignment in terms of the coupling that would be measured for a particular pair of nuclei separated at unit distance with $\theta = 0$, and R is the rhombicity (a measure of the anisotropy in alignment). The five variables are represented by D_{aij}, R, and three Euler angles that orient the molecular frame to the principal alignment frame.

In eqn (4.2) the S_{ij} are order parameters (elements of a 3 by 3 order matrix written in Cartesian coordinates) that define alignment in the molecular frame. The matrix is symmetric and sums to zero on the diagonal. Hence, there are also 5 unknowns. $D_{\max ij}$ is a fundamental constant dependent on the magnetogyric ratios of the nuclei, γ_i and γ_j. Eqn (4.2) can be transformed to the principle alignment frame, once known, in which case the $S_{zz}' D_{\max ij} = 2D_{aij}$ and $(S_{yy}' - S_{xx}')/S_{zz}' = \eta = 3/2R$ where η is the anisotropy.

The presence of 5 unknowns in either formalism means that at least 5 independent RDCs must be collected for each semi-rigid element when attempting to extract structural information from a set of RDC measurements. This may be particularly difficult for carbohydrates where pyranose rings are frequently the most likely semi-rigid element. For some rings, β-Glc, for example, C–H vectors for carbons 1,3 and 5 are nearly parallel, as are vectors for carbons 2 and 4. These two sets will have nearly identical angular dependencies for RDCs, making it necessary to seek a larger semi-rigid structure, or find complementary information, before a definitive structure can be determined.

Residual chemical shift anisotropies (RCSAs) arise from the fact that chemical shielding (equivalently chemical shifts) are orientation dependent. As with RDCs, rapid isotropic sampling of orientations results in reducing this to the scalar quantity that we recognize as the chemical shift in normal solution spectra. The equation describing RCSAs follows:

$$\delta_i^{\mathrm{csa}} = 2/3 \sum_{k,l} \sum_{j=x,y,z} S_{kl} \cos(\theta_{jk}) \cos(\theta_{jl}) \delta_{jj} \tag{4.3}$$

Note the similarity to eqn (4.2). δ_{jj} are principal elements of the anisotropic part of a chemical shift tensor for nucleus i, and the θ_{jk} and θ_{jl} are the angles between the principal axes of the shift tensor for a particular site of interest and the axes of the molecular frame. If the shift tensor happened to be axially symmetric, the equation would reduce to a single double sum and a single shift anisotropy parameter making the form of the equation identical to eqn (4.2).

For carbohydrates the primary sites of interest for RCSAs are the carbonyl carbons that are part of acetyl groups in acetylated sugars. The difference in elements of the shift tensor for a carbonyl group is approximately 200 ppm. This can vary somewhat from site to site, but the effect of these variations is likely to fall below the precision of measurement in most cases. At a magnetic field of 18T (800 MHz for protons) the maximum difference in chemical shift for different orientations would be about 40 000 Hz. This is comparable to a $D_{\max ij}$ for a $^{13}C–^{1}H$ pair of about 60 000 Hz. For similar levels of orientation the offset of a ^{13}C carbonyl resonance would therefore be similar in magnitude to a $^{13}C–^{1}H$ RDC and considerably larger than a $^{15}N–^{1}H$ RDC.

4.2.1.2 Aligning Samples for RDC and RCSA

Achieving partial orientation of samples frequently involves the use of liquid crystals. These materials are made of molecules, or assemblies of molecules, with sufficiently large geometric anisotropies that they do not tumble independently, but form cooperative domains with sufficient anisotropy in magnetic susceptibility that the domains orient in a magnetic field. The domains have sufficient space between constituent molecules or assemblies to allow relatively unfettered rotation of dissolved molecules of interest. Occasional collisions between media molecules and molecules of interest, or weak electrostatic interactions between the two, impart a degree of order that is 10^{-3} to 10^{-4} that of complete order. For carbohydrates some of the water-based liquid crystals used are the same as those used for proteins, namely bacteriophage, bicelles made of phospholipids or alkyl-ethylene-glycol-based (PEG) detergents. More recently special attention to the needs of the carbohydrate community has been given, and new media based on ionic liquid crystals have been introduced.[8] In addition to liquid crystals there are gels (polyacrylamide for example) that can be mechanically distorted.[9,10] See Table 4.1 for more specific references to media used in carbohydrate applications.

It is even possible to capitalize on direct alignment by the magnetic field, when groups with a sufficiently large anisotropy of magnetic susceptibility are added to a protein or carbohydrate. These groups are often chelates capable of binding paramagnetic lanthanide ions. These will be discussed in more detail later when we talk about PCSs, but their simultaneous use to orient a molecule for RDC measurement has the advantage that order

Table 4.1 Aqueous alignment media used with carbohydrates.

Media	Alignment interaction	Comments	Reference
Bicelle (phospholipid)	Steric	DMPC/DHPC	11
Bicelle (phospholipid)	Steric/ electrostatic-negative	DMPC/CHAPSO/DSS	12
Bicelle (phospholipid)	Steric/ electrostatic-positive	DMPC/DHPC/CTAB	13
Bicelle (PEG detergent)	Steric	C12E5/hexanol	14
Ionic liquid crystal	Steric/ electrostatic-positive	*N*-dodecyl-*N*-methyl pyrrolidinium bromide	8
Gelatin	Steric	Stretched gel	15
Bacteriophage	Steric/ electrostatic-negative	Pf1 phage	16,17

parameters or alignment parameters can be rewritten in terms of the same susceptibility parameters that lead to PCSs, and RDC and PCS data can be combined without introducing a second set of unknowns. In all cases one must be mindful of potential signal interference from media components in the spectra of molecules of interest. This is less of an issue for proteins where ^{15}N labeling is often used to generate sites for RDC measurement. N–H moieties in media molecules are less common than C–H moieties, and the natural abundance of ^{15}N is only about 1/3 that of ^{13}C. Not all spectral signals from ^{13}C sites in media will interfere, since more rigid parts of media molecules are highly ordered and have very broad lines, but attention must be given to more mobile parts.

4.2.1.3 Measuring RDC

The spin operator part of the dipolar interaction is identical to that for first order scalar couplings, so RDCs simply add to the scalar couplings normally seen in solution spectra of molecules of interest. This necessitates the separation of scalar and dipolar parts of observed splitting. It is frequently done by comparing measurements in isotropic solution to those made in media that impart partial orientation. As long as dipolar contributions are kept smaller than the scalar couplings, a few tens of Hz for ^{13}C–^{1}H, the signs of dipolar couplings can be extracted. Note that $(3\cos^2\theta - 1)$ can be positive or negative, and note that a positive RDC will add to a ^{13}C–^{1}H splitting, but subtract from a ^{15}N–^{1}H splitting, because of the difference in signs of magnetogyric ratios for ^{15}N and ^{13}C. The actual measurements can be as simple as a direct measure of splittings in non-decoupled HSQC spectra. Measurement in the proton dimension, with the ^{13}C decoupling turned off, is illustrated in Figure 4.2. This avoids the need to collect large numbers of points in the indirect dimension to maintain resolution, however, measurement

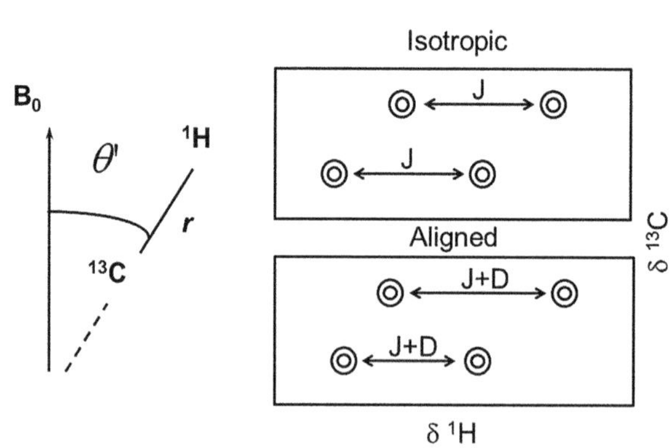

Figure 4.2 RDC dependence on angle θ' relative to the magnetic field, B_0, and illustration of measurement from HSCQ spectra ^{13}C-coupled in the ^{1}H dimension.

precision may be limited by the fine structure caused by multiple ^{1}H–^{1}H couplings. Thus, measurement in the indirect dimension does offer an advantage, and also because several factors that introduce errors in RDC measurement can be minimized (see Section 4.21.4). In case this option is selected, the number of indirect points needed can be reduced somewhat by using pulse sequences that accentuate coupling evolution.[18] It is also possible to introduce a third dimension to HSQC spectra in which modulation of ^{13}C magnetization results only from coupling to protons.[19–22] This dimension can be fit directly in the time domain to yield couplings, along with estimates of errors.

One-bond ^{13}C–^{1}H coupling measurement in the direct proton dimension has seen renewed interest in recent years with the application of broadband homonuclear ^{1}H decoupling, based on bilinear rotation decoupling (BIRD)[23] method, to collapse the multiplet structure of ^{1}H–^{1}H couplings. The corresponding implementations of CLIP/CLAP experiments, called "pure shift" CLIP/CLAP–HSQC[24] and CLIP/CLAP–RESET[15] were shown to deliver increased precision of coupling measurement for CH and CH_3 groups. NMR spectra are acquired in the form of pseudo-3D datasets that are computationally processed to yield decoupled 2D datasets. Since such homonuclear decoupling methods use the directly bonded ^{13}C nucleus to distinguish the target ^{1}H spins from remotely coupled ^{1}H spins, these experiments are only applicable to compounds with carbon at natural abundance. For the same reason they are unable to decouple geminal ^{1}H–^{1}H couplings, unless constant-time frequency labeling is used in the direct dimension. This limitation is avoided in the perfectBIRD CLIP/CLAP–HSQC experiment,[25] which employs an additional echo element to produce singlet lines for all CH_n multiplicities at the cost of a somewhat larger linewidth. These experiments along with references to additional methods that may be useful for work on carbohydrates are included in Table 4.2.

4.2.1.4 Accuracy of RDC Measurement

Several processes can make unwanted contributions to splittings measured for RDC extraction. Measurement in the proton dimension, for example, can be affected by cross-correlation effects that differentially add to the widths of multiplet components. When components are not well resolved, the effects lead to apparent increases or decreases in ^{13}C–^{1}H splittings.[33] These effects are usually smaller in the indirect ^{13}C dimension. However, when field induced alignment of paramagnetically tagged molecules is used, additional electron-nuclear cross correlation effects can produce dynamic frequency shifts in both dimensions.[34] The contributions are often small, less than 1 Hz, but if precise interpretation is planned, this can be a concern. It is important to note that other factors such as accurate representation of bond length and the effects of local motion on the molecular models used to interpret RDCs can also produce variations in predicted RDCs approaching 10%. So, the benefits of precise measurement of an RDC may be lost when using imprecise models.

Table 4.2 Methods for measurement of $^{13}C{-}^{1}H$, $^{1}H{-}^{1}H$, and $^{13}C{-}^{13}C$ RDCs.

RDC	Type of measurement	Comments	Reference
$^{13}C{-}^{1}H$	^{1}H frequency domain HSQC	High resolution, but with significant second order effects	26
$^{13}C{-}^{1}H$	^{13}C frequency domain HSQC	Fewer artifacts, but needs many points in the indirect dimension	12
$^{13}C{-}^{1}H$	^{13}C frequency domain coupling enhanced HSQC	Reduces number of points needed in the indirect domain	18
$^{13}C{-}^{1}H$	J modulation HSQC	Good quantitation, but adds a third dimension	19,27
$^{13}C{-}^{1}H$	Constant time J modulation HSQC	Minimizes effects of proton second order effects	11,22
$^{13}C{-}^{1}H$	IPAP	Separates coupled pairs into two spectra – reduces overlap	28
$^{13}C{-}^{1}H$	IPAP ^{13}C-detected	Separates coupled pairs into two spectra – reduces overlap	29
$^{13}C{-}^{1}H$	Pure shift CLIP/CLAP–HSQC CLIP/CLAP–RESET	Broadband homonuclear ^{1}H decoupling for CH and CH_3 groups	15,24
$^{13}C{-}^{1}H$	perfectBIRD CLIP/CLAP–HSQC	Broadband homonuclear ^{1}H decoupling for CH, CH_2 and CH_3 groups	25
$^{1}H{-}^{1}H$	HSQC–CT–COSY	$^{1}H{-}^{1}H$ couplings in uniformly ^{13}C labeled material are measured	30
$^{1}H{-}^{1}H$	CT–COSY	$^{1}H{-}^{1}H$ couplings from J modulation of CT crosspeaks/autopeak ratios	31
$^{13}C{-}^{13}C$	$^{13}C{-}^{13}C$–CT–COSY	$^{13}C{-}^{13}C$ couplings in uniformly ^{13}C labeled material are measured	30
$^{1}H{-}^{1}H$	ECOSY	Eliminates 2 spin flip multiplet components – eases measurement	12
$^{1}H{-}^{1}H$	J-modulated 1D spectra	Efficient quantitative measurement for smaller systems	32

There are also second order effects that can affect the interpretation of splittings. This is particularly the case for carbohydrates where strong J coupling can exist between vicinal pairs of protons.[11] These effects are very difficult to remove and can affect measurements of both ^{13}C–^{1}H and ^{1}H–^{1}H couplings. Errors up to 7 Hz for ^{13}C–^{1}H have been seen in the proton dimension of coupled HSQC spectra, resulting mostly from peak distortion. Even broadband homonuclear decoupled experiments[15,24,25] are not immune to ^{1}H–^{1}H strong coupling artifacts, although they offer improved accuracy by virtue of enhanced resolution and minimized signal overlap.

Other effects enter directly even to the ^{13}C dimension. They are analogous to virtual coupling effects in second order spectra where strong coupling between a pair of protons results in a splitting by the sum of ^{13}C couplings to both protons. The extra contribution can approach 10%. One solution to the problem is to simply simulate the second order effects on spectra.[11] However, recently pulse sequences based on constant time J-modulation have been proposed that suppress many of the effects.[22]

4.2.1.5 Measuring RCSA

Measurement of RCSAs is simpler in principle, but has its own practical challenges. Upon partial alignment, RCSAs are simply manifested as shifts of resonance lines for spins of interest, such as the carbonyl carbon of acetyl groups on *N*-acetylated sugars. Second order effects and dynamic frequency shifts are of little consequence since they do not change significantly on alignment. But unlike RDCs, which are internally referenced by movement of one line of a multiplet relative to another, RCSAs have no internal reference; a single line position must often be compared in separately acquired isotropic and aligned spectra. Sample conditions must, therefore, be reproduced very precisely to give measurements accurate on the order of 1–2 Hz. Specialized devices are available that allow accurate referencing, such as a two-stage tube that allows measurement of stretched and relaxed polyacrylamide gels using the same sample.[35] A careful study of field dependence of magnetically aligned samples may also prove better than comparison of separate isotropic and media aligned samples. Aside from errors associated with maintaining identical environments, analysis of RCSAs generally assumes a chemical shift tensor for a particular functional group to be independent of local environment. Variations have been studied for carbonyl carbons of amide groups in proteins.[36] These are likely to be similar to variations for carbonyl carbons in carbohydrates and are not likely to dominate other sources of error.

4.2.2 PCS and PRE

Given the likely lack of sufficient independent RDC measurements, as well as the existence of substantial internal motion in carbohydrates, acquisition of data complementary to RDCs is important. Of course, some of the data may come from more traditional NOE and scalar coupling measurements, as discussed in other chapters. But these provide primarily short range information that can restrict a particular pair of glycosidic torsional angles. Uncertainty in these measurements propagates throughout larger carbohydrate structures making it desirable to have some long-range constraints on positions of residues at the extremes of these structures. Also, in the case of a glycan being docked into a binding pocket on a protein, the position of glycan residues relative to points on the protein surface is of interest. The addition of a paramagnetic site to one end of a glycan, or to a particular site on a protein, can generate pseudo contact shifts (PCSs) and paramagnetic relaxation enhancements (PREs) that have the desirable long range effects on NMR observable nuclei.[37,38]

4.2.2.1 PCS Theory and Measurement

PCSs are most frequently generated by attachment of a chelating group such as a DOTA (tetraazacyclododecane-1,4,7,10-tetraacetic acid) derivative or an ion-binding peptide segment in the case of a protein.[39,40] When bound to these entities certain paramagnetic lanthanide ions contribute very large and anisotropic magnetic susceptibilities to the resulting complex. When lifetimes for spin states of the unpaired electrons in these ions are very short, broadening of NMR resonances, at least of sites remote from the ion, is minimal, and chemical shifts change in a distance- and angular-dependent fashion. The functional form of this dependence, written in an arbitrary molecular frame is shown in eqn (4.4).[41]

$$\delta_i^{\text{pcs}} = \frac{1}{4\pi r^3}\sum_{j,k}\cos\theta_{i,j}\cos\theta_{i,k}\chi_{j,k}\times 10^6 \tag{4.4}$$

Here δ_i^{pcs} is the induced chemical shift in ppm, r is the distance between the ion and the nucleus observed, θ_{ij} and θ_{ik} are the angle between the vector connecting the ion and the nucleus and the x, y, or z axes of the molecular frame and the $\chi_{j,k}$ are the elements of the anisotropic part of the susceptibility tensor in this frame. If one knows the orientation of the principle frame of the susceptibility tensor and transforms molecular coordinates of eqn (4.4) to this frame, eqn (4.5) results.

$$\delta^{\text{pcs}} = \frac{1}{12\pi r^3}\left[\Delta\chi'\left(3\cos^2\theta-1\right) + \frac{3}{2}\delta\chi'\sin^2\theta\cos(2\phi)\right]\times 10^6 \tag{4.5}$$

Here θ and ϕ are polar angles in the principal frame, $\Delta\chi' = (\chi_{zz}' - (\chi_{xx}' + \chi_{yy}')/2)$, and $\delta\chi' = \chi_{xx}' - \chi_{yy}'$, where χ_{ii}' are elements of the principal frame susceptibility tensor. Note the similarity to the functional forms of the equations for RDCS and RCSAs. The main difference is that the susceptibility tensor elements have replaced the products of large interaction constants and small order parameters. The χ values for paramagnetic lanthanide ions are of the order of 10^{-31} m^3. This means that at distances of 30 Å (3×10^{-9} m) shifts can be on the order of 0.1 ppm. These are quite easily measured by comparing spectra of complexes with a diamagnetic lanthanide, such as Lu^{3+}, to spectra

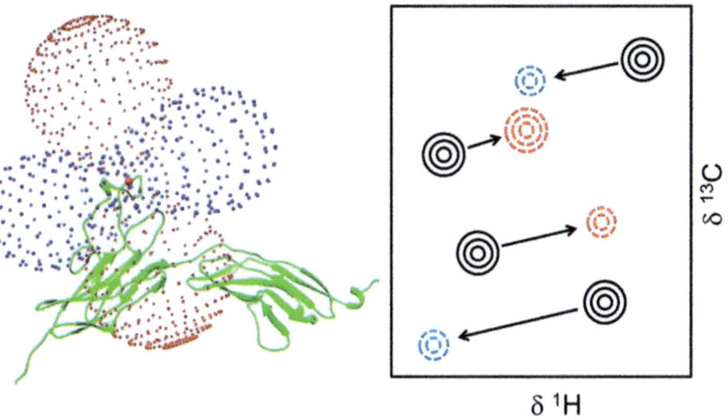

Figure 4.3 Illustration of PCS effects. Left, the blue and red dots superimposed on a ribbon diagram of a protein having a lanthanide binding peptide replacing an internal loop show a surface of constant shift. Red is upfield and blue is downfield. Right, a representation of a ^{13}C–^{1}H HSQC spectrum shows crosspeaks with a diamagnetic lanthanide (black) superimposed on the same segment with a paramagnetic lanthanide (red, upfield shift and blue downfield shift).

of complexes with a paramagnetic lanthanide, such as Dy^{3+} (illustrated in Figure 4.3). Identification of diamagnetic and paramagnetic peak pairs in two-dimensional heteronuclear correlation spectra, such as a 2D ^{13}C–^{1}H HSQC, is facilitated by the fact that at large distances from the lanthanide ion, resonance shifts for directly bonded ^{13}C and ^{1}H nuclei have nearly the same magnitude in ppm with crosspeaks lying on a line with slope equal to the ratio of the operational frequencies for the two nuclei.

4.2.2.2 PRE Theory and Measurement

PREs result from nuclear dipole - electron dipole interaction contributions to nuclear spin relaxation rate, usually the transverse rate, R_{2p}. They are usually measured from the decrease in crosspeaks volumes in 2D HSQC experiments due to increased transverse spin relaxation during the INEPT transfer and refocusing periods of this experiment (see Figure 4.4)[39,42]. The geometric information comes from the $1/r^6$ dependence of this mechanism, as illustrated in eqn (4.6). Here S is the spin of the electronic state, γ_H and γ_e are the magnetogyric ratios of the proton and electron respectively and τ_c is an effective correlation time related to the electron spin lifetime and molecular tumbling time by $(1/\tau_e + 1/\tau_m)^{-1}$.[43] The distance dependence is identical to that for NOEs, but the effective measurement distances are much larger. Both lanthanides with low to zero anisotropies like Gd^{3+} and nitroxide carrying spin labels like TEMPO produce these PREs. For Gd^{3+} the electron lifetime is in most cases shorter (10^{-9} s) than the molecular tumbling time of a protein complex and this will dominate τ_c. For free glycans carrying a paramagnetic tag τ_m will dominate. Operating at high fields (800 MHz for protons) with a τ_c between 0.1 and 1 ns, eqn (4.6) simplifies to eqn (4.7).

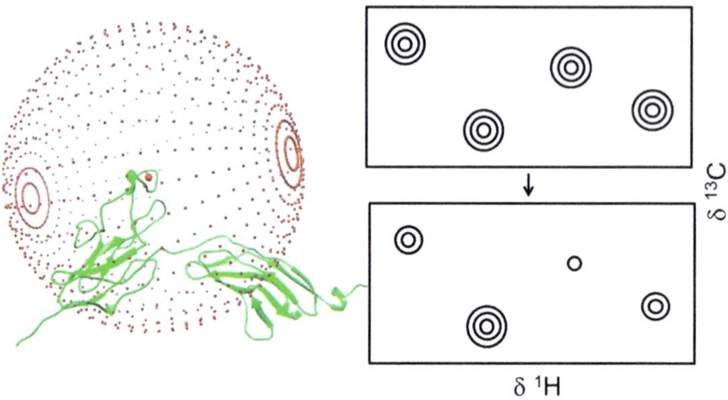

Figure 4.4 Illustration of PRE effects on a segment of a ^{13}C–^{1}H HSQC experiment. Left, red dots show a surface of equal PRE on a lanthanide-tagged protein. Right top, spectrum of a molecule with a diamagnetic ion. Right bottom, spectrum of a molecule with paramagnetic ion. Crosspeaks are reduced in intensity proportional to a $1/r^6$.

$$R_{2P} = \frac{1}{15}\left(\frac{\mu_0}{4\pi}\right)^2 \frac{\gamma_H^2 \gamma_e^2 S(S+1)}{r^6}\left[4\tau_c + \frac{\tau_c}{1+(\omega_H-\omega_S)^2\tau_c^2} + \frac{3\tau_c}{1+\omega_H^2\tau_c^2} + \frac{6\tau_c}{1+(\omega_H+\omega_S)^2\tau_c^2} + \frac{6\tau_c}{1+(\omega_S)^2\tau_c^2}\right] \tag{4.6}$$

$$R_{2P} = \frac{1}{15}\left(\frac{\mu_0}{4\pi}\right)^2 \frac{\gamma_H^2 \gamma_e^2 S(S+1)}{r^6}\left[4\tau_c + \frac{3\tau_c}{1+\omega_H^2\tau_c^2}\right] \tag{4.7}$$

For typical periods of 15 ms where proton magnetization is transverse in a ^{1}H–^{13}C HSQC these mechanisms will lead to a loss of about 30% of an HSQC peak when the proton being observed is 20 Å from the Gd^{3+} site. There are also Curie relaxation contributions that are particularly important for lanthanides with much shorter electron spin lifetimes such as Dy^{3+}. These effects also display a $1/r^6$ distance dependence, but the magnitude of the effect is smaller. They often lead to the undesirable loss of intensity for sites closer than 10 Å when measuring PCSs. The very straightforward dependence of all PRE effects on distance alone makes their measurement a convenient supplement to other data when determining structures of free or protein-bound glycans.

4.3 Structure Determination

The actual process of determining a glycan structure based on the information described can be approached in many ways and there are computational tools to assist in this process. Because of the similarity of functional forms for many of the data types discussed, programs initially written for the use of NOEs and RDCs can be adapted to use other data types. When the molecule is small, it may be possible to iteratively adjust glycosidic torsion angles in a trial structure as observables are back-calculated for comparison with experiment. Tools such as REDCAT provide these calculations for RDCs,[44,45] but REDCAT has also been used to analyze RCSAs and PCSs. It is also possible to break a glycan into semi-rigid segments and use programs like this to extract a set of Euler angles for each segment that allow rotation into a principal alignment frame.[46] Properly oriented segments can then be connected to generate a model for a structure. There is normally a degeneracy in relative orientation of segments due to the insensitivity of RDCs to inversion of axes, but the need to connect segments by glycosidic bonds often removes this degeneracy. Other programs exist that incorporate similar utilities, PALES, for example.[47] Similar approaches are integrated into other systems that allow evaluation of ensemble representations of flexible structures. The "flexible-meccano" package[48] has been used to model flexible peptide segments and it likely can be used for glycan structures.[49]

A more common modeling approach is to use programs originally developed for protein structure determination. These incorporate one or more of the above data types as restraints in molecular dynamics-based simulated annealing, or Monte Carlo sampling searches. The restraints are usually built on a square deviation between experimental and back-calculated data. The similarity in equations given for PCSs, CSAs, and PREs to either RDC equations or NOE equations makes it possible to use existing tools developed for NOE or RDC restraints to handle this broader range of data. The structures that result from the search are then those that best satisfy a combination of experimental data and molecular energies. Commonly used programs include AMBER,[50] XPLOR,[51] CYANA,[52] CNS,[53] and GROMACS.[54] The usual challenge when working with glycans is construction of a topology file that is compatible with force field parameters developed primarily to deal with proteins and nucleic acids. Computational tools are available to aid construction of these files. The GLYCAM website, for example, has an interactive glycan building utility that will produce topology files consistent with AMBER forcefields and parameter modifications that have been made specifically for glycans.[55] PRODRG will convert a variety of small molecule coordinate files to topology files compatible with programs such as GROMACS, X–PLOR/CNS, and AutoDock.[56] There is also the Glycan Reader software for preparing inputs for CHARMM-based simulation programs,[57] and a GLYCAN server that is part of the HADDOCK suite of docking software.[58] Even with these tools it is sometimes necessary to add specific atom types and parameters. To do this properly requires substantial computational effort, but it is often possible to find parameters that may be appropriate for carbohydrates in parameter sets built for other molecular types.

In many cases not only the characterization of the free glycan is of interest, but also the investigation of the glycan as a ligand for a protein. Docking programs often use streamlined scoring functions and search routines to rapidly find appropriate binding sites and dock glycans into those sites. AutoDock Vina[59] and Glide[60] and are examples. Restraints are not often incorporated directly, but they can be used in subsequent scoring of posed structures. It is also common to subject the structures to short periods of MD refinement before final scoring to introduce some level of flexibility into ligand geometry. The program HADDOCK integrates these functions, using restraints directly and adding MD segments that treat ligands and protein segments as flexible entities.[58,61] It has been applied to a number of glycan binding proteins.[62,63]

4.4 Application to Glycan Structure and Dynamics

While applications of RDCs to protein structure determination are more widely recognized today, use of RDCs in the structure determination of glycans actually precedes their use in structure determination of proteins by several years. As early as 1989 ^{13}C–^{1}H and ^{1}H–^{1}H RDCs were combined with deuterium quadrupole coupling to deduce a solution structure of trehalose, a dimer of glucose α-linked reducing end to reducing end.[64] The media used for partial orientation was a bicelle-like system made from an aqueous dispersion of cesium perfluorooctanoate. This replaced earlier media preparations using potassium laurate that worked for deuterium quadrupole coupling observations. It offered the advantage of suppressing media signals from natural abundance ^{13}C at fluorinated sites by detecting only magnetization arising from INEPT transfers from protons. The resulting structure agreed very well with existing crystal and solution structures of trehalose. Additional applications to uniformly ^{13}C-enriched glycolipids that used both ^{13}C–^{1}H and ^{13}C–^{13}C RDCs followed,[65,66] with the orientation media switching to the phospholipid bicelle system that is widely used today.[67] Applications to soluble glycans also became widespread.[12,17,68–79] While most of these treated monosaccharide units in complex glycans as rigid and used RDCs to provide information on ϕ and ψ dihedral angles of glycosidic bonds, some used sufficiently extensive sets of ^{13}C–^{1}H, ^{13}C–^{13}C, and ^{1}H–^{1}H RDCs to determine the conformational preferences of individual sugar rings as well.[80,81]

More recently, PREs and PCSs started to be used in studies of structure and dynamics of complex glycans. PRE effects, caused by a soluble lanthanide chelate added to solvent, were used to guide interpretation of MD simulations of a ganglioside, GM1, integrated into a bicelle system.[82] Covalent attachment of EDTA-derived metal chelates to soluble complex sugars[26,83,84] allowed the measurement of RDCs, PCS and PREs, often within the same sample. For example, RDC and PRE data were measured in lactose,[26] PCS in the GM3 and GM2 saccharides,[85] and PCS in lactose.[86] Attachment of a spin-label allowed measurement of PREs in the large undecasaccharide, M9.[87]

Figure 4.5 illustrates some PCS data acquired using an EDTA-derived chelate attached to a particularly complex system.[88] The molecule depicted in Figure 4.5A is an *N*-glycan commonly found on glycoproteins. The trisaccharide branches have identical composition and normally display overlapping

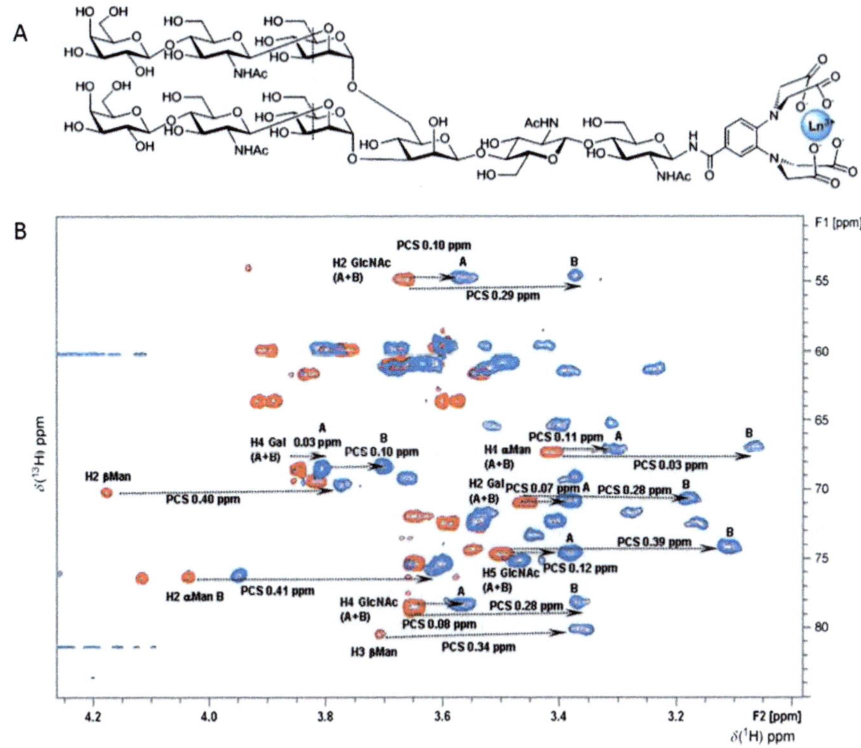

Figure 4.5 PCSs allow resolution of overlapped peaks and structural analysis of a complex *N*-glycan. (A) *N*-glycan with an EDTA derived lanthanide tag. (B) overlaid HSQC spectra with La^{3+} (red) and Dy^{3+} (blue) ions. Reproduced from Canales *et al.*, *Angewandte Chemie-International Edition* 2013 **52** 13789–13793, with permission of John Wiley & Sons. Copyright © 2013 WILEY-VCH Verlag GmbH & Co. KGaA, Weinheim.

resonances in HSQC spectra (red crosspeaks near 3.68 and 55.0 ppm in Figure 4.5B are an example). In the presence of a paramagnetic ion (Dy^{3+} in this case) PCSs (blue crosspeaks) enhance the difference in chemical shift allowing resolution and assignment of crosspeaks. The differences in PCSs for equivalent residues in the two branches also suggest that they adopt quite different geometries. The authors used the shifts to test a set of low energy conformers for consistency with data. While no single conformer produced an ideal fit to all data, a highly populated conformer could be identified.

Other examples of a complex systems arise in the case of oligomeric glycosaminoglycans, including chondroitin sulfate, heparan sulfate, heparin and dermatan sulfate. Fragments from these molecules are very heterogeneous because of varying sulfation patterns, yet there is considerable evidence to suggest that specific patterns are involved in modulating the activity of proteins including growth factors, chemokines and blood coagulation factors. Considerable effort has been devoted to structural studies of fragments from these molecules. One example used ^{13}C–^{1}H and ^{1}H–^{1}H RDCs on a tetrasaccharide from heparin to deduce conformational properties in solution.[14] In heparin, sulfated *N*-acetylglucosamine residues alternate with sulfated iduronic acid residues. *N*-acetylglucosamine residues exist in well-defined chair forms that allowed determination of an alignment frame and alignment parameters for each of these. This indicated an extended conformation with similar degrees of alignment for both rings. The intervening iduronic acid residue exists in a mixture of chair and skew-boat forms. Using RDCs as restraints in an AMBER-based molecular dynamics study allowed an estimate of populations for each of these forms and a conclusion that interconversion did little to alter the overall structure of the tetramer.

The aforementioned study was done with ^{13}C at natural abundance. ^{13}C enrichment obviously improves sensitivity and allows work with smaller amounts of material. Approaches to synthesis of ^{13}C enriched material are evolving.[89–92] However, material isolated from natural sources is still the most readily available form for many glycosaminoglycan oligomers. An alternate approach is to introduce a ^{13}C-enriched group to a natural product, post-isolation. This has been done on a pentasaccharide from chondroitin sulfate, a glycoaminoglycan in which *N*-acetyl-galactosamine residues alternate with glucuronic acid residues.[93] Here acetyl groups on the *N*-acetyl-galactosamine residues have been removed and replaced with ^{13}C enriched groups. The carbonyl and directly bonded methyl group provide a ^{13}C–^{13}C RDC and a ^{13}C–^{1}H RDC, but because of methyl rotation these report on the orientation of the same C–C bond vector. This would leave a minimum of information if it were not for the large RCSA of the carbonyl carbon. This is very complementary to the RDC, because the largest shielding direction of the chemical shift tensor is perpendicular to the plane containing the C–C vector (see Figure 4.6). In the application cited RDC and RCSA data were combined with other NMR data to generate a preferred conformation for an isolated pentamer. This is one of the rare applications of RCSAs, but because of the ease of isotopic labeling and the ease of measurement we expect to see more applications in the future.

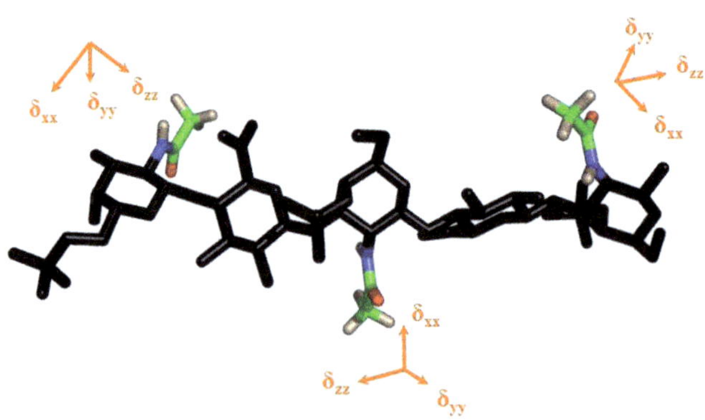

Figure 4.6 6,4,4-sulfated-chondroitin sulfate pentamer. ^{13}C-labeled acetyl groups and the orientation of individual carbonyl chemical shift tensors are shown in color.

One of the major problems with structure determination of free glycans in solution, is that most exhibit a significant level of flexibility about glycosidic bonds, particularly when linkage to an exo-cyclic methylene is involved, leading to variations in ω, as well as ϕ and ψ torsion angles. The extra information coming from RDC and paramagnetic effects allows detection of this flexibility, and in some cases, modeling of this flexibility. An early approach to detecting flexibility relied on comparing order tensors for each monosaccharide of a trimannoside and interpreting difference as arising from internal motion.[46] Subsequently, combined use of experimental RDCs and molecular dynamics (MD), Monte-Carlo (MC), or grid search simulations, allowed interpretation of such flexibility in the form of two or more conformations with different populations in equilibrium.[17,78,94–97] For oligosaccharides aligned *via* steric interactions in neutral anisotropic media such analysis has been greatly facilitated by the methods for prediction of alignment tensor orientation from molecular structure,[98,99] allowing calculation of RDCs averaged over large structure ensembles, such as MD trajectories. Treatment of motion in glycans has become more sophisticated in recent studies relying heavily on computational methods. Examples of the latter would be the APME approach of calculating the φ/ψ torsion angle distribution function,[13,100] or the use of RDC data to filter the snapshots from an MD trajectory.[101,102]

A recent example from the Kato laboratory has used replica-exchange molecular dynamics (REMD) to explore conformational space more efficiently.[103] The study looks at a pair of high-mannose forms of *N*-glycan found on many glycoproteins, Man8 and Man9. The removal of one mannose residue from the central branch in M9 to produce M8 is believed to be a signal targeting glycoproteins for disposal. The two forms were studied using paramagnetically tagged forms similar to that depicted in Figure 4.5. PCSs were collected and compared to averages of PCSs back-calculated using conformers sampled in MD trajectories. Even long trajectories (240 ns) were not capable of reproducing observations. However, use of REMD substantially improved the fit to observation. Examining the conformations in the improved ensemble showed that removal of the single mannose not only exposed underlying residues, but allowed other branches to sample a broader range of conformations, some of which may be recognized in disposal signaling. Further advances along these lines will certainly improve our understanding of how glycans produce their physiological effects.

4.5 Application to Glycan–Protein Complexes

Given the importance of glycan–protein interactions in processes that range from protein homeostasis to cell-adhesion/migration, to the modulation of various signaling events, it is no surprise that structural characterization of these complexes has attracted a great deal of attention. Most of the NMR work employs transferred NOE (trNOE) and saturation transfer difference (STD) spectroscopy to provide information on intra-glycan distance constraints and identify binding epitopes on the glycans. Because of the amplification of cross-relaxation rates in large complexes with rapid glycan exchange rates, it is possible to work at high glycan to protein ratios where high resolution and sensitivity of glycan proton signals is maintained. Despite the quality of the data, it is still too limited for *de novo* structure determination and analysis usually relies on computational methods to generate reliable models. This approach is covered more extensively in another chapter. There are also some excellent reviews that cover the methodology and applications.[104] Here we focus on the use of RDCs, PCSs, and PREs to supplement trNOE and STD data, reducing some of the dependence on computational programs to produce reliable models.

It was not long after the first RDC applications to protein structure appeared that the potential for determination of protein-bound glycan structure was recognized.[105] A ^{13}C enriched glycan (Galα1-4Galβ1-4Glc) was used to study its conformation when bound to the B-subunit homopentamer of a toxin derived from *Escherichia coli* O157. This was not a trivial undertaking because the contributions of bound state RDCs to the averages observed for rapidly exchanging systems are not amplified as those of trNOEs and STDs; for RDCs one measures a simple population average that is usually heavily weighted by the RDCs of the free form. The authors took advantage of the larger level of orientation of the protein in this case (DMPC/DHPC bicelles) as well as a relatively low ligand to protein ratio (5 : 1) to allow extraction of bound state RDCs from a series of HSQC spectra that were ^{13}C coupled in the proton dimension. It nevertheless required a double difference process, measuring splittings of the glycan signals for isotropic and aligned phases, both with and without the protein, and then using a known binding constant to extract bound state RDCs. Several pairs of RDCs in a given sugar ring are also nearly degenerate reducing the number of truly independent pieces of data. Nevertheless, a cluster of single state structures for the trisaccharide which fit the twelve pieces of data acquired was determined. The cluster was fairly tight and in close agreement with the glycan conformation seen in the crystal structure of a similar complex. While the trisaccharide was not actually docked into the protein binding site, the potential for collecting RDC data on the protein and using this to actually restrain alignment of the trisaccharide relative to the protein was clearly recognized.

About this same time a similar set of studies on a trimeric form of mannose-binding protein (MBP) was initiated.[106] MBP is a part of the mammalian innate defense system where it binds to the mannose-terminated glycans frequently found on bacterial pathogens and initiates their destruction. The initial study used just a ^{13}C-enriched monosaccharide, α-methyl-mannoside, and relied on the slight deviations from co-linearity of ^{13}C–^{1}H pairs in the $^{4}C_1$ mannose ring to define its orientation. It was then docked into the protein binding site, not by collecting RDC data on the protein, but by recognizing that the three-fold rotational symmetry of a complex ensures that the alignment in partially ordered media will be axially symmetric with the principal axis being the rotation axis of the protein.[107] Because of uncertainty about the relative domain orientations in a trimeric complex like MBP, RDC data on the protein was eventually collected and the docked structure of a trimannoside determined.[108] The actual docking was accomplished with the AutoDock program without the use of RDC-based constraints. Instead, poses were examined for consistency with the RDC determined principal axis of symmetry. This is not a very stringent criterion since rotation about the axis of symmetry is not defined in an axially symmetric system. However, some useful screening of poses was accomplished.

In the aforementioned cases the extent of orientation for the protein complex proved to be larger than the free glycan, making the extraction of data somewhat easier. However, this will not always be the case. Bicelle-induced alignment is largely steric in nature, and as proteins, or protein complexes, become more spherical, their level of alignment will approach zero. There have, therefore, been several attempts to assure a higher level of orientation for the protein.

One system used to illustrate these methods was the carbohydrate binding domain (CRD) from Galectin-3, a human lectin known to bind galactose residues found at the non-reducing terminus of cell-surface glycans. To mimic those glycans a freely soluble lactose molecule (Galβ1-4Glc) was used. The alignment medium was a bicelle-like medium composed of 3% alkyl-ethylene glycol detergent (C12E5) and a small amount of hexanol in aqueous buffer. The extent of orientation for the Galectin-3-CRD was small, yielding $^{15}N-^{1}H$ RDCs less than 10 Hz. Free lactose $^{13}C-^{1}H$ RDCs in the same medium reached 4.2 Hz, a significant contribution to the average when a ten-fold excess of lactose would reduce the contributions of bound RDCs to less than 1 Hz. One solution illustrated in Figure 4.7A proved to be the addition of a short alkyl chain to the C-terminus of the protein.[109] This was done by mutating the

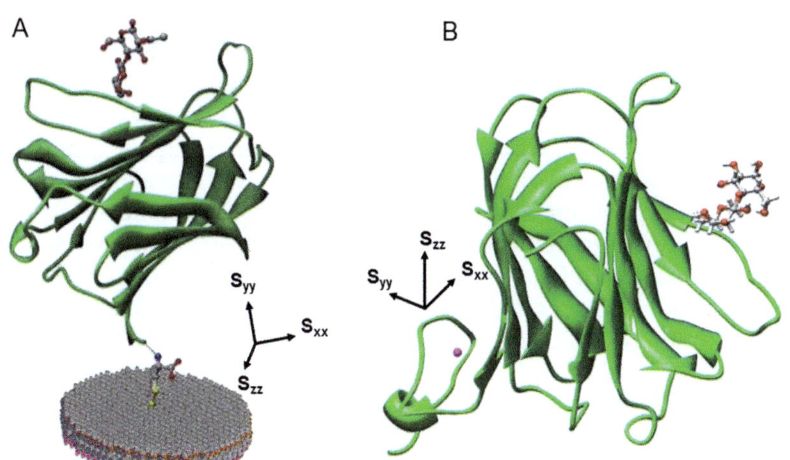

Figure 4.7 Two different ways of accentuating bound-state RDCs for exchanging ligands. (A) Increasing order of the protein by association with orienting media. (B) Preferentially aligning the protein by field-induced alignment of a lanthanide tagged protein. The Galectin-3-CRD-lactose complex is illustrated.

terminal residue to a cysteine and reacting this with propylmethanethiolsulfonate. The micelles formed by C12E5 have a distinctly hydrophobic interior into which an alkyl chain will insert. In the presence of the propyl chain, the protein RDCs rose by a factor of 4 allowing easier separation of bound state RDCs for the ligand. In this case modeling of the complex was accomplished by using an energy-minimized structure for the lactose molecule, finding the principal alignment frames for ligand and protein using the program REDCAT,[45] placing protein and ligand in this frame, and translating the ligand to optimize contacts in the binding site. Additional docking was accomplished with the HADDOCK program.[61]

A subsequent study on the lactose-Galectin3-CRD system was pursued using a His-tagged version of the protein and the C12E5 bicelle system with a lipid-like nickel chelate (DOGS-NTA-Ni, 1,2-dioleoyl-*sn*-glycero-3-{[*N*(5-amino-1-carboxypentyl) iminodiacetic acid]succinyl}(nickel salt)) added to the medium.[110] This bicelle-anchored chelate is capable of binding His-tagged proteins much like the nickel columns used in protein purification. A very similar level of RDC enhancement from bound ligands was attained.

Another completely different approach to making sure that bound-state RDCs of ligands dominate the measured average is to design experiments in which only the protein is aligned. Field-induced alignment of a paramagnetically tagged protein comes very close to this ideal. Alignment of any molecule depends on the anisotropic part of its magnetic susceptibility. This is ordinarily very small for glycans, but it can be made very large for proteins by adding chelates,[111] peptide segments,[112,113] or even whole proteins[114] that can carry metal ions with large paramagnetic susceptibilities. The difference in susceptibilities means that the contribution of free glycan (usually in 5–10 fold excess) to a measured average RDC for a rapidly exchanging glycan can be neglected. One of the first applications of this type was made to the Galectin-3-CRD system previously mentioned. It is illustrated in Figure 4.7B. A seventeen-residue lanthanide binding peptide was added to the C-terminus of Galectin-3-CRD, a point well away from the ligand binding site. The RDCs do not depend on distance from the ion, but broadening of ligand resonances from paramagnetic effects do, and there is some advantage in having the bound ligand far removed from the ion site. In the present case both RDCs and small PCSs were measured at a ligand to protein ratio of 5 to 1. The observed RDCs ranged only up to 3.4 Hz, but all of this comes from the bound state RDC of 17 Hz. Modeling was done with the XPLOR program. Using just RDCs and PCSs the lactose placement gave a 4.3 Å RMSD for heavy atoms from an X-ray structure containing this ligand. Using a single NOE to complement this data reduced the RMSD to 1.9 Å.

Much research since the time of the above application has gone into the design of a wider variety of paramagnetic tags that can be attached to proteins. Putting lanthanide-binding peptides at the N or C termini often leads to large amplitude internal motion that reduces the magnitude of effects and complicates interpretation. Internal motion is also a complication for many of the original chelate tags that employed a disulfide linkage to a single cysteine introduced by mutation for attachment. Insertion of peptides in loops and using pairs of linkages to cysteines are among the approaches used to reduce tag mobility.[115,116] However, modification of native loops in proteins without disrupting structure or function is not always easy, and proteins often contain native cysteines that make selective addition of a tag to mutated sites difficult. New strategies that avoid loop insertion and the introduction of non-native cysteines should expand opportunities for applications similar to that described.

The roles of protein carrying a paramagnetic tag and glycans providing observation sites can be reversed. Some examples of glycans carrying paramagnetic tags were discussed above as an aid in conformational analysis of free glycans. Similar tagged glycans produce perturbations on isotopically labeled protein sites that can be used to locate binding sites and orient the modified glycan in that site. Some early studies of this type were again carried out using the Galectin-3-CRD system and a lactose derivative synthesized by the addition of a carboxy-TEMPO moiety (1-Oxyl-4-carboxyl-2,2,6,6-tetramethylpiperidine) to the amino analog of lactose through an amide bond.[117] TEMPO derivatives produce PREs of protein nuclear sites in a $1/r^6$ dependent fashion. Quantitative comparison of R_2 and R_1 enhancements for sites in the 11–20 Å range were in good agreement with enhancements calculated from a crystal structure with the sugar rings of the TEMPO analog superimposed on the sugar rings of lactose in the crystal. Since that time similar analogs have been used to place ligands in protein complexes for which fewer structural data were available. For example, addition of a 4-amino-TEMPO group has been added to the reducing end of a chondroitin sulfate hexamer using reductive animation and the bound geometries of these analogs determined using PRE-derived distance constraints in docking programs for both chemokine systems (CCL5)[118] and link module systems (LINK–TSG6).[62] These are systems where interactions with chondroitin sulfate and other glycosaminoglycan polymers play important roles in oligomerization and crosslinking elements of the extracellular matrix. The models of complexes were generated using distance restraints derived from PREs using utilities in the AMBER MD module and the HADDOCK docking program, respectively.

Similar PRE effects can be generated by attaching metal-binding chelates to glycans and loading these with ions such as Gd^{3+} or Mn^{2+}. A very recent example has used "click" chemistry to attach an alkyne-containing EDTA derivative to a nona-sulfated hyaluronan-tetrasaccharide carrying an azide at its reducing end.[119] Heavily sulfated hyaluronan oligomers are a derivative of natural hyaluronan that has shown promise as a component of artificial extracellular matrices. Here the tagged version of this molecule was used to demonstrate binding to the cytokine, interleukin-10 (IL-10). IL-10 is known to bind heparin and the modified product was shown to compete with heparin. Intensity changes in the ^{15}N–^{1}H HSQC spectrum of the protein on titration with the Mn^{2+} loaded version of the tagged sulfated hyaluronan oligomer were used to derive ion-to-binding site distances in a manner similar to that used with TEMPO derivatives. Distance restraints were used in the XPLOR package to precisely position a pair of ions in the IL-10 dimer. While the position of the remainder of the hyaluronan oligomer was not as well defined, this clearly provides an important first step in a potentially biomedically important interaction.

The lanthanide chelates offer an additional advantage in providing PCSs, which carry both an angular and a distance dependence. A very nice example comes from the Jimenez-Barbero laboratory.[86] Here a biphenyl EDTA lanthanide-chelating unit was added to an amino analog of lactose *via* an amide bond, and this modified lactose was bound to the CRD of Galectin-3. Figure 4.8 illustrates the complex geometry as well as some of the observed PCSs. PCSs could be observed out to sites at distances in excess of 40 Å. This is about twice the distance for which PRE effects from a TEMPO-tagged lactose bound to the Galectin-3-CRD could be observed. The crystal structure of the complex was used to back-calculate PCSs and these gave excellent agreement with observation (a Q factor of 0.3 and 0.2 for Tm^{3+} and Dy^{3+} respectively).

A similar tag attached to a fucose residue has recently been constructed with the intent to use PCS and PRE perturbations to aid in NMR resonance assignment of a very large protein system.[120] The protein target is the protruding domains of norovirus which are believed to bind to fucose-containing blood group antigens. Expressible constructs exist as 70 kDa dimers, well outside the applicable range of conventional resonance assignment strategies. Several PCS and PRE perturbations were seen in TROSY variants of ^{15}N–^{1}H HSQC spectra that should facilitate assignment of the perturbed resonances.

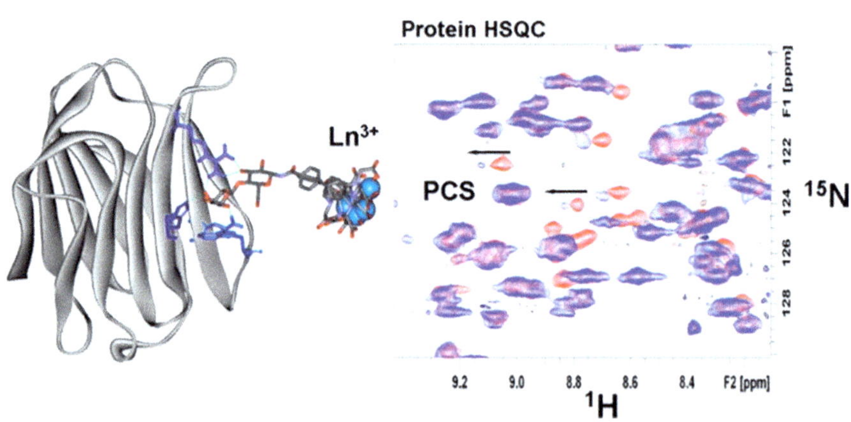

Figure 4.8 PCS effects on Galectin-3 CRD spectra on binding a Lanthanide tagged lactose molecule. Left, ribbon diagram of the protein-glycan complex. Right, segment of a ^{15}N–^{1}H HSQC spectrum with crosspeaks with a diamagnetic lanthanide (La^{3+}) in red and with a paramagnetic lanthanide (Tm^{3+}) in blue. Reprinted with permission from Canales *et al.*, *Journal of the American Chemical Society* 2014, **136**, 8011–8017. Copyright (2014) American Chemical Society.

From the examples presented, it is clear that RDCs, PCSs, and PREs can provide useful descriptions of both the conformational distributions that glycans sample in solution and the conformations that they adopt when bound to proteins. The introduction of lanthanide-carrying chelate tags for both proteins and glycans is having a dramatic effect on the field. RDCs, PCSs and PREs each have distinct geometric dependencies and are highly complementary. The ability to measure all of these from a single construct can provide the wealth of information needed to define the structure and dynamics of the most complex glycan systems.

References

1. K. Kato and T. Yamaguchi, *Glycoconjugate J.*, 2015, **32**, 505–513.
2. M. D. Battistel, H. F. Azurmendi, B. Yu and D. I. Freedberg, *Prog. Nucl. Magn. Reson. Spectrosc.*, 2014, **79**, 48–68.
3. A. Canales, J. Jimenez-Barbero and M. Martin-Pastor, *Magn. Reson. Chem.*, 2012, **50**, S80–S85.
4. D. Diaz, A. Canales-Mayordomo, F. J. Canada and J. Jimenez-Barbero, *Methods Mol Biol.*, 2015, **1273**, 261–287.
5. A. Bax, *Protein Sci.*, 2003, **12**, 1–16.
6. J. H. Prestegard, C. M. Bougault and A. I. Kishore, *Chem. Rev.*, 2004, **104**, 3519–3540.
7. J. R. Tolman, J. M. Flanagan, M. A. Kennedy and J. H. Prestegard, *Proc. Natl. Acad. Sci. U. S. A.*, 1995, **92**, 9279–9283.
8. M. Dama and S. Berger, *Carbohydr. Res.*, 2013, **377**, 44–47.
9. J. J. Chou, S. Gaemers, B. Howder, J. M. Louis and A. Bax, *J. Biomol. NMR*, 2001, **21**, 377–382.
10. H. J. Sass, G. Musco, S. J. Stahl, P. T. Wingfield and S. Grzesiek, *J. Biomol. NMR*, 2000, **18**, 303–309.
11. B. Yu, H. van Ingen, S. Vivekanandan, C. Rademacher, S. E. Norris and D. I. Freedberg, *J. Magn. Reson.*, 2012, **215**, 10–22.
12. H. Neubauer, J. Meiler, W. Peti and C. Griesinger, *Helv. Chim. Acta*, 2001, **84**, 243–258.
13. E. Sawen, B. Stevensson, J. Ostervall, A. Maliniak and G. Widmalm, *J. Phys. Chem. B*, 2011, **115**, 7109–7121.
14. L. Jin, M. Hricovini, J. A. Deakin, M. Lyon and D. Uhrin, *Glycobiology*, 2009, **19**, 1185–1196.
15. T. Reinsperger and B. Luy, *J. Magn. Reson.*, 2014, **239**, 110–120.
16. A. Silipo, Z. Zhang, F. J. Canada, A. Molinaro, R. J. Linhardt and J. Jimenez-Barbero, *ChemBioChem*, 2008, **9**, 240–252.
17. A. Almond and J. O. Duus, *J. Biomol. NMR*, 2001, **20**, 351–363.
18. J. R. Tolman and J. H. Prestegard, *J. Magn. Reson., Ser. B*, 1996, **112**, 269–274.
19. J. R. Tolman and J. H. Prestegard, *J. Magn. Reson., Ser. B*, 1996, **112**, 245–252.

20. N. Tjandra and A. Bax, *J. Magn. Reson.*, 1997, **124**, 512–515.
21. Y. Liu and J. H. Prestegard, *J. Magn. Reson.*, 2009, **200**, 109–118.
22. B. Yu, H. van Ingen and D. I. Freedberg, *J. Magn. Reson.*, 2013, **228**, 159–165.
23. J. R. Garbow, D. P. Weitekamp and A. Pines, *Chem. Phys. Lett.*, 1982, **93**, 504–509.
24. I. Timári, L. Kaltschnee, A. Kolmer, R. W. Adams, M. Nilsson, C. M. Thiele, G. A. Morris and K. E. Kövér, *J. Magn. Reson.*, 2014, **239**, 130–138.
25. L. Kaltschnee, A. Kolmer, I. Timari, V. Schmidts, R. W. Adams, M. Nilsson, K. E. Koever, G. A. Morris and C. M. Thiele, *Chem. Commun.*, 2014, **50**, 15702–15705.
26. M. Erdelyi, E. d'Auvergne, A. Navarro-Vazquez, A. Leonov and C. Griesinger, *Chem.–Eur. J.*, 2011, **17**, 9368–9376.
27. M. Ottiger, F. Delaglio and A. Bax, *J. Magn. Reson.*, 1998, **131**, 373–378.
28. L. Yao, J. Ying and A. Bax, *J. Biomol. NMR*, 2009, **43**, 161–170.
29. L. Jin and D. Uhrin, *Magn. Reson. Chem.*, 2007, **45**, 628–633.
30. M. Martin-Pastor, A. Canales-Mayordomo and J. Jimenez-Barbero, *J. Biomol. NMR*, 2003, **26**, 345–353.
31. F. Tian, P. J. Bolon and J. H. Prestegard, *J. Am. Chem. Soc.*, 1999, **121**, 7712–7713.
32. L. Jin, T. N. Pham and D. Uhrin, *ChemPhysChem*, 2007, **8**, 1228–1235.
33. J. H. Prestegard, H. M. Al-Hashimi and J. R. Tolman, *Q. Rev. Biophys.*, 2000, **33**, 371–424.
34. R. Ghose and J. H. Prestegard, *J. Magn. Reson.*, 1997, **128**, 138–143.
35. Y. Liu and J. H. Prestegard, *J. Biomol. NMR*, 2010, **47**, 249–258.
36. R. A. Burton and N. Tjandra, *J. Am. Chem. Soc.*, 2007, **129**, 1321–1326.
37. W.-M. Liu, M. Overhand and M. Ubbink, *Coord. Chem. Rev.*, 2014, **273**, 2–12.
38. J. Koehler and J. Meiler, *Prog. Nucl. Magn. Reson. Spectrosc.*, 2011, **59**, 360–389.
39. G. Otting, in *Annual Review of Biophysics*, ed. D. C. Rees, K. A. Dill and J. R. Williamson, 2010, vol. 39, pp. 387–405.
40. K. N. Allen and B. Imperiali, *Curr. Opin. Chem. Biol.*, 2010, **14**, 247–254.
41. I. Bertini, C. Luchinat and G. Parigi, *Prog. Nucl. Magn. Reson. Spectrosc.*, 2002, **40**, 249–273.
42. J. L. Battiste and G. Wagner, *Biochemistry*, 2000, **39**, 5355–5365.
43. I. Bertini, C. Luchinat and G. Parigi, *Solution NMR of Paramagnetic Molecules*, Elsevier, 2001.
44. C. Schmidt, S. J. Irausquin and H. Valafar, *BMC Bioinf.*, 2013, **14**, 302.
45. H. Valafar and J. H. Prestegard, *J. Magn. Reson.*, 2004, **167**, 228–241.
46. F. Tian, H. M. Al-Hashimi, J. L. Craighead and J. H. Prestegard, *J. Am. Chem. Soc.*, 2001, **123**, 485–492.
47. M. Zweckstetter and A. Bax, *J. Am. Chem. Soc.*, 2000, **122**, 3791–3792.
48. V. Ozenne, F. Bauer, L. Salmon, J. R. Huang, M. R. Jensen, S. Segard, P. Bernado, C. Charavay and M. Blackledge, *Bioinformatics*, 2012, **28**, 1463–1470.

49. J.-r. Huang, M. Gentner, N. Vajpai, S. Grzesiek and M. Blackledge, *Biochem. Soc. Trans.*, 2012, **40**, 989–994.
50. R. Salomon-Ferrer, D. A. Case and R. C. Walker, *Wiley Interdiscip. Rev.: Comput. Mol. Sci.*, 2013, **3**, 198–210.
51. C. D. Schwieters, J. J. Kuszewski and G. M. Clore, *Prog. Nucl. Magn. Reson. Spectrosc.*, 2006, **48**, 47–62.
52. P. Guntert and L. Buchner, *J. Biomol. NMR*, 2015, **62**, 453–471.
53. A. T. Brunger, *Nat. Protoc.*, 2007, **2**, 2728–2733.
54. S. Pronk, S. Pall, R. Schulz, P. Larsson, P. Bjelkmar, R. Apostolov, M. R. Shirts, J. C. Smith, P. M. Kasson, D. van der Spoel, B. Hess and E. Lindahl, *Bioinformatics*, 2013, **29**, 845–854.
55. K. N. Kirschner, A. B. Yongye, S. M. Tschampel, J. González-Outeiriño, C. R. Daniels, B. Lachele Foley and R. J. Woods, *J. Comput. Chem.*, 2008, **29**, 622–655.
56. A. W. Schuttelkopf and D. M. F. van Aalten, *Acta Crystallogr., Sect. D: Biol. Crystallogr.*, 2004, **60**, 1355–1363.
57. S. Jo, K. C. Song, H. Desaire, A. D. MacKerell and W. Im, *J. Comput. Chem.*, 2011, **32**, 3135–3141.
58. G. C. P. van Zundert, J. P. G. L. M. Rodrigues, M. Trellet, C. Schmitz, P. L. Kastritis, E. Karaca, A. S. J. Melquiond, M. van Dijk, S. J. de Vries and A. M. J. J. Bonvin, *J. Mol. Biol.*, 2016, **428**, 720–725.
59. O. Trott and A. J. Olson, *J. Comput. Chem.*, 2010, **31**, 455–461.
60. R. A. Friesner, J. L. Banks, R. B. Murphy, T. A. Halgren, J. J. Klicic, D. T. Mainz, M. P. Repasky, E. H. Knoll, M. Shelley, J. K. Perry, D. E. Shaw, P. Francis and P. S. Shenkin, *J. Med. Chem.*, 2004, **47**, 1739–1749.
61. C. Dominguez, R. Boelens and A. Bonvin, *J. Am. Chem. Soc.*, 2003, **125**, 1731–1737.
62. Y. Park, T. A. Jowitt, A. J. Day and J. H. Prestegard, *Biochemistry*, 2016, **55**, 262–276.
63. L. Y. Chiu, K. W. Hung, S. C. Tjong, Y. W. Chiang and S. C. Sue, *Biochim. Biophys. Acta, Proteins Proteomics*, 2014, **1844**, 1851–1859.
64. P. Ram, L. Mazzola and J. H. Prestegard, *J. Am. Chem. Soc.*, 1989, **111**, 3176–3182.
65. K. P. Howard and J. H. Prestegard, *J. Am. Chem. Soc.*, 1995, **117**, 5031–5040.
66. K. P. Howard and J. H. Prestegard, *J. Am. Chem. Soc.*, 1996, **118**, 3345–3353.
67. C. R. Sanders, B. J. Hare, K. P. Howard and J. H. Prestegard, *Prog. Nucl. Magn. Reson. Spectrosc.*, 1994, **26**, 421–444.
68. M. Martin-Pastor and C. A. Bush, *Carbohydr. Res.*, 2000, **323**, 147–155.
69. T. Rundlöf, C. Landersjö, K. Lycknert, A. Maliniak and G. Widmalm, *Magn. Reson. Chem.*, 1998, **36**, 773–776.
70. G. R. Kiddle and S. W. Homans, *FEBS Lett.*, 1998, **436**, 128–130.
71. C. Landersjö, C. Höög, A. Maliniak and G. Widmalm, *J. Phys. Chem. B*, 2000, **104**, 5618–5624.
72. M. Martin-Pastor and C. A. Bush, *Biochemistry*, 2000, **39**, 4674–4683.

73. A. Almond, J. Bunkenborg, T. Franch, C. H. Gotfredsen and J. O. Duus, *J. Am. Chem. Soc.*, 2001, **123**, 4792–4802.
74. K. Lycknert, A. Maliniak and G. Widmalm, *J. Phys. Chem. A*, 2001, **105**, 5119–5122.
75. M. Martin-Pastor and C. A. Bush, *J. Biomol. NMR*, 2001, **19**, 125–139.
76. H. F. Azurmendi and C. A. Bush, *Carbohydr. Res.*, 2002, **337**, 905–915.
77. J. Adeyeye, H. F. Azurmendi, C. J. M. Stroop, S. Sozhamannan, A. L. Williams, A. M. Adetumbi, J. A. Johnson and C. A. Bush, *Biochemistry*, 2003, **42**, 3979–3988.
78. C. Landersjo, J. L. M. Jansson, A. Maliniak and G. Widmalm, *J. Phys. Chem. B*, 2005, **109**, 17320–17326.
79. M. Martín-Pastor, A. Canales, F. Corzana, J. L. Asensio and J. Jiménez-Barbero, *J. Am. Chem. Soc.*, 2005, **127**, 3589–3595.
80. D. I. Freedberg, *J. Am. Chem. Soc.*, 2002, **124**, 2358–2362.
81. T. N. Pham, S. L. Hinchley, D. W. H. Rankin, T. Liptaj and D. Uhrín, *J. Am. Chem. Soc.*, 2004, **126**, 13100–13110.
82. M. L. DeMarco, R. J. Woods, J. H. Prestegard and F. Tian, *J. Am. Chem. Soc.*, 2010, **132**, 1334–1338.
83. A. Mallagaray, A. Canales, G. Dominguez, J. Jimenez-Barbero and J. Perez-Castells, *Chem. Comm.*, 2011, **47**, 7179–7181.
84. S. Yamamoto, T. Yamaguchi, M. Erdélyi, C. Griesinger and K. Kato, *Chem.–Eur. J.*, 2011, **17**, 9280–9282.
85. S. Yamamoto, Y. Zhang, T. Yamaguchi, T. Kameda and K. Kato, *Chem. Commun.*, 2012, **48**, 4752–4754.
86. A. Canales, A. Mallagaray, M. Alvaro Berbis, A. Navarro-Vazquez, G. Dominguez, F. Javier Canada, S. Andre, H.-J. Gabius, J. Perez-Castells and J. Jimenez-Barbero, *J. Am. Chem. Soc.*, 2014, **136**, 8011–8017.
87. T. Yamaguchi, Y. Kamiya, Y.-M. Choo, S. Yamamoto and K. Kato, *Chem. Lett.*, 2013, **42**, 544–546.
88. A. Canales, A. Mallagaray, J. Perez-Castells, I. Boos, C. Unverzagt, S. Andre, H.-J. Gabius, F. J. Canada and J. Jimenez-Barbero, *Angew. Chem., Int. Ed.*, 2013, **52**, 13789–13793.
89. C. Laguri, N. Sapay, J.-P. Simorre, B. Brutscher, A. Imberty, P. Gans and H. Lortat-Jacob, *J. Am. Chem. Soc.*, 2011, **133**, 9642–9645.
90. Z. Zhang, S. A. McCallum, J. Xie, L. Nieto, F. Corzana, J. Jimenez-Barbero, M. Chen, J. Liu and R. J. Linhardtt, *J. Am. Chem. Soc.*, 2008, **130**, 12998–13007.
91. S. A. Colebrooke, C. D. Blundell, P. L. DeAngelis, I. D. Campbell and A. Almond, *Magn. Reson. Chem.*, 2005, **43**, 805–815.
92. C. Fontana, H. Kovacs and G. Widmalm, *J. Biomol. NMR*, 2014, **59**, 95–110.
93. F. Yu, J. J. Wolff, I. J. Amster and J. H. Prestegard, *J. Am. Chem. Soc.*, 2007, **129**, 13288–13297.
94. R. M. Venable, F. Delaglio, S. E. Norris and D. I. Freedberg, *Carbohydr. Res.*, 2005, **340**, 863–874.
95. S. Ganguly, J. Xia, C. Margulis, L. Stanwyck and C. A. Bush, *Biopolymers*, 2011, **95**, 39–50.

96. M. Kramer and E. Kleinpeter, *J. Magn. Reson.*, 2011, **212**, 174–185.
97. J. Xia, C. J. Margulis and D. A. Case, *J. Am. Chem. Soc.*, 2011, **133**, 15252–15255.
98. H. F. Azurmendi and C. A. Bush, *J. Am. Chem. Soc.*, 2002, **124**, 2426–2427.
99. A. Almond and J. B. Axelsen, *J. Am. Chem. Soc.*, 2002, **124**, 9986–9987.
100. C. Landersjö, B. Stevensson, R. Eklund, J. Östervall, P. Söderman, G. Widmalm and A. Maliniak, *J. Biomol. NMR*, 2006, **35**, 89–101.
101. V. Gargiulo, M. A. Morando, A. Silipo, A. Nurisso, S. Perez, A. Imberty, F. Javier Canada, M. Parrilli, J. Jimenez-Barbero and C. De Castro, *Glycobiology*, 2010, **20**, 1208–1216.
102. J. Xia and D. A. Case, *Biopolymers*, 2012, **97**, 276–288.
103. T. Yamaguchi, Y. Sakae, Y. Zhang, S. Yamamoto, Y. Okamoto and K. Kato, *Angew. Chem., Int. Ed.*, 2014, **53**, 10941–10944.
104. V. Roldos, F. Javier Canada and J. Jimenez-Barbero, *ChemBioChem*, 2011, **12**, 990–1005.
105. H. Shimizu, A. Donohue-Rolfe and S. W. Homans, *J. Am. Chem. Soc.*, 1999, **121**, 5815–5816.
106. P. J. Bolon, H. M. Al-Hashimi and J. H. Prestegard, *J. Mol. Biol.*, 1999, **293**, 107–115.
107. H. M. Al-Hashimi, P. J. Bolon and J. H. Prestegard, *J. Magn. Reson.*, 2000, **142**, 153–158.
108. N. U. Jain, S. Noble and J. H. Prestegard, *J. Mol. Biol.*, 2003, **328**, 451–462.
109. T. D. Zhuang, H. Leffler and J. H. Prestegard, *Protein Sci.*, 2006, **15**, 1780–1790.
110. R. D. Seidel III, T. Zhuang and J. H. Prestegard, *J. Am. Chem. Soc.*, 2007, **129**, 4834–4839.
111. T. Ikegami, L. Verdier, P. Sakhaii, S. Grimme, B. Pescatore, K. Saxena, K. M. Fiebig and C. Griesinger, *J. Biomol. NMR*, 2004, **29**, 339–349.
112. C. Ma and S. J. Opella, *J. Magn. Reson.*, 2000, **146**, 381–384.
113. J. Wohnert, K. J. Franz, M. Nitz, B. Imperiali and H. Schwalbe, *J. Am. Chem. Soc.*, 2003, **125**, 13338–13339.
114. J. Feeney, B. Birdsall, A. F. Bradbury, R. R. Biekofsky and P. M. Bayley, *J. Biomol. NMR*, 2001, **21**, 41–48.
115. J. D. Swarbrick, P. Ung, X.-C. Su, A. Maleckis, S. Chhabra, T. Huber, G. Otting and B. Graham, *Chem. Commun.*, 2011, **47**, 7368–7370.
116. K. Barthelmes, A. M. Reynolds, E. Peisach, H. R. A. Jonker, N. J. DeNunzio, K. N. Allen, B. Imperiali and H. Schwalbe, *J. Am. Chem. Soc.*, 2011, **133**, 808–819.
117. N. U. Jain, A. Venot, K. Umemoto, H. Leffler and J. H. Prestegard, *Protein Sci.*, 2001, **10**, 2393–2400.
118. C. Deshauer, A. M. Morgan, E. O. Ryan, T. M. Handel, J. H. Prestegard and X. Wang, *Structure*, 2015, **23**, 1066–1077.
119. S. Koehling, G. Kuenze, K. Lemmnitzer, M. Bermudez, G. Wolber, J. Schiller, D. Huster and J. Rademann, *Chem.–Eur. J.*, 2016, **22**, 5563–5574.
120. A. Mallagaray, G. Dominguez, T. Peters and J. Perez-Castells, *Chem. Commun.*, 2016, **52**, 601–604.

CHAPTER 5

Lanthanide-Chelating Carbohydrate Conjugates to Detect Carbohydrate–Protein Interactions

ANGELES CANALES[a] AND JESÚS JIMÉNEZ-BARBERO*[b,c,d]

[a]Departamento de Química Orgánica I, Facultad de Ciencias Químicas, Universidad Complutense de Madrid, E-28040 Madrid, Spain; [b]CIC bioGUNE, Bizkaia Technology Park, Building 801A, 48160 Derio-Bizkaia, Spain; [c]Ikerbasque, Basque Foundation for Science, Bilbao, Spain; [d]Department of Organic Chemistry II, Faculty of Science & Technology, University of the Basque Country, 48940 Leioa, Bizkaia, Spain
*E-mail: jjbarbero@cicbiogune.es

5.1 Introduction

Carbohydrate–protein recognition is a key step in processes of paramount importance for life, such as; cell–cell recognition, inflammatory processes, cell growth or pathogen-host infection.[1] Therefore, the knowledge of the key details of this interaction process is essential to design methods, protocols or molecules able to enhance or disfavor the corresponding event.

Nowadays, a series of biophysical techniques are available to characterize these interactions at different levels of complexity. In this respect, surface plasmon resonance can be used to detect binding and determine

New Developments in NMR No.10
NMR in Glycoscience and Glycotechnology
Edited by Koichi Kato and Thomas Peters

Published by the Royal Society of Chemistry, www.rsc.org

the binding constant,[2] isothermal titration calorimetry might be employed for measuring the thermodynamic parameters of the binding process,[3] analytical centrifugation[4,5] and light scattering[6] shed light on the possibility of the formation large aggregates. From the structural perspective, X-ray crystallography[7] and NMR spectroscopy provide information at the atomic level about the recognition process.[8]

However, despite this variety of available methodologies, it is still needed to develop additional tools to characterize carbohydrate–protein interactions due to the inherent complexity, heterogeneity and flexibility of natural oligosaccharides. These features often preclude the crystallization of carbohydrates both in the free state and in complex with other biomolecules and even in those cases where an oligosaccharide protein complex can be crystallized, it is difficult to observed electron density for the flexible regions of the oligosaccharide, even displaying moderate motion, and therefore the conformation and interaction modes of these regions cannot be inferred.[9]

The access to atomic details of protein–glycan complexes is also very challenging from the NMR perspective. The standard NMR parameters with structural information, NOEs and J couplings, only provide local information. Generally speaking, it is fairly common for complex oligosaccharides to be found in extended shapes, with no long-range NOE contacts. In addition, a limited number of oligosaccharide–protein intermolecular NOEs can usually be measured to derive the relative orientation of the carbohydrate respect to the protein.[8] Moreover, oligosaccharides usually contain repeating units that display isochronous chemical shifts. Therefore, knowledge on key individual residue-specific structural information may frequently be elusive.[10]

Residual dipolar couplings (RDCs) are also difficult to measure in carbohydrates bound to proteins since RDCs of exchanging ligands are dominated by free-state contributions. In order to enhance the RDCs from the bound state, protein modifications have been proposed by addition of a hydrophobic alkyl tail that anchors to the bicelles of the alignment media. This method should be applied when bound ligand geometry, rather than orientation relative to the protein, is of interest. Here, data on the protein do not need to be collected; the protein does not need to be isotopically labeled, and it can be very strongly oriented. In contrast, this approach cannot be applied if the ligands themselves are highly hydrophobic and strongly associate with the orienting medium.[11]

5.2 Lanthanide-Chelating Carbohydrate Conjugates

In this context, the use of lanthanide-chelating carbohydrate conjugates has emerged in the last years as a novel approach to characterize the conformation of complex oligosaccharides[10,12,13] and to access to relevant details of carbohydrate–protein interactions in solution.[10,14]

The physical basis of this application is well documented. The presence of a lanthanide nucleus in a molecule gives rise to large chemical shift changes in the NMR signals of the nuclei in the vicinity of the lanthanide.

These changes are called pseudo-contact shifts (PCS) and are due to the dipolar interaction between the unpaired electron of the lanthanide and the nuclei around. The effect is distance dependent, as can be seen in eqn (5.1).

$$\Delta\delta_{\mathrm{PCS}} = \frac{1}{12\pi r^3}\left[\Delta\chi_{\mathrm{ax}}(3\cos^2\theta - 1) + 3/2\Delta\chi_{\mathrm{rh}}\sin^2\theta\cos^2\varphi)\right] \tag{5.1}$$

Where $\Delta\delta_{\mathrm{PCS}}$ is the difference in chemical shifts measured between diamagnetic and paramagnetic samples, r is the distance between the metal ion and the nuclear spin and θ and ϕ are the polar coordinates describing the position of the nuclear spin with respect to the principal axes of the magnetic susceptibility tensor ($\Delta\chi$). The effect is proportional to the anisotropy of the magnetic susceptibility tensor and independent of the isotropic component of the χ tensor. It is therefore sufficient to describe the PCS as a function of the axial and rhombic components of the $\Delta\chi$ tensor, $\Delta\chi_{\mathrm{ax}}$ and $\Delta\chi_{\mathrm{rh}}$.

The PCS decrease less rapidly with the distance than the nuclear Overhauser effect. Therefore, it is possible to use these parameters to quantitatively obtain medium- and long-range distance constraints.

Moreover, the lanthanide induces paramagnetic relaxation enhancement effects (PRE) in the surrounding nuclei. The relaxation enhancement decreases with r^{-6}, as the NOE, although its effect is more noticeable. In fact, resonances very close to the lanthanide paramagnetic center show dramatic line broadening and may be frequently hidden under the noise level.

These parameters have been widely employed in the structural biology and chemical biology fields. Pioneering applications of this methodology took advantage of the combined PCS and PRE parameters to refine the 3D structure calculation of metalloproteins. Obviously, metalloproteins already contain a metal ion within its structure and provided that the metal is paramagnetic, PCS and/or PRE effects might be observed. Alternatively, the naturally occurring ion might be replaced by a proper lanthanide in order to generate PCS and PRE effects on the protein signals. Then, these parameters are employed for the structural calculation or refinement. For instance, it is well known that many calcium-binding sites might be suitable for lanthanide recognition.[15]

A chemical variation of this methodology has been designed for its application to the study of non-metalloproteins. In this protocol, a lanthanide chelating unit is designed to be covalently attached to the peptide chain at the specific chosen site. Very frequently, a cysteine residue is employed for anchoring the tag through the formation of a disulfide bond.[16] Alternatively, a well characterized lanthanide-binding-peptide can be appended to the protein C-terminus by regular cloning techniques and employed as the tag for generating the PCS and PRE.[17] In favorable cases, also residual dipolar coupling (RDC) information can be obtained, since the paramagnetic metal may favor one particular orientation of the protein with respect to the magnetic field.[16] Therefore, there are certain advantages in the employment of this methodology to measure RDCs with respect to those that use external orienting media.[18]

This type of approach with lanthanide-binding-peptides has been successfully used to study a variety of intermolecular interactions. In particular, protein–peptide[19] and carbohydrate–protein interactions have been deduced using this protocol.[20] As leading example, the interaction mode of lactose with galectin-3 has been deduced by using a galectin-3 variant modified with a lanthanide binding peptide motif attached to the protein.[20,21] Also from the molecular recognition perspective, other approaches have employed TEMPO-like tags attached to Lys residues at the protein surface to generate strong PRE effects that can then easily be used to screen ligand binding to protein surfaces. This method is known as SLAPSTIC (Spin Labels Attached to Protein Side chains a Tool to identify Interacting Compounds).[22]

However, the modification of the protein with these tags may locally modify the structure and dynamics of the receptor and therefore modulate its recognition properties. Therefore, alternative protocols have also been envisaged, attaching the paramagnetic metal to the interacting ligand.

The first examples of modifications of the ligand with paramagnetic labels also employed TEMPO-type tags.[23,24] These fragments are potent enhancers of PRE and permit to detect molecular interactions at long distances and identification of allosteric modulators by using labeled compounds that target the orthoesteric binding site.[23] For instance, adenine derivatives modified with a nitroxide spin labeled have been applied in the identification of allosteric inhibitors of kinases. The adenine derivative targets the ATP binding site therefore if another kinase ligand binds to a second site will feel the paramagnetic relaxation enhancement effects provided that it is bound simultaneously and the allosteric site is within 15–20 Å.

Carbohydrate derivates modified with nitroxide spin labels have also been described and employed in glycan recognition characterization (compound **1**, Figure 5.1).[24,25]

However, although the employment of TEMPO or its variations permit to easily deduce PRE on the signals of the interacting partner, the access to quantitative distance information is not easy. Moreover, PCS are not available using this methodology.

Therefore, the use of lanthanide-binding-tags to be attached to the glycan chain has been proposed by different research groups. The initial designs were based on the tag proposed by C. Griesinger and coworkers for proteins.[16] First examples were described for 1-β-aminochitobiose derivatives[26,27] and lactose derivatives (See compounds **2** and **4**, Figure 5.1).[28]

In this context, and regarding the design of the lanthanide chelating unit, some considerations should be taken into account. First, it is important to use rigid units in order to minimize conformational averaging and to maintain the metal in a well-defined position of the molecule. Depending on the size of the carbohydrate, it is also important to include a proper linker between the chelating unit and the carbohydrate signals to confine the dramatic PRE effect only within the linker signals. In this manner, the resonance signals of the glycan are still accessible and may be used to deduce their PCS. Finally, it is also important to avoid pseudoasymmetric centers

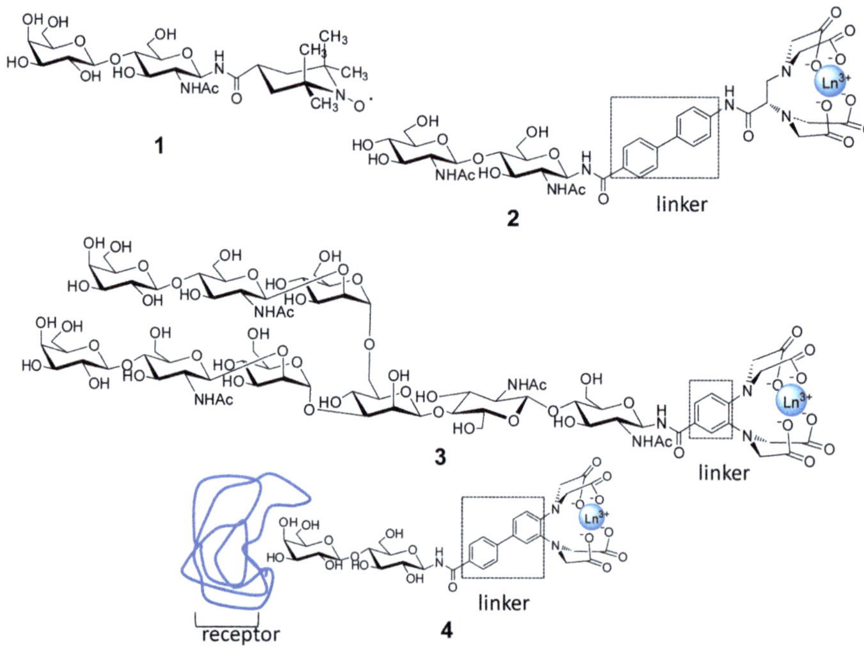

Figure 5.1 Carbohydrate conjugates synthesized for NMR structural characterization. Different linkers were designed depending on the size of the molecule under study. In the case of compound **3**, this conjugate also allows detecting the interaction of lactose derivative with a lactose receptor.

that can give rise to a double set of signals in the NMR spectra of glycans upon addition of the metal ion.[16] Glycans display a tremendous number of stereogenic centers and therefore, the PCS analysis might be impeded.

Initial disaccharide conjugates contain a biphenyl linker between the carbohydrate and the chelating units (compound **2**, Figure 5.1) and a stereogenic center.[26,28] These conjugates allow the validation of lanthanide binding tag methodology in the carbohydrate field showing that it is possible to detect all the carbohydrate signals in the presence of the paramagnetic metal and the experimental PCS values are in good agreement with the values back-calculated from the known geometries of these disaccharides. In addition, one bond $^{1}H-^{13}C$ residual dipolar couplings could be measured for the carbohydrate signals thanks to the partial orientation of the disaccharide induced by the paramagnetic ion.[28]

In a next step, new tags lacking the stereogenic center where designed in order to simply the synthetic process (compounds **3** and **4**, Figure 5.1). This new generation of lanthanide binding tags has shown broad potential in the structural characterization of disaccharides,[14] trisaccharides,[29] tetrasaccharides[13] and even a nonasaccharide (compound **3**, Figure 5.1).[10]

Nowadays, the conformation of both complex type[10] and high mannose *N*-glycans[12] have been successfully studied by using conjugates bearing lanthanide binding tags, a smaller linker was chosen for these molecules due to the higher size of the oligosaccharide (compound **3**, Figure 5.1). In these oligosaccharides the use of lanthanide binding conjugates provide additional benefits due to the possibility of distinguishing all the individual signals of the molecule breaking the natural pseudosymmetry of the oligosaccharides that usually display isochronous chemical shifts for the units located at the end of the arms.[10]

5.3 Characterization of Carbohydrate–Protein Recognition by Using Carbohydrate Conjugates Bearing Lanthanide Binding Tags

The examples mentioned previously have shown that the lanthanide binding tag methodology is a powerful approach in the conformational analysis of carbohydrates. However, lanthanide binding carbohydrate conjugates have further application in the study of carbohydrate–protein interactions.[14]

The advantages of this methodology in the study of carbohydrate recognition rely on a higher sensitivity in the detection of the binding event and the ability of measurement of a higher number of chemical shift perturbations on the protein signals resulting in a better definition of the binding surface. This is especially important in the carbohydrate field since in the absence of multivalent effects the molecules interact with low affinity.[30]

Two approaches can be chosen to study carbohydrate–protein interactions by using paramagnetism, the ligand-based approach or the receptor-based approach. In those cases where the size of the receptor allows direct detection of the protein signals in NMR it is very useful to acquire receptor-based experiments. Isotopic labelled proteins are usually required, at least with ^{15}N, and ^{1}H–^{15}N HSQC experiments adding increasing amounts of carbohydrate conjugate. In order to measure PCSs two experiments are required one with a diamagnetic reference (La^{3+}, Lu^{3+}) and one with a paramagnetic ion (Dy^{3+}, Tb^{3+}, Tm^{3+}), see Figure 5.2c.

This receptor-based approach has been used to characterize the carbohydrate recognition domain (CRD) of human galectin-3 (h-Gal3) in complex with lactose. No modification of the protein is required since the tag is attached to the carbohydrate. This is a great advantage in those cases where proteins cannot be easily being modified for including a cysteine residue. The h-Gal3 was expressed and labeled with ^{15}N, and a series of HSQC spectra were acquired in the presence of an increasing amount of the lactose derivative loaded with dysprosium. Chemical shift changes were detected for residues located in an extended surface of the protein corresponding with the carbohydrate binding surface. It is important to note that a different set of data can be obtained just by changing the paramagnetic ion. The combination of PCS or PRE information with different metals allows a better definition of the system.

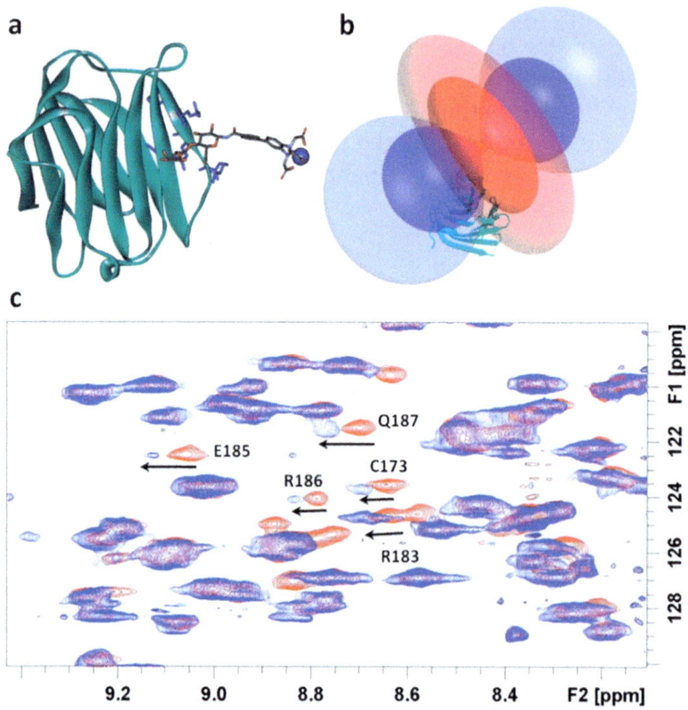

Figure 5.2 (a) Carbohydrate recognition domain of human galectin-3 (h-Gal3) in complex with a lactose derivative bearing a lanthanide binding tag. (b) Isosurfaces of positive (red color) and negative (blue color) PCS expected for h-gal3 in complex with the paramagnetic derivative. (c) Superimposition of the ^{1}H–^{15}N HSQC spectra of h-Gal3 in the presence of lactose conjugate loaded with lanthanum (red color) and h-gal3 in the presence of the lactose conjugate loaded with dysprosium (blue color).[14]

Interestingly, in the h-Gal3/lactose system, PCS values higher than 0.02 were observed for 32 NH signals, chemical shift changes were detected even in residues that are 42 Å apart from the metal core. In addition, an excellent correlation was obtained between experimental data and the PCSs back-calculated from a 3D model built starting from the structure of galectin-3 lactose complex obtained by X-ray crystallography. These data represent an improvement with respect to the results described with nitroxide spin labeled carbohydrate analogs.[24] In that case, only perturbation of residues that fall within a radius of 20 Å from the spin label could be detected and, in addition, only intensity losses could be measured for obtaining structural information. In addition, the PCS and PRE information derived from these experiments can be used to refine the structure of carbohydrate–protein complexes determined by NMR of X-ray methods.

Recognition of complex type biantennary *N*-glycans (compound **2**, Figure 5.3) has been also studied by taking advantage of paramagnetic effects.[10] In this case the system was characterized from the ligand point of view.

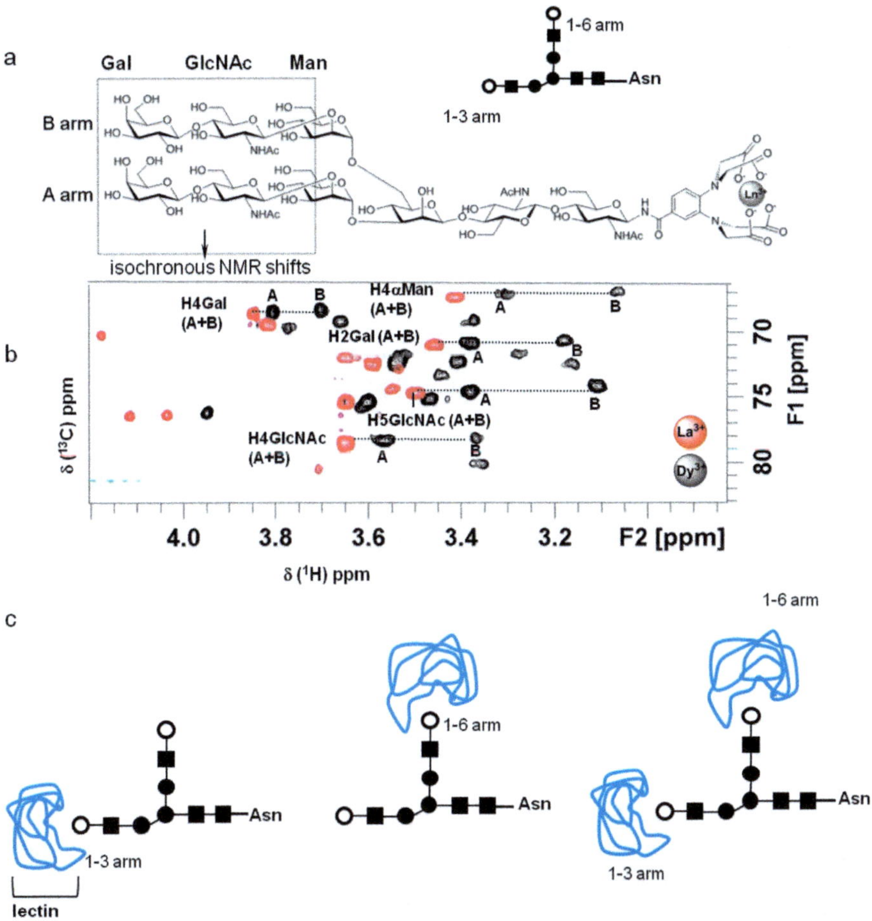

Figure 5.3 (a) Complex type biantennary *N*-glycan structure. The monosaccharide units inside the box display isochronous chemical shifts due to the pseudosymmetry of the molecule. (b) Superimposition of the ^1H–^{13}C HSQC spectra of the compound in the presence of La^{3+} in red and in the presence of Dy^{3+} in black. Several signals are labeled to show that both arms can be individually distinguished in the presence of Dy^{3+}. (c) *N*-glycan–lectin recognition scheme.[10]

Multiantennary *N*-glycans contain an additional point of complexity since the galactose units located in different branches cannot be distinguished with conventional NMR experiments (Figure 5.3). These units display isochronous chemical shifts in NMR due to the similar environment of all of them in the oligosaccharide. The same behavior is observed for the *N*-acetyl-glucosamine units and for part of the alpha-mannose signals. In contrast, the use of the carbohydrate conjugate loaded with a paramagnetic ion allows the unambiguous assignment of all the carbohydrate signals.

Once the assignment is performed it is possible to envision experiments to know which arm is involved in receptor recognition (Figure 5.3C). First experiments have been already described in this line by measuring line broadening of the carbohydrate signals upon addition of h-Gal3.[10] The analysis of the broadening in the HSQC peaks allowed confirming the specific interaction of hGal3 with the Gal and GlcNAc residues at both ends of the A and B arms.

These results open the field to study the recognition of complex type *N*-glycan by other proteins.

5.4 Perspectives

Finally, it is important to remark that the potential of carbohydrates conjugates bearing lanthanide binding tags is under exploration. Different aspects are still under development, such as the design of new tags, the ability to apply this methodology to higher molecular weight oligosaccharides and the synthesis of carbohydrates conjugates suitable for the study of lectins with metal binding sites. This last application is especially important since several lectins with paramount biological relevance, as C-type lectins, contain calcium binding sites.

References

1. H. J. Gabius, *The Sugar Code: Fundamentals of Glycosciences*, Wiley VCH, Weinheim, 2009.
2. E. Munoz, D. Xu, M. Kemp, F. Zhang, J. Liu and R. J. Linhardt, Affinity, kinetic, and structural study of the interaction of 3-O-sulfotransferase isoform 1 with heparan sulfate, *Biochemistry*, 2006, **45**, 5122–5128.
3. A. Viegas, J. Sardinha, F. Freire, D. F. Duarte, A. L. Carvalho, C. M. Fontes, M. J. Romao, A. L. Macedo and E. J. Cabrita, Solution structure, dynamics and binding studies of a family 11 carbohydrate-binding module from Clostridium thermocellum (CtCBM11), *Biochem. J.*, 2013, **451**, 289–300.
4. C. Fernandez-Tornero, R. M. Lozano, G. Rivas, M. A. Jimenez, L. Standker, D. Diaz-Gonzalez, W. G. Forssmann, P. Cuevas, A. Romero and G. Gimenez-Gallego, Synthesis of the blood circulating C-terminal fragment of insulin-like growth factor (IGF)-binding protein-4 in its native conformation. Crystallization, heparin and IGF binding, and osteogenic activity, *J. Biol. Chem.*, 2005, **280**, 18899–18907.
5. D. J. Scott, S. E. Harding and A. J. Rowe, *Analytical Ultracentrifugation: Techniques and Methods*, RSC Publishing Royal Society of Chemistry, 2005.
6. P. L. Abdian, J. J. Caramelo, N. Ausmees and A. Zorreguieta, RapA2 is a calcium-binding lectin composed of two highly conserved cadherin-like domains that specifically recognize Rhizobium leguminosarum acidic exopolysaccharides, *J. Biol. Chem.*, 2013, **288**, 2893–2904.

7. D. Solis, M. J. Mate, M. Lohr, J. P. Ribeiro, L. Lopez-Merino, S. Andre, E. Buzamet, F. J. Canada, H. Kaltner, M. Lensch, F. M. Ruiz, G. Haroske, U. Wollina, M. Kloor, J. Kopitz, J. L. Saiz, M. Menendez, J. Jimenez-Barbero, A. Romero and H. J. Gabius, N-domain of human adhesion/growth-regulatory galectin-9: preference for distinct conformers and non-sialylated N-glycans and detection of ligand-induced structural changes in crystal and solution, *Int. J. Biochem. Cell Biol.*, 2010, **42**, 1019–1029.
8. A. Canales, R. Lozano, B. Lopez-Mendez, J. Angulo, R. Ojeda, P. M. Nieto, M. Martin-Lomas, G. Gimenez-Gallego and J. Jimenez-Barbero, Solution NMR structure of a human FGF-1 monomer, activated by a hexasaccharide heparin-analogue, *FEBS J.*, 2006, **273**, 4716–4727.
9. Y. Ha, D. J. Stevens, J. J. Skehel and D. C. Wiley, X-ray structures of H5 avian and H9 swine influenza virus hemagglutinins bound to avian and human receptor analogs, *Proc. Natl. Acad. Sci. U. S. A.*, 2001, **98**, 11181–11186.
10. A. Canales, A. Mallagaray, J. Perez-Castells, I. Boos, C. Unverzagt, S. Andre, H. J. Gabius, F. J. Canada and J. Jimenez-Barbero, Breaking pseudo-symmetry in multiantennary complex N-glycans using lanthanide-binding tags and NMR pseudo-contact shifts, *Angew. Chem.*, 2013, **52**, 13789–13793.
11. T. Zhuang, H. Leffler and J. H. Prestegard, Enhancement of bound-state residual dipolar couplings: conformational analysis of lactose bound to Galectin-3, *Protein Sci.*, 2006, **15**, 1780–1790.
12. T. Yamaguchi, Y. Sakae, Y. Zhang, S. Yamamoto, Y. Okamoto and K. Kato, Exploration of conformational spaces of high-mannose-type oligosaccharides by an NMR-validated simulation, *Angew. Chem.*, 2014, **53**, 10941–10944.
13. Y. Zhang, S. Yamamoto, T. Yamaguchi and K. Kato, Application of paramagnetic NMR-validated molecular dynamics simulation to the analysis of a conformational ensemble of a branched oligosaccharide, *Molecules*, 2012, **17**, 6658–6671.
14. A. Canales, A. Mallagaray, M. A. Berbis, A. Navarro-Vazquez, G. Dominguez, F. J. Canada, S. Andre, H. J. Gabius, J. Perez-Castells and J. Jimenez-Barbero, Lanthanide-chelating carbohydrate conjugates are useful tools to characterize carbohydrate conformation in solution and sensitive sensors to detect carbohydrate-protein interactions, *J. Am. Chem. Soc.*, 2014, **136**, 8011–8017.
15. K. Barthelmes, A. M. Reynolds, E. Peisach, H. R. Jonker, N. J. DeNunzio, K. N. Allen, B. Imperiali and H. Schwalbe, Engineering encodable lanthanide-binding tags into loop regions of proteins, *J. Am. Chem. Soc.*, 2011, **133**, 808–819.
16. A. Leonov, B. Voigt, F. Rodriguez-Castaneda, P. Sakhaii and C. Griesinger, Convenient synthesis of multifunctional EDTA-based chiral metal chelates substituted with an S-mesylcysteine, *Chem.–Eur. J.*, 2005, **11**, 3342–3348.

17. T. Saio, K. Ogura, K. Shimizu, M. Yokochi, T. R. Burke Jr and F. Inagaki, An NMR strategy for fragment-based ligand screening utilizing a paramagnetic lanthanide probe, *J. Biomol. NMR*, 2011, **51**, 395–408.
18. A. Canales, J. Jimenez-Barbero and M. Martin-Pastor, Review: Use of residual dipolar couplings to determine the structure of carbohydrates, *Magn. Reson. Chem.*, 2012, **50**(suppl. 1), S80–S85.
19. T. Saio, K. Ogura, M. Yokochi, Y. Kobashigawa and F. Inagaki, Two-point anchoring of a lanthanide-binding peptide to a target protein enhances the paramagnetic anisotropic effect, *J. Biomol. NMR*, 2009, **44**, 157–166.
20. T. Zhuang, H. S. Lee, B. Imperiali and J. H. Prestegard, Structure determination of a Galectin-3-carbohydrate complex using paramagnetism-based NMR constraints, *Protein Sci.*, 2008, **17**, 1220–1231.
21. M. Nitz, K. J. Franz, R. L. Maglathlin and B. Imperiali, A powerful combinatorial screen to identify high-affinity terbium(III)-binding peptides, *ChemBioChem*, 2003, **4**, 272–276.
22. W. Jahnke, S. Rudisser and M. Zurini, Spin label enhanced NMR screening, *J. Am. Chem. Soc.*, 2001, **123**, 3149–3150.
23. W. Jahnke, M. J. Blommers, C. Fernandez, C. Zwingelstein and R. Amstutz, Strategies for the NMR-based identification and optimization of allosteric protein kinase inhibitors, *ChemBioChem*, 2005, **6**, 1607–1610.
24. N. U. Jain, A. Venot, K. Umemoto, H. Leffler and J. H. Prestegard, Distance mapping of protein-binding sites using spin-labeled oligosaccharide ligands, *Protein Sci.*, 2001, **10**, 2393–2400.
25. P. E. Johnson, E. Brun, L. F. MacKenzie, S. G. Withers and L. P. McIntosh, The cellulose-binding domains from Cellulomonas fimi beta-1, 4-glucanase CenC bind nitroxide spin-labeled cellooligosaccharides in multiple orientations, *J. Mol. Biol.*, 1999, **287**, 609–625.
26. A. Mallagaray, A. Canales, G. Dominguez, J. Jimenez-Barbero and J. Perez-Castells, A rigid lanthanide binding tag for NMR structural analysis of carbohydrates, *Chem. Commun. (Camb.)*, 2011, **47**, 7179–7181.
27. S. Yamamoto, T. Yamaguchi, M. Erdelyi, C. Griesinger and K. Kato, Paramagnetic lanthanide tagging for NMR conformational analyses of N-linked oligosaccharides, *Chem.–Eur. J.*, 2011, **17**, 9280–9282.
28. M. Erdelyi, E. d'Auvergne, A. Navarro-Vazquez, A. Leonov and C. Griesinger, Dynamics of the glycosidic bond: conformational space of lactose, *Chem.–Eur. J.*, 2011, **17**, 9368–9376.
29. S. Yamamoto, Y. Zhang, T. Yamaguchi, T. Kameda and K. Kato, Lanthanide-assisted NMR evaluation of a dynamic ensemble of oligosaccharide conformations, *Chem. Commun. (Camb.)*, 2012, **48**, 4752–4754.
30. H. J. Gabius, Glycans: bioactive signals decoded by lectins, *Biochem. Soc. Trans.*, 2008, **36**, 1491–1496.

CHAPTER 6

NMR Characterization of the Conformations, Dynamics, and Interactions of Glycosphingolipids

MAHO YAGI-UTSUMI[a,b], TAKUMI YAMAGUCHI[a,b,c], YOSHINORI UEKUSA[a] AND KOICHI KATO*[a,b]

[a]Okazaki Institute for Integrative Bioscience and Institute for Molecular Science, National Institutes of Natural Sciences, Higashiyama 5-1, Myodaiji, Okazaki 444-8787, Japan; [b]Graduate School of Pharmaceutical Sciences, Nagoya City University, Tanabe-dori 3-1, Mizuho-ku, Nagoya 467-8603, Japan; [c]School of Materials Science, Japan Advanced Institute of Science and Technology, 1-1 Asahidai, Nomi 923-1292, Japan
*E-mail: kkatonmr@ims.ac.jp

6.1 Introduction

Molecular interactions involving carbohydrate chains are often characterized by their intrinsically low affinities, which can be avidly enhanced through the formation of multivalent clusters. These cluster effects are best exemplified by glycosphingolipids on cell surfaces.[1] Gangliosides, which are glycosphingolipids containing at least one sialyl group (Figure 6.1), are major components of membrane microdomains and also provide acceptor sites for viruses and bacterial toxins, thereby mediating a variety of physiological and pathological

New Developments in NMR No.10
NMR in Glycoscience and Glycotechnology
Edited by Koichi Kato and Thomas Peters

Published by the Royal Society of Chemistry, www.rsc.org

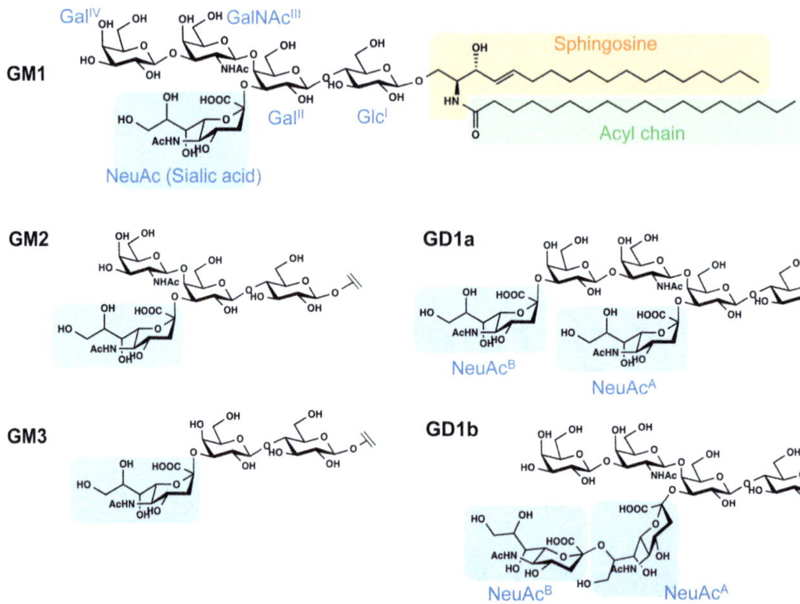

Figure 6.1 Structure of the entire GM1 ganglioside and the carbohydrate structures of GM2, GM3, GD1a, and GD1b gangliosides. Glc, glucose; Gal, galactose; GalNAc, *N*-acetylgalactosamine; NeuAc, *N*-acetylneuraminic acid.

events.[2–5] Accumulating evidence indicates that ganglioside clusters on neuronal cell surfaces provide sites for the association of amyloidogenic proteins involved in neurodegenerative diseases, such as amyloid beta (Aβ), and promote their conformational transition and subsequent deposition.[6,7] To gain deeper insights into the ongoing molecular processes in ganglioside clusters, structural and dynamic views of gangliosides and their assemblies are essential.

Crystallographic approaches to these systems are hampered by the dynamic properties of the oligosaccharide moieties of gangliosides and also the fluidity of their assemblies. In contrast, nuclear magnetic resonance (NMR) techniques have successfully been applied to the conformational characterization of gangliosides in solution and in membrane-like environments.[8] NMR spectroscopy also offers powerful tools for probing dynamic glycolipid clusters and their specific interactions with proteins. This chapter outlines NMR applications for investigating the conformations, dynamics, and interactions of glycosphingolipids represented by gangliosides.

6.2 Sample Preparation for Solution NMR Experiments

Needless to say, specific molecular recognition by gangliosides is exerted primarily through their carbohydrate moieties. Therefore, structural studies on gangliosides have often focused on the conformational properties of their glycan components. For example, the pentasaccharide moiety of GM1 ganglioside carries determinants recognized by cholera toxin and polyoma virus, as visualized by X-ray crystallographic data obtained using a liberated glycan.[9,10] In early studies, conformational analyses of ganglioside glycans were performed using nuclear Overhauser effect (NOE) data as restraints.[11–16] In these solution NMR approaches, the target systems have to be designed to be of appropriate size because gangliosides form huge aggregates in aqueous environments, as exemplified by the 140 kDa micelles of GM1[17,18] (Figure 6.2). The most conventional method is to isolate the carbohydrate moieties, as was done in crystallographic studies. In addition to synthetic methods, the glycan moiety can be prepared by enzymatic or chemical cleavage from gangliosides.[12,14,19–22] An alternative approach is to solubilize the ganglioside aggregates. Organic solvents, such as dimethyl sulfoxide (DMSO), have traditionally been used for this purpose.[13,23,24] Host–guest methods can also be used to isolate gangliosides. For example, GM1 can be isolated even in aqueous solutions by adding α-cyclodextrin, which encapsulates the hydrophobic lipid moieties in its hydrophobic cavity.[17] A variety of smaller membrane-mimic systems have also been used for solution NMR analyses, including mixed micelles[12,19,25] and smaller bicelles[26–28] (*vide infra*). Lysogangliosides lack the acyl chain and can therefore form smaller aqueous micelles that are suitable for high-resolution NMR analyses[29–31] (Figure 6.3).

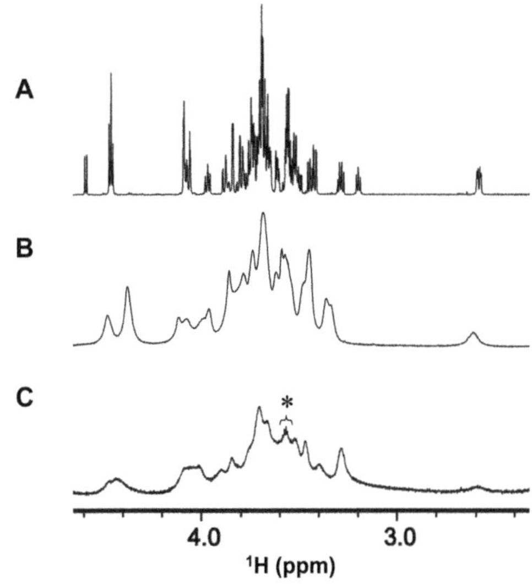

Figure 6.2 One-dimensional ^{1}H-NMR spectra of (A) liberated GM1 pentasaccharide (1.0 kDa), (B) lyso–GM1 micelle (approximate molecular mass, 60 kDa), and (C) GM1 micelle (approximate molecular mass, 140 kDa). Peaks originating from a low molecular weight contaminant are indicated by asterisks. The spectra were recorded at 300 K using a Bruker Avance-800 spectrometer (at Instrument Center, IMS).

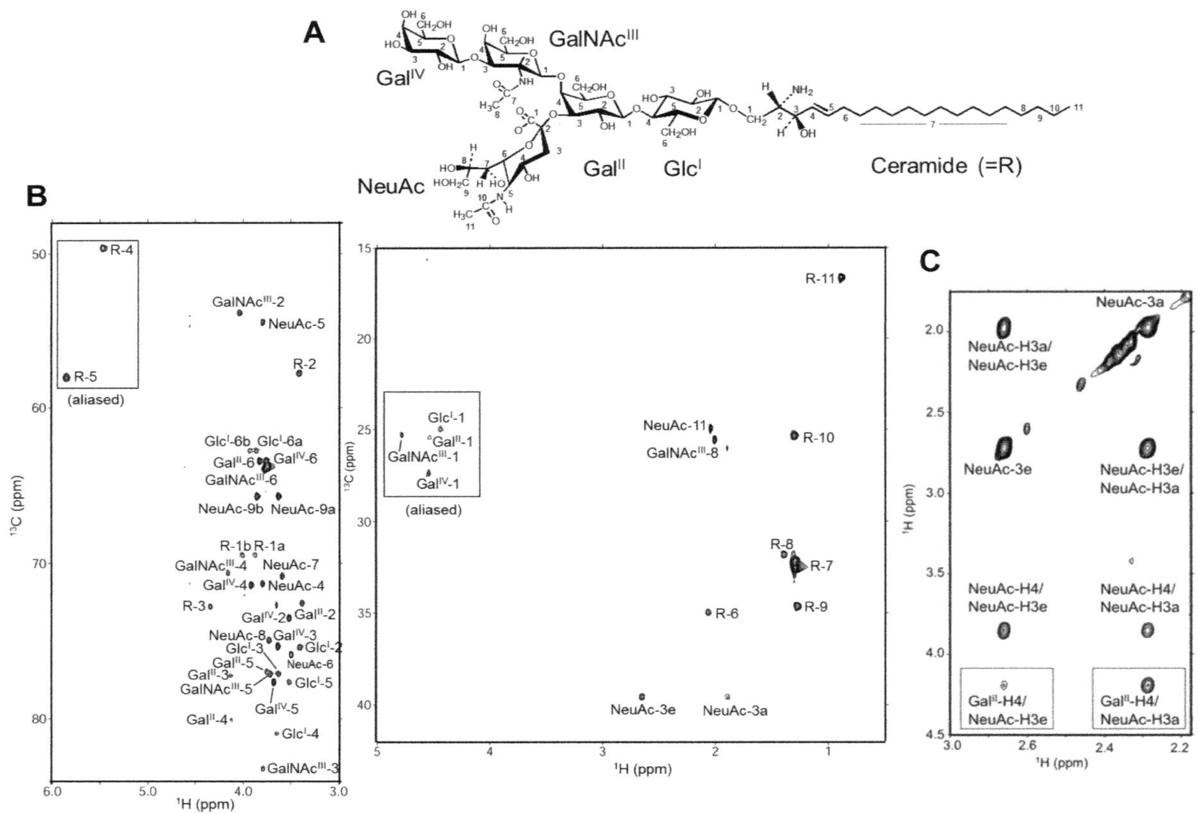

Figure 6.3 (A) The chemical structure of lyso–GM1 ganglioside. ^{1}H–^{13}C HSQC spectra (B) and a part of the ^{1}H–^{1}H NOESY spectrum (C) of aqueous lyso–GM1 micelles. (A) Reprinted from Yagi-Utsumi *et al.*[30] with permissions from John Wiley and Sons. © 2015 Federation of European Biochemical Societies. (B) Reproduced from *Glycobiology of the Nervous System*, Structural Analysis of Oligosaccharides and Glycoconjugates Using NMR, vol. 9, 2014, pp. 165–183, Yoshiki Yamaguchi, Takumi Yamaguchi and Koichi Kato, (Copyright © 2014, Springer Science+Business Media New York). With permission of Springer.

6.3 Dynamic Conformations of Ganglioside Glycans

The carbohydrate moieties of glycoconjugates generally possess substantial degrees of freedom in internal motion, which is, at least partially, associated with their weak interactions due to the entropic penalty upon binding their target molecules. NMR spectroscopy is potentially useful for illustrating the dynamic conformations of the oligosaccharide chains.[32] However, conformational information offered by NMR experiments is averaged over the dynamic conformational ensembles of these flexible biomolecules. In contrast, molecular dynamics (MD) simulations potentially describe individual conformers in the ensembles, possible conformational spaces, or even landscapes of oligosaccharides in solution. However, the simulation results significantly depend on the calculation protocols employed, including force fields and initial states. NMR information is useful for validating the results of MD simulations. Recently developed lanthanide-tagging methods have enabled NMR validation of the conformational spaces of oligosaccharides derived from MD simulation through the observation of pseudocontact shifts (PCSs)[33–35] (Figure 6.4). This method was used to explore the conformational spaces occupied by a series of ganglioside oligosaccharides, *i.e.* GM3 trisaccharide, GM2 tetrasaccharide, and GM1 pentasaccharide[21,22,36] (Figure 6.5). Indeed, 100 ns simulations with the GLYCAM_06 force field were necessary to obtain conformational ensembles of GM2 and GM3 oligosaccharides that satisfy the experimentally observed PCS values. A generalized-ensemble algorithm with a simulation time of 1.9 μs was employed to reproduce the experimental PCS values of the GM1 pentasaccharide. Although the sialyl moieties of the GM1 and GM2 gangliosides shared similar conformations, distinct sialyl linkage conformations were more populated in GM3 because the *N*-acetylgalactosamine residue in GM1 and GM2 restricts the conformational freedom of the sialyl linkage. This suggests that the branched structures of ganglioside glycans affect protein binding processes. Moreover, each of the conformers of the GM1 pentasaccharide determined in the previously reported crystal structures in complex with proteins were found in the conformational space occupied by this glycan in its uncomplexed state. This suggests conformational selection by the proteins during the ganglioside recognition process (Figure 6.4).

6.4 Ganglioside Interactions in Membrane Environments

6.4.1 Lateral Carbohydrate Interactions

Gangliosides and other glycolipids exert their biological functions through cluster formation in membrane environments. Therefore, to better understand the molecular mechanisms of these glycolipid functions, it is essential to develop NMR methods that combine appropriate membrane mimics with sophisticated spectral measurements. Demarco *et al.* applied a paramagnetism-assisted NMR method for the conformational characterization of GM1

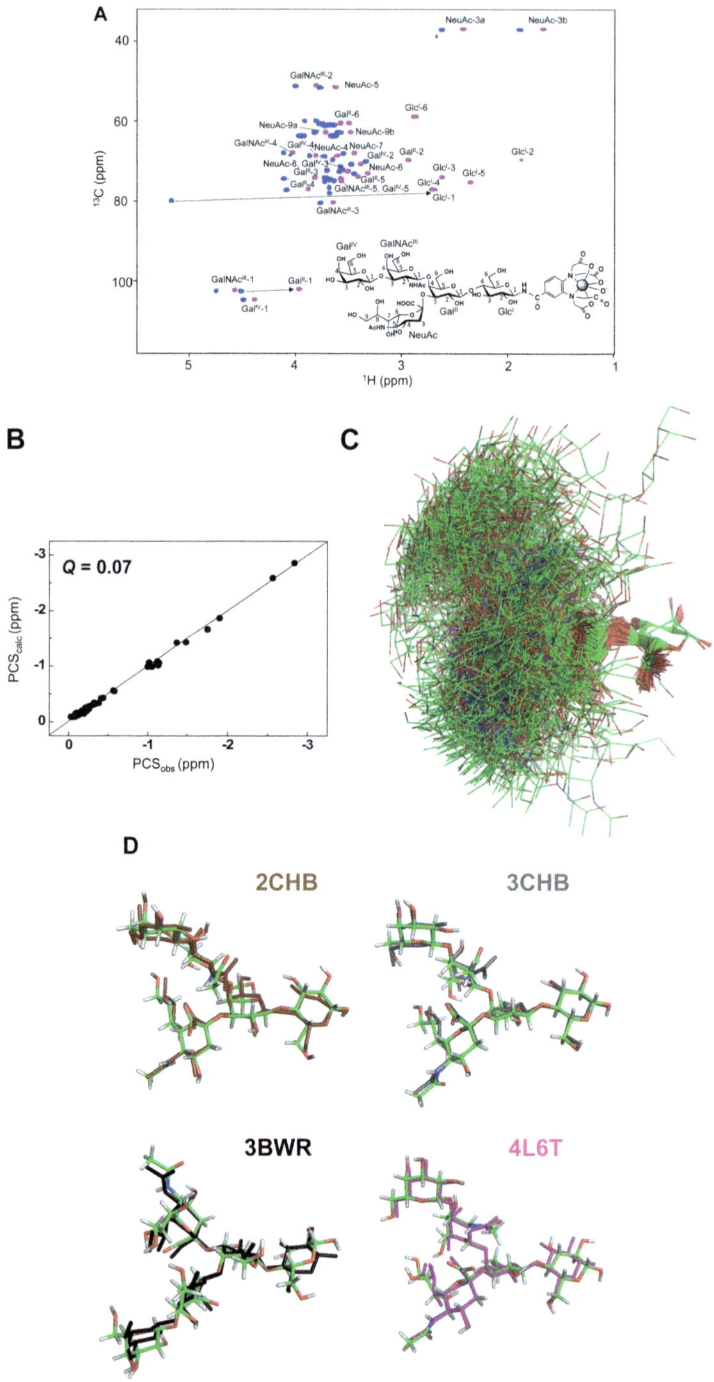

Figure 6.4 Characterization of the dynamic conformations of the GM1 pentasaccharide by paramagnetism-assisted NMR analyses. (A) A phenylenediaminetetraacetic acid-based lanthanide-chelating tag was attached, and

embedded in phospholipid bicelles.[26] The paramagnetic relaxation enhancement (PRE) observed for GM1 using water-soluble and membrane-anchored paramagnetic probes was interpreted by combining the NMR data with MD simulations. The results illustrate the overall structures of GM1 in the membrane in which the outer galactose (Gal) and *N*-acetylneuraminic acid (NeuAc) residues are most accessible. This view is consistent with their involvement in interactions with toxins and viruses as their primary acceptor sites.

It has been supposed that ganglioside molecules in membranes interact with each other and with other membrane components with certain specificities, forming functional clusters exemplified by membrane microdomains.[3,37] Some autoantibodies detected in sera from patients with Guillain–Barré syndrome have epitopes comprising specific combinations of two gangliosides.[38]

Specific ganglioside–ganglioside interactions could be detected on the basis of NMR spectral changes observed on mixing two different gangliosides in a mixed micelle system. Figure 6.6 shows the ^{1}H–^{13}C HSQC spectra of dodecylphosphocholine (DPC) micelles containing a mixture of GD1b and GD1a, GD1b and GM1, or GD1b and asialo-GM1 compared with those containing each single ganglioside species. On mixing with GD1b, GD1a and GM1 exhibited significant chemical shift changes for peaks originating from Gal^{II}–H3 and $NeuAc^{A}$–H8, although no significant spectral changes were induced for asialo-GM1 on mixing GD1b. This implies the successful detection of specific carbohydrate–carbohydrate interactions of GD1b with GD1a and GM1 ganglioside oligosaccharides. Indeed, the chemical shifts of the Gal^{II} and $NeuAc^{A}$ moieties of GD1a and GM1 were perturbed upon their interactions with GD1b. These results indicate that the sialyl moiety of GD1a and GM1 are involved in their specific interactions with GD1b.

The lateral carbohydrate–carbohydrate interactions between gangliosides are generally weak and transient, but they can be enhanced by connecting the ganglioside molecules with a covalent linker. These synthetic

---

^{1}H–^{13}C HSQC spectra of 1 : 1 complexes of the pentasaccharide with paramagnetic Tm^{3+} (*magenta*) or diamagnetic La^{3+} (*blue*) were recorded. Chemical shift differences of the anomeric CH groups induced by PCS are indicated by arrows. (B) Correlation between the experimentally observed PCS values and back-calculated PCS data. The PCS values were back-calculated from the replica-exchange MD-derived ensemble models. (C) Superimposition of 200 conformers extracted from the NMR-validated simulation results. (D) The lowest-RMSD structure selected from the conformational ensemble is superposed on the 3D structures of the carbohydrate moieties of GM1 obtained by crystallographic analyses [PDB codes: 2CHB (brown), 3CHB (gray), 3BWR (black), and 4L6T (magenta)]. Reproduced from Springer *Biochemical Roles of Eukaryotic Cell Surface Macromolecules*, Conformational Dynamics of Oligosaccharides Characterized by Paramagnetism-Assisted NMR Spectroscopy in Conjunction with Molecular Dynamics Simulation, vol. 842, 2015, pp. 217–230, Y. Zhang, T. Yamaguchi, T. Satoh, M. Yagi-Utsumi, Y. Kamiya, Y. Sakae, Y. Okamoto and K. Kato, (© Springer International Publishing Switzerland 2015) and *Glycoconjugate Journal*, Structural and dynamic views of GM1 ganglioside, vol. 32, pp. 105–112, M. Yagi-Utsumi, K. Kato, (© Springer Science+Business Media New York 2015). With permission of Springer.

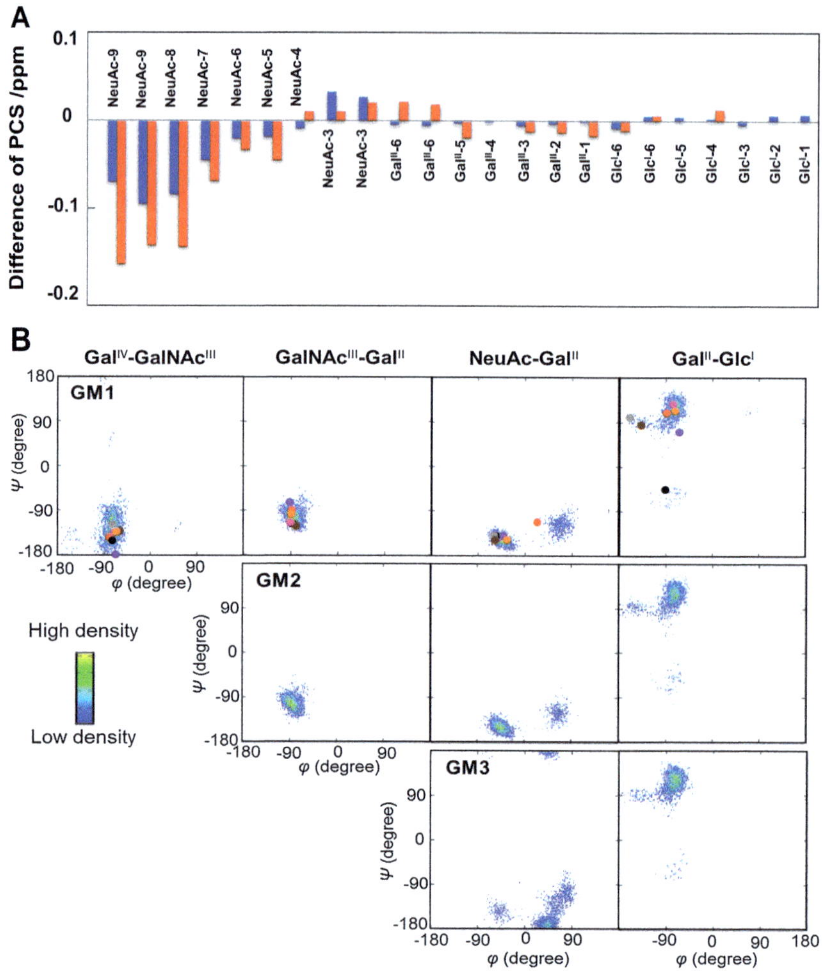

Figure 6.5 (A) The differences in proton PCS values shown as a subtraction of $\Delta\delta^1$H of the GM2 tetrasaccharide from $\Delta\delta^1$H of the GM3 trisaccharide tagged with Tm^{3+} (blue) or Tb^{3+} (red). n.d., not detected due to extensive line broadening. (B) Density map of glycosidic linkage torsion angles in the experimentally validated MD trajectories of the GM1, GM2, and GM3 oligosaccharides. The glycosidic torsions of the carbohydrate moieties of GM1 in DMSO calculated using NOE-based molecular mechanics[12] (●), the carbohydrate moieties of GM1 in aqueous solution calculated by NOE-based molecular modeling[11] (●), and GM1–acetyl in aqueous solution calculated by MD simulation[16] (●) are plotted along with those of the GM1 pentasaccharide in complex with cholera toxin [PDB code: 2CHB (●) and 3CHB (●)], polyoma virus [3BWR (●)], and AB_5 toxin [4L6T (●)]. The definitions of φ and ψ used for the Gal^{IV}–GalNAc^{III}, GalNAc^{III}–Gal^{II}, and Gal^{II}–Glc^{I} linkage were (φ = O5–C1–O1–C′n, ψ = C1–O1–C′n–C′n−1), and those for the NeuAc–Gal^{II} linkage were (φ = O6–C2–O1–C′n, ψ = C2–O1–C′n–C′n−1). Reproduced with permission from Zhang *et al.*[22] (© 2012 by the authors); Biochemical Roles of Eukaryotic Cell Surface Macromolecules, Conformational Dynamics of Oligosaccharides Characterized by Paramagnetism-Assisted NMR Spectroscopy in Conjunction with Molecular Dynamics Simulation, vol. 842, 2015, pp. 217–230, Y. Zhang, T. Yamaguchi, T. Satoh, M. Yagi-Utsumi, Y. Kamiya, Y. Sakae, Y. Okamoto and K. Kato, (© Springer International Publishing Switzerland 2015) with permission of Springer;[8] and Glycoconjugate Journal, Structural and dynamic views of GM1 ganglioside, vol. 32, pp. 105–112, M. Yagi-Utsumi, K. Kato, (© Springer Science+Business Media New York 2015). With permission of Springer.[36]

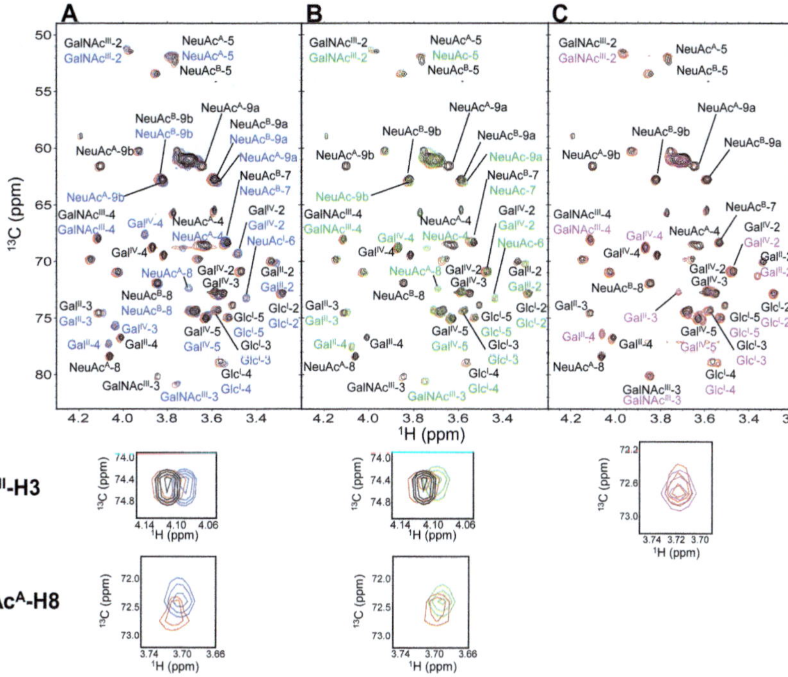

Figure 6.6 ^{1}H–^{13}C HSQC spectra of GD1b (black), GD1a (blue), GM1 (green), asialo-GM1 (magenta) (2 mM each) or their mixtures (GD1b and GD1a (A), GD1b and GM1 (B), and GD1b and asialo-GM1 (C)) embedded in DPC-d_{38} (104 mM) micelles in 10 mM phosphate-buffered (pH 7.0) D_2O solution (red). The spectra were recorded at 303 K using a JEOL ECA-920 spectrometer (at Instrument Center, IMS). The peaks originating from Gal^{II}–H3 of GD1a and GM1 became overlapped with that of GD1b after mixing.

hybrid gangliosides, containing two or more oligosaccharide chains, mimic the cluster of glycolipids on cell membranes and can represent a tool for characterizing carbohydrate–carbohydrate interaction using solution NMR techniques.[39]

6.4.2 Lateral Ceramide Interactions

Gangliosides can form certain ordered phases known as microdomains in the presence of phospholipids, sphingomyelin (SM), and cholesterol (Chol).[40–42] Recent studies have underscored the functional importance of microdomains involving gangliosides in physiological and pathological contexts.[43] Although SM and Chol are considered essential for the formation of microdomains, the types of molecular interactions involved in this process remain to be completely elucidated. One of the most promising approaches to address this issue is the application of solid-state NMR techniques assisted by site-specific isotope labeling.[44]

Matsumori and co-workers successfully characterized the conformation of SM in a membrane environment using NMR spectroscopy. On the basis of coupling constants and NOE correlations observed in small bicelles, it was revealed that SM molecules adopt a relatively stable conformation.[45] Moreover, the orientation and order of SM in bilayers were determined based on C–N dipole couplings and chemical-shift anisotropies obtained from solid-state NMR techniques, including CP–MAS and REDOR spectral measurements.[46] These measurements suggest possible intermolecular hydrogen bond formation involving SM's amide group. They also revealed that the orientation of the amide group of SM was minimally affected by Chol; however, its order parameter significantly increased in the presence of Chol. Their group synthesized a series of site-specific deuterated SM analogs and thereby observed the ^{2}H quadrupole splitting of stearoyl–SM in a systematic manner, indicating that Chol causes ordering of the middle portions of the sphingosine and the acyl chains[47,48] (Figure 6.7). These findings imply that the alicyclic skeleton of Chol

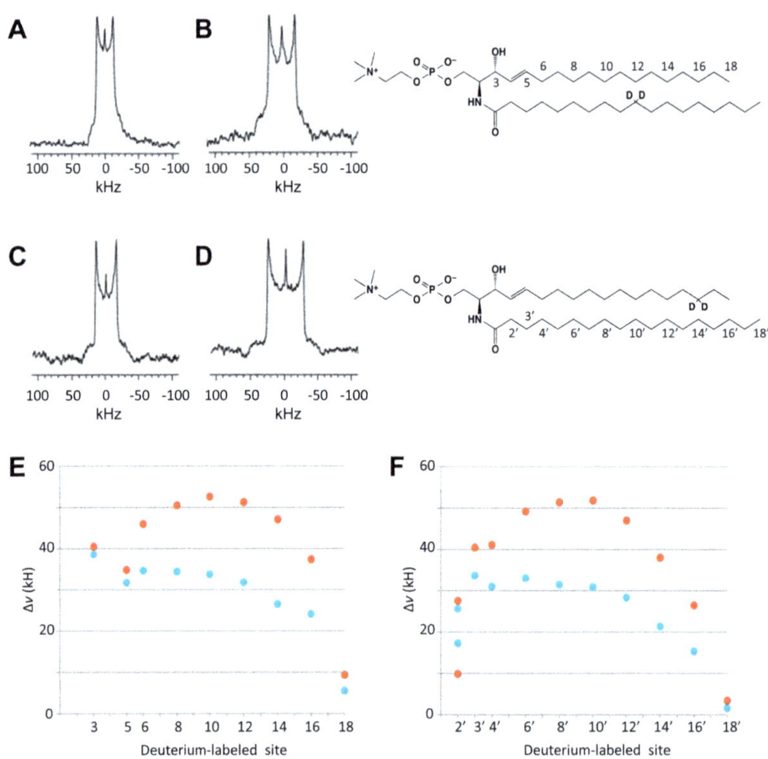

Figure 6.7 (A–D) ^{2}H-NMR spectra of stearoyl-SM derivatives deuterated at C16 of the sphingosine chain (A and B) and at C10' of stearoyl (C and D) chain in the presence (A and C) or absence (B and D) of Chol (50 mol%) at 45 °C, (E and F). The profile of quadrupole splitting magnitude obtained from a series of site-specifically deuterium-labeled SM derivatives deuterated on the sphingosine (E) and stearoyl (F) chains in the absence (blue circles) and presence (red ones) of Chol. Reprinted with permission from Matsumori *et al.*[47] Copyright (2012) American Chemical Society.

is preferentially located near the middle portions of the stearoyl–SM hydrocarbon chains, thereby decreasing their motional freedom.

Thus, NMR provides detailed insights into the formation of glycosphingolipid clusters in membrane environments. The ceramide parts have an intrinsic affinity because of intermolecular hydrogen bonding at the amide group, which can be promoted through Chol-induced ordering of the alkyl chains coupled with enhanced membrane packing. The specific carbohydrate–carbohydrate interactions between gangliosides may contribute to their cluster formation as well as to the consequent biological functions of membrane microdomains.

6.5 Ganglioside Clusters as a Platform for Protein Interactions

Accumulating evidence has indicated that ganglioside clusters on neuronal cell surfaces act as platforms for amyloid formation by the intrinsically disordered proteins involved in neurodegenerative diseases, such as Aβ in Alzheimer's disease and α-synuclein (αSN) in Parkinson's disease.[6,7,49,50] These proteins interact with specific ganglioside clusters and subsequently undergo conformational changes, resulting in the formation of their toxic, aggregated forms. Therefore, the ganglioside cluster is a potential therapeutic target for various intractable diseases.[51]

A series of NMR studies have characterized the interactions of Aβ(1–40) with GM1 clusters using aqueous micelles composed of GM1 or lyso–GM1 as membrane mimics.[30,52,53] The PRE and chemical shift perturbation data highlight the importance of the carbohydrate–ceramide interface of the ganglioside clusters for accommodating Aβ.[30] On addition a spin-labeled Aβ(1–40) analog, which contained an extra C-terminal cysteine residue modified with a nitroxide radical, the peaks originating from the hydrophilic–hydrophobic interface of the micelles exhibited significant attenuation due to PRE (Figure 6.8). This result is consistent with the formation of the amphiphilic up-and-down topology in Aβ–GM1 clusters, which has been characterized by chemical shift perturbation and transverse relaxation-optimized spectroscopy-based saturation transfer data[52] (Figure 6.9). It was further demonstrated that a thioflavin T-reactive β-structure is formed at the C-terminal segment of Aβ(1–40) when present at higher density on GM1 micelles.[53] These findings suggest that the ganglioside clusters offer a stage at the interface between the hydrophilic glycan and the hydrophobic ceramide for binding coupled with α-helix formation by Aβ molecules. Consequently, their spatial rearrangements are restricted to promote specific intermolecular interactions leading to amyloid formation.

Moreover, these NMR approaches can also provide detailed information regarding the interactions of hereditary Aβ mutants, which possess a single amino acid substitution and thereby exhibit different physicochemical and pathological properties, on ganglioside clusters, including ganglioside

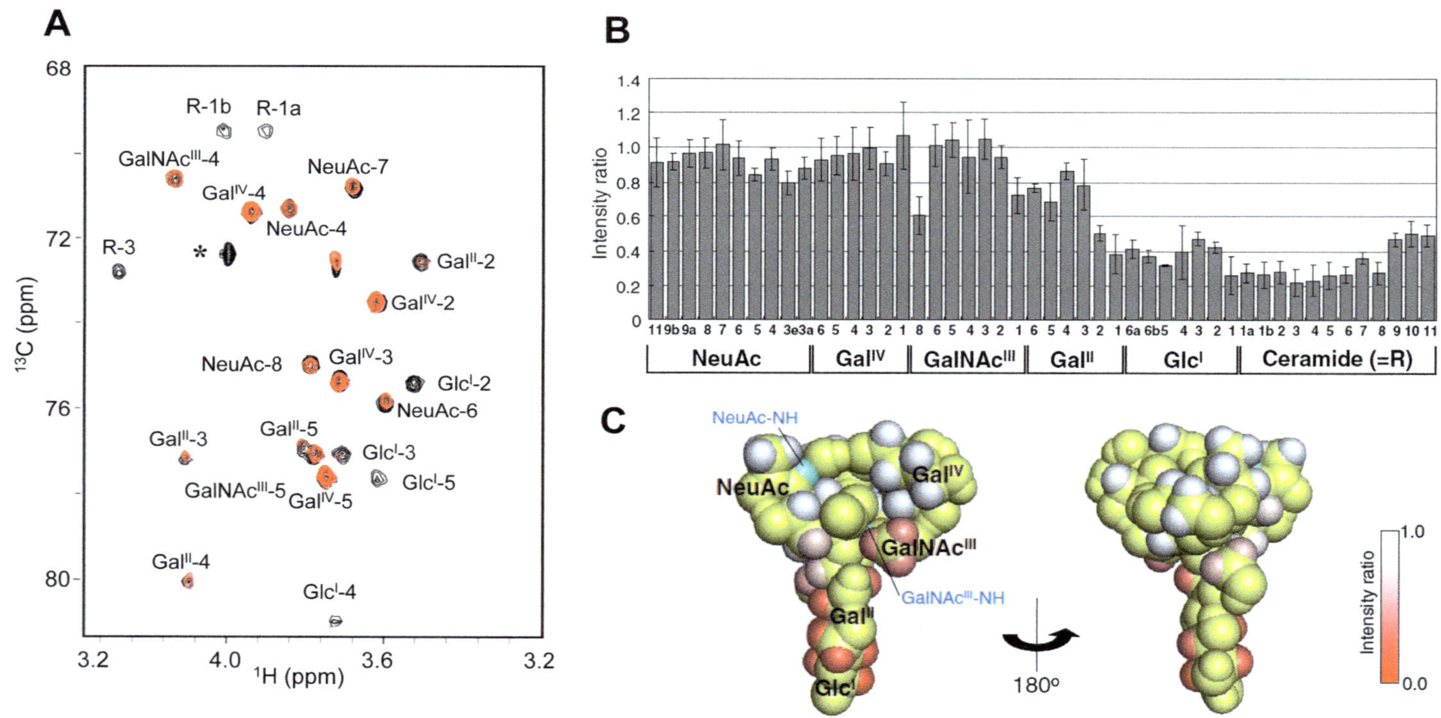

Figure 6.8 PRE effects observed for lyso–GM1 micelles in the presence of a spin-labeled Aβ(1–40) analog. (A) A part of the ^{1}H–^{13}C HSQC spectra of aqueous lyso–GM1 micelles in the presence of spin-labeled Aβ(1–40) before (black) and after (red) radical quenching. Asterisk indicates the peak originating from ascorbic acid. (B) Plots of the intensity ratios of the CH peaks of lyso–GM1 in the presence of spin-labeled Aβ(1–40) before and after radical quenching. The lower intensity ratios are indicative of closer distance to the unpaired electron of the spin label. Intensity ratios are the mean ± S.D. of three independent experiments. (C) Mapping of non-exchangeable hydrogens showing the PRE effects on the lowest-penalty 3D model of the carbohydrate moiety of lyso–GM1 with a linear color gradient (from red to white). Other atoms are colored yellow. Parts of this figure Reprinted from FEBS Letters, vol. 548, M. Yagi-Utsumi, T. Kameda, Y. Yamaguchi and K. Kato, NMR characterization of the interactions between lyso-GM1 aqueous micelles and amyloid β, pp. 831–836, Copyright (2010), with permission from Elsevier.

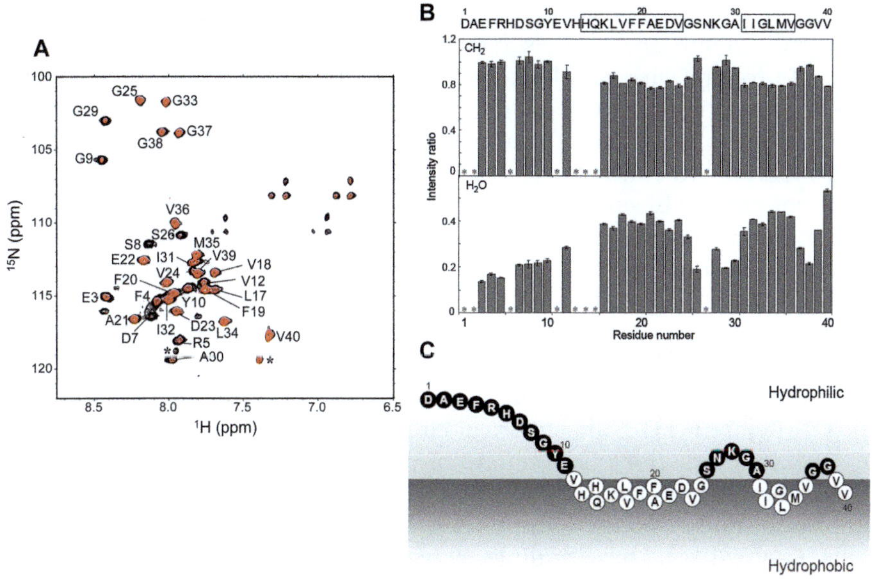

Figure 6.9 (A) 21.6 T TROSY-based ^1H–^{15}N HSQC spectra of U-[^{13}C, ^{15}N, ^2H] Aβ(1–40) in the presence of lyso–GM1 micelles with on-resonance (red) and off-resonance (black) irradiation of H_2O. Asterisk indicates the peak originating from lyso–GM1 micelles. (B) Plots of the intensity ratios of the backbone amide peaks of Aβ(1–40) with on-resonance and off-resonance irradiation of the acyl (CH_2) groups of lyso–GM1 (upper) and H_2O (lower). The primary structure of Aβ(1–40) peptide. Boxes indicating the α-helical regions are presented at the top of the plots. Asterisk indicates the amino acid residue that did not exhibit an observable peak in the spectrum due to severe broadening. Intensity ratios are presented as the mean ± S.D. of three independent experiments. (C) A topological model of Aβ(1–40) lying on ganglioside clusters based on the present NMR data. The amino acid residues exposed to the hydrophilic and hydrophobic milieus are represented by closed and open circles, respectively, with single-letter codes. Parts of this figure are reprinted from Glycoconjugate Journal, Up-and-down topological mode of amyloid β-peptide lying on hydrophilic/hydrophobic interface of ganglioside clusters, vol. 26, 2009, pp. 999–1006, M. Utsumi, Y. Yamaguchi, H. Sasakawa, N. Yamamoto, K. Yanagisawa and K. Kato, (© Springer Science+Business Media, LLC 2008). With permission of Springer.

specificities. Indeed, NMR data from Flemish-type (A21G) Aβ(1–40) bound to lyso–GM1 micelles indicated that the N-terminal segment of this mutant is in an intermediate conformational exchange between different conformational states on the NMR time scale, whereas the C-terminal α-helical segment remains tightly bound and acts as a membrane anchor.[31] These data reveal that the A21G mutation disrupts the first α-helix in the membrane-bound Aβ(1–40) molecule and suggest that this destabilization of α-helical structure is related to the altered fibril formation compared with that of wild-type Aβ, which is promoted through its interaction with ganglioside clusters.

Small bicelles with embedded gangliosides can be used as standardized nanoscale membrane models for detailed NMR analyses of the interactions of proteins with ganglioside clusters.[27,28,54] For example, intrinsically unstructured neuropeptides bind to the lipid bilayers and thereby attain a well-ordered secondary structure on bicelles containing GM1 or other gangliosides.[28] In contrast, ^{1}H–^{15}N HSQC experiments using small ganglioside-containing bicelles demonstrated specific interactions between the N-terminal segment of αSN and GM1 or GM2 but not GM3, without the formation of any secondary structure[27] (Figure 6.10). Previous NMR studies have shown that αSN can form a long α-helix in larger anionic membranes.[55–58] However, this long α-helix cannot be accommodated in the limited sizes of the bicelles.[59–62] A well-defined sugar cluster was recently constructed by modifying a self-assembled $M_{12}L_{24}$ sphere with the sugar moiety of GM1[63] (Figure 6.11). This spherical complex has enabled the observation of specific interactions between Aβ or αSN and the GM1 pentasaccharide cluster because of the lack of the hydrophobic ceramide, which would otherwise stably trap the hydrophobic segments of these pathogenic proteins (Figure 6.11). These findings suggest that the creation of gangliosidic

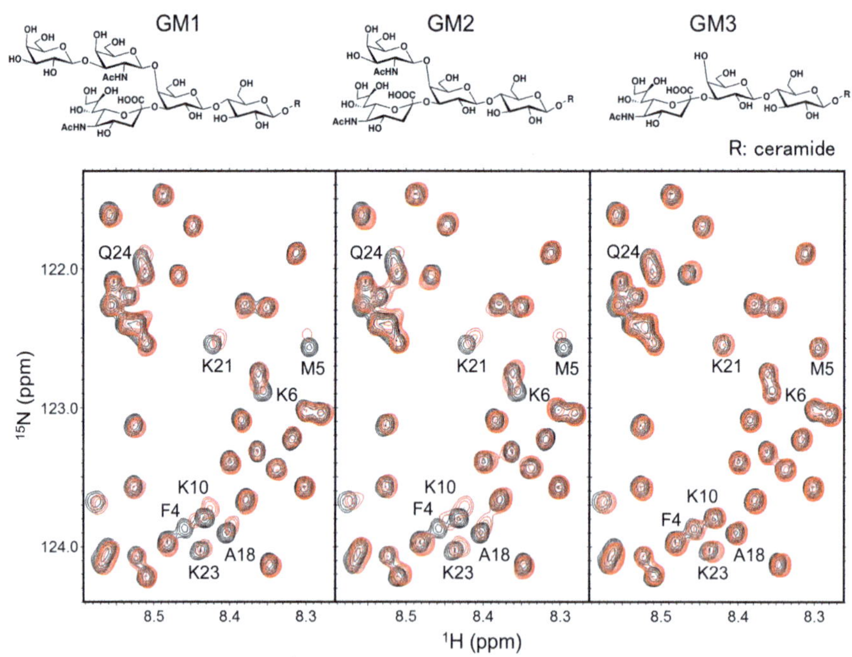

Figure 6.10 NMR characterization of the interactions of αSN with the ganglioside-embedding small bicelles as nanoscale standardized membrane mimics for NMR analyses. ^{1}H–^{15}N HSQC spectral changes in αSN induced by interactions with gangliosides GM1 (left), GM2 (middle), and GM3 (right) embedded in bicelles. Parts of this figure are reproduced from ref. 27 with permission from The Royal Society of Chemistry.

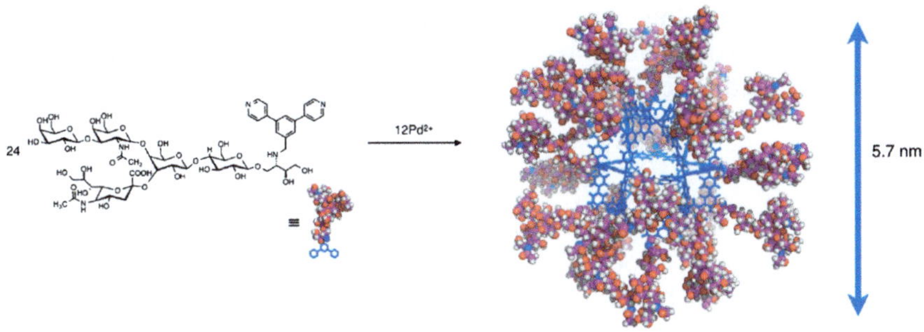

Figure 6.11 Schematic of the synthesis of a carbohydrate cluster from Pd^{2+} ions and a bidentate ligand-bearing GM1 sugar moiety. The molecular structure of the spherical GM1-sugar cluster was optimized using MM calculations (Pd, yellow; C, light blue; N, blue; O, red; H, gray). Reprinted from Sato *et al.*[63] with the permission of John Wiley and Sons. © 2015 WILEY-VCH Verlag GmbH & Co. KGaA, Weinheim.

oligosaccharide clusters having little or no hydrophobic interior will open up new possibilities for characterizing the early stages of ganglioside-philic protein interactions.

6.6 Concluding Remarks

Gangliosides promote their functions through the formation of ordered clusters that are mediated by the hydrophobic ceramide moiety and display dynamic glycans, which mediate non-stoichiometric interactions with proteins. Accumulating evidence has indicated that the size and curvature of the ganglioside clusters are determining factors in cluster–protein interactions and subsequent aggregate formation.[41,53,64,65] This highlights the need to develop gangliosidic assemblies as model systems that are systematically controllable in terms of these factors. A variety of ganglioside clusters with different sizes and curvatures would then be accessible for systematic studies using a series of self-assembled spherical complexes as the scaffolds.[66–68] Bicelles and nanodiscs can be further developed for embedding a variety of gangliosides along with Chol and SM, enabling detailed characterization of the *cis*- and *trans*-modes of biomolecular interactions on glycosphingolipid clusters using solution and solid-state NMR spectroscopy.

Acknowledgements

We thank Dr Nobuaki Matsumori (Kyushu University) for the useful discussion. We also thank Dr Saeko Yanaka (IMS) for NMR data collection. This study was partly supported by a JSPS/MEXT KAKENHI Grant-in-Aid for Yong Scientists (B) (JP15K21680 to M.Y.-U. and JP15K17889 to T.Y.), Grant-in-Aid for Scientific Research on Innovation Areas (JP25102001 and JP25102008 to K.K.), Grant-in-Aid for Challenging Exploratory Research (JP26560451 to

K.K.), Research Funding for Longevity Sciences (25-19) from the National Center for Geriatrics and Gerontology, the Nanotechnology Platform Program of MEXT, and the Okazaki ORION project.

References

1. S. Hakomori, *Curr. Opin. Hematol.*, 2003, **10**, 16–24.
2. K. Simons and D. Toomre, *Nat. Rev. Mol. Cell Biol.*, 2000, **1**, 31–39.
3. A. Regina Todeschini and S. I. Hakomori, *Biochim. Biophys. Acta*, 2008, **1780**, 421–433.
4. S. I. Hakomori, *Biochim. Biophys. Acta*, 2008, **1780**, 325–346.
5. P. H. Lopez and R. L. Schnaar, *Curr. Opin. Struct. Biol.*, 2009, **19**, 549–557.
6. T. Ariga, M. P. McDonald and R. K. Yu, *J. Lipid Res.*, 2008, **49**, 1157–1175.
7. K. Matsuzaki, K. Kato and K. Yanagisawa, *Biochim. Biophys. Acta*, 2010, **1801**, 868–877.
8. M. Yagi-Utsumi and K. Kato, *Glycoconjugate J.*, 2015, **32**, 105–112.
9. E. A. Merritt, P. Kuhn, S. Sarfaty, J. L. Erbe, R. K. Holmes and W. G. Hol, *J. Mol. Biol.*, 1998, **282**, 1043–1059.
10. U. Neu, K. Woellner, G. Gauglitz and T. Stehle, *Proc. Natl. Acad. Sci. U. S. A.*, 2008, **105**, 5219–5224.
11. J. N. Scarsdale, J. H. Prestegard and R. K. Yu, *Biochemistry*, 1990, **29**, 9843–9855.
12. D. Acquotti, L. Poppe, J. Dabrowski, C. Wilhelm Von der Lieth, S. Sonnino and G. Tettamanti, *J. Am. Chem. Soc.*, 1990, **112**, 7772–7778.
13. D. Acquotti, G. Fronza, E. Ragg and S. Sonnino, *Chem. Phys. Lipids*, 1991, **59**, 107–125.
14. S. Sabesan, J. O. Duus, T. Fukunaga, K. Bock and S. Ludvigsen, *J. Am. Chem. Soc.*, 1991, **113**, 3236–3246.
15. H. C. Siebert, G. Reuter, R. Schauer, C. W. von der Lieth and J. Dabrowski, *Biochemistry*, 1992, **31**, 6962–6971.
16. P. Brocca, A. Bernardi, L. Raimondi and S. Sonnino, *Glycoconjugate J.*, 2000, **17**, 283–299.
17. K. Kato, H. Sasakawa, Y. Kamiya, M. Utsumi, M. Nakano, N. Takahashi and Y. Yamaguchi, *Biochim. Biophys. Acta*, 2008, **1780**, 619–625.
18. Y. Yamaguchi, T. Yamaguchi and K. Kato, *Adv. Neurobiol.*, 2014, **9**, 165–183.
19. P. Brocca, P. Berthault and S. Sonnino, *Biophys. J.*, 1998, **74**, 309–318.
20. H. C. Siebert, K. Born, S. André, M. Frank, H. Kaltner, C. W. von der Lieth, A. J. Heck, J. Jiménez-Barbero, J. Kopitz and H. J. Gabius, *Chemistry*, 2005, **12**, 388–402.
21. S. Yamamoto, Y. Zhang, T. Yamaguchi, T. Kameda and K. Kato, *Chem. Commun. (Camb.)*, 2012, **48**, 4752–4754.
22. Y. Zhang, S. Yamamoto, T. Yamaguchi and K. Kato, *Molecules*, 2012, **17**, 6658–6671.
23. L. O. Sillerud, R. K. Yu and D. E. Schafer, *Biochemistry*, 1982, **21**, 1260–1271.
24. T. A. Koerner Jr, J. H. Prestegard, P. C. Demou and R. K. Yu, *Biochemistry*, 1983, **22**, 2676–2687.

25. L. Poppe, H. van Halbeek, D. Acquotti and S. Sonnino, *Biophys. J.*, 1994, **66**, 1642–1652.
26. M. L. Demarco, R. J. Woods, J. H. Prestegard and F. Tian, *J. Am. Chem. Soc.*, 2010, **132**, 1334–1338.
27. T. Yamaguchi, T. Uno, Y. Uekusa, M. Yagi-Utsumi and K. Kato, *Chem. Commun. (Camb.)*, 2013, **49**, 1235–1237.
28. U. L. Khatun, A. Gayen and C. Mukhopadhyay, *Glycoconjugate J.*, 2014, **31**, 435–447.
29. M. Utsumi, Y. Yamaguchi, H. Sasakawa, N. Yamamoto, K. Yanagisawa and K. Kato, *Glycoconjugate J.*, 2009, **26**, 999–1006.
30. M. Yagi-Utsumi, T. Kameda, Y. Yamaguchi and K. Kato, *FEBS Lett.*, 2010, **584**, 831–836.
31. M. Yagi-Utsumi and C. M. Dobson, *Biol. Pharm. Bull.*, 2015, **38**, 1668–1672.
32. Y. Zhang, T. Yamaguchi and K. Kato, *Chem. Lett.*, 2013, **42**, 1455–1462.
33. A. Canales, A. Mallagaray, J. Perez-Castells, I. Boos, C. Unverzagt, S. Andre, H. J. Gabius, F. J. Canada and J. Jimenez-Barbero, *Angew. Chem., Int. Ed. Engl.*, 2013, **52**, 13789–13793.
34. T. Yamaguchi, Y. Sakae, Y. Zhang, S. Yamamoto, Y. Okamoto and K. Kato, *Angew. Chem., Int. Ed. Engl.*, 2014, **53**, 10941–10944.
35. K. Kato and T. Yamaguchi, *Glycoconjugate J.*, 2015, **32**, 505–513.
36. Y. Zhang, T. Yamaguchi, T. Satoh, M. Yagi-Utsumi, Y. Kamiya, Y. Sakae, Y. Okamoto and K. Kato, *Adv. Exp. Med. Biol.*, 2015, **842**, 217–230.
37. S. Hakomori, *Glycoconjugate J.*, 2004, **21**, 125–137.
38. S. Kusunoki, K. Kaida and M. Ueda, *Biochim. Biophys. Acta*, 2008, **1780**, 441–444.
39. L. Mauri, R. Casellato, M. G. Ciampa, Y. Uekusa, K. Kato, K. Kaida, M. Motoyama, S. Kusunoki and S. Sonnino, *Glycobiology*, 2012, **22**, 352–360.
40. K. A. Vyas, H. V. Patel, A. A. Vyas and R. L. Schnaar, *Biol. Chem.*, 2001, **382**, 241–250.
41. Y. Mao, Z. Shang, Y. Imai, T. Hoshino, R. Tero, M. Tanaka, N. Yamamoto, K. Yanagisawa and T. Urisu, *Biochim. Biophys. Acta*, 2010, **1798**, 1090–1099.
42. M. M. Lozano, Z. Liu, E. Sunnick, A. Janshoff, K. Kumar and S. G. Boxer, *J. Am. Chem. Soc.*, 2013, **135**, 5620–5630.
43. V. Michel and M. Bakovic, *Biol. Cell*, 2007, **99**, 129–140.
44. M. Murata, S. Sugiyama, S. Matsuoka and N. Matsumori, *Chem. Rec.*, 2015, **15**, 675–690.
45. T. Yamaguchi, T. Suzuki, T. Yasuda, T. Oishi, N. Matsumori and M. Murata, *Bioorg. Med. Chem.*, 2012, **20**, 270–278.
46. N. Matsumori, T. Yamaguchi, Y. Maeta and M. Murata, *Biophys. J.*, 2015, **108**, 2816–2824.
47. N. Matsumori, T. Yasuda, H. Okazaki, T. Suzuki, T. Yamaguchi, H. Tsuchikawa, M. Doi, T. Oishi and M. Murata, *Biochemistry*, 2012, **51**, 8363–8370.
48. T. Yasuda, M. Kinoshita, M. Murata and N. Matsumori, *Biophys. J.*, 2014, **106**, 631–638.

49. M. Piccinini, F. Scandroglio, S. Prioni, B. Buccinnà, N. Loberto, M. Aureli, V. Chigorno, E. Lupino, G. DeMarco, A. Lomartire, M. T. Rinaudo, S. Sonnino and A. Prinetti, *Mol. Neurobiol.*, 2010, **41**, 314–340.
50. J. Fantini and N. Yahi, *J. Mol. Biol.*, 2011, **408**, 654–669.
51. Y. Kamiya, M. Yagi-Utsumi, H. Yagi and K. Kato, *Curr. Pharm. Des.*, 2011, **17**, 1672–1684.
52. M. Utsumi, Y. Yamaguchi, H. Sasakawa, N. Yamamoto, K. Yanagisawa and K. Kato, *Glycoconjugate J.*, 2009, **26**, 999–1006.
53. M. Yagi-Utsumi, K. Matsuo, K. Yanagisawa, K. Gekko and K. Kato, *Int. J. Alzheimers Dis.*, 2010, **2011**, e925073.
54. U. L. Khatun, S. K. Goswami and C. Mukhopadhyay, *Biophys. Chem.*, 2012, **168-169**, 48–59.
55. Z. Martinez, M. Zhu, S. Han and A. L. Fink, *Biochemistry*, 2007, **46**, 1868–1877.
56. C. R. Bodner, C. M. Dobson and A. Bax, *J. Mol. Biol.*, 2009, **390**, 775–790.
57. C. R. Bodner, A. S. Maltsev, C. M. Dobson and A. Bax, *Biochemistry*, 2010, **49**, 862–871.
58. G. Comellas, L. R. Lemkau, D. H. Zhou, J. M. George and C. M. Rienstra, *J. Am. Chem. Soc.*, 2012, **134**, 5090–5099.
59. T. S. Ulmer, A. Bax, N. B. Cole and R. L. Nussbaum, *J. Biol. Chem.*, 2005, **280**, 9595–9603.
60. C. R. Bodner, C. M. Dobson and A. Bax, *J. Mol. Biol.*, 2009, **390**, 775–790.
61. C. R. Bodner, A. S. Maltsev, C. M. Dobson and A. Bax, *Biochemistry*, 2010, **49**, 862–871.
62. T. Bartels, L. S. Ahlstrom, A. Leftin, F. Kamp, C. Haass, M. F. Brown and K. Beyer, *Biophys. J.*, 2010, **99**, 2116–2124.
63. S. Sato, Y. Yoshimasa, D. Fujita, M. Yagi-Utsumi, T. Yamaguchi, K. Kato and M. Fujita, *Angew. Chem., Int. Ed. Engl.*, 2015, **54**, 8435–8439.
64. J. Fantini, N. Yahi and N. Garmy, *Front. Physiol.*, 2013, **4**, 120.
65. L. Botto, D. Cunati, S. Coco, S. Sesana, A. Bulbarelli, E. Biasini, L. Colombo, A. Negro, R. Chiesa, M. Masserini and P. Palestini, *PLoS One*, 2014, **9**, e98344.
66. K. Suzuki, S. Sato and M. Fujita, *Nat. Chem.*, 2010, **2**, 25–29.
67. Q. F. Sun, J. Iwasa, D. Ogawa, Y. Ishido, S. Sato, T. Ozeki, Y. Sei, K. Yamaguchi and M. Fujita, *Science*, 2010, **328**, 1144–1147.
68. D. Fujita, H. Yokoyama, Y. Ueda, S. Sato and M. Fujita, *Angew. Chem., Int. Ed. Engl.*, 2015, **54**, 155–158.

CHAPTER 7

NMR Analysis of Glycosyltransferases

THOMAS PETERS[a]

[a]University of Lübeck, Institute of Chemistry, Ratzeburger Allee 160, 23562 Lübeck, Germany
*E-mail: thomas.peters@chemie.uni-luebeck.de

Glycosylation is a highly abundant and by far the most complex post-translational modification.[1] In mammalian organisms about 80% of all proteins are glycosylated leading to myriads of different glycoforms. Analysis of these different species and linking glycosylation patterns to disease states is a challenge.[2,3] Certainly, diagnosis and therapy would significantly benefit from such knowledge.[4] Especially in the emerging field of stratified medicine this would be an essential addition to a growing set of novel tools. Therefore, it is very important to understand the biosynthesis of glycoproteins, and its regulation at molecular level. Biosynthesis of glycoproteins involves complex biological machinery located mainly in the ER and in the Golgi apparatus employing membrane-bound glycosyltransferases as key players.[5] The concerted action of these glycosyltransferases leads to the final decoration of cell surfaces with complex glycans. Understanding the structural and dynamical basis of the enzymatic mechanisms of glycosyltransferases[5–8] would provide novel approaches toward modulation of glycosyltransferase function. For instance, selective inhibition of individual glycosyltransferases would pave the way for the development of novel pharmaceuticals, *e.g.* a novel class of antibiotics or anti-cancer drugs.[9–15]

New Developments in NMR No.10
NMR in Glycoscience and Glycotechnology
Edited by Koichi Kato and Thomas Peters

Published by the Royal Society of Chemistry, www.rsc.org

As glycosyltransferases are rather large proteins and mostly membrane-bound they are difficult targets for protein NMR studies. For NMR studies only soluble catalytic domains have been utilized so far. Often these enzymes assemble in quaternary structures further complicating NMR experiments. For instance, human blood group A and B glycosyltransferases form stable homodimers with a molecular weight around 70 kDa. As NMR analysis of proteins larger than about 25 to 30 kDa is far from being routine[16,17] NMR studies often take advantage of structural data from crystallography. NMR experiments then extend the static picture from crystallography by providing information about protein dynamics,[18–20] binding thermodynamics and kinetics of ligand binding,[21–23] as well as identification and characterization of binding of ligands.[24–26] In the following it will be summarized how NMR studies have contributed to a better understanding of dynamical properties of glycosyltransferases.

7.1 Static Structures of Glycosyltransferases

7.1.1 Structures from X-Ray Crystallography

Crystallography has delivered a large number of three-dimensional structures of glycosyltransferases of bacterial as well as of mammalian origin. These structures can be assigned to one of three folds, GT-A, GT-B, or GT-C.[27,28] The majority of cases falls into the families GT-A or GT-B. Enzymes adopting the GT-A fold are characterized by two tightly associated abutting β/α/β domains. The GT-B fold is also characterized by two associated β/α/β domains but in this case the two domains face each other creating an active site along the cleft generated by this domain arrangement.[5] Another classification can be made according to the stereochemical features of the glycosyltransfer reaction.[5,8] Transfer of a donor sugar from a donor substrate to an acceptor sugar can either occur with retention of configuration of the anomeric center of the donor sugar (retaining GTs), or with inversion of configuration of the anomeric C-atom (inverting GTs). An example for the latter class of GTs is β-1,4-galactosyltransferase that synthesizes *N*-acetyllactosamine units using UDP-Gal as a donor sugar. This GT is among the first having been successfully crystallized and structurally analyzed.[29] It is widely agreed that inverting GTs follow a direct-displacement S_N2-like mechanism whereas the mechanism for retaining GTs is still disputed, and eventually there is no general mechanism that applies to retaining GTs.[5] Experimental and theoretical evidence has been found for a broad spectrum of mechanisms ranging from double-displacement S_N2 like mechanisms to so-called S_Ni mechanisms.[6,7,30–33]

A particularly well-studied case is provided by human blood group A and B glycosyltransferases, GTA and GTB, respectively. A breakthrough in the field came with the first crystal structure analysis of these enzymes.[34] These enzymes belong to the GT-A fold and are retaining glycosyltransferases. GTA and GTB differ by only four out of 354 amino acids: Arg176/Gly176,

Gly235/Ser235, Leu266/Met266, and Gly268/Ala268 in GTA and GTB, respectively. A comprehensive crystallographic study delivered high-resolution structures of GTA, GTB and of mutants varying the four distinguishing amino acids.[35] A special nomenclature is used to distinguish the mutants: AAAA translates into GTA; BBBB translates into GTB; AABB translates into a mutant where the first two amino acids that distinguish A and B are Arg176 (A) and Gly235 (A) whereas the second two amino acids are Met266 (B) and Ala268 (B); *etc.* Crystals were soaked with a variety of donor and acceptor substrates demonstrating that GTA and GTB come in three different conformations: an open conformation, a semi-closed conformation, and a closed conformation. Without any ligand, or in the presence of acceptor substrate only, the enzymes are open.[35] This means an internal loop comprising amino acids 175–195 is only partially ordered and the C-term consisting of amino acids 346–351 is completely disordered. Binding of UDP leads to partial closing of the active site by the internal loop but disorder is increasing with the C-term still being completely disordered. When donor- and acceptor substrate or analogs thereof are present the internal loop adopts one helical conformation and the C-term becomes ordered closing completely that part of the active site where the activated sugar UDP-Gal, or UDP-GlaNAc binds. It is an open question on which timescale the interconversion of different states of the enzymes occurs. Figure 7.1 shows an example for the open, semi-closed and closed conformations. The challenge for the NMR spectroscopist is the identification of the timescale of exchange. It will be shown below that NMR experiments give insight in such motions, and in fortunate cases deliver quantitative information about the kinetics of the conformational exchange process.

7.1.2 Structures from NMR Spectroscopy

The large size, and the tendency to form multimers has precluded NMR-based structural analysis of GTs except in one case, an *N*-acetylglucosaminyltransferase Alg13 from yeast.[36] Alg13 belongs to the GT-B fold and represents the donor substrate-binding domain. For GT activity association with Alg14 a membrane-bound yeast protein. Alg13 and Alg14 are thought to form a complex on the surface of the ER that then exhibits catalytic activity. Fortunately, Alg13 is only about half the molecular weight of a typical intact GT, and therefore lends itself well to NMR analysis. An almost complete backbone and side chain assignment was possible. Using NOE restraints and RDCs structure refinement was possible showing the Alg13 has some structural peculiarities compared to other GTs adopting a GT-B fold. Chemical shift perturbation experiments in combination with PRE experiments then allowed identification of the donor substrate-binding pocket. The measurement of motional correlation times τ_c using relaxation interference experiments[37] indicated significant local mobility at the C- and N-terminus as well as in the donor-substrate binding pocket. Further studies using relaxation dispersion experiments did not reveal any motions on the micro- to millisecond timescale.

7.2 Dynamic Structures of Glycosyltransferases

7.2.1 Ligand-Based NMR Experiments

NMR provides a number of experimental approaches to analyze binding thermodynamics and kinetics indirectly reflecting underlying protein dynamics. Especially the on- and off-rates for ligand binding contain information about associated protein movements in the corresponding binding site. For instance, the off-rate cannot be faster than opening and closing of a binding pocket, and thus would give a lower limit for the rate at which *e.g.* a loop closes a binding pocket.

7.2.1.1 Binding Kinetics from NMR Experiments

In a ligand-based NMR study the binding of donor and acceptor substrates to GTB (see above) had been studied using STD NMR experiments.[38] Saturation transfer critically depends on the exchange kinetics, and therefore contains all information required for the determination of corresponding rate constants. Unfortunately, saturation transfer also strongly depends on the chemical (proton) environment in the binding pocket, and in order to quantitatively evaluate rate constants one requires access to 3D structural data of the corresponding complex. At the time of the study only data for the apo-form of GTB were available and the UDP-Gal and H-antigen had to be modeled.[38] Then, STD NMR buildup curves were measured and a full relaxation matrix simulation was performed to match the experimental data. Theoretical STD buildup curves were fitted to the experimental data yielding off-rates for UDP-Gal and H-disaccharide of around 10 Hz and larger than 100 Hz, respectively (Figure 7.1).

If the opening and closing of the donor site of GTB is rate-limiting then this means that the exchange rate constant for this conformational process cannot be slower than 10 Hz under these conditions. The situation may alter in the presence of H-disaccharide. Thus monitoring the binding of donor substrate UDP-Gal to GTB had allowed insight into the protein movements associated with the binding process.

7.2.1.2 Enzyme Kinetics from NMR Experiments

Very much like binding kinetics enzyme kinetics provides indirect insight into protein movement associated with catalysis.[39,40] Traditionally, enzyme kinetics of GTs is done with radioactively labeled donor substrates. Lately, covalently attached fluorescence labels have been used especially for applications in GT-inhibitor design.[41] GT-catalyzed reactions are usually slower than many other enzymatic transformations. For such cases it had been shown that the analysis of progression curves detected with simple 1D ^{1}H NMR experiments is a fast and robust tool not only delivering kinetic parameters but at the same time providing information about the stereochemistry of the reaction and the structure of products formed.[42] This approach makes

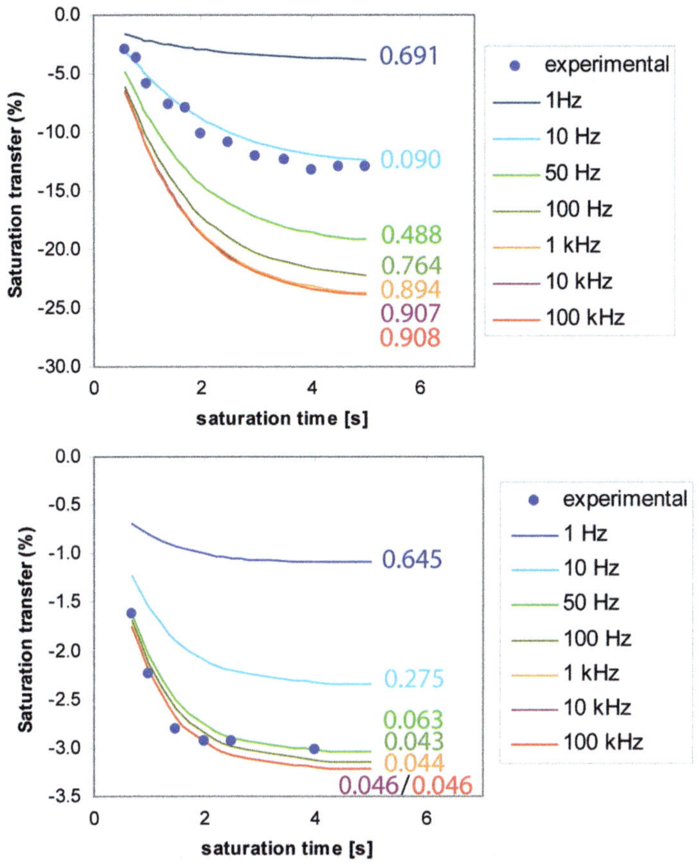

Figure 7.1 Different kinetics of the binding of donors and acceptor ligands to GTB. The figure shows the experimental build-up curves (dots) for the most intense STD signals of UDP-Gal (H1′ of ribose; top) and the H-antigen (H4′ of the galactose; bottom), as well as the STD curves predicted from full relaxation rate matrix calculations with the model of the complex, using different values of the dissociation rate constant (k_{off}). Color-coded are shown the R-factors for each proton as a function of k_{off}. For UDP-Gal, an excellent match is obtained for k_{off} = 10 Hz. For the H-antigen the best fits are obtained for a range of k_{off} values between 100 and up to 100.000 Hz. Reprinted with permission from J. Angulo, B. Langpap, A. Blume, T. Biet, B. Meyer, N. R. Krishna, H. Peters, M. M. Palcic and T. Peters, *J. Am. Chem. Soc.*, 2006, **128**, 13 529–13 538.[38] Copyright (2006) American Chemical Society.

uses of the so-called Lambert W solution of the Michaelis–Menten equation. Non-linear regression data analysis of progress curves delivers k_{cat} and K_M. Progress curves are obtained by adding access donor-substrate into a solution of GT enzyme and acceptor substrate, and then monitoring the disappearance of educt signals and the appearance of product signals in ^{1}H NMR spectra with time.

A beautiful example has been provided by the analysis of the enzyme kinetics of a human fucosyltransferase FUT8.[43] This inverting GT transfers an L-fucose residue from GDP-Fuc to the reducing GlcNAc residue of *N*-glycan chains to form so-called core-fucosylated *N*-glycans. The absence of this core-fucose residue plays an important role in a variety of biological recognition reactions. For example, therapeutic monoclonal antibodies carrying *N*-glycans with no or depleted core-fucosylation display massively increased Fc-γ mediated cytotoxicity. Therefore, it is important to develop strategies to modulate the activity of FUT8, *e.g.* by designing synthetic inhibitors.[44] For this to achieve a thorough understanding of binding to natural substrates is essential. The study describes a ligand–NMR-based approach using STD NMR spectroscopy to analyze recognition of GDP-Fuc by FUT8.[43] STD epitope mapping combined with information from crystal structure data of a related fucosyltransferase, and MD simulations led to a binding model that well explains the high specificity of the transfer reaction, and suggests critical amino acids essential for catalysis. Then, using the Lambert W solution of the Michaelis–Menten equation the study delivers kinetic constants from analysis of full progression curves (Figure 7.2).

It is found that FUT8 is a "slow" enzyme with k_{cat} = 0.45 Hz, and relatively tight binding of acceptor substrate reflected by a $K_{M,acceptor}$ of 12 μM. Especially the k_{cat} value is about one order of magnitude lower than corresponding k_{cat} values for GTA and GTB. Although not derived from NMR data but from surface plasmon resonance (SPR) studies a value is also given for the off-rate of GDP-Fuc. With a value of k_{off} of 0.2 Hz the dissociation reaction of a complex of FUT8 and GDP-Fuc is much slower than observed for GTB where k_{off} was found to be around 10 Hz as determined from STD buildup curves[38] (see above). This comparison of an inverting GT (FUT8) to a retaining GT (GTB) shows that there must be a wide variety of binding kinetics for GTs that should also reflect in corresponding protein dynamics. As will become clearer below not much is known about this at present.

7.2.2 Protein-Based NMR Experiments

Using protein NMR techniques direct insight into protein dynamics can be obtained by a variety of techniques.[45] For enzymes monitoring the influence of binding of substrates is of special interest aiming at understanding what happens in the protein during catalysis.[46] Impressive results have been achieved over the years by applying so-called relaxation dispersion experiments to this set of problems. A key paper in this respect describes intrinsic motions of prolyl *cis–trans* isomerase Cyclophilin A (CypA) underlying catalysis.[47] Compared to GTs this class of enzymes is very fast with exchange rates of substrate binding of about 2500 Hz as determined from simultaneously fitting backbone ^{15}N amide and ^{13}C methyl relaxation dispersion data. Interestingly, here the timescale of conformational changes coincides with substrate turnover. Obviously, for GTs quite another timescale applies for substrate turnover. In the following the focus is on a retaining bacterial

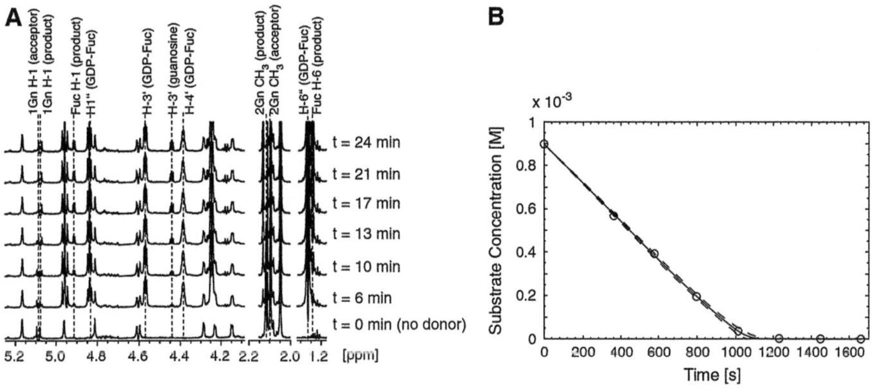

Figure 7.2 *Upper part:* Enzymatic reaction catalyzed by FUT8. A fucosyl moiety is transferred from the activated sugar nucleotide GDP-Fuc to the 6-hydroxy group of the proximal GlcNAc of the acceptor with inversion of the anomeric configuration. The acceptor, a branched *N*-type oligosaccharide, has to exhibit an unsubstituted GlcNAc residue at the 3-mannose to be recognized by the enzyme. In contrast, the linkage of the proximal GlcNAc to the asparagine side chain is not essential. *Lower part:* Enzyme kinetic characterization of FUT8 by means of progress curve analysis using NMR detection of products and substrates. A: Selected sections of 1H NMR spectra recorded at different time points of the reaction (for clarity, the spectrum at 27 min is not shown). Resonances that were used to determine substrate and product concentrations in the region between 1.0 and 5.2 ppm are indicated by dotted lines. The well-dispersed resonances of GDP-Fuc (H-8, H-1′, H-3′, H-4′, H-1″, H-6″), guanosine (H-8, H-1′, H-3′), heptasaccharide 1 (GlcNAc1 H-1, GlcNAc2 CH3) and the fucosylated product 2 (GlcNAc1 H-1, Fuc H-1,GlcNAc2 CH3, Fuc H-6) were integrated and used for the determination of substrate concentrations. The sample contained 2 μM FUT8, a starting concentration of 0.9 mM acceptor substrate and 3 mM donor substrate, respectively, 10 U alkaline phosphatase and 1 mg mL^{-1} BSA in 160 μL deuterated Mes–NaOH buffer, pH 7.0. B: Plot of the concentration of acceptor substrate *versus* time, *i.e.* progress curve, of the FUT8 reaction using the integral of the 2 *N*-acetyl resonance of GlcNAc1. The Lambert-W fit yields the enzyme kinetic constants $K_{M,acceptor} = 12$ μM and $k_{cat} = 0.45\ s^{-1}$. The dashed lines give the confidence interval obtained from the fit. Reprinted from Kotzler, M.P, Blank, S., Bantleon, F.I., Spillner, E. and Meyer, B. (2012) Donor substrate binding and enzymatic mechanism of human core alpha1,6-fucosyltransferase (FUT8). *Biochim Biophys Acta*, **1820**, 1915–1925.[43] Copyright (2012), with permission from Elsevier.

glycosyltransferase LgtC (Lipopolysaccharide α-1,4-galactosyltransferase C). Here, a partial backbone and side chain assignment was achieved, and based on this studies into the dynamics were possible. Other GTs have been targeted with protein NMR, too but only in the cases of Alg13[36] (see Section 7.1.2) and LgtC,[48,49] both of them monomeric GTs, this approach was really successful. For a bacterial sialyltransferase CstII from *Campylobacter jejuni* a number of point mutations had been necessary in order to turn this enzyme from the natural tetrameric form into a monomeric state making it more tractable by NMR.[50] Even in the monomeric state CstII displayed severe conformational exchange causing about half of the expected NH backbone signals to disappear. Further studies into the protein dynamics were not performed yet.

7.2.2.1 Backbone Assignments of Glycosyltransferases

As alluded to above, the size and multimeric character of GTs renders protein NMR experiments challenging. LgtC, a bacterial GT from *Neisseria meningitidis*, is one of the few GTs staying monomeric in solution. Like GTA and GTB LgtC is a retaining GT and belongs to the GT-A fold. Being a 292 amino acid residue protein with a molecular weight of 32 kDa LgtC still is a challenge for protein NMR spectroscopy.[48,49] About 70% of the expected backbone NH resonances could be identified, and about 50% were assigned.[49] Exchange broadening caused the other peaks to disappear as this had been observed before for the sialyltransferase CstII.[50] Notably, signals in the substrate-binding pocket could not be observed. It was found that addition of UDP-2FGal, a non-hydrolysable analog of the donor substrate UDP-Gal, to the apo form of LgtC gave rise to a distinct set of peaks in the $^{1}H,^{15}N$ TROSY HSQC spectra indicating slow exchange between free and bound state. The partial backbone assignment was utilized as a basis for NMR experiments into the protein dynamics of LgtC.[48,49]

7.2.2.2 Methyl Labeling of Glycosyltransferases

As many of the backbone NH resonances of LgtC had not been assigned an alternative approach was taken utilizing ^{13}C-methyl labeled samples. ILV labeled samples were prepared and subjected to methyl TROSY[51] experiments. It was immediately obvious that more peaks than expected were observed. The study focused on the Ile cross peaks. An assignment of methyl cross peaks was achieved by combining NMR experiments that took advantage of known backbone assignments and site-directed mutagenesis of ILV residues. All Ile residues were assigned and a clear distinction between two forms of the enzyme called "a" and "b" was possible. Many, though not all, Ile residues showed a splitting into two peaks termed "a" and "b" (Figure 7.3). Temperature dependent measurements clearly showed intensity changes of the two states with temperature strongly suggesting that the two forms represent a conformational equilibrium. Raising the temperature favors state "a". Titration of δ_1-^{13}C-methyl Ile labeled LgtC with UDP-2FGal led to a decrease of intensities of the "a" peaks with a concomitant increase of intensities of the "b" peaks. At the same time the chemical shift of only the "b" peaks changed. This suggests that only the "b" form of LgtC binds donor substrate and thus is catalytically competent.

Although crystal structure data are available for LgtC and its complexes with substrates there is no indication yet what the difference between the "a" and the "b" form is in terms of *e.g.* different conformational states. The NMR

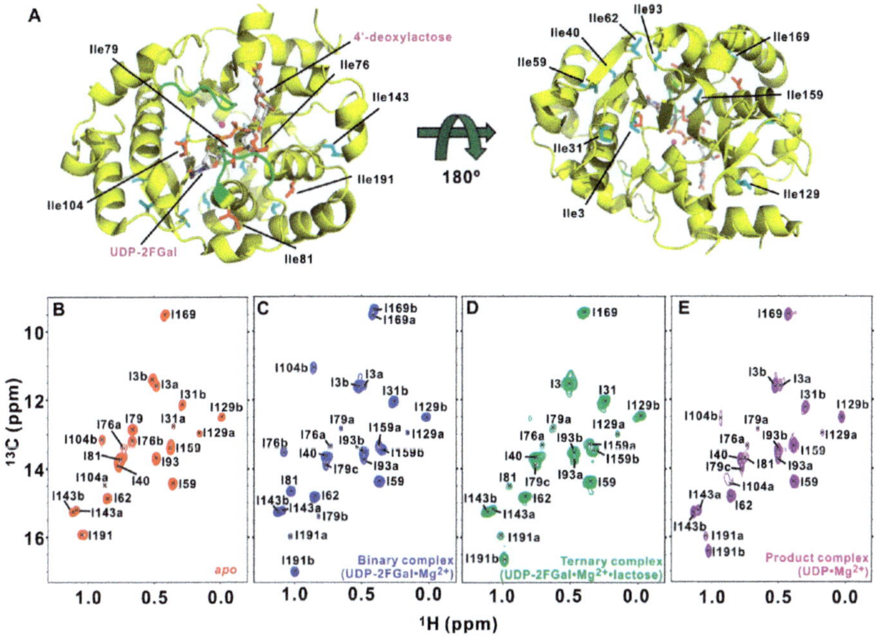

Figure 7.3 (A) Crystal structure of the LgtC ternary complex showing the isoleucine residues close to (red) or distal from (cyan) the active site. Ile76, Ile79, and Ile81 are within or flanking the flexible loops (green) covering the active site. Ile104 belongs to the Asp–X–Asp motif used for metal (Mn^{2+} or Mg^{2+}) binding. Ile191 is located in the α-helix that also contains Gln189 and Asp190. (B–E) methyl-TROSY spectra of the isoleucine residues of the uniformly deuterated and selectively [^{1}H,^{13}C]methyl-labeled (B) apo LgtC, (C) LgtC binary complex with 10 mM Mg^{2+} and 1 mM UDP-2FGal, (D) LgtC ternary complex with 10 mM Mg^{2+}, 1 mM UDP-2FGal, and 300 mM lactose, and (E) LgtC product complex with 20 mM Mg^{2+} and 1 mM UDP. The spectra (B–E) show the presence of two conformational states of LgtC, termed "a" and "b". Exchange between the two states is slow on the chemical shift timescale (*cf.* Figure 7.4). Reprinted with permission from P. H. Chan, A. H. Cheung, M. Okon, H. M. Chen, S. G. Withers and L. P. McIntosh, *Biochemistry*, 2013, **52**, 320–332.[48] Copyright (2013) American Chemical Society.

experiments nicely demonstrate that this enzyme exhibits significantly more plasticity than static pictures from crystal structure analysis suggest.

7.2.2.3 Glycosyltransferase Dynamics from Relaxation Dispersion

Relaxation dispersion experiments were used to probe the dynamics of δ_1-^{13}C-methyl Ile labeled LgtC. Both forms "a" and "b" showed relaxation dispersion of a number of Ile methyl cross peaks indicating molecular motion on the microsecond to millisecond timescale. The largest changes of $R_{2,\mathrm{eff}}$ were observed for Ile104 located in the donor binding pocket. Global fitting of five

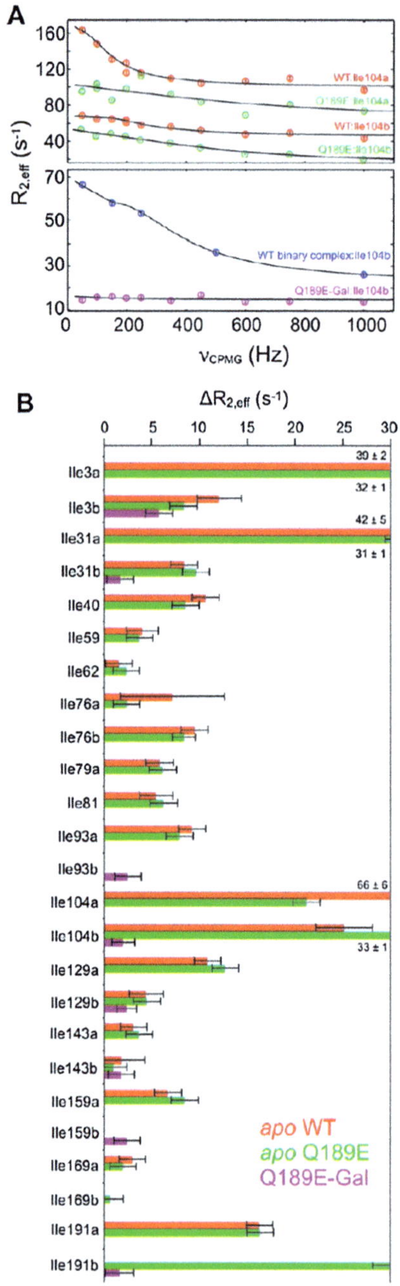

Figure 7.4 Methyl relaxation dispersion studies of LgtC–Q189E reveal millisecond timescale conformational exchange that is dampened upon formation of the covalent glycosyl–enzyme intermediate, LgtC–Q189E–Gal. (A) Complete 600 MHz relaxation dispersion profiles (top) of Ile104 in the "a" and "b" states of apo wild-type LgtC (red) and the "a" and "b" states of apo LgtC–Q189E (green). (A) Relaxation dispersion profiles (bottom) of state "b" of the wild-type binary complex with UDP-2FGal·Mg^{2+}

relaxation dispersion curves for Ile residues 3, 31, 76, 104, and 129 yielded exchange rate constants k_{ex} of 880 ± 40 and 780 ± 80 Hz for the "a" and the "b" form, respectively. These values are about one order of magnitude larger than the k_{cat} value of 46 Hz. The structural or conformational changes associated with these motions are unknown at present. It would be very interesting to uncover whether these motions are a prerequisite for catalysis, and whether and how they are linked to k_{cat}.

A rather interesting observation was made when studying relaxation dispersion for the Q189E mutant of LgtC. This mutant is known to form a covalent intermediate with Galactose that is transferred from UDP-Gal to Asp190. The mechanism of how this occurs is unknown as the carboxylate of Asp190 is about 9 Å away from the donor site. It was observed that in LgtC with Galactose covalently attached to Asp190 relaxation dispersion is almost switched off as this is convincingly seen for Ile104 (Figure 7.4). Obviously, formation of this covalent intermediate significantly dampens millisecond motions of the enzyme.

7.2.2.4 *Glycosyltransferase Dynamics from ZZ-Exchange Experiments*

Overlaid to these millisecond motions LgtC displays exchange between two forms "a" and "b" as discussed in section 7.2.2.3. The NMR study on LgtC also addresses this slow exchange problem. Employing zz-exchange NMR experiments[52,53] the authors were able to estimate the exchange rate between "a" and "b" in the apo form of LgtC (Figure 7.5). As a consequence of severe signal overlap only two Ile residues could be used for analyzing the corresponding exchange rate, which was found to be about 0.15 Hz.

7.3 Perspectives for the Investigation of Glycosyltransferase Dynamics Using NMR Experiments

A picture emerges that describes GTs as highly flexible enzymes with molecular motions on a variety of timescales. We are only beginning to understand how these motions are coupled to catalysis. So far very limited data material

(blue) and the covalently modified LgtC–Q189E–Gal (magenta). Error bars are smaller than the symbols, and the lines are best fits to a two-state exchange model. (B) Histogram showing the experimental $\Delta R_{2,\mathrm{eff}} = R_{2,\mathrm{eff}}(\nu_{\mathrm{CPMG\ 50\ Hz}}) - R_{2,\mathrm{eff}}(\nu_{\mathrm{CPMG\ 1000\ Hz}})$ values for apo LgtC (red), apo LgtC–Q189E (green), and LgtC–Q189E–Gal (magenta) obtained using a 600 MHz spectrometer. The numbers are $\Delta R_{2,\mathrm{eff}}$ values for residues with clipped histogram bars. Blanks indicate peaks with insufficient signal-to-noise ratios for reliable analysis. Reprinted with permission from P. H. Chan, A. H. Cheung, M. Okon, H. M. Chen, S. G. Withers and L. P. McIntosh, *Biochemistry*, 2013, **52**, 320–332.[48] Copyright (2013) American Chemical Society.

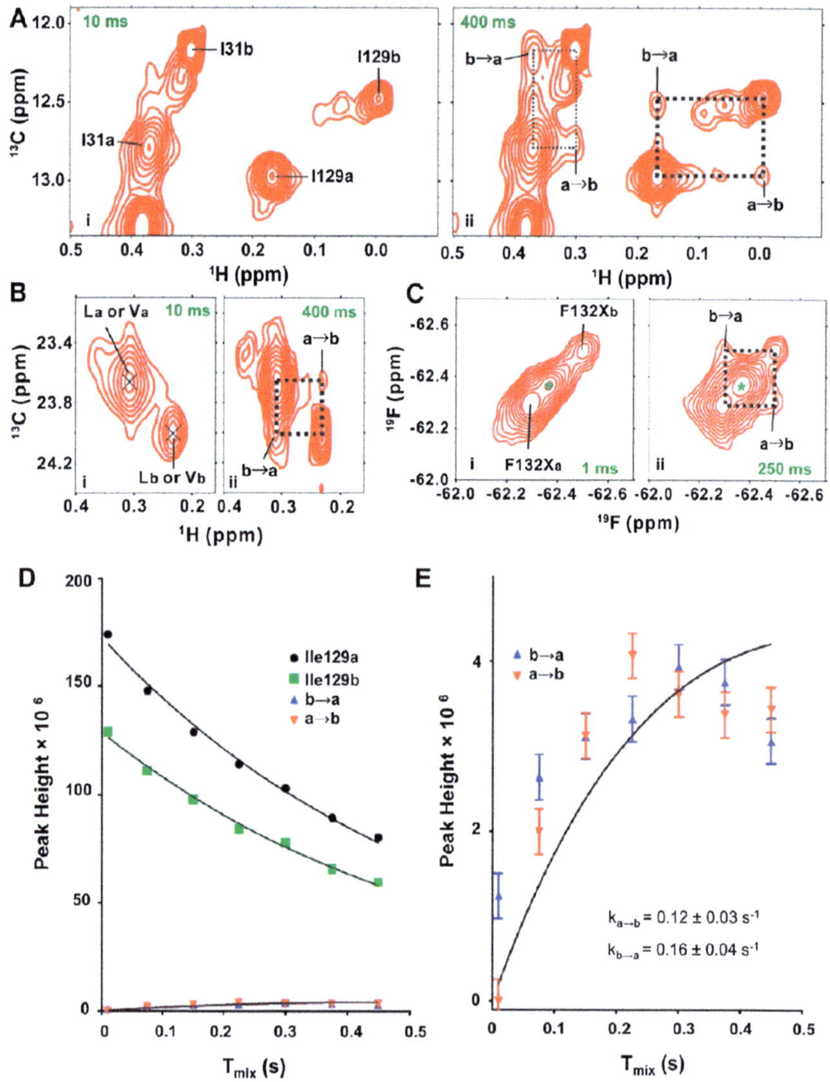

Figure 7.5 Magnetization exchange experiments reveal the slow interconversion between conformational states of apo LgtC at 25 °C. Exchange cross-peaks were detected in the form of methyl-TROSY spectra for (A) Ile129 and Ile31 and (B) a pair of unassigned leucines/valines. Panels i show spectra with a 10 ms transfer delay, and panels ii show spectra with a 400 ms transfer delay. (C) Exchange was also detected for CF_3Phe-labeled LgtC–F132X using a ^{19}F-NOESY experiment with (i) 1 or (ii) 250 ms transfer times. The additional diagonal signal (asterisk) was not present in the initial spectra recorded with fresh protein and is attributed to sample degradation. (D and E) The conformational exchange rate constants for Ile129 were extracted by fitting the time-dependent heights of the auto and exchange peaks. The root-mean-square errors of the peak heights are smaller than the symbols in panel D. Panel E is an expanded view showing the "a → b" and "b → a" exchange peaks, for which the best-fit curves overlap. The modest agreement of the fit curves with the data points reflects the relatively slow conformational exchange and hence weak exchange peaks for Ile129. Because of limited spectral dispersion, it was not possible to reliably analyze the exchange data for the other residues shown in panels A–C. Reprinted with permission from P. H. Chan, A. H. Cheung, M. Okon, H. M. Chen, S. G. Withers and L. P. McIntosh, *Biochemistry*, 2013, **52**, 320–332.[48] Copyright (2013) American Chemical Society.

indicates that this class of enzymes requires a combination of different timescale motions in order to function properly. Certainly, a picture where the rate of substrate turnover coincides with micro- or millisecond motions as observed for CypA is too simple. So far crystallography fell short in delivering answers. Therefore, a major challenge is to link the observation of a multitude of conformational states by NMR to three-dimensional pictures.

References

1. R. Hurtado-Guerrero and G. J. Davies, *Curr. Opin. Chem. Biol.*, 2012, **16**, 479–487.
2. R. D. Cummings and J. M. Pierce, *Chem. Biol.*, 2014, **21**, 1–15.
3. C. D. Rillahan, A. Antonopoulos, C. T. Lefort, R. Sonon, P. Azadi, K. Ley, A. Dell, S. M. Haslam and J. C. Paulson, *Nat. Chem. Biol.*, 2012, **8**, 661–668.
4. M. Dalziel, M. Crispin, C. N. Scanlan, N. Zitzmann and R. A. Dwek, *Science*, 2014, **343**, 1235681.
5. L. L. Lairson, B. Henrissat, G. J. Davies and S. G. Withers, *Annu. Rev. Biochem.*, 2008, **77**, 521–555.
6. A. Ardevol, J. Iglesias-Fernandez, V. Rojas-Cervellera and C. Rovira, *Biochem. Soc. Trans.*, 2016, **44**, 51–60.
7. B. Schuman, S. V. Evans and T. M. Fyles, *PLoS One*, 2013, **8**, e71077.
8. C. Breton, S. Fournel-Gigleux and M. M. Palcic, *Curr. Opin. Struct. Biol.*, 2012, **22**, 540–549.
9. M. M. Palcic, L. D. Heerze, O. P. Srivastava and O. Hindsgaul, *J. Biol. Chem.*, 1989, **264**, 17174–17181.
10. B. Müller, C. Schaub and R. R. Schmidt, *Angew. Chem., Int. Ed.*, 1998, **37**, 2893–2897.
11. K. Schaefer, N. Sindhuwinata, T. Hackl, M. P. Kotzler, F. C. Niemeyer, M. M. Palcic, T. Peters and B. Meyer, *J. Med. Chem.*, 2013, **56**, 2150–2154.
12. K. Schaefer, J. Albers, N. Sindhuwinata, T. Peters and B. Meyer, *ChemBioChem*, 2012, **13**, 443–450.
13. Z. Tu, Y. N. Lin and C. H. Lin, *Chem. Soc. Rev.*, 2013, **42**, 4459–4475.
14. R. Jorgensen, L. L. Grimm, N. Sindhuwinata, T. Peters and M. M. Palcic, *Angew. Chem.*, 2012, **51**, 4171–4175.
15. T. Pesnot, R. Jorgensen, M. M. Palcic and G. K. Wagner, *Nat. Chem. Biol.*, 2010, **6**, 321–323.
16. D. P. Frueh, *Prog. Nucl. Magn. Reson. Spectrosc.*, 2014, **78**, 47–75.
17. Y. Xu and S. Matthews, *Top. Curr. Chem.*, 2013, **335**, 97–119.
18. R. Rosenzweig and L. E. Kay, *Annu. Rev. Biochem.*, 2014, **83**, 291–315.
19. R. Sprangers and L. E. Kay, *Nature*, 2007, **445**, 618–622.
20. I. R. Kleckner and M. P. Foster, *Biochim. Biophys. Acta*, 2011, **1814**, 942–968.
21. M. P. Williamson, *Prog. Nucl. Magn. Reson. Spectrosc.*, 2013, **73**, 1–16.
22. L. Fielding, *Prog. Nucl. Magn. Reson. Spectrosc.*, 2007, **51**, 219–242.
23. S. W. Homans, *Top. Curr. Chem.*, 2007, **272**, 51–82.

24. T. Carlomagno, *Annu. Rev. Biophys. Biomol. Struct.*, 2005, **34**, 245–266.
25. B. Meyer and T. Peters, *Angew. Chem., Int. Ed. Engl.*, 2003, **42**, 864–890.
26. M. del Carmen Fernandez-Alonso, D. Diaz, M. A. Berbis, F. Marcelo, J. Canada and J. Jimenez-Barbero, *Curr. Protein Pept. Sci.*, 2012, **13**, 816–830.
27. B. Henrissat, G. Sulzenbacher and Y. Bourne, *Curr. Opin. Struct. Biol.*, 2008, **18**, 527–533.
28. Y. Bourne and B. Henrissat, *Curr. Opin. Struct. Biol.*, 2001, **11**, 593–600.
29. L. N. Gastinel, C. Cambillau and Y. Bourne, *EMBO J.*, 1999, **18**, 3546–3557.
30. D. Albesa-Jove, F. Mendoza, A. Rodrigo-Unzueta, F. Gomollon-Bel, J. O. Cifuente, S. Urresti, N. Comino, H. Gomez, J. Romero-Garcia, J. M. Lluch, E. Sancho-Vaello, X. Biarnes, A. Planas, P. Merino, L. Masgrau and M. E. Guerin, *Angew. Chem., Int. Ed. Engl.*, 2015, **54**, 9898–9902.
31. A. Monegal and A. Planas, *J. Am. Chem. Soc.*, 2006, **128**, 16030–16031.
32. N. Soya, Y. Fang, M. M. Palcic and J. S. Klassen, *Glycobiology*, 2011, **21**, 547–552.
33. A. Bobovska, I. Tvaroska and J. Kona, *Glycobiology*, 2015, **25**, 3–7.
34. S. I. Patenaude, N. O. Seto, S. N. Borisova, A. Szpacenko, S. L. Marcus, M. M. Palcic and S. V. Evans, *Nat. Struct. Biol.*, 2002, **9**, 685–690.
35. J. A. Alfaro, R. B. Zheng, M. Persson, J. A. Letts, R. Polakowski, Y. Bai, S. N. Borisova, N. O. Seto, T. L. Lowary, M. M. Palcic and S. V. Evans, *J. Biol. Chem.*, 2008, **283**, 10097–10108.
36. X. Wang, T. Weldeghiorghis, G. Zhang, B. Imperiali and J. H. Prestegard, *Structure*, 2008, **16**, 965–975.
37. D. Lee, C. Hilty, G. Wider and K. Wüthrich, *J. Magn. Reson.*, 2006, **178**, 72–76.
38. J. Angulo, B. Langpap, A. Blume, T. Biet, B. Meyer, N. R. Krishna, H. Peters, M. M. Palcic and T. Peters, *J. Am. Chem. Soc.*, 2006, **128**, 13529–13538.
39. G. Bhabha, J. Lee, D. C. Ekiert, J. Gam, I. A. Wilson, H. J. Dyson, S. J. Benkovic and P. E. Wright, *Science*, 2011, **332**, 234–238.
40. D. D. Boehr, H. J. Dyson and P. E. Wright, *Chem. Rev.*, 2006, **106**, 3055–3079.
41. T. Pesnot, M. M. Palcic and G. K. Wagner, *ChemBioChem*, 2010, **11**, 1392–1398.
42. F. Exnowitz, B. Meyer and T. Hackl, *Biochim. Biophys. Acta*, 2012, **1824**, 443–449.
43. M. P. Kotzler, S. Blank, F. I. Bantleon, E. Spillner and B. Meyer, *Biochim. Biophys. Acta*, 2012, **1820**, 1915–1925.
44. M. P. Kotzler, S. Blank, F. I. Bantleon, M. Wienke, E. Spillner and B. Meyer, *ACS Chem. Biol.*, 2013, **8**, 1830–1840.
45. A. Mittermaier and L. E. Kay, *Science*, 2006, **312**, 224–228.
46. K. Henzler-Wildman and D. Kern, *Nature*, 2007, **450**, 964–972.
47. E. Z. Eisenmesser, O. Millet and W. Labeikovsky, *et al., Nature*, 2005, **483**, 117–121.
48. P. H. Chan, A. H. Cheung, M. Okon, H. M. Chen, S. G. Withers and L. P. McIntosh, *Biochemistry*, 2013, **52**, 320–332.

49. P. H. Chan, S. Weissbach, M. Okon, S. G. Withers and L. P. McIntosh, *Biochemistry*, 2012, **51**, 8278–8292.
50. P. H. Chan, L. L. Lairson, H. J. Lee, W. W. Wakarchuk, N. C. Strynadka, S. G. Withers and L. P. McIntosh, *Biochemistry*, 2009, **48**, 11220–11230.
51. J. E. Ollerenshaw, V. Tugarinov and L. E. Kay, *Magn. Reson. Chem.*, 2003, **41**, 843–852.
52. K. Kloiber, R. Spitzer, S. Grutsch, C. Kreutz and M. Tollinger, *J. Biomol. NMR*, 2011, **51**, 123–129.
53. G. T. Montelione and G. Wagner, *J. Am. Chem. Soc.*, 1989, **111**, 3096–3098.

CHAPTER 8

Stable Isotope Labeling of Glycoproteins for NMR Study

YOSHIKI YAMAGUCHI*[a], HIROKAZU YAGI[b] AND KOICHI KATO[b,c]

[a]Structural Glycobiology Team, Systems Glycobiology Research Group, RIKEN-Max Planck Joint Research Center, RIKEN Global Research Cluster, 2-1 Hirosawa, Wako, Saitama 351-0198, Japan; [b]Graduate School of Pharmaceutical Sciences, Nagoya City University, 3-1 Tanabe-dori, Mizuho-ku, Nagoya, 467-8603, Japan; [c]Institute for Molecular Science and Okazaki Institute for Integrative Bioscience, National Institutes of National Sciences, 5-1 Higashimaya Myodaiji, Okazaki 444-8787, Japan
*E-mail: yyoshiki@riken.jp

8.1 Introduction

Glycosylation is one of the most complex co-/post-translational modifications of proteins. In general, glycan structures of proteins are heterogeneous because the biosynthesis of carbohydrate chains does not proceed in a template-driven manner as in the translation of polypeptide chains.[1] In addition, glycosylation profiles are significantly species dependent, and even reflect physiological and pathological conditions.[2,3] Despite such complexity, there has been rapid progress in elucidating the physiological roles of the glycoprotein glycans in biological systems.[4–6] In the pharmaceutical context, protein glycosylation has been considered a key factor in developing and producing biopharmaceuticals.[7] One prominent example is the therapeutic application

New Developments in NMR No.10
NMR in Glycoscience and Glycotechnology
Edited by Koichi Kato and Thomas Peters

Published by the Royal Society of Chemistry, www.rsc.org

of glycoprotein antibodies, where glycan structure greatly influences effector function and immunogenicity.[8–10] Thus, it is increasingly important to elucidate the detailed relationship between glycan structure and glycoprotein function. Another well-studied example is the heavily glycosylated human immunodeficiency virus (HIV) protein gp120, one of the targets in the development of drugs and vaccines against HIV.[11,12] The antigenicity of the gp120 polypeptide is reduced by its extensive glycosylation, indicating that attention should be paid to the structure and dynamics of glycans when designing neutralizing antibodies.[13–15]

X-ray crystallography has been extensively applied to the 3D structural analysis of biomolecules at atomic-level resolution. Unfortunately, however, the glycans attached to proteins are seldom visualized by X-ray crystallography. Their electron densities are often missing because of their inherent flexibility.[16–18] In contrast, NMR spectroscopy plays a significant role in analyzing the structures, dynamics and interactions of glycoproteins. Site-specific ^{13}C labeling by synthetic approaches has provided valuable information for conformational study of mono- and oligo-saccharides[19–21] (see Chapter 2). Metabolic or enzymatic isotope labeling methods have opened new opportunities to study dynamic structures and interactions of glycoproteins.[22,23] The usefulness of NMR in structural glycoscience has been dramatically enhanced by the development of stable isotope labeling techniques. This chapter will outline the state of the art in isotope labeling of glycoproteins.

8.2 Isotope Labeling of Glycoproteins Using Mammalian Expression Systems

Clearly, bacterial expression systems such as *Escherichia coli* cannot produce glycoproteins unless glycosylation pathways are genetically introduced. Eukaryotic glycosylated proteins have been produced in extensively engineered *E. coli* containing a set of heterologous genes required for *N*-glycosylation.[24] At present, however, it is not practical to produce a protein with mammalian-type glycans in bacterial systems. Hence, appropriate non-bacterial expression systems have to be chosen to produce a glycoprotein for NMR study. Among the various glycoprotein-producing systems, mammalian cells are frequently used as vehicles for production of biologically functional glycoproteins with stable isotope labeling.[22,25] The cultivation of mammalian cells, however, does require rich media, which must include most amino acids, especially a considerable quantity (typically 450 mg L^{-1}) of L-glutamine, and metabolic carbon sources, *i.e.* glucose, pyruvate, and succinate, which makes production quite costly. In their pioneering work, Lustbader *et al.* successfully obtained heteronuclear NMR spectral data of uniformly ^{15}N-labeled and $^{13}C/^{15}N$-labeled human chorionic gonadotropin (hCG), prepared using Chinese hamster ovary (CHO) cells.[26,27] The stably transfected CHO cells were grown in a medium containing algal hydrolysates composed of isotope-labeled amino acids. The yield of isotopically labeled hCG was approximately 10 mg L^{-1} with greater than 90% enrichment of ^{13}C and ^{15}N.

Using hybridoma and CHO cells cultivated in synthetic media containing $^{15}N/^{13}C$-labeled algal amino acid mixture and additional components, uniformly labeled mouse IgG2b and recombinant mouse–human chimeric IgG1 glycoproteins have been prepared, respectively.[22,25] Backbone assignments of a human IgG1 Fc fragment with molecular mass 53 kDa were attained partially (66%)[28] and then almost completely (99%).[29] Intriguingly, a cluster of amino acid residues (Gln295–Thr299) in the vicinity of the Asn297-linked *N*-glycan exhibited significant chemical shift differences in comparison with the non-glycosylated Fc protein produced in *E. coli*.[30] A detailed protocol of stable isotope labeling of IgG has been published.[25]

The use of transient transfection systems is an alternative for producing glycosylated proteins in mammalian cells. Sastry *et al.* employed an adenovirus vector-based mammalian expression system to produce a ^{15}N or $^{15}N/^{13}C$-enriched recombinant glycoprotein corresponding to the outer domain variant of HIV-1 gp120 envelope proteins with 15 *N*-glycosylation sites.[31] Commercially available media (BioExpress® 6000, ^{15}N- and $^{15}N/^{13}C$-labeled) yielded 45 and 44 mg L^{-1} for the ^{15}N- and $^{15}N/^{13}C$-labeled glycoproteins, respectively, with an average of over 80% isotope incorporation. A notable point of the adenoviral expression system is that the level of expression was maintained even in 70% D_2O, which makes it promising for NMR analysis of larger glycoproteins (>20–30 kDa) with fractional deuteration.

8.3 Glycan Labeling of Glycoproteins

In addition to uniform labeling, selective labeling of the glycan moieties of glycoproteins can be achieved by metabolic or enzymatic means. D-Glucose is an essential metabolic precursor that reaches every sugar[32,33] (Figure 8.1), and the use of ^{13}C-labeled glucose in the cultivation of glycoprotein-producing mammalian cells metabolically labels all glycans[34] (Figure 8.2). Selective labeling of sugar residues can be achieved with isotopically labeled glucosamine, which is cell-permeable and metabolically converted to *N*-acetylglucosamine.[34] During the metabolic conversion of deuterated glucose to other monosaccharides, hydrogen can replace deuterium in the sugar rings to a varying extent depending on position and residue type. In fact metabolic deuterium-hydrogen exchange ratios allow mannose and fucose signals to be distinguished.[35] Such techniques facilitate spectral assignments of glycoprotein glycans.

Isotope-labeled sugar residues can be enzymatically attached onto the non-reducing termini of glycans using glycosyltransferases in *in vitro* reactions, and has been successful in the selective NMR observation of the non-reducing terminal residues of IgG–Fc.[34,36,37]

Metabolic ^{13}C labeling of glycoprotein glycans widens applications of ^{13}C-detection NMR measurements. In contrast to carbohydrate proton signals, which, in general, overlap considerably, ^{13}C signals of carbohydrates are moderately separated, which is advantageous for spectral editing. Figure 8.2E shows a 2D ^{13}C–^{13}C NOESY spectrum of the IgG–Fc glycoprotein with

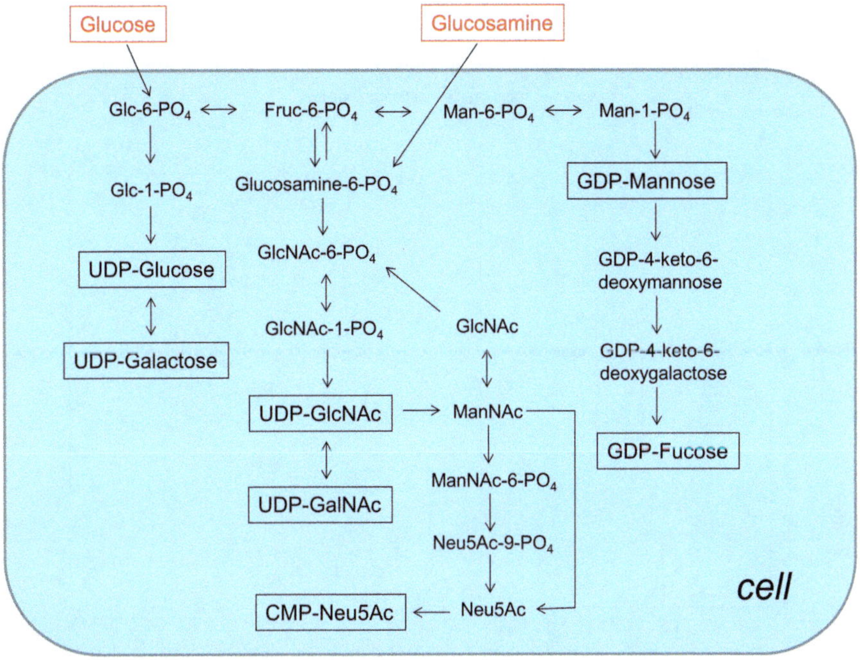

Figure 8.1 Metabolic conversion of sugar nucleotides in mammalian cells.

^{13}C-labeled glycans.[38] The peak patterns of intra-residue ^{13}C–^{13}C connectivity are useful for the classification of monosaccharide types with specific linkage configurations (Figure 8.2D). ^{13}C-detected 2D NMR spectra are especially helpful in analyses of larger glycoproteins with extensive overlap of proton signals.

8.4 Isotope Labeling of Glycoproteins Using Non-mammalian Expression Systems

8.4.1 Insect Cell–Silkworm Expression Systems

Non-mammalian expression systems have also been developed to overcome the disadvantages of mammalian systems, which include cost, time, and low yield. Insect cells are an alternative and are often used for the production of glycoproteins for structural studies. Walton *et al.* have developed a less costly approach using Sf9 cells.[39] Key strategies are to reduce the amount of glutamine in the medium (2 g L^{-1} to 500 mg L^{-1}) and to supply glucose and algal-derived amino acids. The yield of Thy-1 protein was 95 μg L^{-1} and roughly 80% of the expected peaks were observed in the ^{1}H–^{15}N HSQC spectrum. Furthermore, carbohydrate signals were observed in the 1D ^{13}C-edited ^{1}H spectrum, indicating direct incorporation of ^{13}C-labeled glucose into the glycan.

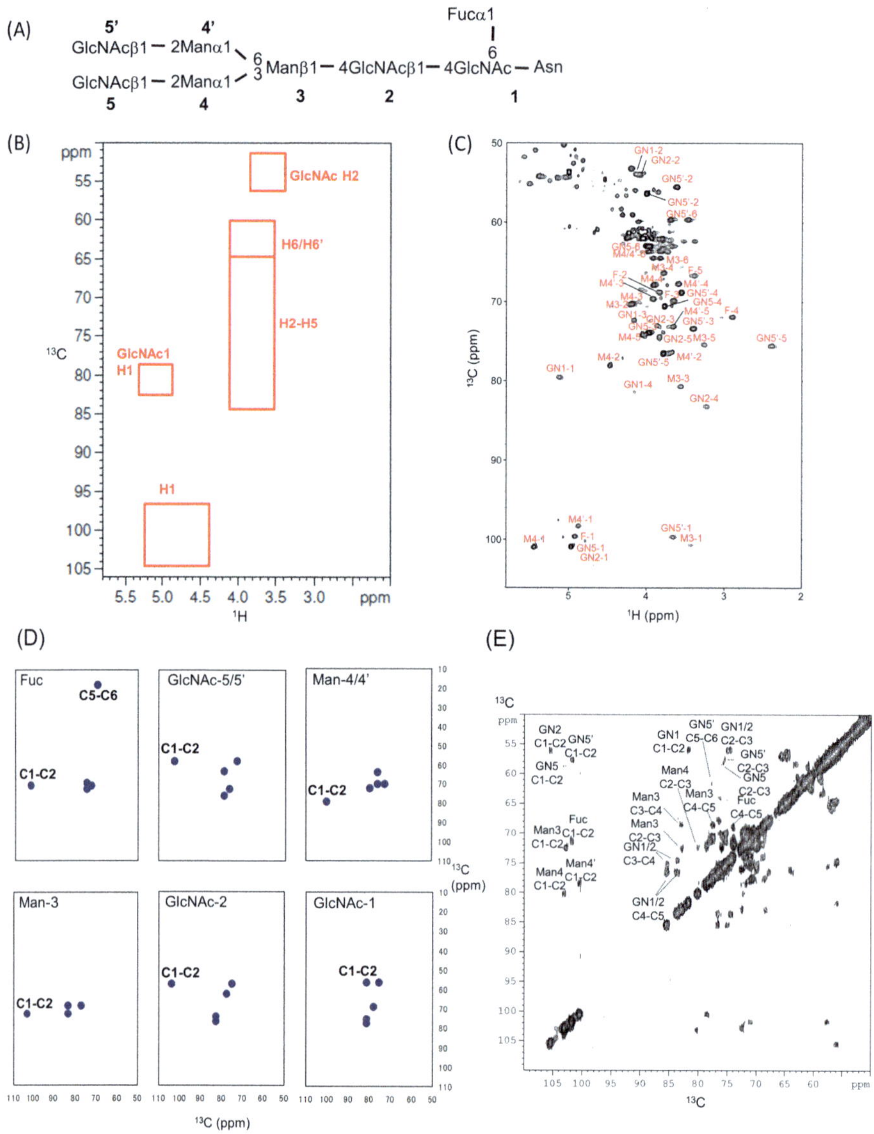

Figure 8.2 NMR analysis of glycans. (A) Structure of bi-antennary *N*-glycan. (B) Characteristic chemical shift values (^{1}H and ^{13}C) of carbohydrate signals on ^{1}H–^{13}C HSQC spectrum. (C) ^{1}H–^{13}C HSQC spectrum of [$^{13}C_6$] glucose-labeled murine IgG2b–Fc. The figure is adapted from Kato *et al.*[22] (D) Two-dimensional ^{13}C–^{13}C NOESY patterns expected for the sugar residues in a bi-antennary complex-type oligosaccharide. (E) Oligosaccharide region of the two-dimensional ^{13}C–^{13}C NOESY spectrum of ^{13}C-labeled murine IgG2b–Fc. This figure is adapted from Yamaguchi *et al.* with permission.[38]

Baculovirus-infected silkworm larvae have been developed as an expression vehicle for recombinant proteins.[40–42] Recently, living silkworm larvae reared with a ^{15}N-labeled artificial diet were used for the production of isotopically labeled IgG glycoprotein.[43] In this system, calnexin, a glycoprotein chaperone in the endoplasmic reticulum, was co-expressed to assist in the correct folding of the recombinant IgG molecules. The larvae were fed an isotope-labeled artificial diet, which included a protein mixture derived from the yeast strain *Candida utilis*. The yeast cells were grown in a culture medium containing ^{15}N-labeled ammonium sulfate as the nitrogen source. Baculovirus-containing hemolymph was collected from the larvae injected with individual bacmid-derived baculovirus expression vectors with IgG and calnexin cDNAs, and then injected into fifth-instar silkworm larvae for IgG glycoprotein expression. The larvae were reared on a ^{15}N-labeled artificial diet for 96 h, and their hemolymph then collected. From a single larva, 0.1 mg of IgG was harvested with ^{15}N-enrichment ratio of 80%, enabling ^{1}H–^{15}N HSQC spectral measurements. The major (90%) *N*-glycan glycoform of this IgG–Fc was of fucosylated paucimannose-type with $Man_{2\text{-}3}GlcNAc_2Fuc$ structures (Table 8.1). Heterologous expression vehicles do not necessarily produce the

Table 8.1 The structure and incidence of *N*-glycans derived from IgGs produced in silkworms, tobacco, and CHO cells.[28,43,46] The ratio (mol %) was calculated from the peak areas in HPLC profiles of pyridylaminated derivatives of *N*-glycans from IgGs. Key: ● Glucose, ● Mannose, ● Galactose, ▲ α1,6-Fucose, ▼ α1,3-Fucose, ■ *N*-acetylglucosamine, ★ Xylose.

Structures	mol%		
	Silkworm	Tobacco	CHO cell
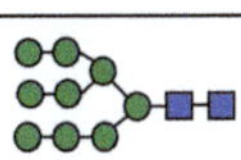	-	1.1	-
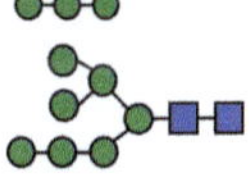	-	1.4	-
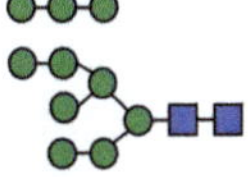	-	4.3	-
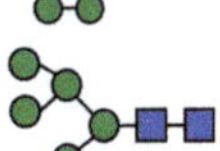	5.1	-	-
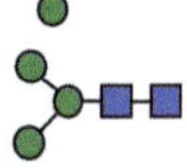	5.8	5.8	-
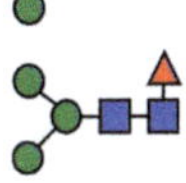	35.2	-	-
	52.2	-	-
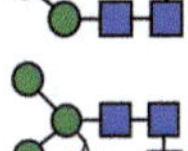	-	6.2	-

(continued)

Table 8.1 (*continued*)

Structures	mol%		
	Silkworm	Tobacco	CHO cell
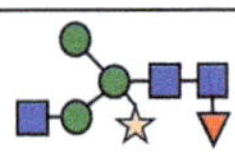	-	1.7	-
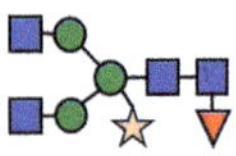	-	69.4	-
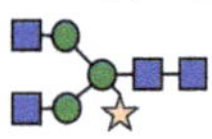	-	7.1	-
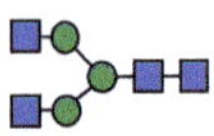	-	3.6	-
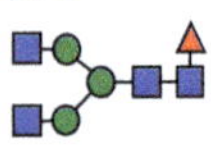	-	-	31.4
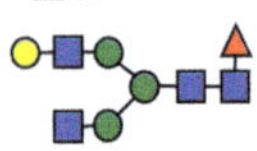	-	-	42.1
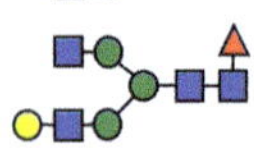	-	-	12.3
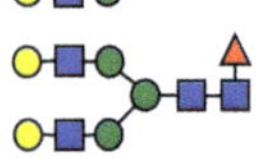	-	-	14.2
Others	1.0	5.2	-

desired result, although, in this instance, the NMR data indicated that the altered glycoforms had limited impact on Fc conformation in proximity to the *N*-glycans[43] (Figure 8.3).

8.4.2 Plant Expression Systems

Plant expression is now attracting attention from pharmaceutical fields.[44,45] A transgenic tobacco (*Nicotiana benthamiana*) has been tested for the expression of ^{15}N-labeled IgG.[46] Four-week-old seedlings were placed in a drip hydroponic system and cultivated for 49 days with an isotopically labeled Murashige and Skoog medium containing ^{15}N-labeled potassium nitrate and ^{15}N-labeled ammonium nitrate as major nitrogen sources. The yield of IgG from 1 g of the fresh leaves was 0.1 mg, which was commensurate with unlabeled medium cultivation, and ^{15}N-enrichment was in the 52–58% range. The *N*-glycans derived from the recombinant IgG glycoprotein produced in this protocol were plant-specific, mostly containing β1,2-xylose and α1,3-fucose (Table 8.1), and affected a limited region of the Fc glycoprotein[46] (Figure 8.3).

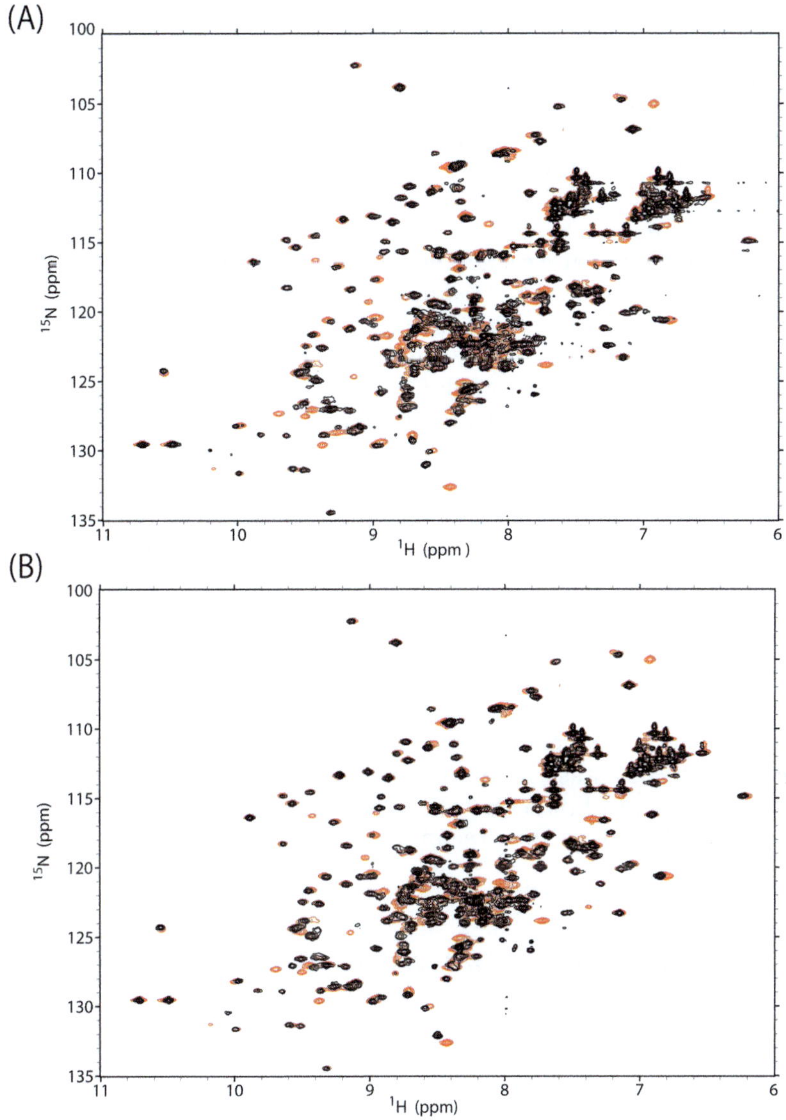

Figure 8.3 Expression of ^{15}N-labeled IgG in transgenic tobacco and silkworm. ^{1}H–^{15}N HSQC spectra of ^{15}N-labeled IgG–Fc are shown, which were obtained from transgenic tobacco plants (A, in black) and from silkworms (B, in black). For comparison, ^{1}H–^{15}H HSQC spectra of ^{15}N-labeled IgG–Fc produced in CHO cells are overlaid (A and B, in red).

8.4.3 Yeast Expression Systems

Yeast cells often produce large amounts of protein, rather like *E. coli*, and are therefore another candidate for glycoprotein-expression for cost-effective uniform isotope labeling. *Pichia pastoris* strain GS115 has been used for producing ^{15}N-labeled hCG and human follicle-stimulating hormone.[47] One 3L culture of *P. pastoris* using $^{15}NH_4Cl$/glucose-glycerol-methanol as nitrogen–carbon sources produced 25 mg of hCG with a ^{15}N enrichment of 70%. The glycan structure was of yeast-strain specific high-mannose type ranging from $Man_8GlcNAc_2$ to $Man_{11}GlcNAc_2$.

With accumulating knowledge of the glycan processing pathway in yeast cells, it is evident that genetic engineering is a feasible method to attain desired glycan structures.[48–50] Kamiya *et al.* describe a method for metabolic production of uniformly ^{13}C-labeled oligosaccharides using genetically engineered *Saccharomyces cerevisiae* cells. For isotope labeling, the yeast cells were grown in the medium containing ^{13}C-labeled glucose. In the engineered cells, homogeneous high-mannose-type oligosaccharide accumulates because of deletions of genes encoding three or four proteins involved in the processing pathway of *N*-linked oligosaccharides.[51,52] Position-selective labeling of the oligosaccharide using site-specific ^{13}C-labeled glucose isotopomers as metabolic precursors is a direct way to assign the non-anomeric sugar signals (Figure 8.4). Following this approach, analyses of high-mannose glycan conformations were performed using long-range NOE[52] and pseudo-contact shift data.[53]

Furthermore, a hybrid strategy has been developed that combines yeast genetic engineering and chemo-enzymatic synthesis to produce uniformly and selectively labeled high-mannose glycan (Figure 8.5).[54] Donor substrate UDP–[$^{13}C_6$]glucose was chemically synthesized from D-[$^{13}C_6$]glucose and the enzyme uridine diphosphate (UDP)-glucose:glycoprotein glucosyltransferase (UGGT) produced as a recombinant protein in *E. coli* on a milligram scale.

8.5 Concluding Remarks and Perspective

A variety of production vehicles is now available for metabolic stable isotope labeling of glycoproteins for NMR studies.[16] However, cost reduction, effective deuteration, and glycoform control remain challenges. Bacterial expression and cell-free expression of glycosylated proteins are formidable hurdles,[24,55,56] but will open up new horizons for NMR structural glycobiology dealing with these issues. In addition, sophisticated synthetic approaches in conjunction with *in vitro* enzymatic reaction[57–59] will facilitate the production of glycoproteins with tailored or even artificial glycoforms with highly specific isotope labeling. Development of these techniques will provide insights into basic and applied glycobiology, not only through the resulting NMR data but also through knowledge acquired during the processes of technical development.

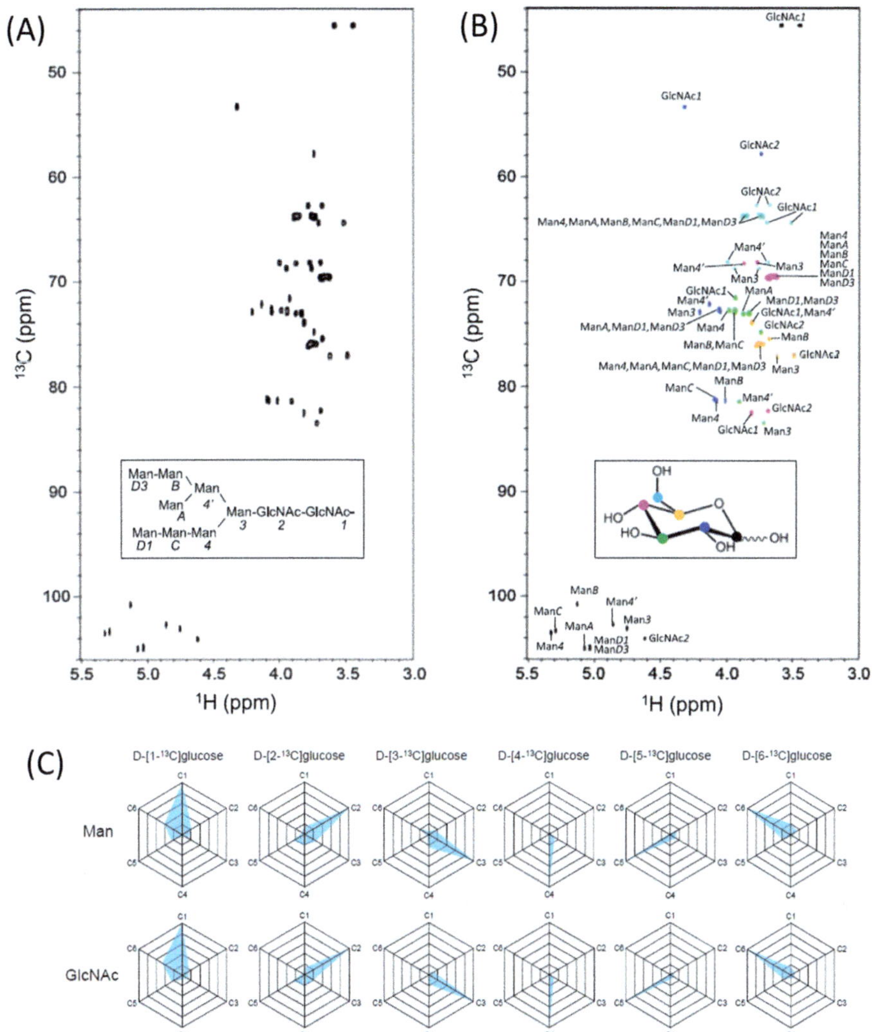

Figure 8.4 ^{1}H–^{13}C HSQC spectra of a fluorescently labeled $Man_8GlcNAc_2$ glycan, which is metabolically labeled with [$^{13}C_6$]glucose (A) or [x-^{13}C]glucose ($x = 1, 2, 3, 4, 5$ and 6) (B). In (B), isotope enrichment ratio was estimated at each position (C). These figures are adapted from Kamiya *et al.* with permission.[51]

Acknowledgements

This work was supported in part by the Program for the Promotion of Fundamental Studies in Health Sciences of the National Institute of Biomedical Innovation (NIBIO), the Japan Agency for Medical Research and Development, AMED (Project for utilizing glycans in the development of innovative drug discovery technologies), and Grants-in-Aids for Scientific Research (C) (JP25460054

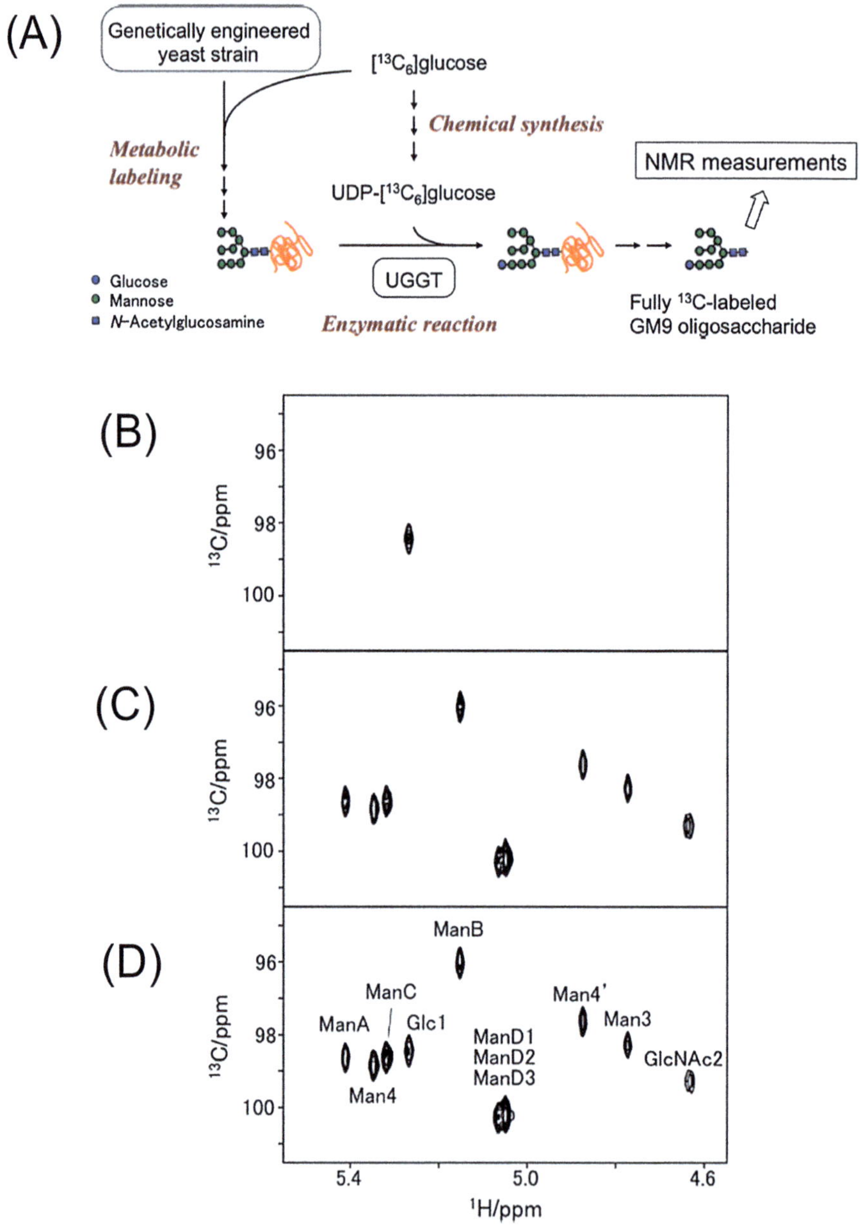

Figure 8.5 (A) A hybrid approach combining yeast engineering and chemo-enzymatic synthesis using a glucosyltransferase UGGT to produce uniformly and selectively ^{13}C-labeled dodecasaccharide. (B–D) ^{1}H–^{13}C HSQC spectra (anomeric region) of fluorescently labeled Glc_1Man_9, (B) only the terminal glucose residue was labeled with ^{13}C, (C) $Man_9GlcNAc_2$ part was labeled with ^{13}C, and (D) both glucose and $Man_9GlcNAc_2$ part were ^{13}C-labeled. This figure was adapted from Zhu *et al.* with permission.[54]

to Y. Y. and JP15K07935 to H. Y.), Scientific Research (B) (JP16H04758 to Y. Y.), Scientific Research (A) (JP24249002 to K. K.) and Scientific Research on Innovative Areas (JP20211849 to K. K.) from the Ministry of Education, Culture, Sports, Science and Technology (MEXT) of Japan.

References

1. P. M. Rudd and R. A. Dwek, *Crit. Rev. Biochem. Mol. Biol.*, 1997, **32**, 1.
2. K. Ohtsubo and J. D. Marth, *Cell*, 2006, **126**, 855.
3. K. Marino, J. Bones, J. J. Kattla and P. M. Rudd, *Nat. Chem. Biol.*, 2010, **6**, 713.
4. K. W. Moremen, M. Tiemeyer and A. V. Nairn, *Nat. Rev. Mol. Cell Biol.*, 2012, **13**, 448.
5. P. M. Rudd, T. Elliott, P. Cresswell, I. A. Wilson and R. A. Dwek, *Science*, 2001, **291**, 2370.
6. R. S. Haltiwanger and J. B. Lowe, *Annu. Rev. Biochem.*, 2004, **73**, 491.
7. H. Li and M. d'Anjou, *Curr. Opin. Biotechnol.*, 2009, **20**, 678.
8. Y. Yamaguchi, N. Takahashi and K. Kato, in *Comprehensive Glycoscience*, ed. J. P. Kamerling, Elsevier, Oxford, 2007, vol. 3, p. 745.
9. L. Liu, *J. Pharm. Sci.*, 2015, **104**, 1866.
10. R. Jefferis, *Nat. Rev. Drug Discovery*, 2009, **8**, 226.
11. P. D. Kwong, R. Wyatt, J. Robinson, R. W. Sweet, J. Sodroski and W. A. Hendrickson, *Nature*, 1998, **393**, 648.
12. R. Wyatt, P. D. Kwong, E. Desjardins, R. W. Sweet, J. Robinson, W. A. Hendrickson and J. G. Sodroski, *Nature*, 1998, **393**, 705.
13. K. J. Doores, *FEBS J.*, 2015, **282**, 4679.
14. S. Horiya, I. S. MacPherson and I. J. Krauss, *Nat. Chem. Biol.*, 2014, **10**, 990.
15. R. Pejchal, K. J. Doores, L. M. Walker, R. Khayat, P. S. Huang, S. K. Wang, R. L. Stanfield, J. P. Julien, A. Ramos, M. Crispin, R. Depetris, U. Katpally, A. Marozsan, A. Cupo, S. Maloveste, Y. Liu, R. McBride, Y. Ito, R. W. Sanders, C. Ogohara, J. C. Paulson, T. Feizi, C. N. Scanlan, C. H. Wong, J. P. Moore, W. C. Olson, A. B. Ward, P. Poignard, W. R. Schief, D. R. Burton and I. A. Wilson, *Science*, 2011, **334**, 1097.
16. Y. Kamiya, T. Satoh and K. Kato, *Curr. Opin. Struct. Biol.*, 2014, **26**, 44.
17. M. Nagae and Y. Yamaguchi, *Int. J. Mol. Sci.*, 2012, **13**, 8398.
18. M. R. Wormald, A. J. Petrescu, Y. L. Pao, A. Glithero, T. Elliott and R. A. Dwek, *Chem. Rev.*, 2002, **102**, 371.
19. W. Zhang, S. Zhao and A. S. Serianni, *Methods Enzymol.*, 2015, **565**, 423.
20. B. Bose-Basu, T. Klepach, G. Bondo, P. B. Bondo, W. Zhang, I. Carmichael and A. S. Serianni, *J. Org. Chem.*, 2007, **72**, 7511.
21. T. E. Klepach, I. Carmichael and A. S. Serianni, *J. Am. Chem. Soc.*, 2005, **127**, 9781.
22. K. Kato, Y. Yamaguchi and Y. Arata, *Prog. Nucl. Magn. Reson. Spectrosc.*, 2010, **56**, 346.

23. Y. Yamaguchi and K. Kato, in *Modern Magnetic Resonance*, ed. G. A. Webb, Springer-Verlag, The Netherlands, 2007, vol. 1, p. 219.
24. J. D. Valderrama-Rincon, A. C. Fisher, J. H. Merritt, Y. Y. Fan, C. A. Reading, K. Chhiba, C. Heiss, P. Azadi, M. Aebi and M. P. DeLisa, *Nat. Chem. Biol.*, 2012, **8**, 434.
25. Y. Yamaguchi and K. Kato, *Methods Enzymol.*, 2010, **478**, 305.
26. J. W. Lustbader, S. Birken, S. Pollak, A. Pound, B. T. Chait, U. A. Mirza, S. Ramnarain, R. E. Canfield and J. M. Brown, *J. Biomol. NMR*, 1996, **7**, 295.
27. C. T. Weller, J. Lustbader, K. Seshadri, J. M. Brown, C. A. Chadwick, C. E. Kolthoff, S. Ramnarain, S. Pollak, R. Canfield and S. W. Homans, *Biochemistry*, 1996, **35**, 8815.
28. Y. Yamaguchi, M. Nishimura, M. Nagano, H. Yagi, H. Sasakawa, K. Uchida, K. Shitara and K. Kato, *Biochim. Biophys. Acta*, 2006, **1760**, 693.
29. H. Yagi, Y. Zhang, M. Yagi-Utsumi, T. Yamaguchi, S. Iida, Y. Yamaguchi and K. Kato, *Biomol. NMR Assignments*, 2015, **9**, 257.
30. D. Liu, M. J. Cocco, R. Rosenfied, J. K. Lewis, D. Ren, L. Li, R. L. Remmele Jr and D. N. Brems, *Biomol. NMR Assignments*, 2007, **1**, 233.
31. M. Sastry, L. Xu, I. S. Georgiev, C. A. Bewley, G. J. Nabel and P. D. Kwong, *J. Biomol. NMR*, 2011, **50**, 197.
32. A. Varki, *FASEB J.*, 1991, **5**, 226.
33. A. Varki, *Methods Enzymol.*, 1994, **230**, 16.
34. Y. Yamaguchi, K. Kato, M. Shindo, S. Aoki, K. Furusho, K. Koga, N. Takahashi, Y. Arata and I. Shimada, *J. Biomol. NMR*, 1998, **12**, 385.
35. Y. Yamaguchi, T. Takizawa, K. Kato, Y. Arata and I. Shimada, *J. Biomol. NMR*, 2000, **18**, 357.
36. A. M. Gilhespy-Muskett, J. Partridge, R. Jefferis and S. W. Homans, *Glycobiology*, 1994, **4**, 485.
37. A. W. Barb, L. Meng, Z. Gao, R. W. Johnson, K. W. Moremen and J. H. Prestegard, *Biochemistry*, 2012, **51**, 4618.
38. Y. Yamaguchi, M. Walchli, M. Nagano and K. Kato, *Carbohydr. Res.*, 2009, **344**, 535.
39. W. J. Walton, A. J. Kasprzak, J. T. Hare and T. M. Logan, *J. Biomol. NMR*, 2006, **36**, 225.
40. M. Hiyoshi, A. Kageshima, T. Kato and E. Y. Park, *J. Virol. Methods*, 2007, **144**, 91.
41. T. Motohashi, T. Shimojima, T. Fukagawa, K. Maenaka and E. Y. Park, *Biochem. Biophys. Res. Commun.*, 2005, **326**, 564.
42. T. Kato, M. Kajikawa, K. Maenaka and E. Y. Park, *Appl. Microbiol. Biotechnol.*, 2010, **85**, 459.
43. H. Yagi, M. Nakamura, J. Yokoyama, Y. Zhang, T. Yamaguchi, S. Kondo, J. Kobayashi, T. Kato, E. Y. Park, S. Nakazawa, N. Hashii, N. Kawasaki and K. Kato, *J. Biomol. NMR*, 2015, **62**, 157.
44. V. Gomord, P. Chamberlain, R. Jefferis and L. Faye, *Trends Biotechnol.*, 2005, **23**, 559.
45. L. Joshi and L. C. Lopez, *Curr. Opin. Plant Biol.*, 2005, **8**, 223.

46. H. Yagi, N. Fukuzawa, Y. Tasaka, K. Matsuo, Y. Zhang, T. Yamaguchi, S. Kondo, S. Nakazawa, N. Hashii, N. Kawasaki, T. Matsumura and K. Kato, *Plant Cell Rep.*, 2015, **34**, 959.
47. V. Blanchard, R. A. Gadkari, A. V. George, S. Roy, G. J. Gerwig, B. R. Leeflang, R. R. Dighe, R. Boelens and J. P. Kamerling, *Glycoconjugate J.*, 2008, **25**, 245.
48. S. R. Hamilton and T. U. Gerngross, *Curr. Opin. Biotechnol.*, 2007, **18**, 387.
49. S. Wildt and T. U. Gerngross, *Nat. Rev. Microbiol.*, 2005, **3**, 119.
50. Y. Chiba and Y. Jigami, *Curr. Opin. Chem. Biol.*, 2007, **11**, 670.
51. Y. Kamiya, S. Yamamoto, Y. Chiba, Y. Jigami and K. Kato, *J. Biomol. NMR*, 2011, **50**, 397.
52. Y. Kamiya, K. Yanagi, T. Kitajima, T. Yamaguchi, Y. Chiba and K. Kato, *Biomolecules*, 2013, **3**, 108.
53. T. Yamaguchi, Y. Sakae, Y. Zhang, S. Yamamoto, Y. Okamoto and K. Kato, *Angew. Chem., Int. Ed. Engl.*, 2014, **53**, 10941.
54. T. Zhu, T. Yamaguchi, T. Satoh and K. Kato, *Chem. Lett.*, 2015, **44**, 1744.
55. C. Guarino and M. P. DeLisa, *Glycobiology*, 2012, **22**, 596.
56. S. I. van Kasteren, H. B. Kramer, H. H. Jensen, S. J. Campbell, J. Kirkpatrick, N. J. Oldham, D. C. Anthony and B. G. Davis, *Nature*, 2007, **446**, 1105.
57. Y. Kajihara, Y. Tanabe, S. Sasaoka and R. Okamoto, *Chemistry*, 2012, **18**, 5944.
58. M. Umekawa, T. Higashiyama, Y. Koga, T. Tanaka, M. Noguchi, A. Kobayashi, S. Shoda, W. Huang, L. X. Wang, H. Ashida and K. Yamamoto, *Biochim. Biophys. Acta*, 2010, **1800**, 1203.
59. W. Huang, J. Giddens, S. Q. Fan, C. Toonstra and L. X. Wang, *J. Am. Chem. Soc.*, 2012, **134**, 12308.

CHAPTER 9

Quantifying Carbohydrate Motions Through Solution Measurements: Applications to Immunoglobulin G Fc

ADAM W. BARB[a]

[a]Roy J. Carver Department of Biochemistry, Biophysics and Molecular Biology, Iowa State University, Ames, IA 50011, USA
*E-mail: abarb@iastate.edu

9.1 Introduction

9.1.1 General Glycoprotein

Glycoproteins are polypeptides covalently modified with a carbohydrate and are predicted to represent at least 50% of the human proteome.[1] The most common form is asparagine(*N*)-linked glycosylation that occurs when the polypeptide chain is exported from the ribosome and imported into the lumen of the endoplasmic reticulum (ER). An amino acid sequence of three residues, Asn-X-(Ser/Thr) where X is any residue other than P, signals the oligosaccharyl transferase to transfer *en bloc* a fourteen residue carbohydrate chain from a dolichol-linked donor to the Asn sidechain amide nitrogen (see ref. 2). Following proper folding of the polypeptide, the *N*-glycan is remodeled through the action of glycosylhydrolases and glycosyltransferases

New Developments in NMR No.10
NMR in Glycoscience and Glycotechnology
Edited by Koichi Kato and Thomas Peters

Published by the Royal Society of Chemistry, www.rsc.org

during transit of the protein through the secretory pathway of the ER and Golgi. While *N*-glycosylation sites are genetically encoded, glycan remodeling in the secretory pathway occurs as a template-independent process, introducing significant heterogeneity. Though this heterogeneity may represent an obstacle to researchers, there is a growing body of evidence that the body controls cell–cell communication by tuning the glycan distribution itself.

Mature *N*-glycans on mammalian proteins are often of a complex type with two, three, four or more branches (see Figure 9.1C for a diagram of biantennary *N*-glycan) and contain at least *N*-acetylglucosamine (GlcNAc) and mannose with galactose, fucose and *N*-acetylneuraminic acid residues as common modifications. The number of potential glycoforms is enormous considering the variation in linkage position and stereochemistry as well as a number of potential modifications, though the distribution of any single *N*-glycan at a single protein site is thought to be relatively restricted due to local epitope environment factors that influence *N*-glycan processing.

9.1.2 Glycoprotein Challenges

N-glycans are known to contribute to protein function, protein thermodynamic stability, stability in the serum and harbor epitopes for receptor binding (see ref. 3). Identifying the structural and functional role of *N*-glycans is often challenging due to conformational and configurational heterogeneity. Dynamic moieties like *N*-glycans inhibit crystallization, limiting studies of intact glycoproteins by X-ray crystallography. In fact, the carbohydrates are often removed entirely or dramatically shortened to enhance crystallization. While this is important for describing the structure of the underlying polypeptide, these studies provide little structural insight into the diversity of *N*-glycan roles. Solution nuclear magnetic resonance (NMR) spectroscopy is uniquely suited to structure-based studies of dynamic molecules including glycoproteins and does not require *N*-glycan cleavage. This chapter is devoted to describing the preparation and characterization of *N*-glycoproteins using solution NMR spectroscopy, though the expression and remodeling techniques described are widely suited to protein preparation for many other techniques as well. The discussion is focused on studies of a model glycoprotein, the Fragment crystallizable (Fc) region of human immunoglobulin G1 (IgG1).

9.1.3 IgG1 Fc

IgG1 Fc is the stalwart target of many investigations into the impact of glycosylation on the structure/function relationship. IgG1 is produced by the immune system to specifically recognize foreign material in the body and target this material for destruction.[4] Once bound to a pathogen epitope, the Fc of IgG1 is recognized by a cell surface Fc γ receptor (FcγR) in one, of multiple, potential signaling cascades (Figure 9.1). This interaction and

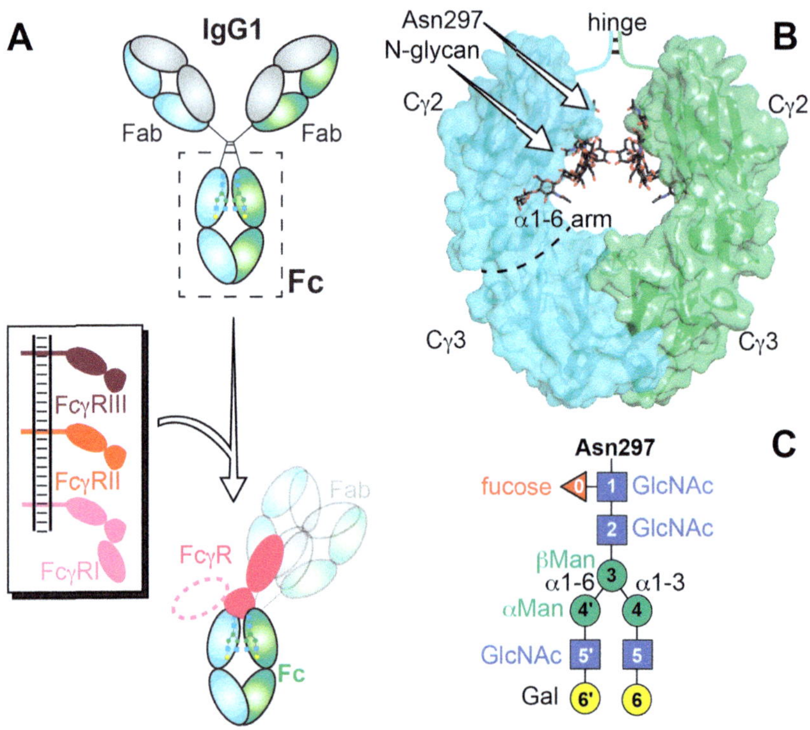

Figure 9.1 Structure and function of IgG1 Fc. (**A**) IgG1 binds to multiple receptors through the Fc region. (**B**) The Fc *N*-glycan is required for interactions with many receptors and is found buried between two polypeptide domains in models by X-ray crystallography. This model is based on PDB 4kui. (**C**) The most abundant Fc *N*-glycans are core-fucosylated and of a biantennary complex-type. The immunoglobulin G heavy chains are shown in *green* or *cyan* and the light chains in *gray*.

subsequent signaling event require the presence of an *N*-glycan at Asn297 of the IgG1 Fc. Removing the *N*-glycan abrogates signaling, a strategy utilized by *Streptococcus* pathogens that enzymatically cleave the *N*-glycan of serum IgG.[5] Furthermore, the activity of therapeutic monoclonal antibodies requires *N*-glycosylation for full activity to treat many cancers, autoimmune disorders and other diseases.

IgG1 is formed by two longer "heavy chain" and two light chain polypeptides (Figure 9.1). The Fc is a homodimer formed by the C-terminal portion of the two heavy chains and can be proteolytically cleaved from the antigen-binding Fab domains to retain full receptor binding capacity in an *in vitro* Fc: FcγR binding assay. The first high-resolution structure of the Fc was solved forty years ago and surprisingly revealed near complete resolution of the *N*-glycan in electron density maps.[6] The *N*-glycan forms an interface with the surface of the Fc polypeptide and occupies a space formed between the two chains of the Fc homodimer (Figure 9.1B). As will

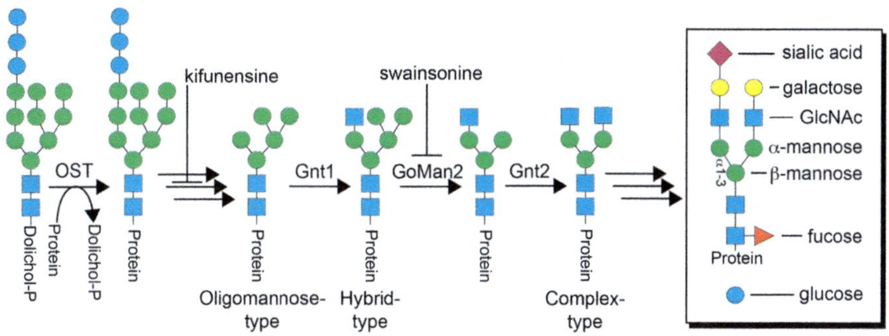

Figure 9.2 The *N*-glycan remodeling pathway. Abbreviations: GoMan2, Golgi mannosidase 2; Gnt 1/2, *N*-acetylglucosaminyltransferase 1/2; OST-oligosaccharyltransferase.

be discussed, despite its static appearance in X-ray crystallography data, the *N*-glycan is quite mobile in solution and contributes directly to Fc structure and function.

The predominant IgG1 Fc glycoforms are of a core fucosylated, biantennary complex-type with zero (~30–35%), one (~35%) or two (~20%) terminal galactose residues (See Figure 9.2). Sialylated forms account for 5–10% of the pool with other minor forms (including those lacking fucose) accounting for <1% each. It was found that the distribution of glycoforms shifts to under-galactosylated forms in rheumatoid arthritis patients[7] and changes to the glycan composition affect Fc affinity for at least one FcγR (IIIa).[8,9] Many groups have focused their efforts on characterizing the structure and motion of the Fc *N*-glycan. This review aims to describe some of the techniques used in these studies and the results.

9.2 Methods to Prepare Labeled Glycoproteins

9.2.1 Glycan Homogeneity is Crucial

A significant obstacle prohibiting many structure-based studies of glycoproteins is preparing samples with limited chemical heterogeneity; a high degree of homogeneity is optimal. Methods to express *N*-glycosylated proteins generally require the use of mammalian cells to generate complex-type *N*-glycosylation and are covered in detail elsewhere,[10,11] though an engineered baculovirus system,[12] a chemoenzymatic method using protein expression in *Escherichia coli* with a transglycosidase-catalyzed step[13] and purely synthetic approach are described.[14] High potential protein yields are an advantage of the mammalian expression systems, though the standard cell lines result in a distribution of glycoforms due to the template-independent synthesis of complex-type *N*-glycans in the Golgi.[2] Glycan heterogeneity can be addressed with engineered cell lines or *in vitro* enzymatic remodeling of the glycans post-purification.

9.2.1.1 Genetic Manipulation of Mammalian Cells for Homogenous Glycoprotein Expression

Gnt1$^{-/-}$ cells are deficient in the key step in the pathway that converts oligomannose-type glycans to complex-type structures as shown in Figure 9.2 and were first identified in a screen of mutagenized cells resistant to the lectin phytohemagglutinin.[15] *Gnt1*$^{-/-}$ cells are particularly advantageous for glycoprotein expression and express proteins with predominantly oligomannose (Man5) glycan structures that have been utilized directly or enzymatically converted with a single GlcNAc residue at the α1-3Man branch.[16] Furthermore, oligomannose glycans are sensitive to a range of endoglycosidase enzymes that remove the glycan leaving only a single GlcNAc residue (EndoH/F) or removing it entirely and converting the linking asparagine residue to an aspartate (PNGaseF). The presence of a single GlcNAc residue is often sufficient to preserve the stabilizing properties of the entire *N*-glycan[17] and is used to prepare samples for crystallization.[18] Other examples of engineering the *N*-glycan remodeling pathway[19] or co-expression with glycan-modifying enzymes[20,21] are also applicable to generating homogenous glycoproteins.

9.2.1.2 Chemical Manipulation of Mammalian Cells for Homogenous Glycoprotein Expression

Many commercially available chemical inhibitors of *N*-glycan processing are bioactive and applicable to glycoprotein expression. Kifunensine and swainsonine are potent natural products that inhibit Golgi α-mannosidases and result in proteins expressed with glycans similar to the Man5 structure previously discussed[22–24] and shown in Figure 9.2.

Fluorinated sugar analogs are also potent inhibitors of *N*-glycan maturation. 2-deoxy-2-fluoro fucose inhibits protein fucosylation and was used to prepare immunoglobulin G with <5% fucosylation.[10,25] Fluorine substitution of the axial proton at the 3 position of *N*-acetylneuraminic acid resulted in the complete loss of sialylation in HL-60 cells and is likely applicable to glycoprotein expression.[26] Unfortunately the scope of fluorinated sugars as modulators of *N*-glycan remodeling appears limited as 2-deoxy-2-fluoro analogs of mannose and galactose and 4-deoxy-4-fluoro GlcNAc inhibit general glycoprotein production.[27–29]

9.2.2 Expression-Based Methods for Polypeptide and Glycan Labeling

Incorporating stable isotope labels into glycoproteins represents a significant challenge. Standard multidimensional solution NMR experiments of large proteins require high levels of ^{13}C and ^{15}N labeling, which is supported by culturing *Escherichia coli* in the presence of the metabolic precursors [U–^{13}C] glucose and $^{15}NH_4Cl$ in a chemically defined medium, however, this system is not suitable for expressing glycoproteins with mammalian-type *N*-glycosylation.

9.2.2.1 *Amino Acid Labeling Strategies*

The use of a chemically defined medium is essential for high levels of isotope incorporation; this is in contrast to many mammalian cell culture medium preparations with significant serum content. [amide–^{15}N]lysine labeling of the human CD2 glycoprotein expressed using Chinese hamster ovary (CHO) cells in MEM medium[30] and the later analyses of CD2 structure and conformation represent a significant historical achievement in glycoprotein studies.[31,32] Selective or complete amino acid labeling in mammalian cell expression systems is achieved by supplementing an amino-acid-free medium with labeled amino acids. Kato and coworkers prepared human [^{13}C/^{15}N]IgG1 Fc for NMR experiments and reported the near-complete assignment of backbone resonances.[9,33] Needless to say, medium supplementation with 20 ^{13}C/^{15}N-labeled amino acids and other metabolites can be cost-prohibitive when considering the cost of human cell tissue culture; however, selective amino acid ^{15}N-labeling is quite cost effective, and minimal amide scrambling occurs with mammalian cultures in an otherwise ^{14}N medium (unenriched).[34–36]

Mammalian systems are not expected to support uniform ^{2}H labeling and might limit NMR studies on very large glycoproteins and glycoprotein complexes. In select cases the yeast *Pichia pastoris* may represent an economical option for protein expression in a system that supports high levels of ^{2}H, ^{13}C(from deuterated methanol) and ^{15}N(NH_4Cl) labeling.[37,38] Unfortunately, fungi including *Pichia*, synthesize high-mannose *N*-glycans rather than the complex or hybrid types found on human proteins though some success with engineered strains is reported.[39,40]

9.2.2.2 *Carbohydrate Labeling During Protein Expression*

Supplementing the expression medium as described by Kato and coworkers will lead to uniform carbohydrate labeling.[9] Alternatively, the same laboratory also determined that specific and uniform *N*-glycan labeling is achievable by adding [U–^{13}C]glucose to the medium.[41] In our laboratory, incorporating 3 gL^{-1} glucose concentrations was sufficient to label carbohydrate residues with few observable polypeptide signals.[42] Furthermore, protein expression in the presence of [1–^{13}C]glucose led to anomeric labeling of each *N*-glycan residue, dramatically reducing spectral complexity and revealing the potential for selective labeling at other positions.[43]

9.2.3 ***In vitro* Methods for Glycan Labeling**

Labeling specific carbohydrate residues provides high sensitivity and specificity to measurements of glycoprotein glycan structure and motion. Incorporating glycan labels post-purification extends NMR capabilities to both recombinant proteins and glycoproteins purified from natural sources. This section will describe aspects of substrate preparation and the enzyme-catalyzed labeling reactions. Techniques to assess the efficiency of labeling reactions through reliable chromatography and mass spectrometric methods are described elsewhere.[44,45]

9.2.3.1 Preparing Labeled Sugar Donor Substrates

Sugar nucleotides donate residues to *N*-glycan chains in reactions catalyzed by glycosyltransferase enzymes. Schemes to prepare isotope-enriched donor molecules for *in vitro* reactions are shown in Figure 9.3.

The six-step synthesis of UDP-[U–^{13}C,^{15}N]*N*-acetylglucosamine (GlcNAc) from [U–^{13}C]glucose and [amido–^{15}N]glutamine was demonstrated using a one-pot reaction catalyzed by five enzymes and is shown in Figure 9.3.[16] The first two enzymes of the pathway are commercially available, and the last three must be prepared though expression clones for these enzymes are accessible. This strategy recapitulates the bacterial pathway for UDP-GlcNAc synthesis and resulted in the recovery of 17% of the starting glucose labels in the form of UDP-[U–^{13}C,^{15}N]GlcNAc. Alternatively, labeled GlcNAc can be purchased and converted to GlcNAc-1-P with a GlcNAc kinase and then to UDP-GlcNAc with a uridylyltransferase.[46–50] Chemical routes are also described.[51] In a further manipulation, an epimerase efficiently converted UDP-[U–^{13}C,^{15}N]GlcNAc to UDP-[U–^{13}C,^{15}N]*N*-acetylgalactosamine for use in labeling studies of glycosaminoglycans, O-linked glycans and glycosphingolipids.[52]

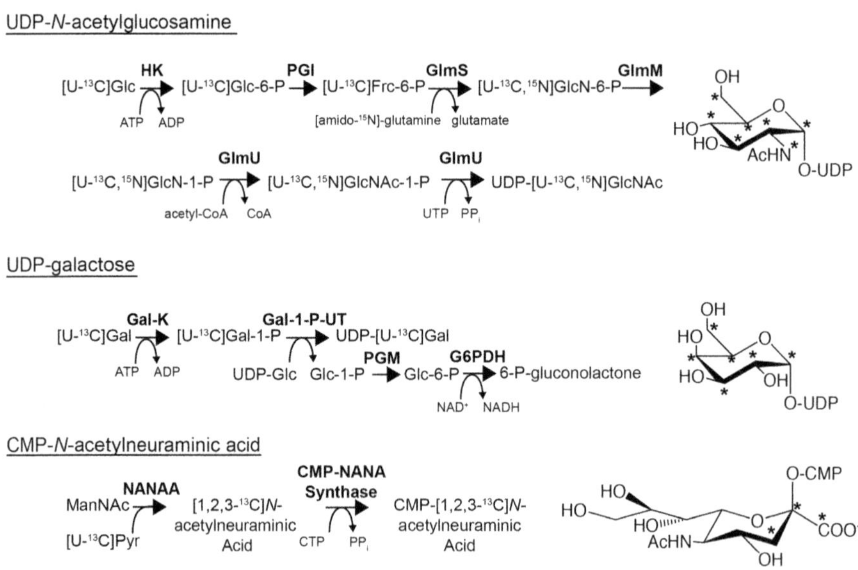

Figure 9.3 Strategies for enzymatically preparing labeled sugar nucleotide donors. Abbreviations: CMP-NANA synthase- CMP *N*-acetylneuraminic acid synthase, G6PDH- glucose-6-phosphate dehydrogenase, Gal-1-P-UT- galactose-1-phosphate uridylyltransferase, GalK- galactokinase, GlmM- phosphoglucosamine mutase, GlmS- glutamine-fructose-6-phosphate aminotransferase, GlmU- bifunctional glucosamine-1-phosphate *N*-acetyltransferase/UDP-*N*-acetylglucosamine pyrophosphorylase, HK- hexokinase, NANA-*N*-acetylneuraminic acid aldolase, PGI-phosphoglucose isomerase, PGM- phosphoglucomutase.

UDP-[U–^{13}C]galactose can be synthesized by utilizing the first two enzymes of the Leloir galactose metabolism pathway as shown in Figure 9.3[53] starting with [U–^{13}C]galactose.[54] The first enzyme in the pathway, galactokinase, may be difficult to source commercially though it can be prepared as described.[55] The second step catalyzed by galactose-1-phosphate uridylyltransferase requires UDP-Glc and is governed by a $K_{eq} \sim 1$. Glc-1-P is efficiently removed by phosphoglucomutase and an NAD$^+$-dependent glucose-6-P-dehydrogenase to drive the synthesis of UDP-[U–^{13}C]galactose to completion.

CMP-[1,2,3–^{13}C]*N*-acetylneuraminic acid preparation is relatively straightforward and only requires one or two enzymes. [1,2,3–^{13}C]*N*-acetylneuraminic acid can be obtained in one of two ways. Forcing the degradative *N*-acetylneuraminic acid aldolase in reverse will condense pyruvate with *N*-acetylmannosamine.[56,57] Alternatively, digesting a polysialic acid polymer excreted by certain *E. coli* strains grown in labeling medium will generate [U–^{13}C,^{15}N]*N*-acetylneuraminic acid.[58] A single synthetase efficiently converts CTP and *N*-acetylneuraminic acid to CMP-*N*-acetylneuraminic acid and inorganic pyrophosphate.

9.2.3.2 *Glycosylhydrolases and Transferases*

Enzymes are powerful tools for *in vitro* glycan remodeling due to high efficiency and specificity. In principle, the distribution of mammalian glycoforms on a single purified protein can be remodeled *in vitro* to a single species due to the conserved pentasaccharide core structure. In practice, however, enzymes to remove each modification from the expression are not uniformly available and active toward protein-bounds substrates. Here the discussion is focused on enzymes used to remodel the IgG Fc *N*-glycan to homogeneity, a notoriously difficult remodeling target, with a brief discussion of applications to more elaborate glycan structures found on other proteins. There are many good commercial sources for glycan modifying enzymes, particularly endoglycosidases. Glycosyltransferases can be harder to find, though many expression clones for soluble versions of glycan-modifying enzymes are freely available from an NIH-funded resource (http://glycoenzymes.ccrc.uga.edu).

Methods to add and remove galactose and *N*-acetylneuraminic acid residues are well described.[41,57,59,60] It is interesting to note that bovine lactose synthase (a β1-4 galactosyltransferase) is widely available and capable of modifying unsubstituted β-linked GlcNAc residues. Branch specificities of both the β1-4 galactosyltransferase and α2-6 sialyltransferase have been reported though the bovine and human forms will transfer to both branches of the biantennary IgG1 Fc *N*-glycan.[61,62] It should be noted that while addition of an α2-6-linked *N*-acetylneuraminic acid to the (a1-3Man-linked)branch of the IgG1 Fc *N*-glycan is reasonably efficient, addition of the second residue is extremely slow and requires large quantities of enzyme and CMP-*N*-acetylneuraminic acid.[60]

Though removing GlcNAc from *N*-glycans is straightforward with an *N*-acetylhexosaminidase, adding GlcNAc residues to acceptor glycoproteins requires position-specific GlcNAc transferases. Gnt1 (also called lec1 or MGAT1) transfers a GlcNAc residue to the 2 hydroxyl of the α1-3mannose residue of a Man5 *N*-glycan. The human Gnt1 enzyme will also recognize a Man3 paucimannose structure with high efficiency.[16] A second GlcNAc can be efficiently added using the human Gnt2 (MGAT2) after Gnt1 to form the core of a complex-type, biantennary *N*-glycan. Remaining GlcNAc transferases likely will serve as efficient catalysts to generate high levels of bisected (Gnt3; MGAT3), and tri-or tetra-antennary glycans (Gnt4/5; MGAT4/5). Additional modifications include fucosylation and sialylation.

Chemoenzymatic methods of glycan remodeling permit one-step *en bloc* remodeling reactions that may prove useful for incorporating labeled residues.[63,64] Recent applications of this technology utilize an oxazaline donor molecule and are applicable to at least core fucosylated, disialylated complex-type biantennary and Man9 structures.[13,65]

9.3 NMR Techniques to Probe Glycan Mobility

Solution NMR spectroscopy is particularly well suited to studies of dynamic molecules to atomic resolution. Signals in NMR spectra reflect the atom-specific chemical environment as well motion on multiple time scales, and NMR experiments can be used to characterize these molecular features with high precision. NMR spectroscopy of glycoproteins is highly similar to studies of proteins or nucleic acids, and many resources are available that cover the theoretical and practical aspects of characterizing these phenomena.[66,67] This section seeks not to replicate these texts, but rather offer a concise introduction to certain approaches used with glycoprotein glycans.

9.3.1 Resonance Frequencies are Sensitive to Motion

The features observed in a simple one-dimensional NMR experiment are applicable to the higher two, three and four-dimensional experiments commonly used to characterize biomolecules. In a one-dimensional experiment, an NMR signal is observed for a single type of nucleus in a molecule when the orientation of the bulk magnetic vector that describes the average orientation of all like nuclei in the sample is perturbed from its equilibrium position parallel to the external magnetic field (z axis). It is important to note that the size of this vector, and thus the NMR signal, is dependent upon the strength of the external magnetic field. If this vector is rotated 90° using a radiofrequency pulse, the vector now resides in the transverse x–y plane as shown in Figure 9.4A. Following this rotation the magnetization vector begins to precess in the x–y plane and simultaneously return to the z axis. The rate of precession is sensitive to the local chemical environment, and is observed in the NMR spectrometer as an oscillating wave with a regular

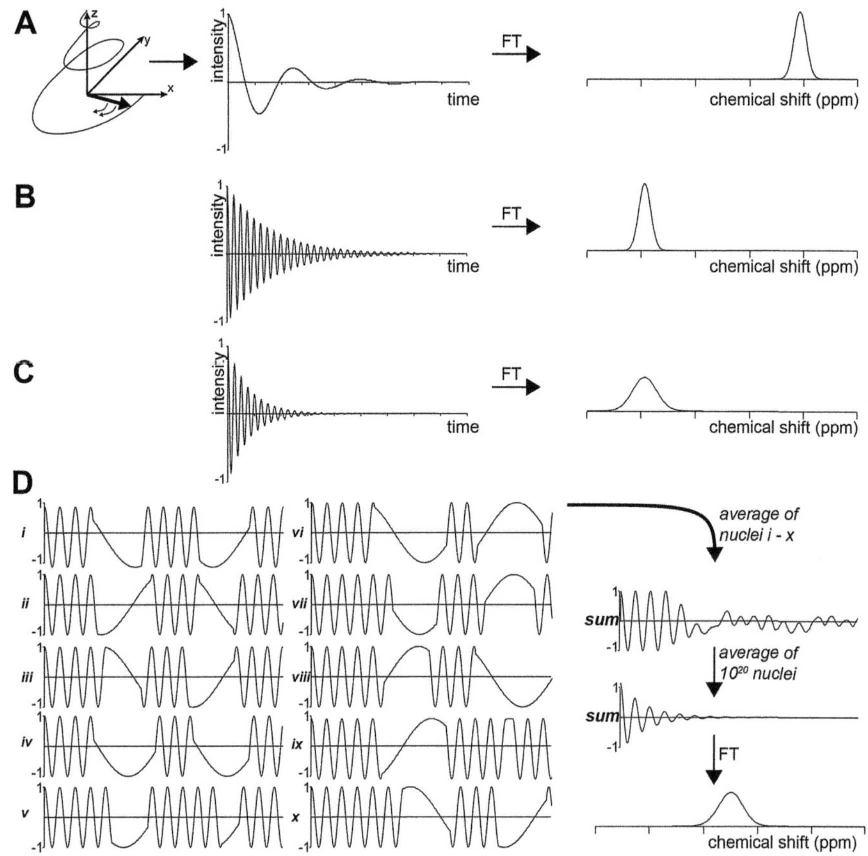

Figure 9.4 Basic concepts in NMR spectroscopy. (**A**) A simple experiment to observe the resonance frequency of a nucleus starts with rotation of the nuclear magnetization vector (*bold* arrow) to the transverse plane that induces an oscillating current in the NMR receiver coil. Fourier transform (FT) of the oscillation with result in a frequency-domain spectrum. Oscillation frequency (**B**) and signal decay (**C**) effect and observed NMR spectrum. (**D**) Exchange of a nucleus between two states with different resonance frequencies leads to decay of the observed signal in a manner independent of molecular tumbling.

period. The rate with which the vector shrinks, or relaxes, is sensitive to multiple factors including the rate at which that nucleus tumbles in the solution. As the signal relaxes, the intensity of transverse magnetization decreases, leading to decay of the oscillating signal.

Modern NMR spectroscopy relies upon a mathematical calculation, the Fourier Transform (FT), to convert the oscillating signal to a single peak representing this frequency. The resonance frequency following the FT is denoted as a "chemical shift" along the *x* axis in units of parts-per-million (ppm), referring to the difference in frequency of the observed signal when compared to a reference signal. The FT procedure is particularly powerful when measuring multiple signals with distinct, overlapping oscillations. A signal with a different resonance frequency, or oscillation rate in the transverse *x–y* plane, has a different position following the FT as shown when comparing Figure 9.4A and 9.4B. The position of these peaks in a spectrum following the FT is dependent upon the chemical properties of the nucleus as well as the local bonded and non-bonded chemical environment. For example, the H2 proton from a galactose residue that is β1-4 linked to a GlcNAc residue is different from the H2 proton from a galactose residue that is β1-4 linked to a glucose residue.

9.3.2 Lineshape Analysis

NMR signals are sensitive to motion in addition to reporting on local chemical environment. Note that the signals in Figure 9.4A and 9.4B decay at the same rate, despite oscillating at different frequencies, and as a result show the same lineshape following FT. A comparison of the signals in Figure 9.4B and 9.4C oscillate at identical frequencies but show the effect of relaxation on the signals. The signal in Figure 9.4C decays at twice the rate of the signal in 4B and shows a broader peak with less height (the total volume of the peaks in this experiment are identical; rates of signal decay are expressed in units of s^{-1}). The mass of a molecule is one primary factor that contributes to the rate of signal relaxation. Larger molecules generally have signals with broader lines, and the larger the molecule, generally, the broader the line. With very large molecules, the signals are often not observed because the peak height does not rise above the noise level in the spectrum. These effects are largely due to *transverse* relaxation. This relates to the shrinking of the magnetic orientation vector in the transverse *x–y* plane. Nuclei in large molecules generally have larger transverse relaxation rates than nuclei in smaller molecules. Transverse relaxation is also called R_2 relaxation where the observed linewidth measured at one half of the peak height is roughly equal to R_2/π.

The measurements of lineshapes in NMR spectra can be informative of molecular motion and have been used to interpret motion of the IgG1 Fc *N*-glycan termini.[41,68] The width of the NMR signal is dependent upon the magnitude of R_2 rate, which may be due to molecular motion. This type of relaxation rate reports on motion that occurs as fast as or faster than the rotational correlation time of the molecule (τ_c; around ~20 ns for a 55 kDa protein like Fc) as well as motion that occurs on the μs–ms timescale. Another type of relaxation relates to how fast the magnetization returns to the *z* axis following perturbation. This is termed R_1 relaxation and is only sensitive to motion that occurs as fast as or faster than τ_c. By comparing R_1 and R_2 relaxation rates, it is possible to estimate the relative motion of a single nucleus on ps–τ_c and (100 s of μs)–ms timescales. By comparing these rates with many nuclei in a protein, it is possible to estimate not only the τ_c of the protein, but also to probe whether certain regions of the protein experience greater motion, such as mobile loops or unconstrained N and C-termini, relative to the median value that reports global protein tumbling in the case of a structured protein. Though R_2 relaxation rates can be measured from an NMR spectrum directly, R_1 cannot. The next section covers methods to measure R_1 and R_2 precisely.

9.3.3 Experiments to Measure Relaxation Rates

Relaxation is measured using experiments designed to characterize the rate at which the bulk nuclear magnetization vector for a single type of nucleus returns to an equilibrium state, which is parallel to the $+z$ axis. These decays should fit to a single exponential equation. To measure R_2 relaxation, an experiment similar to the simple one-dimensional experiment diagramed in Figure 9.4A is performed with rotation of the magnetization vector to the transverse x–y plane. Pulse sequences for these types of experiments are diagramed in multiple resources.[67,69] R_2 relaxation rates increase with molecular weight. Occasionally, the rate of relaxation is too large to measure using standard multi-dimensional pulse sequences, and modifications must be made to preserve signal intensity as much as possible. Such simplifications were required to measure relaxation of galactose C2 nuclei incorporated into the IgG1 Fc glycoprotein *N*-glycan.[60] Phenomena other than molecular weight can also lead to large R_2 rates and are discussed in the next section.

To measure R_1 relaxation, a radio frequency pulse inverts the magnetization vector so that it is aligned with the $-z$ axis, and NMR spectra are observed at varying time intervals. The rate at which the signal in the serial NMR spectra returns to normal strength can be fitted directly to reveal the R_1 rate. A nucleus with a large R_1 rate would return to equilibrium more quickly than a nucleus with a small R_1 rate.

9.3.4 Relaxation Dispersion Methods

Conformational exchange on the μs–ms timescale represents one potential contribution to the observed R_2 relaxation rate. If a nucleus experiences multiple chemical environments, and thus exhibits more than one distinct resonance frequency over the course of the NMR experiment (often 10–10 000 exchanges between states every s), the observed lineshape and R_2 relaxation rates will be larger than expected based upon the molecular weight of the molecule and the τ_c estimate from the R_1 relaxation rate. Additional broadening occurs because of exchange between multiple states and is independent of τ_c.

Figure 9.4D shows the effect of conformational exchange on the observed NMR signal, *in the absence of relaxation* during the observation period. The oscillations of ten nuclei are shown (*i*–*x*). Note that the nuclei start with an identical phase and frequency but switch to a second frequency randomly, and alternate between these two frequencies throughout the observation period. This represents each member of the ensemble of like nuclei randomly exchanging between state A and state B. Once these individual oscillations are summed, the result is a time-dependent decrease in signal intensity as a result of the nuclei losing phase coherence. When 10^{20} nuclei are added to the ensemble the effect looks very much like signal decay observed in Figure 9.4A, but is *fundamentally different*. Thus, both molecular weight and chemical exchange between two states with distinct resonance frequencies as described here contribute to signal decay in an NMR experiment.

Methods to characterize conformational heterogeneity are capable of determining the resonance frequencies for each state and the rate at which states exchange with one another. Relaxation dispersion methods in NMR spectroscopy function by controlling the rate at which the ensemble of nuclei becomes dephased. These methods do not alter the intrinsic rate of molecule conformational exchange (often this can be perturbed by changing temperature), but only the effect of exchange on the intensity of the observed signal, and are described in detail elsewhere.[69] These methods have revealed multiple conformations sampled by the IgG1 Fc *N*-glycan in solution.

If a nucleus experiences two states, A and B, and exchanges between these rapidly on the NMR timescale, a single peak will be observed in an NMR spectrum at the resonance frequency that is the population weighted average of the two states.

9.4 Structure/Function Relationship in IgG1 Fc

The historical development of the theory covering structure, function and receptor binding properties of IgG is covered in detail in a recent review.[70] This chapter will focus on highlights of IgG1 glycan and polypeptide motion with an emphasis on the relaxation rates and rates of motion revealed by solution NMR spectroscopy. These studies utilized proteins with *N*-glycans remodeled to near homogeneity.

9.4.1 The IgG1 Fc *N*-Glycan is Mobile

Complex behavior of the IgG1 Fc *N*-glycan termini was observed in spectra of a mouse IgG2b Fc labeled with [1–^{13}C]galactose.[41] Yamaguchi *et al.* observed two different signals with different resonance frequencies and lineshapes. One fundamental challenge was assigning the galactose resonances to either the (α1-3Man)linked or (α1-6Man)linked branch of the *N*-glycan (See Figure 9.1). The researchers used the inherent preference of a galactosyltransferase for the (α1-6Man)linked GlcNAc residue over the (α1-3Man)linked GlcNAc to assign the spectrum and then determined that the signal from the (α1-6Man) linked galactose C1 nucleus was three times broader (39 Hz) than the (α1-3Man)linked galactose C1 nucleus (13 Hz). This translates into R_2 values of 120 s^{-1} and 40 s^{-1}, respectively at 9.4 T and indicated a significant difference in the chemical environment experienced by each galactose residue.

A study in 2011 determined that the difference in behavior was due to a transient interaction between the (α1-6Man)linked galactose residue and the surface of the Fc polypeptide that was not observed for the (α1-3Man) linked galactose residue.[68] This interaction was characterized following measurements of the R_1 and R_2 relaxation rates for human IgG1 Fc labeled with [2-^{13}C]galactose. A similar relationship of R_2 rates to that reported by Yamaguchi *et al.*[41] was observed with the C2 nuclei of the (α1-6Man)linked (150 s^{-1}) and (α1-3Man)linked galactose residues (40 s^{-1}) at 21.1 T. However, the R_1 rates for both galactose residues gave indistinguishable values (1.2 ± 0.2 s^{-1} and 1.2 ± 0.1 s^{-1}, respectively). The R_1 rates report on motion as fast or faster than global tumbling of the Fc, and it was calculated that the R_1 rates for the C2 nucleus, if bound solidly to the surface of the Fc, should be on the order of 0.17 s^{-1} (note that for large molecules, R_1 rates *decrease* with increased rotational correlation times). Therefore, both termini experience motion greater than that afforded only by rotational tumbling of the Fc molecule in solution. The similarity for R_1 rates and differences in R_2 rates indicated that the (α1-6Man)linked galactose residue exchanged between multiple conformation states.

Relaxation dispersion experiments revealed that the (α1-6Man)linked galactose residue exchanged between a polypeptide bound state and a free, unrestricted state at a frequency of 5300 s^{-1}, or with an average lifetime of 190 μs. Estimates of the population distribution revealed a roughly 1 : 1 proportion of bound and unbound conformations in solution at 37 °C. Later computational simulations of the Fc glycoprotein supported the exchange of the (α1–6Man)linked galactose residue between these two states.[71]

9.4.2 IgG1 Fc *N*-Glycan Motion is Linked to Receptor Binding Affinity

It was long known that the Fc *N*-glycan was required for antibody binding to FcγRs, particularly FcγR IIa, IIb, IIc, IIIa, and IIIb;[72] however, it was unclear why. A structure of the Fc: FcγRIIIa complex solved in 2000 failed to explain this effect because the Fc *N*-glycan did not contact the receptor directly;[73] see Figure 9.5. In 2006 Yamaguchi *et al.*[9] built upon work by Krapp *et al.* (2003)[74] suggesting the Fc quaternary conformation was influenced by the *N*-glycan composition. With the identification of Fc *N*-glycan motion, Subedi and coworkers probed whether or not Fc *N*-glycan motion was related to affinity for FcγRIIIa.[36] Substituting Phe residues at the *N*-glycan/polypeptide interface on IgG1 Fc led to range of different properties (as highlighted Figure 9.5). First, these Fc variants showed a range of glycan mobility as indicated by resonance frequencies and R_2 rates for the galactose C2 nuclei. Variants with single Phe241 or Phe243 substitutions saw moderate decreases in R_2 rates for the (α1-6Man)linked galactose C2 nucleus of ~35 s^{-1} when compared to the wild-type protein with 64 s^{-1} at 16.4 T. Variants with double Phe241–Phe243 substitutions saw further decreases in R_2 rates to ~17 s^{-1} alleviating the predominant galactose-stabilizing interactions (the galactose C2 on an Fc-derived *N*-glycopeptide revealed an R_2 rate of 11 s^{-1}). Second, each Fc variant showed a different affinity for the receptor, and, it was noted, those with the greatest increase in glycan mobility resulting from amino acid substitution

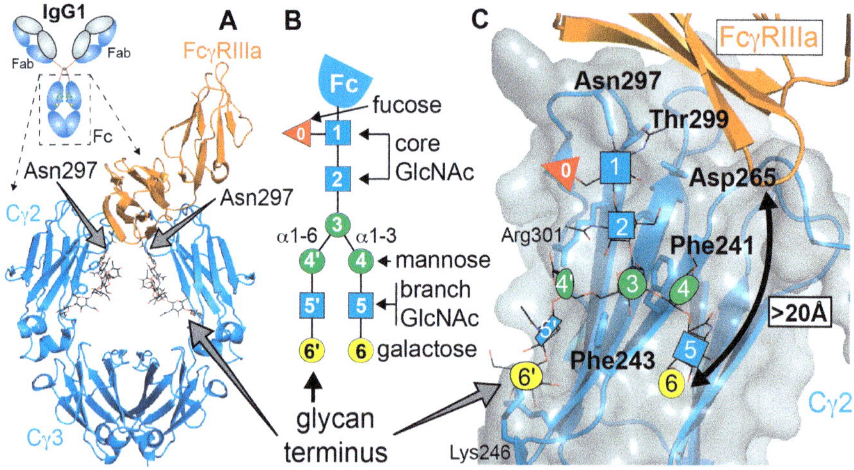

Figure 9.5 The Fc *N*-glycan in receptor interactions. (**A**) The FcγRs (*orange*) bind to IgG1 Fc (*blue*) near the site of *N*-glycan attachment at Asn297 and break rotational symmetry of the Fc dimer. The branch residues of the Fc *N*-glycan chain point away from the receptor as shown in a cartoon (**B**) and atomic-level detail (**C**). This model is based on PDB 3ave.

showed the weakest affinity for FcγRIIIa (wild-type > single substitution > double substitution). This led to the hypothesis that the Fc *N*-glycan contributes to receptor binding by stabilizing the Fc in a conformation with the greatest possible affinity and represents an allosteric mechanism by which the *N*-glycan contributes to antibody function.

9.4.3 Antibody *N*-Glycosylation Stabilizes a Critical Polypeptide Loop

Though previous studies provided a considerable amount of information regarding the behavior of antibody *N*-glycans, it was still unclear what parts of the Fc molecule the *N*-glycan stabilized to promote receptor binding. Multiple groups suggested *N*-glycosylation stabilized IgG1 Fc quaternary structure[9,74,75] and IgG1 Fc conformational heterogeneity ([76–78] among others). Subedi and Barb investigated what portions of IgG1 Fc were stabilized by *N*-glycosylation. They determined that only resonances from Tyr and Lys residues near Asn297 shifted upon *N*-glycan removal. Furthermore, measurements of amide nitrogen R_1 and R_2 relaxation rates showed the primary difference upon deglycosylation in Tyr and Lys residues adjacent to Asn297. The largest of these was a ~25 s^{-1} increase in the R_2 of the Tyr300 amide nitrogen following *N*-glycan removal and no change in the R_1, indicating *N*-glycosylation stabilized the residues near Asn297 on a μs–ms timescale. This region also forms a critical polypeptide/polypeptide interface with FcγRIIIa;[42] see Figure 9.5C. Furthermore, using measurements of another NMR phenomenon called residual dipolar couplings (discussed in Chapter 4), the authors showed the predominant Fc quaternary structure is unchanged by removing the *N*-glycan, suggesting the stabilizing effect of *N*-glycosylation are limited to residues near Asn297.

Further insight into the effect of *N*-glycosylation on antibody function was found by probing the residues at both reducing and non-reducing termini of the *N*-glycan.[42] An Fc variant that removes a critical glycan stabilizing interaction between the (1)GlcNAc residue (Figure 9.5) eliminated FcγRIIIa and dramatically altered the conformation of the carbohydrate and amino acid residues in the vicinity as measured in 2d NMR spectra. Interestingly, Fc Asp265Ala displayed strong interactions between the (α1-6Man)linked galactose residue and the polypeptide surface, indicating carbohydrate/polypeptide interactions proximal to Asn297 were predominant for producing a high-affinity FcγRIIIa binding site and although distal interactions contributed to binding affinity, distal interactions alone proved insufficient. This was exemplified by the discovery that Fc with only a single GlcNAc residue could bind FcγRIIIa with surprisingly minimal 10-fold reduction in affinity when compared to Fc with a full-length *N*-glycan.

9.4.4 Fc *N*-Glycan Residues Distal to Asn297 Stabilize Fc

It was known that GlcNAc and galactose residues on the Fc *N*-glycan enhanced affinity for FcγRIIIa[9] likely by increasing the interface between the glycan and polypeptide. In addition to the studies already introduced, further evidence from structure-based studies supports the role of *N*-glycan composition on IgG1 Fc stabilization. The addition of α2-6 *N*-acetylneuraminic acid to a galactose-terminated biantennary and fucosylated *N*-glycan slightly decreased the R_1 relaxation rate for the (α1-6Man)linked galactose C2 nucleus from $1.2 \pm 0.2\ s^{-1}$ to $0.9 \pm 0.1\ s^{-1}$ and increased the R_2 from 150 to 170 s^{-1}[60]. Both of these values are consistent with the (α1-6Man)linked galactose C2 resonance shifting slightly downfield in the spectrum and indicate a small increase in the interaction between *N*-glycan and polypeptide residues upon addition of *N*-acetylneuraminic acid. Interestingly, the *N*-acetylneuraminic acid residues themselves did not appear to directly interact with the polypeptide and have greater R_1 rates (3.6–3.8 s^{-1}) and smaller R_2 rates (46–59 s^{-1}) for the C3 nuclei when compared to the galactose C2.

A recent study focused on the interaction of the *N*-glycan with the polypeptide at earlier stages in the *N*-glycan processing pathway and identified that each modification, from the intermediate Man5 *N*-glycan (the substrate for Gnt1, Figure 9.2) to the complex type form with two branch GlcNAc residues, sequentially increases the polypeptide/glycan interaction.[16] This was surprising given that two potentially-stabilizing Man residues are removed during the processing step catalyzed by the GoMan2 mannosidase. However, the *N*-glycan proved more stable in their absence based on an analysis of GlcNAc resonance frequencies observed in 2d ^{1}H–^{13}C HSQC experiments.

9.5 Conclusions and Future Prospects

It is clear the remodeling and characterization techniques as applied to antibodies and antibody fragments revealed a wealth of information regarding the properties of glycoproteins and *N*-glycans. Though there are still outstanding questions with respect to IgG1 Fc, the more exciting future discoveries of *N*-glycan structure and motion will come from other protein systems, in particular those with multiple *N*-glycans. Though this represents a technical challenge to prepare materials and assign resonances with multiple glycans present, it is likely the types of interactions that modulate antibody function will be identified as equally critical components in other systems as well. It does not take much searching of the literature to identify proteins with function that depends on *N*-glycosylation and likely intramolecular glycan/polypeptide interactions. Now with the tools available, a new world within structural biology is opening.

Acknowledgements

This work was financially supported by the grant R01GM115489(NIGMS) from the National Institutes of Health, and by funds from the Roy J. Carver Department of Biochemistry, Biophysics & Molecular Biology at Iowa State University. The content of this work is solely the responsibility of the authors and does not necessarily represent the official views of the NIH.

References

1. R. Apweiler, H. Hermjakob and N. Sharon, *Biochim. Biophys. Acta*, 1999, **1473**, 4–8.
2. K. W. Moremen, M. Tiemeyer and A. V. Nairn, *Nat. Rev. Mol. Cell Biol.*, 2012, **13**, 448–462.
3. A. W. Barb, A. J. Borgert, M. Liu, G. Barany and D. Live, *Methods Enzymol.*, 2010, **478**, 365–388.
4. K. Murphy, P. Travers, M. Walport and C. Janeway, *Janeway's Immunobiology*, Garland Science, New York, 8th edn, 2012.
5. M. Collin and A. Olsen, *EMBO J.*, 2001, **20**, 3046–3055.
6. R. Huber, J. Deisenhofer, P. M. Colman, M. Matsushima and W. Palm, *Nature*, 1976, **264**, 415–420.
7. R. B. Parekh, R. A. Dwek, B. J. Sutton, D. L. Fernandes, A. Leung, D. Stanworth, T. W. Rademacher, T. Mizuochi, T. Taniguchi and K. Matsuta, *et al.*, *Nature*, 1985, **316**, 452–457.
8. Y. Kaneko, F. Nimmerjahn and J. V. Ravetch, *Science*, 2006, **313**, 670–673.
9. Y. Yamaguchi, M. Nishimura, M. Nagano, H. Yagi, H. Sasakawa, K. Uchida, K. Shitara and K. Kato, *Biochim. Biophys. Acta*, 2006, **1760**, 693–700.
10. G. P. Subedi, R. W. Johnson, H. A. Moniz, K. W. Moremen and A. W. Barb, *J. Visual Edu.*, 2015, **106**, e53568.

11. G. Backliwal, M. Hildinger, S. Chenuet, S. Wulhfard, M. De Jesus and F. M. Wurm, *Nucleic Acids Res.*, 2008, **36**, e96.
12. A. M. Toth, C. W. Kuo, K. H. Khoo and D. L. Jarvis, *J. Biotechnol.*, 2014, **182–183**, 19–29.
13. J. P. Giddens and L. X. Wang, *Methods Mol. Biol.*, 2015, **1321**, 375–387.
14. P. Wang, S. Dong, J. H. Shieh, E. Peguero, R. Hendrickson, M. A. Moore and S. J. Danishefsky, *Science*, 2013, **342**, 1357–1360.
15. P. Stanley, S. Narasimhan, L. Siminovitch and H. Schachter, *Proc. Natl. Acad. Sci. U. S. A.*, 1975, **72**, 3323–3327.
16. A. W. Barb, *Biochemistry*, 2015, **54**, 313–322.
17. S. R. Hanson, E. K. Culyba, T. L. Hsu, C. H. Wong, J. W. Kelly and E. T. Powers, *Proc. Natl. Acad. Sci. U. S. A.*, 2009, **106**, 3131–3136.
18. P. J. Reeves, N. Callewaert, R. Contreras and H. G. Khorana, *Proc. Natl. Acad. Sci. U. S. A.*, 2002, **99**, 13419–13424.
19. L. Meuris, F. Santens, G. Elson, N. Festjens, M. Boone, A. Dos Santos, S. Devos, F. Rousseau, E. Plets, E. Houthuys, P. Malinge, G. Magistrelli, L. Cons, L. Chatel, B. Devreese and N. Callewaert, *Nat. Biotechnol.*, 2014, **32**, 485–489.
20. H. Salo, E. Sievi, T. Suntio, M. Mecklin, P. Mattila, R. Renkonen and M. Makarow, *FEMS Yeast Res.*, 2005, **5**, 341–350.
21. H. Mabashi-Asazuma, C. W. Kuo, K. H. Khoo and D. L. Jarvis, *ACS Chem. Biol.*, 2015, **10**, 2199–2208.
22. M. Iwami, O. Nakayama, H. Terano, M. Kohsaka, H. Aoki and H. Imanaka, *J. Antibiot.*, 1987, **40**, 612–622.
23. A. D. Elbein, J. E. Tropea, M. Mitchell and G. P. Kaushal, *J. Biol. Chem.*, 1990, **265**, 15599–15605.
24. P. R. Dorling, C. R. Huxtable and S. M. Colegate, *Biochem. J.*, 1980, **191**, 649–651.
25. N. M. Okeley, S. C. Alley, M. E. Anderson, T. E. Boursalian, P. J. Burke, K. M. Emmerton, S. C. Jeffrey, K. Klussman, C. L. Law, D. Sussman, B. E. Toki, L. Westendorf, W. Zeng, X. Zhang, D. R. Benjamin and P. D. Senter, *Proc. Natl. Acad. Sci. U. S. A.*, 2013, **110**, 5404–5409.
26. C. D. Rillahan, A. Antonopoulos, C. T. Lefort, R. Sonon, P. Azadi, K. Ley, A. Dell, S. M. Haslam and J. C. Paulson, *Nat. Chem. Biol.*, 2012, **8**, 661–668.
27. W. McDowell, R. Datema, P. A. Romero and R. T. Schwarz, *Biochemistry*, 1985, **24**, 8145–8152.
28. N. Loch, C. C. Geilen, I. Sporndle, F. Oberdorfer, D. Keppler, R. Tauber and W. Reutter, *FEBS Lett.*, 1991, **294**, 217–220.
29. S. R. Barthel, A. Antonopoulos, F. Cedeno-Laurent, L. Schaffer, G. Hernandez, S. A. Patil, S. J. North, A. Dell, K. L. Matta, S. Neelamegham, S. M. Haslam and C. J. Dimitroff, *J. Biol. Chem.*, 2011, **286**, 21717–21731.
30. J. M. Withka, D. F. Wyss, G. Wagner, A. R. Arulanandam, E. L. Reinherz and M. A. Recny, *Structure*, 1993, **1**, 69–81.
31. D. F. Wyss, J. S. Choi, J. Li, M. H. Knoppers, K. J. Willis, A. R. Arulanandam, A. Smolyar, E. L. Reinherz and G. Wagner, *Science*, 1995, **269**, 1273–1278.

32. D. F. Wyss, J. M. Withka, M. H. Knoppers, K. A. Sterne, M. A. Recny and G. Wagner, *Biochemistry*, 1993, **32**, 10995–11006.
33. H. Yagi, Y. Zhang, M. Yagi-Utsumi, T. Yamaguchi, S. Iida, Y. Yamaguchi and K. Kato, *Biomol. NMR Assignments*, 2015, **9**, 257–260.
34. S. Matsumiya, Y. Yamaguchi, J. Saito, M. Nagano, H. Sasakawa, S. Otaki, M. Satoh, K. Shitara and K. Kato, *J. Mol. Biol.*, 2007, **368**, 767–779.
35. S. Liu, A. Venot, L. Meng, F. Tian, K. W. Moremen, G. J. Boons and J. H. Prestegard, *Chem. Biol.*, 2007, **14**, 409–418.
36. G. P. Subedi, Q. M. Hanson and A. W. Barb, *Structure*, 2014, **22**, 1478–1488.
37. W. D. Morgan, A. Kragt and J. Feeney, *J. Biomol. NMR*, 2000, **17**, 337–347.
38. S. Albright, B. Chen, K. Holbrook and N. U. Jain, *Biochem. Biophys. Res. Commun.*, 2008, **368**, 231–237.
39. B. Laukens, C. De Wachter and N. Callewaert, *Methods Mol. Biol.*, 2015, **1321**, 103–122.
40. Y. Kamiya, S. Yamamoto, Y. Chiba, Y. Jigami and K. Kato, *J. Biomol. NMR*, 2011, **50**, 397–401.
41. Y. Yamaguchi, K. Kato, M. Shindo, S. Aoki, K. Furusho, K. Koga, N. Takahashi, Y. Arata and I. Shimada, *J. Biomol. NMR*, 1998, **12**, 385–394.
42. G. P. Subedi and A. W. Barb, *Structure*, 2015, **23**, 1573–1583.
43. K. Kato, Y. Yamaguchi and Y. Arata, *Prog. Nucl. Magn. Reson. Spectrosc.*, 2010, **56**, 346–359.
44. S. J. North, P. G. Hitchen, S. M. Haslam and A. Dell, *Curr. Opin. Struct. Biol.*, 2009, **19**, 498–506.
45. P. Azadi and C. Heiss, *Methods Mol. Biol.*, 2009, **534**, 37–51.
46. G. Zhao, W. Guan, L. Cai and P. G. Wang, *Nat. Protoc.*, 2010, **5**, 636–646.
47. J. Zhou, L. Fan, P. Wei, L. Huang, J. Cai and Z. Xu, *Prep. Biochem. Biotechnol.*, 2010, **40**, 294–304.
48. K. Okuyama, T. Hamamoto, K. Ishige, K. Takenouchi and T. Noguchi, *Biosci., Biotechnol., Biochem.*, 2000, **64**, 386–392.
49. K. Inoue, M. Nishimoto and M. Kitaoka, *Carbohydr. Res.*, 2011, **346**, 2432–2436.
50. Y. Zhai, M. Liang, J. Fang, X. Wang, W. Guan, X. W. Liu, P. Wang and F. Wang, *Biotechnol. Lett.*, 2012, **34**, 1321–1326.
51. J. E. Heidlas, W. J. Lees, P. Pale and G. M. Whitesides, *J. Org. Chem.*, 1992, **57**, 146–151.
52. A. Varki, *Essentials of Glycobiology*, Cold Spring Harbor Laboratory Press, Cold Spring Harbor, N.Y., 2nd edn, 2009.
53. P. A. Frey, *FASEB J.*, 1996, **10**, 461–470.
54. A. M. Gilhespy-Muskett, J. Partridge, R. Jefferis and S. W. Homans, *Glycobiology*, 1994, **4**, 485–489.
55. T. Yang, L. Bar-Peled, L. Gebhart, S. G. Lee and M. Bar-Peled, *J. Biol. Chem.*, 2009, **284**, 21526–21535.
56. E. S. Simon, M. D. Bednarski and G. M. Whitesides, *J. Am. Chem. Soc.*, 1988, **110**, 7159–7163.
57. M. A. Macnaughtan, F. Tian, S. Liu, L. Meng, S. Park, P. Azadi, K. W. Moremen and J. H. Prestegard, *J. Am. Chem. Soc.*, 2008, **130**, 11864–11865.

58. H. F. Azurmendi, J. Vionnet, L. Wrightson, L. B. Trinh, J. Shiloach and D. I. Freedberg, *Proc. Natl. Acad. Sci. U. S. A.*, 2007, **104**, 11557–11561.
59. A. Kobata, *Biochim. Biophys. Acta*, 2008, **1780**, 472–478.
60. A. W. Barb, L. Meng, Z. Gao, R. W. Johnson, K. W. Moremen and J. H. Prestegard, *Biochemistry*, 2012, **51**, 4618–4626.
61. S. Fujii, T. Nishiura, A. Nishikawa, R. Miura and N. Taniguchi, *J. Biol. Chem.*, 1990, **265**, 6009–6018.
62. A. W. Barb, E. K. Brady and J. H. Prestegard, *Biochemistry*, 2009, **48**, 9705–9707.
63. M. Umekawa, C. Li, T. Higashiyama, W. Huang, H. Ashida, K. Yamamoto and L. X. Wang, *J. Biol. Chem.*, 2010, **285**, 511–521.
64. S. Q. Fan, W. Huang and L. X. Wang, *J. Biol. Chem.*, 2012, **287**, 11272–11281.
65. M. N. Amin, J. S. McLellan, W. Huang, J. Orwenyo, D. R. Burton, W. C. Koff, P. D. Kwong and L. X. Wang, *Nat. Chem. Biol.*, 2013, **9**, 521–526.
66. J. Cavanagh, *Protein NMR Spectroscopy: Principles and Practice*, Academic Press, Amsterdam, Boston, 2nd edn, 2007.
67. M. H. Levitt, *Spin Dynamics: Basics of Nuclear Magnetic Resonance*, John Wiley & Sons, Chichester, England, Hoboken, NJ, 2nd edn, 2008.
68. A. W. Barb and J. H. Prestegard, *Nat. Chem. Biol.*, 2011, **7**, 147–153.
69. A. G. Palmer 3rd, *J. Magn. Reson.*, 2014, **241**, 3–17.
70. Q. M. Hanson and A. W. Barb, *Biochemistry*, 2015, **54**, 2931–2942.
71. M. Frank, R. C. Walker, W. N. Lanzilotta, J. H. Prestegard and A. W. Barb, *J. Mol. Biol.*, 2014, **426**, 1799–1811.
72. M. Nose and H. Wigzell, *Proc. Natl. Acad. Sci. U. S. A.*, 1983, **80**, 6632–6636.
73. P. Sondermann, R. Huber, V. Oosthuizen and U. Jacob, *Nature*, 2000, **406**, 267–273.
74. S. Krapp, Y. Mimura, R. Jefferis, R. Huber and P. Sondermann, *J. Mol. Biol.*, 2003, **325**, 979–989.
75. M. J. Borrok, S. T. Jung, T. H. Kang, A. F. Monzingo and G. Georgiou, *ACS Chem. Biol.*, 2012, **7**, 1596–1602.
76. D. Houde, Y. Peng, S. A. Berkowitz and J. R. Engen, *Mol. Cell. Proteomics*, 2010, **9**, 1716–1728.
77. T. Dashivets, M. Thomann, P. Rueger, A. Knaupp, J. Buchner and T. Schlothauer, *PLoS One*, 2015, **10**, e0143520.
78. M. A. Alsenaidy, J. H. Kim, R. Majumdar, D. D. Weis, S. B. Joshi, T. J. Tolbert, C. R. Middaugh and D. B. Volkin, *J. Pharm. Sci.*, 2013, **102**, 3942–3956.

CHAPTER 10

Analysis of Glycosaminoglycans by ^{15}N-NMR Spectroscopy

VITOR H. POMIN[a,b]

[a]Program of Glycobiology, Institute of Medical Biochemistry Leopoldo de Meis, University Hospital Clementino Fraga Filho, Federal University of Rio de Janeiro, Rio de Janeiro, RJ, 21941-913, Brazil; [b]Complex Carbohydrate Research Center, University of Georgia, Athens, 30602, United States of America
*E-mail: pominvh@bioqmed.ufrj.br, vhpomin@gmail.com, pomin27@uga.edu

10.1 Introduction

Glycosaminoglycans (GAGs) are glycans of diverse functions and complex structures, although composed of disaccharide building blocks. These repeating blocks are in turn composed of alternating uronic acid and hexosamine moieties, which can be differently sulfated depending on the GAG families. Members of the same GAG family share common structural features. For example, heparan sulfate (HS) and heparin (Hp) are composed of α-D-glucosamine (GlcN) units as hexosamine type (Figure 10.1A). While HS is less modified by *N*- and *O*-sulfation and uronic acid C5 epimerization which converts the composing β-D-glucuronic acid (GlcA) into α-L-iduronic acid (IdoA); Hp is highly processed by these modifications (Figure 10.1A).[1,2] Other biologically active GAG families are the chondroitin sulfate (CS), dermatan sulfate (DS), keratan sulfate (KS) and hyaluronic acid (HA). Although CS and DS are both composed of *N*-acetyl-β-D-galactosamine (GalNAc) units, the former is made up primarily of GlcA whereas IdoA composes the latter

New Developments in NMR No.10
NMR in Glycoscience and Glycotechnology
Edited by Koichi Kato and Thomas Peters

Published by the Royal Society of Chemistry, www.rsc.org

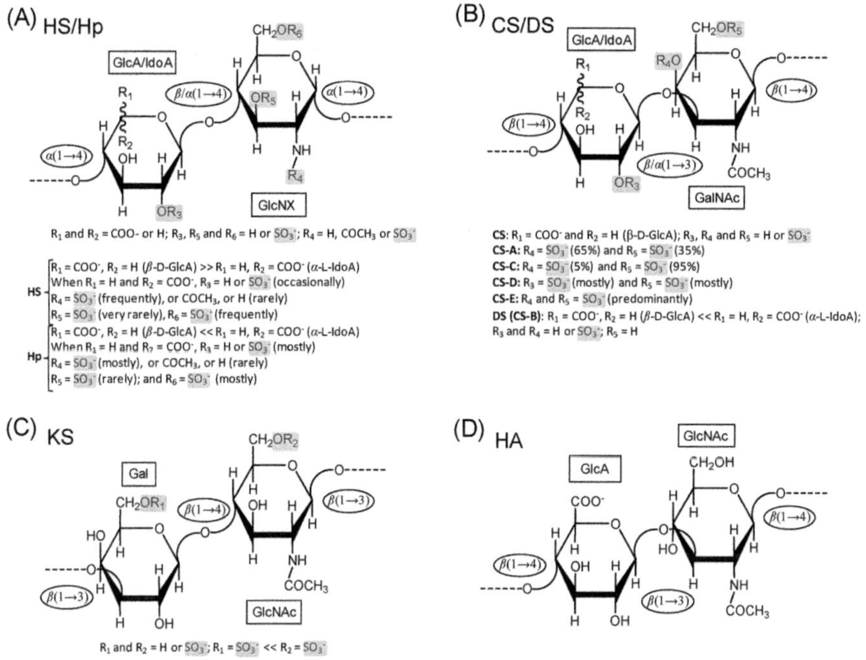

Figure 10.1 Schematic representation of the disaccharide structures of the commonest GAGs normally found in the body and explored in the global pharmaceutical market. (**A**) Heparan sulfate (HS), and heparin (Hp) structures. Both are composed of alternating 4-linked uronic acid and 4-linked *α*-D-glucosamine (GlcN) units. HS has *β*-D-glucuronic acid (GlcA) as its major uronic acid type, whereas Hp has *α*-L-iduronic acid (IdoA). While HS is frequently *N*-sulfated at the GlcN unit (*N*-sulfated glucosamine, GlcNS), with less extension of *N*-acetylation in *N*-acetyl-*α*-D-glucosamine (GlcNAc) and just occasional amino NH_2 groups (*N*-free); Hp is predominantly composed of 2-sulfated IdoA together with *N*,6-di-sulfated GlcN units. Although the 3-*O*-sulfation at the GlcN units occurs more often at Hp than HS, it still happens in the latter but quite rarely. (**B**) Chondroitin sulfate (CS) and dermatan sulfate (DS) structures. CSs are composed of alternating 4-linked *β*-D-GlcA and 3-linked *N*-acetyl-*β*-D-galactosamine (GalNAc) units. CS-A is mostly 4-sulfated at the GalNAc units while CS-C is predominantly 6-sulfated. CS-B, best-known as DS, has *α*-L-IdoA units rather than *β*-GlcA as in CS. The IdoA units in DS may contain 2-sulfation while the GalNAc units are mostly 4-sulfated. While CS-D is mostly composed of 2-sulfated GlcA and 6-sulfated GalNAc units, CS-E is mostly composed of 4,6-di-sulfated GalNAc units. (**C**) Keratan sulfate (KS) structure. It is composed of alternating 3-linked *β*-galactose (Gal) and 4-linked *α*-D-GlcNAc units. KS can bear sulfation at the 6-position of each unit, although sulfation at GlcNAc occurs more often than Gal. (**D**) Hyaluronic acid (HA) structure. It is a non-sulfated polymer of alternating 4-linked *β*-D-GlcA and 3-linked *β*-D-GlcNAc units. In all panels the glycosidic bonds are indicated by ellipses, whereas monosaccharide types are indicated with black rectangles. For fast notation of the sulfation patterns, sulfation sites are highlighted by light-grey rectangles.

(Figure 10.1B).[3] Sulfation can occur at different patterns and this leads to the different CS subfamilies as shown in Figure 10.1B. KS is the only GAG type, which has the neutral sugar galactose (Gal) instead of an uronic acid residue. Although sulfation occurs at the 6-position of any monosaccharide, it occurs more often at the GlcNAc units (Figure 10.1C). HA is the only non-sulfated GAG (Figure 10.1D). Its structure resembles HS but all HA monosaccharides are at their α-configuration. Based on the general structural representation of Figure 10.1, it is clear that hexosamine units in GAGs, regardless of the families, are usually more sulfated or chemically modified than the composing uronic acid units.

Due to the great structural variety and presence of sulfation, GAGs have the capacity to bind to an incredible number of functional proteins, including, but not limited to, inflammatory cytokines and chemokines,[4] angiogenesis-related growth factors like fibroblast growth factors 1,[5] and 2,[6] and plasma cofactors involved in blood coagulation like thrombin,[7] and antithrombin.[7,8] Because of these and other unmentioned roles, GAGs are ubiquitously distributed in all vertebrates, usually linked to a protein core forming thus the well-known proteoglycans. Worldwide, structural and functional analysis of GAGs from mammals is in focus due to the fact that these molecules play key roles in multiple pathophysiological processes such as cancer,[9,10] inflammation,[11] thrombosis,[12] and neovascularization,[13] besides being explored in the global market as nutraceutical or pharmaceutical ingredients. The mechanisms of action of GAGs in these events are directly related to their structural features. Hence, structural analysis is a key procedure to the progress of GAG-related science. Liquid-state nuclear magnetic resonance (NMR) spectroscopy has been playing a pivotal role in this regard.

NMR-based analysis of GAGs is usually performed on ^{1}H and ^{13}C nuclei of anomeric and ring atoms of both uronic acid and hexosamine, *via* both 1D and 2D NMR experiments.[14] The reason for the intense use of ^{1}H and ^{13}C isotopes in NMR GAG analysis is mainly the abundance and magnetic sensitivity

Table 10.1 Some important properties for NMR detection and sensitivity of the commonest nuclei in biomolecules.[32]

Nuclide	Spin	Natural abundance	Gyromagnetic ratio γ [10^7 rad T^{-1} s^{-1}]	NMR Frequency at 18.8 Tesla (ratio related to ^{1}H)	Relative receptivity	Usual chemical shift range
Proton (^{1}H)	1/2	99.985	26.7522	799.734 (1)	1.00	12 ppm
Carbon-12 (^{12}C)	0	98.9	-	-	-	-
Carbon-13 (^{13}C)	1/2	1.108	6.7283	201.133 (1/3.976)	1.76×10^{-4}	220 ppm
Nitrogen-14 (^{14}N)	1	99.63	1.9338	57.820 (1/13.831)	1.00×10^{-3}	-
Nitrogen-15 (^{15}N)	1/2	0.37	−2.7126	81.093 (1/9.861)	3.85×10^{-6}	300 ppm

of these nuclei (Table 10.1). The main difficulty in interpreting 1D or multidimensional NMR spectra of GAGs, involving one or both of these two isotopes, is perhaps the low dispersion of NMR signals (high chemical shift degeneracy) of GAGs aggravated by the inherent structural heterogeneity and complexity of these glycans.[14] For example, in ^{1}H-NMR spectroscopy of the major naturally occurring GAGs, with the exception of the most upfield methyl ^{1}H signal from the acetyl groups around 2.0 ppm, most signals including those from the anomerics and those from the sugar rings are often squeezed between 3.0 and 6.0 ppm on the proton chemical shift scale.[14] Even in 2D NMR spectra (used to enhance dispersion of possible resonance overlap through the addition of another dimension) such as the mostly exploited ^{1}H–^{13}C heteronuclear single quantum coherence (HSQC) experiments in which some structural GAG components (methyl groups, anomerics, sulfation and glycosylation sites) become straightforwardly identifiable because of the specific regions of the resultant cross-peaks, most of the generated ^{1}H–^{13}C resonances from HSQC spectra of GAGs are still condensed in a very limited spectral window.[14,15]

Nowadays, however, NMR analysis of GAGs based on the magnetically less sensitive and less abundant ^{15}N nucleus (Table 10.1) has become more and more common.[16–31] It has been proven that ^{15}N can be successfully used for characterization,[29] dynamical[21] and conformational[31] analyses of GAGs. Although ^{15}N-NMR can be employed in conjunction to ^{1}H and ^{13}C NMR, a new set of structural information can also be retrieved from analyses based solely on the ^{15}N nucleus. It is of note that chemical shift degeneracy can be really reduced when ^{15}N-related analysis is employed since only a few ^{15}N-related resonances are produced. These few resonances facilitate straightforward assignment and structural characterization of mixed GAG species in a single sample or of the very heterogeneous GAG materials. The benefits of this novel ^{15}N-NMR method are obtained however when experimental and instrumental conditions are sufficient to enable reasonable signal detection (adequate spectral acquisition) because of the low NMR sensitivity of ^{15}N and its low natural abundance (Table 10.1). The advent of high-magnetic-field instruments, cryoprobe technology, nuclear polarizer devices for dynamic nuclear polarization (DNP) analysis, implementation of isotopic labeling techniques and novel ^{15}N-related pulse sequences in 2D NMR experiments are all important contributing factors to the proper use of ^{15}N in NMR analyses. This chapter aims at describing the recent structural information retrieved from this less frequently used nucleus in NMR analyses of GAGs.

10.2 Characterization and Recognition of Native GAGs at Natural Abundance

As seen from Figure 10.1, GAG structures vary primarily on (1) sulfation patterns mostly at the hexosamine units; (2) different *N*-substitutions like amine (NH_2), *N*-acetyl ($NHCOCH_3$) and *N*-sulfonation ($NHSO_3^-$); (3) hexosamine types (GlcN *versus* GalNAc); and (4) adjacent uronic acid type (IdoA *versus*

GlcA). Hence, ^{15}N in hexosamines can be used as a structural probe to investigate these features in GAGs once all the afore-mentioned 4 structural variations in GAGs are related with the hexosamines. Despite the low nuclear magnetic sensitivity and low natural abundance of the ^{15}N nucleus (Table 10.1), reliable ^1H–^{15}N spectra of GAGs of different species can be properly achieved when sufficient concentration within a sample is subjected to analysis. Figure 10.2 displays ^1H–^{15}N HSQC spectra of a series of GAG polysaccharides of natural occurrence at the concentration of 15 mg mL^{-1} each sample. Availability of material is not an issue for GAGs derived from commercial sources. Magnetization sensitivity is also enhanced by the INEPT (insensitive nuclei enhanced by polarization transfer) method as seen in common heteronuclear experiments like ^1H–^{15}N HSQC.

Based on Figure 10.2, all GAG types give resonances in distinct regions of the ^1H–^{15}N HSQC spectrum, varying accordingly to their inherent structural features. For instance, all materials made of GalNAc (galactosaminoglycans, GalAGs) give upfield resonances, somewhere between δ_H of 7.94 and 8.09 ppm, (panels A–D) in the ^1H chemical shift scale as compared to the materials made of GlcNAc units (δ_H of 8.36 ppm for HS) (panel E). In terms of sulfation patterns, 4-sulfation results into ^1H–^{15}N cross-peaks with upfield ^{15}N chemical shifts (δ_N of 120.8 ppm) as opposed to 6-sulfation-related ^1H–^{15}N cross-peaks with downfield ^1H chemical shifts (δ_N of 121.7 ppm). Curiously, 4,6-di-sulfation, which can be seen in CS-E (Figure 10.1B) and oversulfated chondroitin sulfate (OSCS), results into a resonance in the midway (δ_N of 121.2 ppm) of the two single sulfation types (panel C).

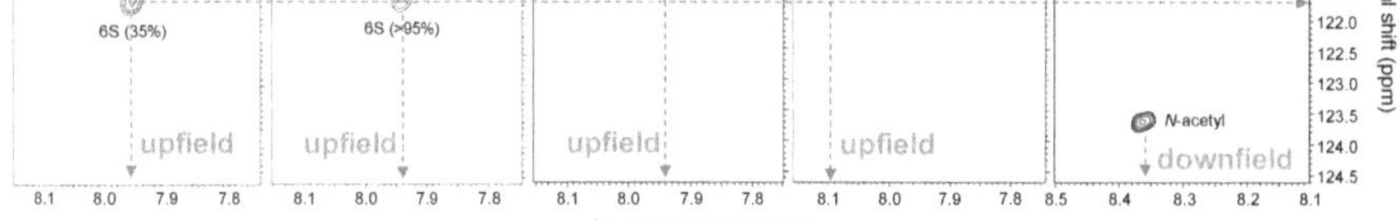

Figure 10.2 Fully assigned ^1H–^{15}N HSQC spectra of the most common native CS polymers: **(A)** CS-A, **(B)** CS-C, **(C)** OSCS, **(D)** CS-B (also known as DS), and **(E)** HS. All spectra were recorded at pH 4.5, 298 K, on a Varian 800 MHz NMR instrument, with samples dissolved in 85/15% H_2O/D_2O. Note that all resonances show at least one distinct chemical shift. 4-sulfation-related signals are upfield at the ^{15}N chemical shift scale in relation to the downfield 6-sulfation-related signals. Galactosaminoglycans (**A–D**) composed of GalNAc units like CS/DS display ^{15}N-related signals at upfield ^1H chemical shifts as compared to the downfield signals of glycosaminoglycans (**E**) composed of GlcNA units as HS.

The rule for the upfield ^{1}H chemical shifts of GalNAc units within the glycans as opposed to the downfield ^{1}H chemical shifts of GlcNAc units (GalAGs *versus* GAGs) can be assured *via* the ^{1}H–^{15}N HSQC spectra of these individual monosaccharides in solution when α and β-configurations occur from the anomeric mutarotation process (Figure 10.3). The amounts of each conformer as well as their structures are shown on the bottom of the figure. These amounts can be deduced from Figure 10.4, in which the kinetics of the anomeric α:β ratio from mutarotation of glucose (Glc) and its derivatives is depicted. Each anomeric is assigned in spectra of Figure 10.3 based on the peak intensity and correlation with the anomeric amounts observed at the equilibrium in the kinetic analysis (Figure 10.4).

In terms of IdoA:GlcA ratios in HS and Hp, they can only be measured in samples at normal pH and room temperature when they are linked to *N*-acetylglucosamine (GlcNAc) units (Figure 10.5). This is because the *N*-sulfated glucosamine (GlcNS) units have their sulfamate protons exchanging with the bulk water very fast at the NMR timescale. Since ^{1}H–^{15}N cross-peaks must be coupled for signal detection, fast chemical exchange in sulfamates of GlcNS must be controlled for signal assessment as discussed in Section 10.4.

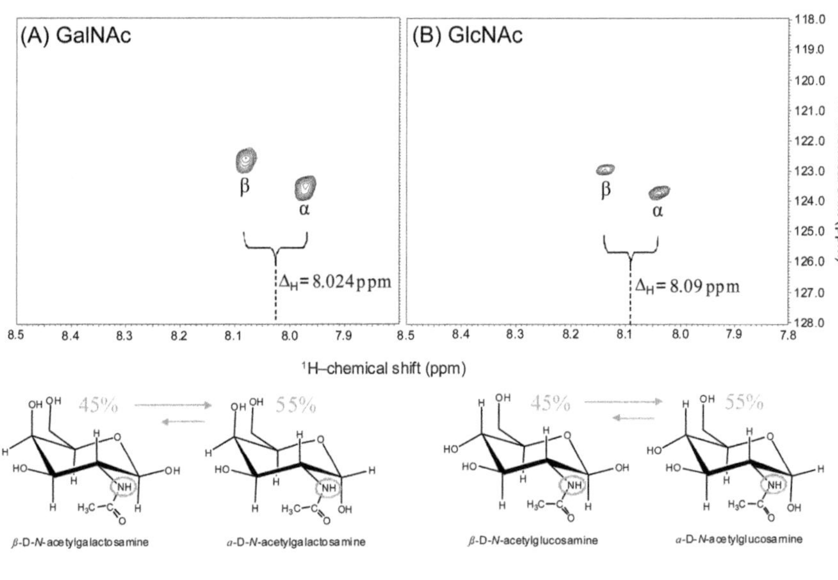

Figure 10.3 Fully assigned ^{1}H–^{15}N HSQC spectra (expansions from δ_H 8.5 to 7.8 ppm, and δ_N 128.0 to 118.0 ppm, which cover all signals) of the composing amino sugars (hexosamines) of GalAGs and GAGs (**A**) *N*-acetyl-D-galactosamine (GalNAc) and (**B**) *N*-acetyl-D-glucosamine (GlcNAc), and structures at the bottom where the resonant NH groups are highlighted with circles. Both spectra were recorded at pH 4.5, 298 K, on a Varian 800 MHz NMR instrument, with samples dissolved in 85/15% H_2O/D_2O. Note the slight chemical shift displacement towards downfield in the ^{1}H scale of the average anomeric signals from GlcNAc as compared to the one from GalNAc unit. The α/β-signals arise from the mutarotation of anomerics in aqueous solution, in which the final equilibrium is 55/45 ratio.

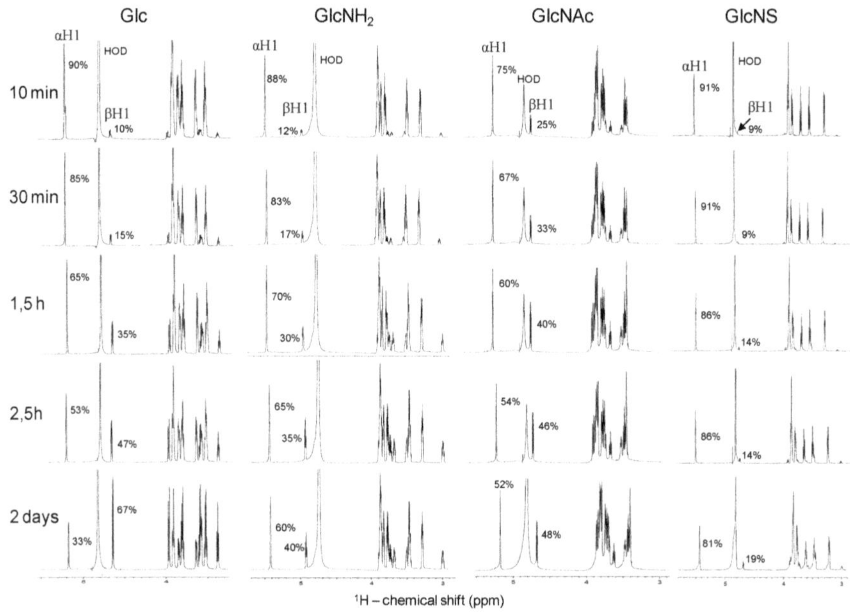

Figure 10.4 1D ^{1}H NMR spectra of glucose (Glc), and its 2-positioned *N*-substituted derivatives: glucosamine ($GlcNH_2$), *N*-acetylglucosamine (GlcNAc), and *N*-sulfated glucosamine (GlcNS), all in ~100% D_2O-solution. The spectra were recorded on a Varian 800 MHz spectrometer at 25 °C and at different time courses after dissolution, as indicated in the left of the panel. The percentage of the α- and β-^{1}H-anomerics changes as a function of time at different rates accordingly to the monosaccharide type. The water (HOD) and anomeric signals are indicated at the top of the panel. Reproduced from ref. 32 with permission from InTech Open Access Publisher.

The impossibility to observe the *N*-sulfated amide groups in GlcNS at normal conditions is one of the major downsides of the ^{15}N-NMR method in analyses of GAGs. Based on Figure 10.5, it is clear to note the higher content of GlcA-linked GlcNAc units in HS (panels A and C) as opposed to Hp which has a higher content of IdoA-attached GlcNAc residues (panel B). The assignments of these ^{1}H–^{15}N signals have been supported by the structural characterization previously achieved through the regular ^{13}C-HSQC experiment (panels D and E).

10.3 Identification of Sulfation Patterns and Position of Composing Hexosamines in GAG Disaccharides and Oligosaccharides of Different Lengths

^{15}N-NMR spectroscopy can also be used for analyses of GAG di-, and oligosaccharides. For instance, unsaturated unreduced CS disaccharides of different sulfation patterns (produced from lyase digestions) were submitted to analysis *via* ^{1}H–^{15}N HSQC experiment (Figure 10.6). As shown in Figure 10.3

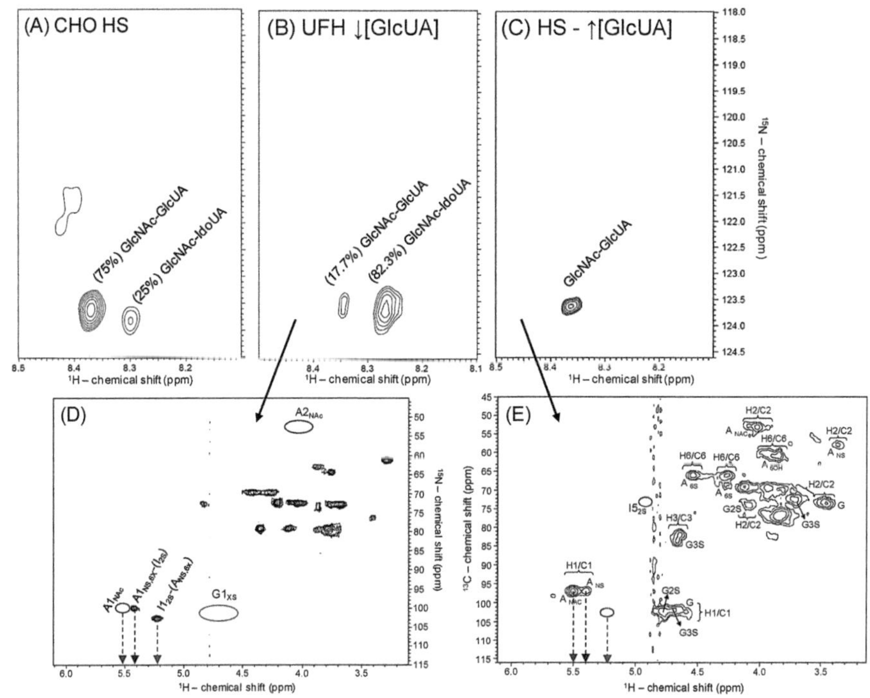

Figure 10.5 ^{15}N- (A–C) and ^{13}C- (D and E) HSQC spectra of heparan sulfate (HS) (A, C, and E) and unfractionated heparin (UFH) (B, D), recorded at Varian 800 MHz (A–C) and Bruker 400 MHz (D, E). The HS in panel A was from cultures of Chinese hamster ovarian (CHO) cells,[29] the UFH in panel B was from a pharmaceutical supplier, and the HS in panel C was extracted from the bivalve *Nodipecten nodosus*.[33] Note that the (B) UFH and (C) HS have respectively low and high amounts of glucuronic acids (GlcUA), as demonstrated by the tiny or missing peaks in their respective ^{13}C–HSQC spectra of panels D and E. In these panels, the letters used are A for glucosamine, I for iduronic acid, G for glucuronic acid, NAc for *N*-acetyl, NS for *N*-sulfation, 6X for 6-(un)substituted, 2S for 2-sulfation, XS for 2- and/or 3-sulfation, and 3S for 3-sulfation. The number right after the capital letters designate the ring position, for example, I_52 s means a cross-peak of the ^{1}H–^{13}C pair at the 5-position of a 2-sulfated iduronic acid ring. Reproduced from ref. 32 with permission from InTech Open Access Publisher.

these disaccharides also showed α and β anomericity in which the cross-peak from the former configuration is upfield in ^{1}H chemical shifts and also shows the major population in solution. When sulfation is absent (like ΔC0S of Figure 10.3A) or far from the structural probe (amide group placed at the C2 position) like the ΔC6S (Figure 10.3B), the average resonances are downfield at the ^{15}N scale ($\delta_N \sim$ 122.5 ppm) and upfield at the ^{1}H scale (δ_H ~8.18 ppm). When sulfation content is higher like the ΔC2;6S (Figure 10.3D) or gets closer to the probe like the ΔC4S (Figure 10.3C) the average resonances are upfield at the ^{15}N scale ($\delta_N \sim$ 120.5 ppm) and downfield at the ^{1}H scale (δ_H ~ 8.2 ppm). This signal profile can be clearly used for diagnostic of sulfation

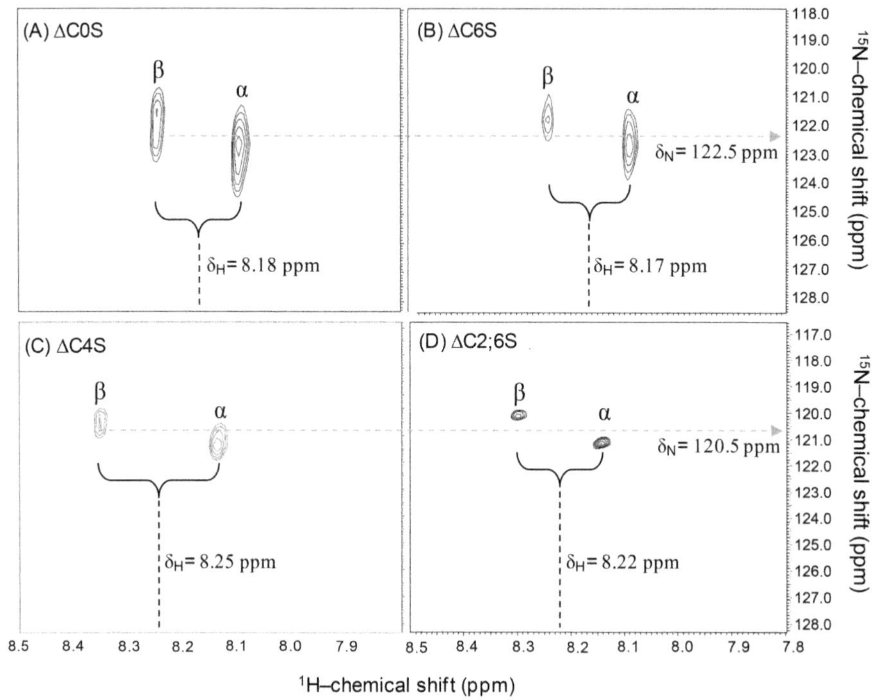

Figure 10.6 ^{15}N-HSQC spectra (800 MHz) of unreduced CS disaccharides with different sulfation patterns. The structures are the following: (A) ΔC0S as ΔUA-(β1 → 3)-GalNAc; (B) ΔC6S as ΔUA-(β1 → 3)-GalNAc-6S; (C) ΔC4S as ΔUA-(β1 → 3)-GalNAc-4S; and (D) ΔC2;6S as ΔUA-2S-(β1 → 3)-GalNAc-6S, where ΔUA, Δ4,5-unsaturated uronic acid; GalNAc, *N*-acetyl galactosamine; and S, sulfation groups in which the numbers before "S" represent the ring position. The dashed arrows at 122.5 and 120.5 ppm indicate the signals with less and more influence caused by sulfation-related deshielding effects, respectively. This influence can also be seen in the ^{1}H chemical shift. α and β anomers are assigned respectively to each CS disaccharide.

patterns in GAG disaccharides; for example, in the illustrative cases of CS derivatives (Figure 10.6).

This method is also useful on medium-sized GAG oligosaccharides, for example, the (un)saturated reduced CS hexasaccharides of different sulfation patterns (Figure 10.7). From this picture, we can clearly note that all composing hexosamines give rise to ^{1}H–^{15}N cross-peaks at different places in the HSQC spectrum. They are labeled as -ol, mid and non-red for the reduced GalNAc, the middle GalNAc, and the GalNAc placed at the non-reducing end. From Figure 10.7 we can also observe that 6-sulfated GalNAc units give resonances at ^{15}N downfield regions (δ_N over 121 ppm) as opposed to 4-sulfated GalNAc units which resonate with δ_H values somewhere close to 120 ppm. Based on Figure 10.7, all sulfation types and units' position within the oligosaccharide chain can be properly assigned as exemplified by the cases of CS hexasaccharides.

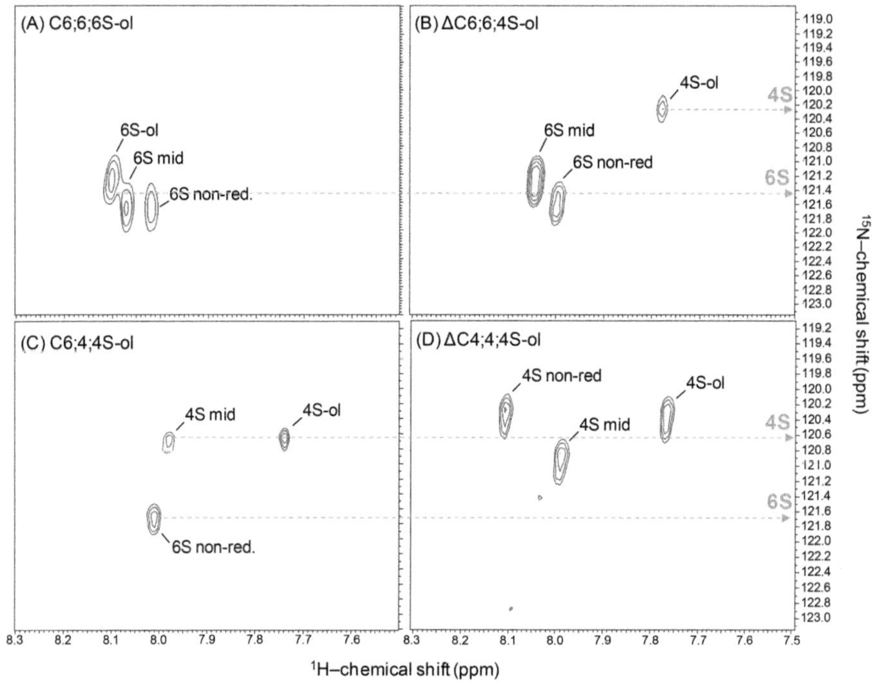

Figure 10.7 Fully assigned ^{1}H–^{15}N HSQC spectra of reduced CS hexasaccharides with different sulfation patterns. They have been obtained from either hyaluronidase (panels A and C) or chondroitnase (panels B, and D) digestions. All spectra were recorded at pH 4.5, 298 K, on a Varian 800 MHz NMR instrument, with samples dissolved in 85/15% H_2O/D_2O. The structures are the following: (A) C666S-ol as GlcA-(β1 → 3)-GalNAc-6S-(β1 → 4)-GlcA-(β1 → 3)-GalNAc-6S-(β1 → 4)-GlcA-(β1 → 3)-GalNAc-6S-ol; (B) ΔC664S-ol as ΔUA-(β1 → 3)-GalNAc-6S-(β1 → 4)-GlcA-(β1 → 3)-GalNAc-6S-(β1 → 4)-GlcA-(β1 → 3)-GalNAc-4S-ol; (C) C644S-ol as GlcA-(β1 → 3)-GalNAc-6S-(β1 → 4)-GlcA-(β1 → 3)-GalNAc-4S-(β1 → 4)-GlcA-(β1 → 3)-GalNAc-4S-ol; and (D) ΔC444S-ol as ΔUA-(β1 → 3)-GalNAc-4S-(β1 → 4)-GlcA-(β1 → 3)-GalNAc-4S-(β1 → 4)-GlcA-(β1 → 3)-GalNAc-4S-ol, where ΔUA, Δ4,5-unsaturated uronic acid; GalNAc, *N*-acetylgalactosamine; GlcA, glucuronic acid; S, sulfation groups where the numbers before "S" represent the ring position; -ol stands for reduced sugars (open rings at the reducing-end termini).[29,30] In the panels, 6S non-red., 6S mid, 6S-ol, 4S non-red., 4S mid, and 4S-ol belong respectively to the assignments of the amide ^{1}H–^{15}N pairs of 6-sulfated and 4-sulfated GalNAc from the non-reducing, middle, and reduced disaccharide units in the CS hexasaccharides. The peaks with ^{15}N-chemical shifts around 120 and over 121 ppm are respectively from 4- and 6-sulfated GalNAc units.

In addition, position of units within oligosaccharide chains can be also recognized by ^{1}H–^{15}N HSQC experiment for the case of the composing GlcNAcs in unreduced HA oligosaccharides of different lengths such as 4-mer, 6-mer and 8-mer as shown at Figure 10.8. The assignment can be even feasible for the α and β conformers placed at the reducing end.

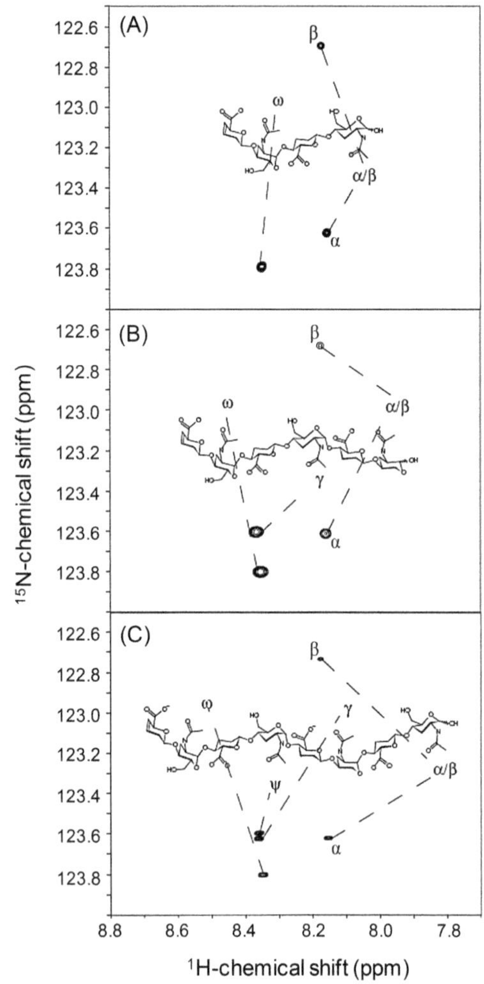

Figure 10.8 ^1H–^{15}N HSQC spectra (900 MHz) of HA-derived (A) tetra-, (B) hexa-, and (C) octasaccharides. Labels on the spectrum correspond to amide groups, according to their positions within the oligosaccharide chains, using Greek letters. They are α/β for the anomeric conformers at the reducing-end terminus, γ for the internal disaccharide adjacent to the reducing-end terminal disaccharide unit, and Ψ for the disaccharide adjacent to the non-reducing end terminal disaccharide unit labeled as ω. Data modified with permission.[21]

10.4 ^{15}N-Based NMR Analysis of *N*-Sulfonate Groups

As already mentioned, protons from *N*-sulfonate groups in GlcNS units exchange very fast with the solvent and thus ^1H–^{15}N cross-peaks from these groups are hard to be seen at the ^{15}N HSQC spectrum of the GlcNS

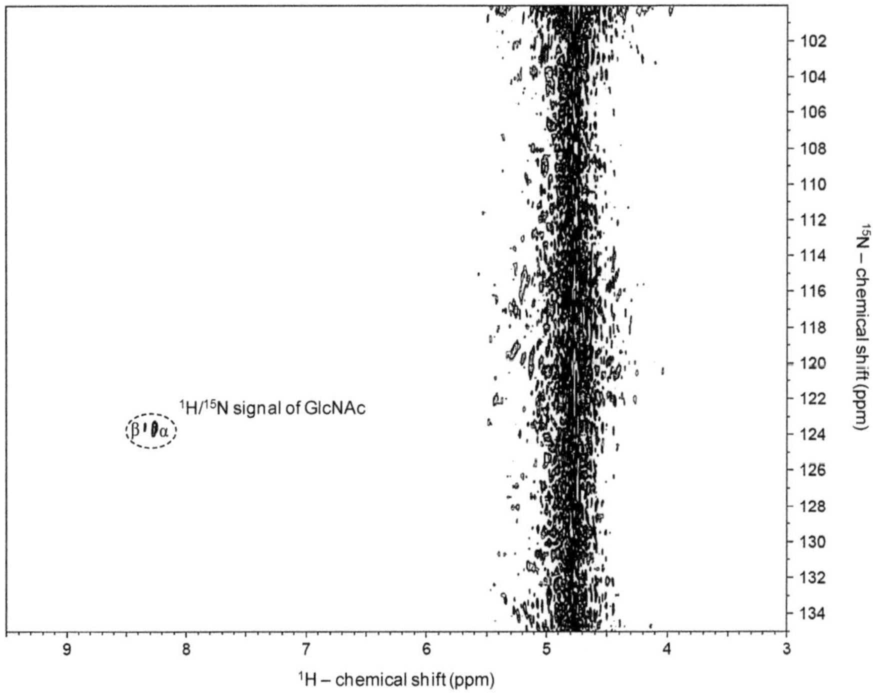

Figure 10.9 ^{1}H–^{15}N HSQC spectra (900 MHz) of Hp dissolved in 10:20:70% D_2O:acetone:H_2O (pH 4.5) at 25 °C. Labels on the spectrum correspond to amide groups of the *α* and *β*-conformers of the composing GlcNAc units.

monosaccharide or from this unit in Hp or HS samples analyzed at regular conditions. On the other hand, the amide groups from their GlcNAc units can be fairly detected, inclusively their anomeric ratios (Figure 10.9). The ^{15}N HSQC spectrum of Figure 10.9 displays a large window in order to shown that only residual water and the ^{1}H–^{15}N cross-peaks of α-/β-GlcNAcs are easily seen. However, when GlcNS units are investigated by NMR at very low temperature (3 °C, for example) all exchangeable protons become detectable (Figure 10.10a). Figure 10.10b *versus* Figure 10.10c shows the unprotonated and protonated conditions of the amide group respectively at room and low temperature, as measured by the αH2 multiplets. When the amide group (also linked at C2) is unprotonated, the αH2 peak gives to a doublet of doublets due to the coupling constant of H2 with H1, and H3. When the amide group is protonated, the αH2 peak gives rise to a triplet of doublets, because of the additional coupling with the NH. The low temperature decreases the fast exchange of the protons in amides of GlcNS and the resultant ^{1}H-NMR signal of the H2 proves the protonation (Figure 10.10a and c). The ^{1}H assignments shown at Figure 10.10a were obtained from the ^{1}H–^{1}H total correlation spectroscopy (TOCSY) spectrum of GlcNS at low temperature (Figure 10.11a). The ^{13}C-HSQC spectrum of this sample is also shown at Figure 10.11b, however, as predicted based on

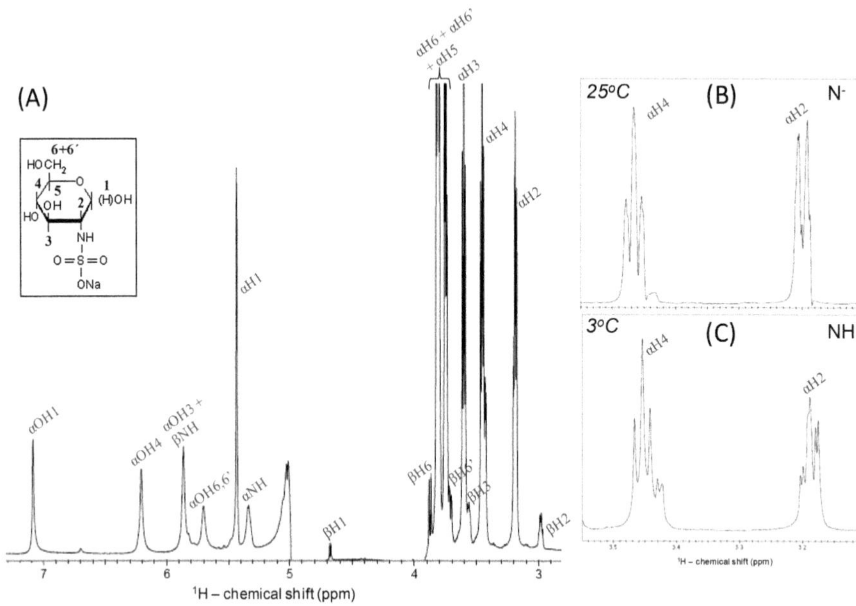

Figure 10.10 1D ^{1}H NMR expansions (A) 2.8–7.3 ppm and (B and C) 3.10–3.55 ppm of 5 mg mL^{-1} GlcNS in 10:20:70% D_2O:acetone:H_2O (pH 4.5) at 3 °C (A and C) and at 25 °C (B) on spectrometer Varian 800 MHz. Note that in panel B, the resonance of αH2 splits in a doublet of doublets due to $^3J_{H2\text{-}H1}$ and $^3J_{H2\text{-}H3}$, while in panel C there are a triplet of doublets for the additional splitting of αH2 with NH ($^3J_{H2\text{-}HN}$) since in the former case the amide is deprotonated because of the higher temperature, but in case of panel C, the amide is protonated because of lower temperature. The structure of the GlcNS in its protonated and sodiated state is shown in an insert on the top left side of the panel A, with protons assigned accordingly. The acetone solvent in the solution was added to achieve completely liquid phase in the sample at very low temperature (3 °C). Reproduced from ref. 32 with permission from InTech Open Access Publisher.

the mutarotation kinetics of Figure 10.4, *α* is the major configuration (the spin system belonging to *β* is not shown at the ^{13}C-HSQC spectrum).

Since the low temperature of the ^{15}N-NMR method can slow down the fast chemical exchange of the amide protons in GlcNS, ^{1}H–^{15}N cross-peaks of its *α* and *β* forms can become detectable *via* the ^{15}N-HSQC spectrum (Figure 10.12). The spectrum of panel A displays a higher threshold in order to clearly differentiate the α^{1}H–^{15}N cross-peak from the nearby residual water signals. The spectrum of panel B at a lower counter level shows both anomeric forms of the sulfated amide in GlcNS monosaccharide. As expected based on anomeric kinetics of Figure 10.4, at equilibrium the signal belonging to the *α* form in GlcNS is much more populated (Figure 10.12B). Langeslay and co-authors have studied in details the protonation of the sulfamates in GlcNS units of Hp and derivatives as a function of pH and temperature.[26–28]

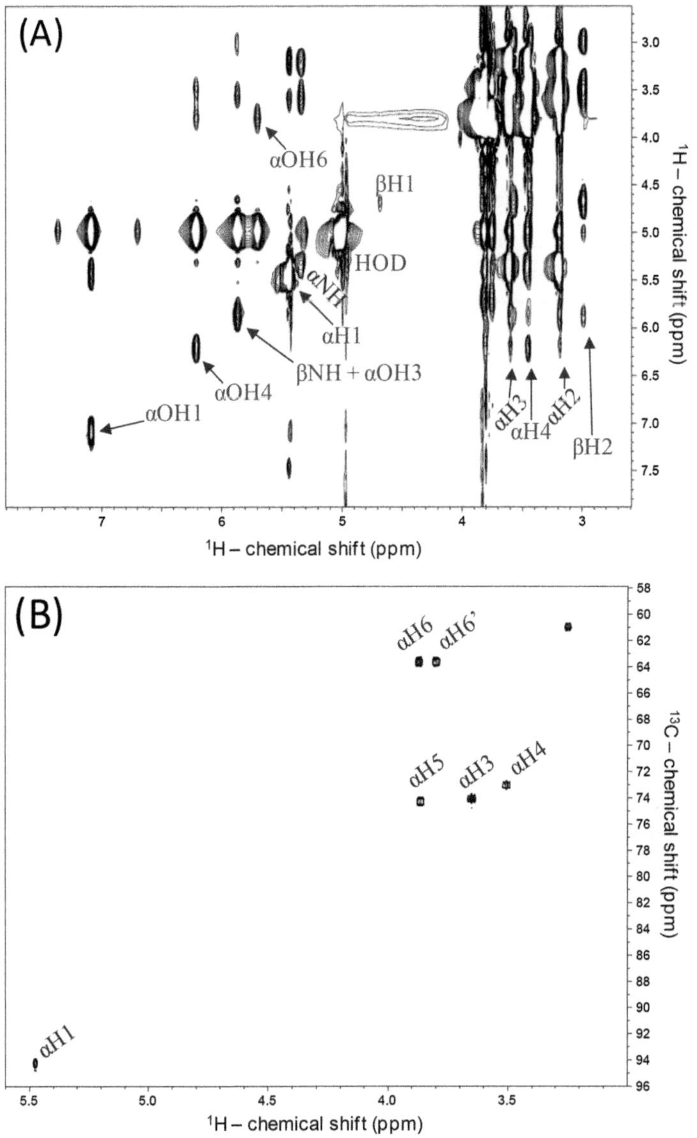

Figure 10.11 (A) TOCSY and (B) ^{13}C–HSQC spectra of 5 mg mL^{-1} GlcNS in 10:20:70% D_2O:acetone:H_2O (pH 4.5) at 3 °C on a Varian 800 MHz. The exchangeable protons from hydroxyl and amide groups are labeled accordingly. The population of the α-anomeric form is more abundant than the β-one as seen in Figure 10.4. Only the spin system of the α-configuration is shown at panel B. Reproduced from ref. 32 with permission from InTech Open Access Publisher.

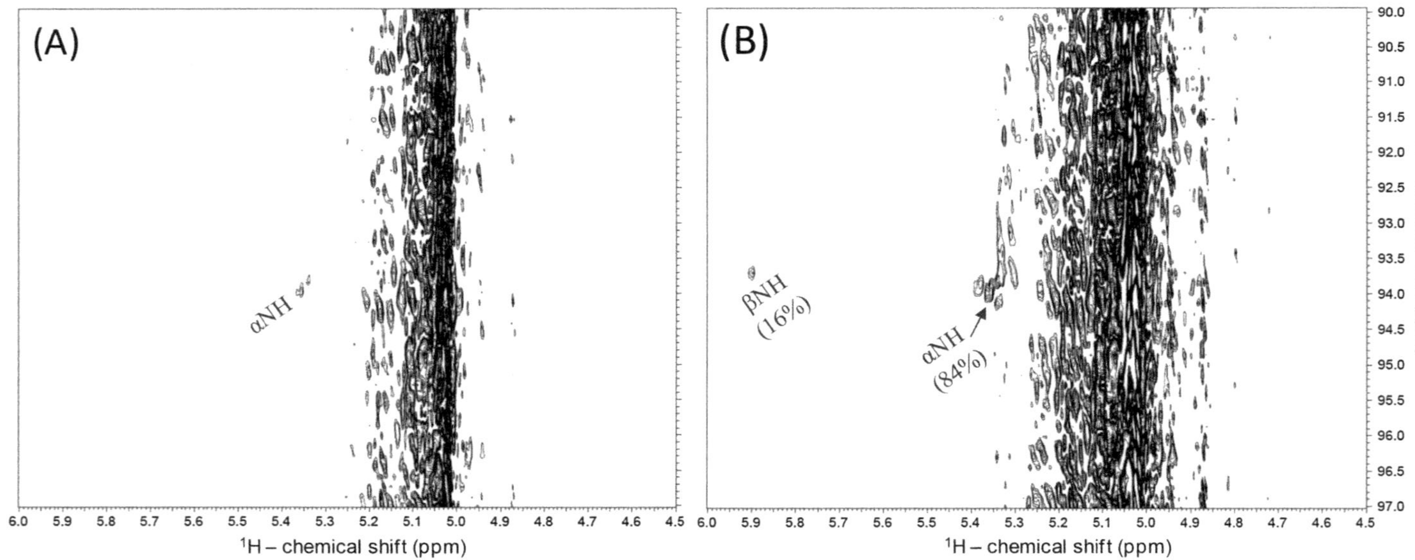

Figure 10.12 (A and B) ^{15}N–HSQC spectra of 5 mg mL^{-1} GlcNS in 10:20:70% D_2O:Acetone:H_2O (pH 4.5) at 3 °C on a Varian 800 MHz at higher (A) and lower (B) counter levels in order to see the βNH and to distinguish the αNH from the nearby water–noise signals. The amide groups are labeled accordingly to the anomeric configurations. The population of the α-anomers of the GlcNS is more abundant than the β-one as seen in Figure 10.4. Reproduced from ref. 32 with permission from InTech Open Access Publisher.

Direct observation of ^{15}N *via* 1D NMR analyses is difficult due to low NMR sensitivity and abundance of this nucleus, even considering ^{15}N-labeled samples like glutamine (Gln) ^{15}N-labeled at its side chain amino group (Figure 10.13A). See that a resultant 1D ^{15}N-NMR spectrum of this compound at 23 mM concentration at 800 MHz instruments submitted for signal averaging through tens of hours for free-induction decay acquisition, results into a ^{15}N signal very close to the noise level. Improvement in signal-to-noise ratio and peak resolution can be achieved (something close to 4000 enhancement factor) in case this same compound is submitted to

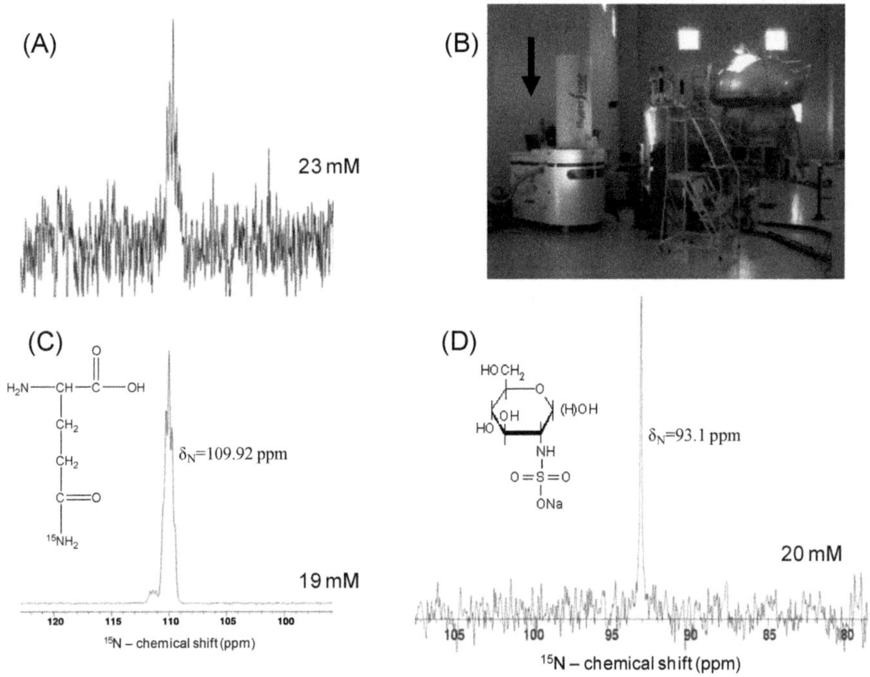

Figure 10.13 (A) ^{15}N-direct observe spectrum of 23 mM ^{15}N-labeled (at lateral amide) glutamine (Gln) at Varian 500 MHz. (B) HiperSense™ polarizer (indicated by arrow) coupled to a Varian 500 MHz NMR instrument (right side) used at Complex Carbohydrate Research Center, University of Georgia, for dynamic nuclear polarization (DNP). See Varian 900 MHz at the far right. (C) ^{15}N-direct observe spectrum of 19 mM ^{15}N-labeled Gln at Varian 500 MHz after 4h-submission to DNP in the HyperSense™. Enhancement of ~4000 in signal-to-noise ratio is achieved as compared to the spectrum of the unpolarized samples (panel A). The same acquisition would take approximately one month *via* the unpolarized method. Note the triplet distribution in the resonance meaning coupling with a single proton at the lateral amide (N–H). (D) ^{15}N-direct observe spectrum of ^{15}N-labeled *N*-sulfated glucosamine (GlcNS) at 20 mM at Varian 500 MHz after 4h-submission to DNP in the HyperSense™.

4h-DNP procedure for magnetization increment in a HyperSense Polarizer (Figure 10.13B) and following submission to a 1D ^{15}N-NMR spectral acquisition (Figure 10.13C) at close sample concentration. The quality of this spectrum could take approximately 1 month at the 800 MHz without the DNP procedure. Note that ^{15}N-Gln results into a triplet in the 1D ^{15}N peak, meaning that Gln side chain amino group has a proton coupled to the ^{15}N nucleus (Figure 10.13C). GlcNS gives a singlet peak at the DNP-assisted 1D ^{15}N-NMR spectrum (Figure 10.13D) collected at room temperature and at nearly equal sample concentration of Gln. This means that ^{15}N nucleus is uncoupled to protons. This explains the impairment in observing the ^{1}H–^{15}N cross-peak of GlcNS *via* ^{15}N-HSQC experiment at room temperature (Figure 10.9).

10.5 ^{15}N-NMR Analysis of Cellular GAGs

All analyses up to now have been done on GAGs or GAG-composing hexosamine units from commercial sources and abundance of material is adequate for those cases in order to overcome the low sensitivity and abundance of ^{15}N. However, when GAGs of cellular origins are under question, amounts are very limited and thus, ^{15}N-NMR analysis at natural abundance will be difficult in those cases. Hence, a ^{15}N-labeling protocol for GAGs of cells is necessary and has been properly implemented (Figure 10.14).[29] This protocol is based on incubation of ^{15}N-labeled side chain Gln in cell cultures (Figure 10.14). This will lead to synthesis of ^{15}N-labeled UDP-hexosamine products in the cytosol of cells in culture that will be further used inside the Golgi apparatus for biosynthesis of GAG chains. Figure 10.15 shows ^{15}N-HSQC spectra of total ^{15}N-labeled GAGs of two types of cells submitted to the isotope labeling procedure: mouse lung endothelium (panel A) and Chinese hamster ovarian (CHO) (panel B). The ^{15}N-labeled GAG profile of these two cells was observed quite different. Endothelium shows essentially more CS types, including CS-A and CS-C and few amounts of HS, whereas CHO expresses great amounts of HS and little CS-A. These CS and HS peaks (Figure 10.15) were assigned based on results of Figure 10.2. The two peaks at the HS region (Figure 10.15) can be clearly assigned as GlcNAc units liked to GlcA (δ_H of 8.37 ppm) or linked to IdoA (δ of 8.32 ppm ^{1}H upfield) as seen on Figure 10.5. Since HS has more GlcA content rather than IdoA, the higher intensity of the downfield ^{1}H–^{15}N cross-peak in Figure 10.15 is expected. Hence, ^{15}N-labeling protocol together with ^{15}N-HSQC NMR experiment is a successful method to investigate the distribution and chemical structures of GAGs derived from cell preparations, even though the yield during the cell GAG extractions are in the range of hundreds of micrograms. The amounts of cell ^{15}N-labeled GAG polysaccharides examined in Figure 10.15 were around 600 micrograms of purified materials.

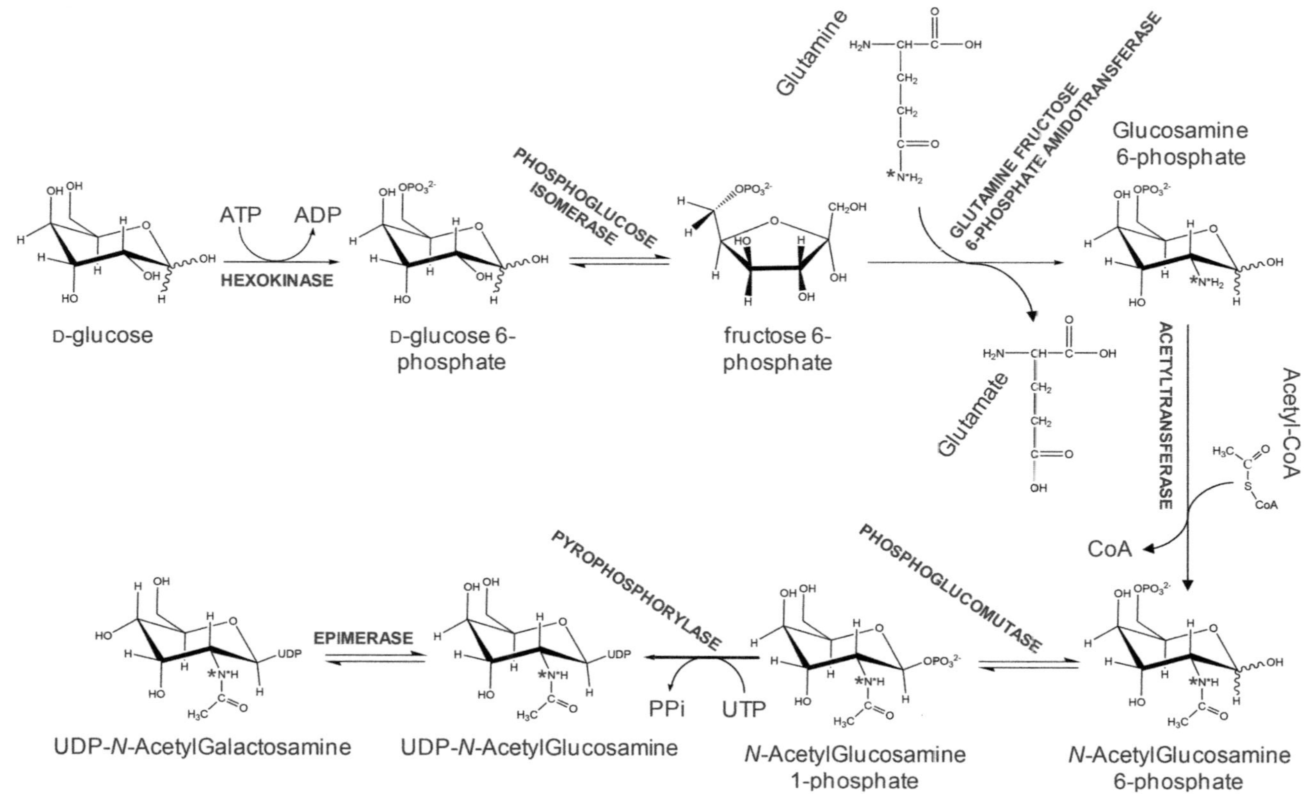

Figure 10.14 Schematic representation of the cytosolic biosynthesis of ^{15}N-labeled *N*-acetyl-hexosamines that will be further used inside the Golgi apparatus for building up the GAG backbones in proteoglycans. The names of the enzymes are fully written in upper cases while the substrates and products are mostly written in lower cases. The ^{15}N-labeled sites are indicated with asterisks. Modified with permission from ref. 29. Copyright (2010) American Chemical Society. For more details about GAG biosynthesis see also ref. 34,35.

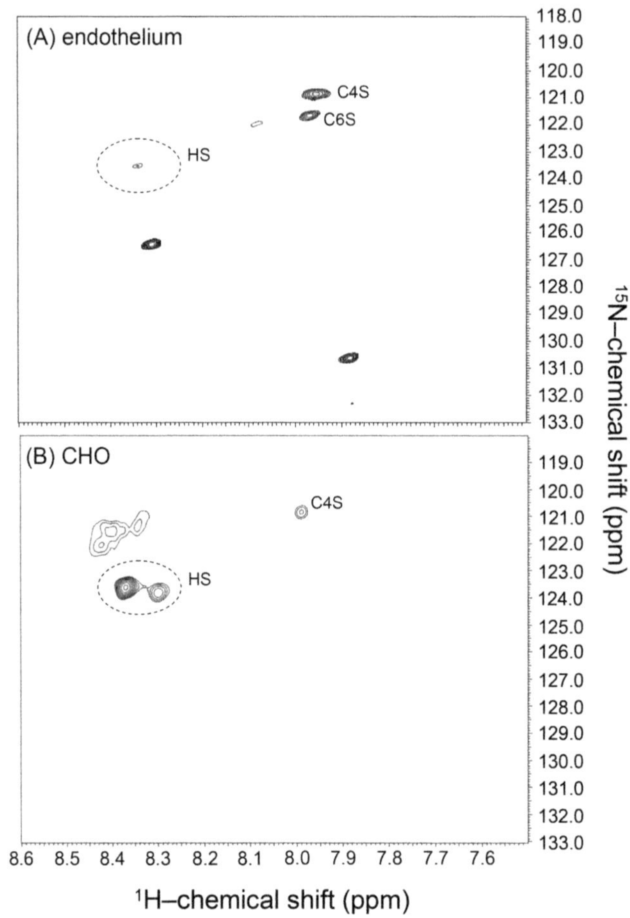

Figure 10.15 $^1H–^{15}N$ HSQC spectra of ^{15}N-labeled negatively charged molecules extracted from the mouse lung endothelial cells (A), and (B) CHO cells.[29]

10.6 Concluding Remarks

Despite the less frequent use of ^{15}N in NMR analyses of GAGs due to its low magnetic sensitivity and natural abundance, unlabeled GAGs when examined at sufficient concentrations or labeled GAGs from cell extracts at very low amounts can be successfully investigated in terms of their structural features. Each GAG family and subfamilies give rise to a specific $^1H–^{15}N$ cross-peak in the ^{15}N-HSQC spectrum (Figure 10.16). Once the resultant ^{15}N-related peaks are well resolved and superimposition is infrequent as opposed to ^1H- and ^{13}C-related resonances, the ^{15}N-NMR technique can be easily employed to recognize the types or subtypes of GAGs in a sample as well as the respective composing units (primarily hexosamines but also uronic acids) and their amounts. Application of ^{15}N-NMR is also quite valuable for recognition

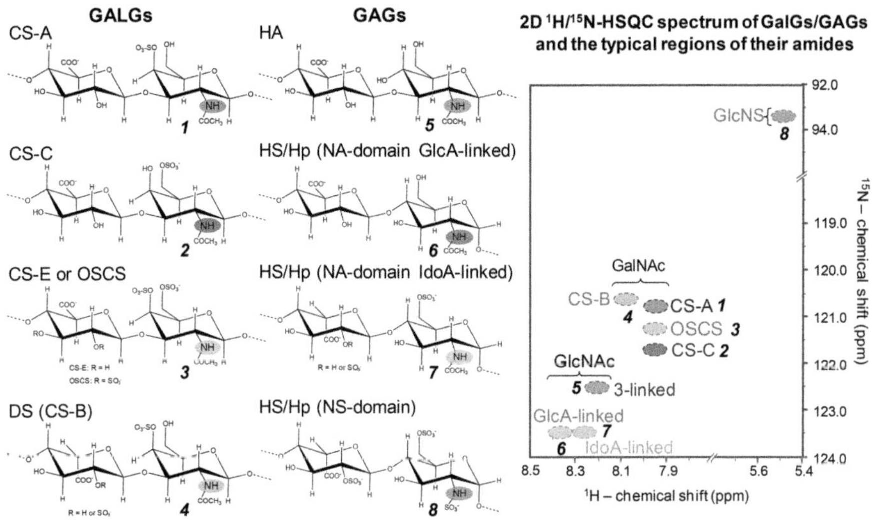

Figure 10.16 Structure of the *N*-acetyl-galactosamine (GalNAc)-composing galactosaminoglycans (GalGs) and *N*-acetyl-glucosamine (GlcNAc)-composing glycosaminoglycans (GAGs) with numbered and highlighted amide groups (left), and their typical positions on a hypothetical ^{1}H–^{15}N HSQC spectrum (right). Note that all GalG and GAG types can be clearly differentiated since their resultant ^{1}H–^{15}N signals do not overlap in the spectrum. Reprinted with permission from ref. 29. Copyright (2010) American Chemical Society.

of sulfation pattern and residue position within a chain of GAG oligosaccharide. The only downside in the application of this novel ^{15}N-based NMR method in structural analysis of GAGs is the difficulties in observing the peaks coming from the amide groups in GlcNS units. In order to overcome this obstacle, controlled conditions like low temperature, proton-rich solution and controlled pH ranges in samples are necessary for proper detection of the peaks coming from the GlcNSs.

References

1. R. Sasisekharan and G. Venkataraman, *Curr. Opin. Chem. Biol.*, 2000, **4**, 626.
2. D. L. Rabestein, *Nat. Prod. Rep.*, 2002, **19**, 312.
3. K. Sugahara, T. Mikami, T. Uyama, S. Mizuguchi, K. Nomura and H. Kitagawa, *Curr. Opin. Struct. Biol.*, 2003, **13**, 612–620.
4. T. M. Handel, Z. Johnson, S. E. Crown, E. K. Lau and A. E. Proudfoot, *Annu. Rev. Biochem.*, 2005, **74**, 385.
5. J. Angulo, R. Ojeda, J. L. de Paz, R. Lucas, P. M. Nieto, R. M. Lozano, M. Redondo-Horcajo, G. Giménez-Gallego and M. Martín-Lomas, *ChemBioChem*, 2004, **5**, 55.

6. R. Pandera, G. Venkataraman, D. Berry, R. Godavarti and R. Sasisekharan, *FASEB J.*, 1999, **13**, 1677.
7. W. Li, D. J. Johnson, C. T. Esmon and J. A. Huntington, *Nat. Struct. Mol. Biol.*, 2004, **11**, 857.
8. I. Capilla and R. J. Linhardt, *Angew. Chem., Int. Ed. Engl.*, 2002, **41**, 391.
9. N. Afratis, C. Gialeli, D. Nikitovic, T. Tsegenidis, E. Karousou, A. D. Theocharis, M. S. Pavão, G. N. Tzanakakis and N. K. Karamanos, *FEBS J.*, 2012, **279**, 1177.
10. G. W. Yip, M. Smollich and M. Götte, *Mol. Cancer Ther.*, 2006, **5**, 2139.
11. K. R. Taylor and R. L. Gallo, *FASEB J.*, 2006, **20**, 9.
12. M. M. Kooman, P. Prandoni, F. Piovella, P. A. Ockelford and D. P. Brandjes, *N. Engl. J. Med.*, 1996, **334**, 682.
13. M. Fujita, *Eur. Heart J.*, 2000, **21**, 270.
14. V. H. Pomin, *Anal. Chem.*, 2014, **86**, 65.
15. V. H. Pomin, Y. Park, R. Huang, C. Heiss, J. S. Sharp, P. Azadi and J. H. Prestegard, *Glycobiology*, 2012, **22**, 826.
16. C. D. Blundell, P. L. DeAngelis, A. J. Day and A. Almond, *Glycobiology*, 2004, **14**, 999.
17. A. Almond, P. L. DeAngelis and C. D. Blundell, *J. Am. Chem. Soc.*, 2005, **127**, 1086.
18. A. Almond, P. L. DeAngelis and C. D. Blundell, *J. Mol. Biol.*, 2006, **358**, 1256.
19. M. Mobli and A. Almond, *Org. Biomol. Chem.*, 2007, **5**, 2243.
20. C. D. Blundell and A. Almond, *Magn. Reson. Chem.*, 2007, **45**, 430.
21. M. Mobli, M. Nilsson and A. Almond, *Glycoconjugate J.*, 2008, **25**, 401.
22. Z. Zhang, S. A. McCallum, J. Xie, L. Nieto, F. Corzana, J. Jiménez-Barbero, M. Chen, J. Liu and R. J. Linhardt, *J. Am. Chem. Soc.*, 2008, **130**, 12998.
23. B. M. Sattelle, J. Shakeri, I. S. Roberts and A. Almond, *Carbohydr. Res.*, 2010, **345**, 291.
24. V. M. Tran, T. K. Nguyen, K. Raman and B. Kuberan, *Anal. Bioanal. Chem.*, 2011, **399**, 559.
25. J. F. K. Limtiaco, D. J. Langeslay, S. Beni and C. K. Larive, *J. Magn. Reson.*, 2011, **209**, 323.
26. D. J. Langeslay, S. Beni and C. K. Larive, *Anal. Chem.*, 2011, **83**, 8006.
27. D. J. Langeslay, S. Beni and C. K. Larive, *J. Magn. Reson.*, 2012, **216**, 169.
28. D. J. Langeslay, R. P. Young, S. Beni, C. N. Beecher, L. J. Mueller and C. K. Larive, *Glycobiology*, 2012, **22**, 1173.
29. V. H. Pomin, J. S. Sharp, X. Li, L. Wang and J. H. Prestegard, *Anal. Chem.*, 2010, **82**, 4078.
30. V. H. Pomin, *Anal. Bioanal. Chem.*, 2013, **405**, 3035.
31. C. D. Blundell, P. L. Deangelis and A. Almond, *Biochem. J.*, 2006, **396**, 487–498.
32. V. H. Pomin, in *Glycosylation*, ed. S. Petrescu, InTech, Rijeka, Croatia, 1st edn, 2012, Chapter 4, pp. 3–11.

33. A. M. Gomes, E. O. Kozlowski, V. H. Pomin, C. M. de Barros, J. L. Zaganeli and M. S. Pavão, *J. Biol. Chem.*, 2010, **285**, 7312–7323.
34. K. Sugahara, T. Mikami, T. Uyama, S. Mizuguchi, K. Nomura and H. Kitagawa, *Curr. Opin. Struct. Biol.*, 2003, **13**, 612–620.
35. J. D. Esko and U. Lindahl, *J. Clin. Invest.*, 2001, **108**, 169–173.

CHAPTER 11

NMR Studies of Protein–Glycosaminoglycan Interactions

XU WANG[a]

[a]School of Molecular Sciences, Arizona State University, Tempe, Arizona, USA
*E-mail: xuwang@asu.edu

11.1 Biology of Glycosaminoglycan

Glycosaminoglycan (GAG) is a class of linear, sulfated extracellular polysaccharide found in eukaryotic organisms. The majority of GAGs are attached to proteins. These GAG-containing glycoproteins are commonly referred to as proteoglycans and are usually anchored to the cell surface or released into the extracellular matrix. The size of GAG chains vary greatly, but can contain tens of monosaccharide to several hundred monosaccharides, making them the largest glycan in mammals. Structurally, GAG is made up of repeating disaccharide units. One of the residues in the disaccharide is always an amino sugar in the form of either *N*-acetylglucosamine (GlcNAc) or *N*-acetylgalactosamine (GalNAc). The other residue can be either an uronic acid sugar or galactose. GAGs are classified according to the identity of their disaccharide units (Figure 11.1). Heparin and heparan sulfate (HS) contain glucosamine (GlcN) and uronic acids, a majority of which is in the form of iduronic acid (IdoA). Chondroitin sulfate (CS) and Dermatan sulfate (DS) contain

New Developments in NMR No.10
NMR in Glycoscience and Glycotechnology
Edited by Koichi Kato and Thomas Peters

Published by the Royal Society of Chemistry, www.rsc.org

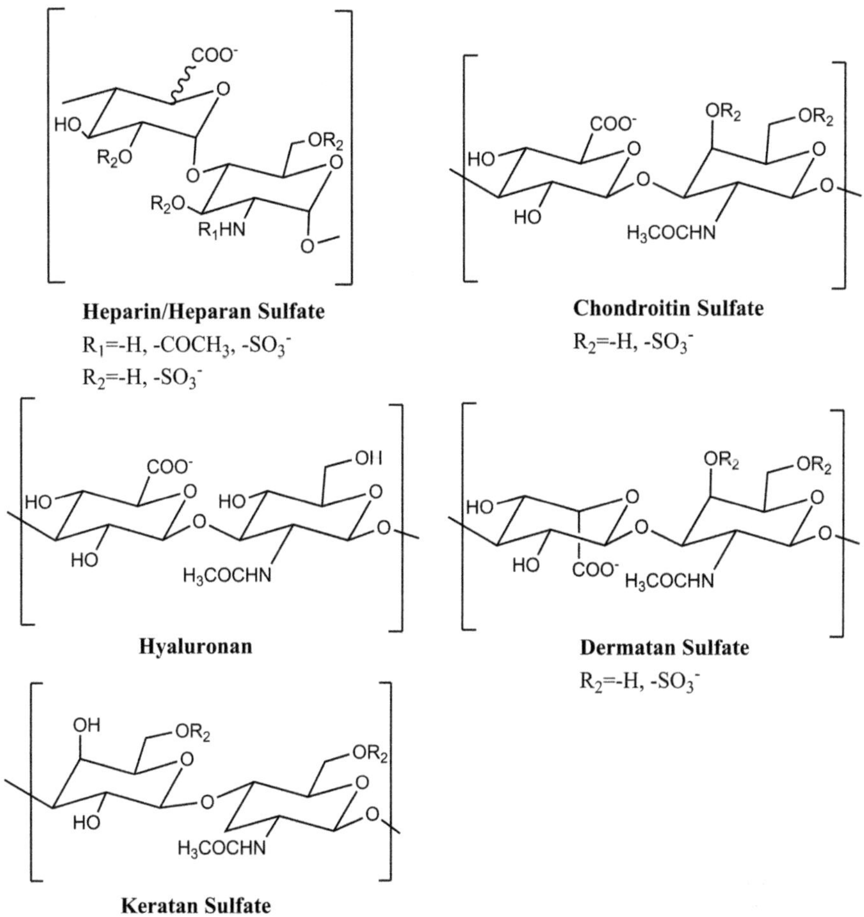

Figure 11.1 Structures of disaccharide units found in different types of GAGs.

GalNAc and uronic acids. All uronic acids in CS are glucuronic acid (GlcA) whereas most of the uronic acids in DS are IdoA. Keratan sulfate is another GalNAc-containing GAG. It is also the only GAG that uses galactose in place of uronic acid residues. Finally, hyaluronan, the only GAG not synthesized as a proteoglycan, contains only GalNAc and GlcA.

Despite their simple carbohydrate backbone, GAGs are among the most complex biomolecules in existence. The complexity originates from two post-synthetic modifications of the carbohydrate backbone. First, IdoA in GAGs is derived from GlcA through C-5 epimerization. However, the conversion is applied semi-randomly and often not completely, producing a mixture of GlcA and IdoA along the GAG chain. Because of IdoA's additional ring flexibility,[1] the IdoA content modulates flexibility of the GAG chain and affect GAG's affinity and specificity for proteins. In addition, all GAGs except hyaluronan are subject to sulfation by various sulfotransferases. Heparin and HS contain the most complex sulfation patterns. GlcNAc in HS and heparin are extensively deacetylated and then *N*-sulfated, producing *N*-sulfoglucosamine (GlcNS). GlcN in these two types of GAGs can also be 6-*O*-sulfated. Epimerization of GlcA and 2-*O*-sulfation of the IdoA can also take place. 3-*O*-sulfation also occurs on some GlcNS and is known to be important in interaction with some proteins.[2] CS and DS undergo similar sulfation modifications. In particular, GalNAc of CS and DS can be sulfated on carbons 4 and 6. 2-*O*-sulfated IdoA has also been detected in these GalNAc-containing GAGs. However, GalNAc in GAGs is never deacetylated. The only type of GAG not known to undergo any post-synthesis modification is hyaluronan.

GAGs have been found in echinoderms, fishes, and mammals. Functions of GAGs in most species are still being actively researched, but biological activities of mammalian GAGs are much better understood. GAGs in mammals are known to regulate development, tissue regeneration, immune response, pathogen adhesion and blood coagulation.[3,4] These activities have made GAGs prominent modulators of cancers, autoimmune and inflammatory disorders, and even Alzheimer's Disease.[4] Most of the activities of GAGs are the result of their interactions with proteins, and GAG-based mimetics have been envisioned as treatments for many pathologies. However, the efficacy of these proposed therapies relies on specifically targeting proteins using GAG mimetics, yet understanding of how protein–GAGs interact has been primitive, and prediction of proteins' GAG specificity is not possible in most cases.

GAG's interactions with proteins are unique among glycans. Because of the high density of sulfates and carboxylates in GAGs, GAGs interact with proteins predominantly through electrostatic interactions, but other weak forces, such as hydrogen bonding and van der Waals interactions, are also present and can exert significant influences on the specificity of the interactions.[5] In some cases, such as the binding of Antithrombin to heparin, the interaction between protein and GAG is so specific that existence of one sulfate group at a particular position can have significant influences on binding affinity.[2,6] However, many GAG-binding proteins are more promiscuous, often having similar affinities for GAGs with different sulfation motifs. The same GAG ligand can even form different complexes with the same protein.[7,8] Further complicating the situation is the fact that GAG can induce protein oligomerization, leading to differences in protein–GAG interactions between monomeric and oligomeric forms of proteins. These dynamic and heterogeneous interactions greatly increase the difficulty of these studies. Investigations of structure–activity relationships between proteins and GAGs are also hampered by the fact that a library of structurally homogeneous GAG ligands is not available. Existing structural data of these systems show that basic amino acid clusters on GAG-binding proteins are crucial in facilitating protein–GAG interactions. However, sequence homology between the clusters is low. In some cases, they are composed of contiguous sequence elements, but they can also be comprised of disparate elements in different parts of the sequence. As a result, identifying GAG-binding sites through primary sequences alone can be challenging.[5]

Study of protein–GAG interactions by NMR has a long history.[9] The versatility of solution NMR and its high-resolution nature has allowed structural information to be obtained easily. Because interactions of proteins with small GAG ligands usually fall into the fast exchange regime on the NMR time scale, heterogeneous samples of GAGs are usually not a barrier to NMR studies since only the average perturbation on the system is measured. In addition, NMR techniques for observing both ligands and proteins exist, allowing interactions to be examined from both perspectives. However, NMR has had limited success obtaining intermolecular contacts between proteins and GAGs. This may be the result of both the dynamics of protein–GAG interactions as well as limitations in NMR methodology used to detect these interactions. Fortunately, computational modeling of the protein–GAG complexes is possible and widely used. In addition, new techniques designed to increase the amount of structural information obtainable are now being actively developed. This chapter reviews the most common NMR technique used in the study of protein–GAG interactions and previews some techniques in development that can further increase our understanding of protein–GAG interactions.

11.2 Preparing GAG Ligands for NMR Studies

Availability of homogeneous and isotopically labeled GAG ligands is one of the biggest bottlenecks in the study of protein–GAG interactions. Natural GAG polysaccharides are heterogeneous and large. Their interactions with proteins often result in protein aggregation, rendering them unsuitable for high-resolution study of protein–GAG interactions. Size-defined fragments are available commercially. However, structurally homogenous GAG ligands beyond disaccharides are usually hard to come by. This is partly because chemical synthesis of GAGs is still beyond the reach of all but the most skilled carbohydrate chemists.[10,11] In comparison, purification of GAG ligands from partial depolymerization of native polymers, although still difficult, is possible. Structurally homogeneous GAG ligands can be obtained by using a three-step procedure. First, native polymers are degraded using GAG-digesting enzymes such as heparinase or chondroitinase. The partially digested mixture is then separated based on size using size exclusion chromatography. Finally, fragments of uniform size are separated based on charge using SAX–HPLC. Because the fragments are the result of enzymatic digestion, sulfation patterns on these ligands are controlled by the specificity of the enzyme, therefore, may not always reflect the distribution of sulfation motifs seen *in vivo*. However, currently, this method remains the most practical way of obtaining structurally homogeneous ligands. Chemoenzymatic synthesis of GAGs has been in development and cloning of a number of important GAG modifying enzymes has greatly enhanced the progress of these endeavors.[12] It is hopeful that such developments will increase availability of homogeneous GAG ligands in the future.

Another bottleneck in NMR study of protein–GAG interactions is the lack of isotopically labeled GAG ligands. This severely limits the number of NMR experiments applicable to protein–GAG investigations and renders investigation of these systems in the presence of low GAG concentrations difficult. Difficulty in chemical synthesis of GAGs render isotope incorporation impractical during synthesis. Because GAGs are not directly produced by prokaryotic organisms capable of growing in minimal media, biosynthesis of isotope labeled GAGs cannot be accomplished easily either. Fortunately, GAG-like polysaccharides are produced by *E. coli* K5 in the form of (GlcAβ1-4GlcNAcα)n, commonly referred to as heparosan. Heparosan can be chemically or enzymatically sulfated and epimerized to yield heparin and HS like molecules.[13] Because growth of *E. coli* K5 in minimal media supplemented with ^{13}C-glucose and ^{15}N-$(NH_4)_2SO_4$ is robust, ^{13}C–^{15}N labeled heparosan can be obtained in sufficient quantities. These isotopically labeled heparosan can then be chemically or enzymatically *O*- and *N*-sulfated at different positions, yielding isotopically labeled heparin ligands. Such a scheme was used by Laguri *et al.* to generate uniformly ^{13}C–^{15}N-labeled GAGs for NMR studies. Currently, there are no naturally occurring prokaryotic systems that produce CS precursors, but work is underway to engineer *E. coli* that can produce such precursors.[14] Another approach is to isotopically label GAGs directly in mammalian cell culture. One such example is the utilization of ^{15}N-glutamine to ^{15}N label amino sugars in Chinese Hamster Ovary cells, leading to labeling of both CS and HS.[15] Interestingly, the authors found that the chemical shifts of amide proton and nitrogen in these *N*-acetylated sugars are sensitive to sulfation pattern in the sugar, therefore, allows NMR to quantify amounts of each sulfated species. However, it should be pointed out that mammalian cell culture labeling schemes are not yet viable sources of GAGs for structural studies.

11.3 Studying Protein-Bound GAGs

The use of NMR in investigations of carbohydrate conformations can be traced back to the initial application of NMR in chemistry when $^3J_{HH}$ were used to determine the anomericity of monosaccharides.[16] NMR has also seen widely used in analysis of protein-bound GAG ligands. Two of the most popular experiments for this application are saturation transfer difference (STD) spectroscopy[17] and transfer NOE (trNOE) spectroscopy.[18] Both techniques utilize dipole–dipole interactions between proximate protons to obtain on protein–GAG contacts. Furthermore, both techniques require only a small amount of unlabeled protein. However, these techniques do require structurally homogeneous GAG ligands, and, because both techniques rely on fast exchange kinetics between protein and GAGs to amplify the effect of protein–GAG interactions on ligands without significantly broadening signals of the ligand, effective applications of these techniques to any system are dependent on the kinetics of interactions between protein and GAGs.

In STD, protein proton signals are specifically saturated without perturbation of the ligand signal. However, the dipole–dipole interactions between the protein and ligand protons allow transfer of the saturation energy from protein to the ligand, producing decreases in signals of ligand protons at the interface.[17] In practice, the ratio of ligand-to-protein must be carefully optimized. Too little protein would result in a small saturation effect that may be difficult to detect, whereas too much protein would lead to excessive broadening of ligand signals, and difficulties in suppressing protein signals. The kinetics of the interaction can also be important. Interactions that fall in the slow exchange regime on the NMR time scale are usually not suitable for STD because the amount of ligands interacting with the protein is similar to the concentration of the protein in the sample, leading to only small changes in ligand signals. Conversely, interactions that fall in the fast exchange kinetics regime will allow multiple ligand molecules to interact with the same protein during the duration of the saturation, therefore amplifying the effect of the saturation and producing a stronger STD signal. STD has been successfully applied to the study of numerous protein-ligand systems.[19–22] An example of the application of STD to protein-GAG interactions can be found in a study by Deshauer *et al.* on interactions of CCL5 with CS hexasaccharides. STD revealed that a wide variety of CS protons, including protons on both GalNAc and GlcA, can be found at the interface (Figure 11.2). However, because of

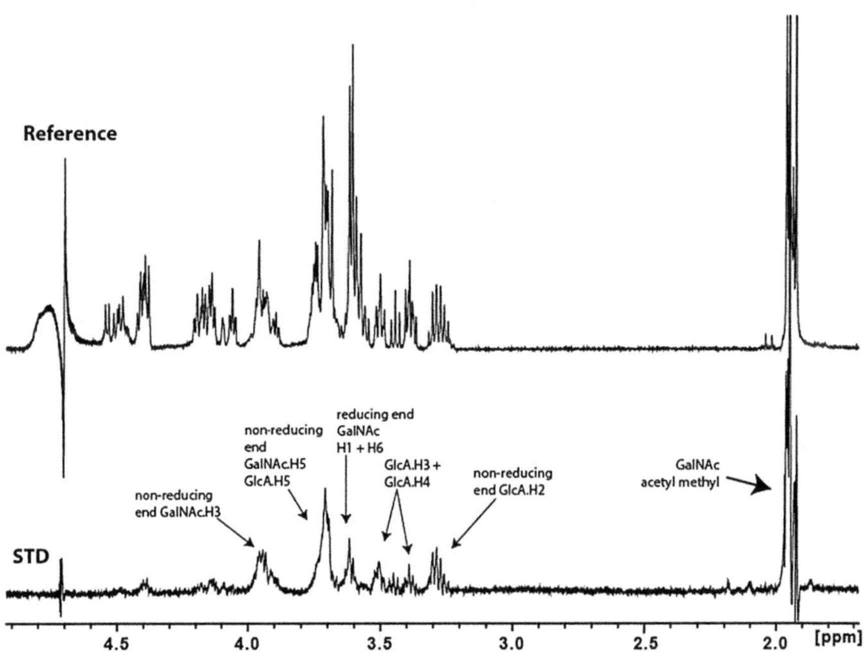

Figure 11.2 STD spectrum of CS hexasaccharides in the presence of the chemokine RANTES. Top spectrum is that of the CS hexasaccharide. The bottom is the STD spectrum of 1 mM hexasaccharide in the presence of 50 μM RANTES. Adapted from Structure, 23, Deshauer *et al.*, Interactions of the Chemokine CCL5/RANTES with medium-sized chondroitin sulfate ligands, 1066–1077, Copyright 2015 with permission from Elsevier.[19]

the ambiguity in ^{1}H chemical shifts, the exact GalNAc at the interface cannot be determined from 1D-STD alone. Kunze *et al.* overcame this limitation by extending the 1D-STD to a natural abundance ^{13}C-edited STD-heteronuclear single quantum coherence (HSQC) experiment, which allowed saturated protons to be determined with higher certainty.[20] This approach is only applicable to systems where ligand signals have much higher signal-to-noise ratio than protein. Yu *et al.* also used STD to determine interactions of several heparin octasaccharides with FGF2 and FGF10.[22] The authors have found that, as a result of inefficient spin diffusion in some proteins, the saturation frequency must also be carefully optimized to obtain the maximum sensitivity. Specifically, frequencies in both the aliphatic and aromatic regions of the spectrum may be saturated to produce the best STD signal. In many cases, temperature, ionic strength and pH should also be optimized to achieve the maximum STD signal in the study of protein–GAG systems. For instance, several studies have shown that lower temperatures help to both stabilize the interaction and enhance the NOE development.[20,21]

Information derived from STD can also be used quantitatively to validate models of protein–glycan complexes using the CORCEMA approach.[23] The CORCEMA model allows STD effects on a ligand to be predicted from a model and when used as a potential in iterative structure refinement procedures, can produce the right structure for protein–glycan complexes. For instance, Tjong *et al.* used CORCEMA to verify that a heparin-binding protein from cobra venom bound heparin in multiple conformations.[24]

Similar to STD, trNOE relies on energy transfer through dipole–dipole interactions between adjacent protons and requires only small amounts of protein. However, unlike STD, the transfer is strictly intramolecular to the ligand, thus no intermolecular contact information can be obtained. Instead, trNOE allows the conformation of the bound ligand to be determined. The technique is made possible by the fact that NOE develops more rapidly in large molecular complexes than in small molecules.[18] Therefore, most of the ligand NOE crosspeaks originated from protein-bound ligands rather than free ligands, allowing us to determine the conformation of the bound ligand. Similar to STD, the amount of protein in the sample and kinetics of interactions has great bearing on the quality of the data and should be optimized for each system. The amount of protein should be low enough to minimize the broadening of ligand signals but sufficiently high to produce a significant trNOE effect. Kinetics of the interaction should be in the fast exchange regime to allow each protein to interact with multiple ligand molecules during the NOE mixing time, thus amplifying the effect. Hricovini used trNOE to determine the bound conformation of the heparin pentasaccharide on antithrombin III.[25] Through quantitative analysis of NOE peak intensities, the authors were able to accurately determine the glycosidic torsion angles between the residues in the bound pentasaccharide and noted that significant differences existed between conformations of the bound and free pentasaccharides.

Because GAGs often induce protein aggregation and bind proteins non-specifically, use of high concentrations of GAGs, as required by STD and trNOE, may not be applicable to every system. In those cases, the approach of Canales *et al.* can be used if isotopically labeled protein is available.[26] In that study, the authors investigated the structure of FGF1-bound heparin hexasaccharide by removing the signal of ^{13}C-labeled protein using a ^{13}C-filtered NOESY. Because the protein-to-ligand ratio is 1 and the equilibrium dissociation constant (K_D) is small, the signals from the ligand originates mostly from the bound fraction. Such an approach can yield data that are easier to interpret than trNOE and is applicable to all systems irrespective of binding affinity or kinetics.

As isotopic labeling of GAG becomes easier, use of ^{13}C–^{15}N-labeled GAG in the study of protein–GAG interactions will be more common, and the introduction of an extra dimension will greatly improve signal dispersion and information content of the data. At the moment, there is only one published report of ^{13}C-labeled GAG being used in NMR investigation of protein–GAG interactions. Laguri *et al.* used a ^{13}C-labeled heparin mimic to study interactions of the chemokine CXCL12α with the heparin octasaccharide.[27] They reported that H4 of GlcNS and H2 of GlcA experienced significant saturation in the presence of chemokine CXCL12α.

11.4 NMR Studies of GAG-Binding Proteins

Solution NMR is a powerful and flexible tool for studying protein structures. This is largely because proteins can be easily isotopically enriched and development of multidimensional heteronuclear experiments that allow high-resolution analysis of proteins as large as 70 kDa. Studies of GAG–protein systems have also benefited greatly from these developments. Since ligands are not observed in these experiments, homogeneity of the ligand sample is not crucial.

The technique used most often in protein–ligand interaction studies is chemical shift perturbation analysis. This technique relies on changes in the backbone amide proton and nitrogen chemical shifts to identify residues involved in ligand binding. The method uses the conventional ^{1}H–^{15}N HSQC experiment to identify signals from ^{1}H–^{15}N bonded pairs in proteins and track changes in their chemical shifts at different GAG ligand concentrations. One advantage of using ^{1}H–^{15}N pair as the reporter atoms is that chemical shifts of amide proton and nitrogen are sensitive to minute differences in magnetic and chemical environment. This reduces signal overlap and allows unambiguous signal identification. They also experience significant changes in response to ligand binding, and magnitudes of these ligand-induced changes often correlate with the proximity of the ligand on the protein. In addition, the magnitude of change also reflects the fraction of protein that is ligand bound. For a 2-state model, the K_D of interaction can be extracted accurately by fitting the plot of chemical shift change and ligand concentration to eqn (11.1), where $\Delta\delta$ is the observed chemical shift change in the protein at the total ligand concentration L, P is the total protein concentration, δ_{max} is the maximum chemical shift change possible and K is the equilibrium dissociation constant.

$$\Delta\delta = \frac{\delta_{max}}{P \times L} \frac{(P + L + K) - \sqrt{(P + L + K)^2 - 4P \times L}}{2} \tag{11.1}$$

This technique has been a workhorse for NMR protein–GAG studies because it can be easily applied as long as assignments for the signals in the ^{1}H–^{15}N HSQC are available. It has been used to determine GAG-binding sites of many GAG-binding proteins, including chemokines, growth factors, bacterial adhesins and other secreted proteins. Figure 11.3 shows chemical shift perturbations induced in the chemokine RANTES by CS hexasaccharides. Not surprisingly, residues exhibiting large GAG-induced chemical shift changes are also located in regions enriched in basic amino acids, providing important clues to the identity of the GAG-binding sites. However, it should be noted that chemical shift changes reflect changes in magnetic environment, which are usually, but not always, the result of ligand binding. For instance, if ligand binding induces large conformation change, then magnitudes of chemical shift change are more likely to reflect conformation changes rather than ligand binding. Fortunately, GAG's interactions with proteins rarely produce large conformation changes, simplifying data interpretation in most cases. It should also be noted that residues involved in directly binding GAGs may not have the largest change in chemical shifts, or any change at all, if the magnetic environment does not change significantly around these residues. Besides chemical shift changes, GAG can also change the intensities of signals by either changing the dynamics of the atoms or the solvent exchange rates of the amide proton.[28] This information can also be used to infer GAG binding.

Residues in GAG-binding sites can also be identified using STD if ^{2}H, ^{15}N-labeled protein is available. The technique was originally developed by Sakakura *et al.* to determine lectin residues responsible for binding *N*-glycans.[29] The experiment is made possible by the fact that complete deuteration of the protein allowed glycan protons to be saturated without disturbing the proton signals of the protein, which are entirely in the amide region. Cross-saturation effects on protein amide protons are then used to identify residues involved in glycan binding. This technique was used by Deshauer *et al.* to confirm the location of the GAG-binding sites in the chemokine RANTES.[19] One weakness of the experiment is that, cross-saturation effect on small complexes are often not strong enough to produce signal intensity changes in amide protons unless they are directly involved in hydrogen-bonding interactions with GAG ligands. However, side chain amide protons of glutamine and asparagine are known to hydrogen bond with GAG glycans, and therefore are likely to be detected by cross-saturation experiments.

^{15}N-labeled proteins are also widely used to study dynamics of proteins. Theoretical treatments on the effect of amide nitrogen dynamics on NMR

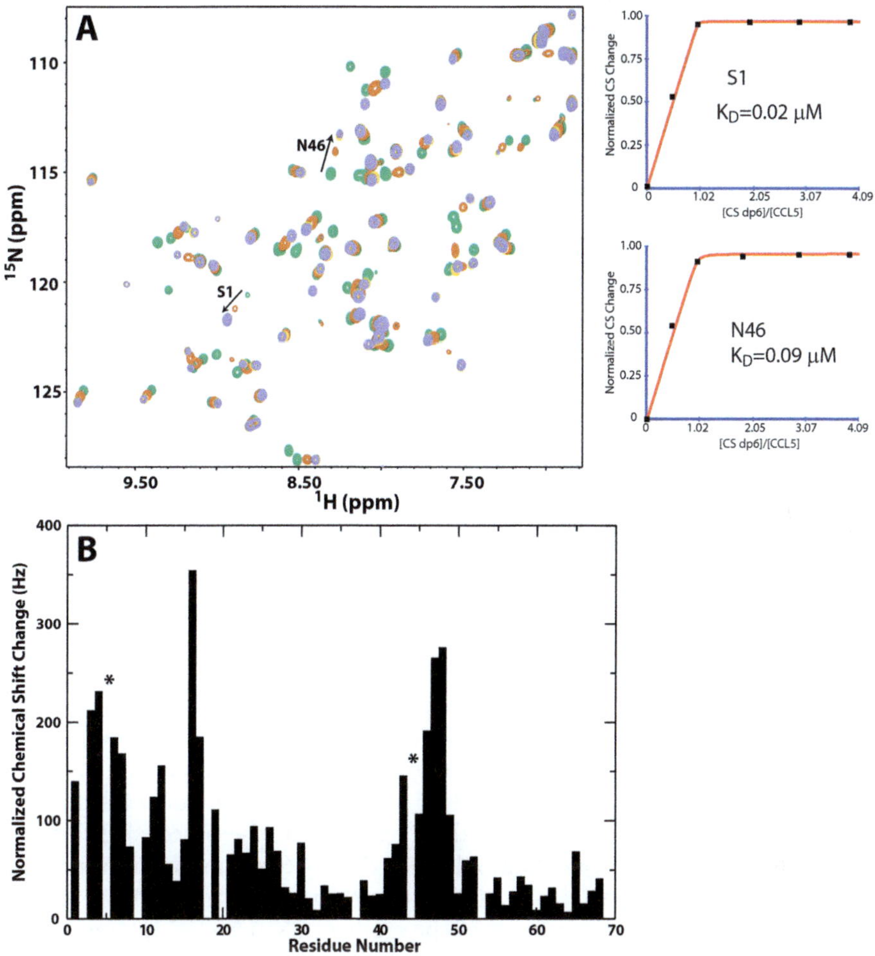

Figure 11.3 Chemical shift perturbation analysis of CS hexasaccharide induced chemical shift changes in the protein RANTES. (A) ^{15}N-edited HSQC spectra of 40 µM RANTES at different concentrations of CS hexasaccharides. The colors of the contours of HSQCs represent CS hexasaccharide concentrations of 0 (dark green), 20 (brown), 40 (yellow), 80 (pink), 120 (light green), 160 (light blue) µM. Chemical shift changes of S1 and N46 were used to construct the binding curves shown on the right and K_D of interaction was calculated by fitting the curve. (B) Scaled and combined chemical shift changes of each residue. The combined chemical shift changes were calculated using the formula $\Delta\delta_S = \sqrt{\Delta\delta^2_{H,Hz} + \left(1.7\Delta\delta_{N,Hz}\right)^2}$. Positions of S5 and R45, which broadened beyond detection during the titration, are indicated by stars. Other residues missing from the graph are prolines. Adapted from Structure, 23, Deshauer *et al.*, Interactions of the Chemokine CCL5/RANTES with medium-sized chondroitin sulfate ligands, 1066–1077, Copyright 2015 with permission from Elsevier.[19]

relaxation rates of the atom is well developed,[30] and experiments used to accurately measure the observables required to calculate the dynamics information are very robust. To characterize motions at pico-and nano-second time scales, the longitudinal relaxation rate, transverse relaxation rate and heteronuclear NOE buildup of amide nitrogen atoms are measured. These values can then be used to calculate the rate of the global tumbling of the protein as well as the internal motion using the Model-Free formalism.[31,32] To detect changes in dynamics that occur on a time scale of micro- to milli-seconds, CPMG-based relaxation dispersion experiments can be used.[33] Feng *et al.* used these dynamic experiments to study the interactions of GAG ligands with a flexible segment in decorin-binding protein A–B (DBPA & DBPB), a bacterial adhesin from the Lyme disease bacterium Borrelia burgdorferi. They showed that presence of heparin ligands did not change the nano- and pico-second time scale motion of the atoms in this flexible segment. However, DS, but not heparin, was able to increase the micro- and milli-second time scale motion of a linker in the protein dramatically.[34] On the other hand, Kim and co-workers also used CPMG-base relaxation dispersion experiments to study interactions of the heparin binding domain of $VEGF_{165}$ with heparin octasaccharide and discovered that GAG ligands reduced the micro- and milli-second time scale motion of the C-terminus of the domain. These results show GAG binding can have significant and varied effects on protein dynamics.

GAG-induced protein oligomerization studies may also be carried out easily using NMR. NMR has been widely used to study dimerization of chemokines.[19,35,36] These studies took advantage of the fact that monomeric species of most chemokines are in slow exchange with the dimers on the NMR time scale, giving the spectra two distinct sets of resonances from each species. Utilizing this, Deshauer *et al.* showed that presence of CS hexasaccharide ligands eliminated all traces of the chemokine RANTES–CCL5 monomer, indicating GAGs can induce dimerization of this chemokine.[19] (Figure 11.4). Similar observations have also been made in the study of CXCL8–GAG interactions.[37] However, oligomerization of other proteins may not be distinguished so easily as chemokines. To detect oligomerization in those cases, changes in the hydrodynamic radius of the protein can be used to infer oligomerization state. NMR can estimate the hydrodynamic radius of the complex using either rotational correlation time of the molecules,[38] which can be derived from transverse and longitudinal relaxation rates, or it can be measured using pulse field gradient translational diffusion experiments. Mayo and co-workers used the pulse field gradient experiments to determine that platelet factor-4 forms a dimer in the presence of heparin dodecasaccharide ligands.[39]

11.5 Obtaining Intermolecular Contacts Between Protein and GAGs

Although NMR has played a significant role in studying glycans and proteins involved in protein–GAG interactions, structures of only a handful of protein–GAG complexes have been determined by NMR.[26,40] The main

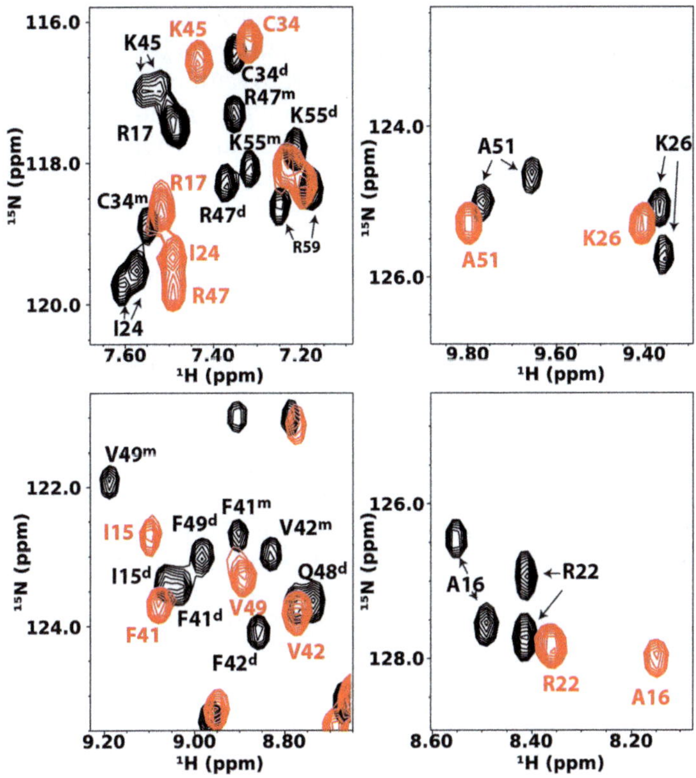

Figure 11.4 Sections of ^{15}N–HSQCs of 30 μM E66S-CCL5 at 40 °C. Black: E66S-CCL5 without CS444, both monomer and dimer species are visible. Peaks representing monomeric species are labeled with the letter m. Red: E66S-CCL5 with two molar equivalents of CS444. Adapted from Structure, 23, Deshauer *et al.*, Interactions of the Chemokine CCL5/RANTES with medium-sized chondroitin sulfate ligands, 1066–1077, Copyright 2015 with permission from Elsevier.[19]

difficulty in such studies is that detecting intermolecular contacts between proteins and GAGs using NOE has proven to be difficult.[41] Several factors contribute to this limitation. First, the interactions between protein and GAG ligands in NMR studies are usually weak and dynamic. This is because NMR studies often employ small GAG ligands ranging from tetrasaccharide to decasaccharide. These ligands can be purified to homogeneity relatively easily and form reasonably sized protein–GAG complexes that are more amenable to NMR investigations. However, the small sizes of the ligands mean their interactions with the protein are also weaker, reducing intensity of intermolecular NOE. These weak interactions are also dynamic, allowing the ligand to interact with GAGs in multiple binding conformations. These dynamic interactions can also induce signal broadening and prevent intermolecular NOE from being observed at all. The dynamic nature of these interactions stems partly from the fact that protein–GAG interactions are usually mediated by electrostatic attraction between the basic amino acids lysine/arginine and the acidic sulfate/carboxylate on GAGs. These interactions are inherently dynamic and weak under physiological conditions. Studies have shown that lysine side chains remain dynamic even when ion pairing with carboxylates, implying that interactions between them are transient in solution.[42,43] Others have shown that complete methylation of lysines has little or no effect on the affinity of proteins for GAGs.[44] This implies lysine's interactions with GAGs can be exerted over a long range, and hydrogen bonding between lysine and the acidic groups may not be necessary. This coupled with the fact that sulfates are usually far removed from the non-labile protons on carbohydrates means the average distance between lysine/arginine side chains and non-labile protons in GAG ligands may be out of the effective range of NOE. In fact, many intermolecular NOEs observed at protein–GAG interfaces involve not lysines and arginines but polar or even hydrophobic amino acids that have either hydrogen bonding or van der Waals interactions with GAGs. Nevertheless, intermolecular NOE has remained the de-facto method for determining structures of protein–GAG complexes. It should be noted that these complications also affect other techniques: electron densities of GAG ligands are often missing or incomplete in crystal structures of protein–GAG complexes.[45,46]

Given these challenges, newer NMR techniques are now being developed to increase the range of intermolecular NOEs. Paramagnetically labeled GAG ligands, in particular, have proven to be applicable to a number of systems.[34,47,48] Paramagnetic effects in NMR originate from dipole–dipole interactions between unpaired electrons and nearby atoms. Such interactions can produce additional relaxation effects on atoms, the extent of this effect on transverse relaxation of the proton can be described by eqn (11.2):

$$R_{2,\text{pre}} = \frac{1}{15}\left(\frac{\mu_0}{4\pi}\right)^2 \frac{\gamma_H^2 g_S^2 \mu_B^2 S(S+1)}{r^6}\left(4\tau_c + \frac{3\tau_c}{1+\omega_H^2\tau_c^2}\right) \tag{11.2}$$

where $R_{2,\text{pre}}$ is the increase in transverse relaxation rate due to the paramagnetic center, r is the distance between the paramagnetic center and the proton, γ_H is the gyromagnetic ratio of proton, g_s is the electron g factor for the unpaired electron, $\omega_H/2\pi$ is the resonance frequency of the proton, τ_c is the combined rotational correlation time for the entire system. This paramagnetic relaxation enhancement (PRE) on protons can be accurately quantified using well-established experiments.[49,50] Because PRE is dependent on the average distance between the paramagnetic center and the atom, its effect on nearby atoms is a more reliable gauge of the location of the ligand than chemical shift perturbation alone. Another important advantage of the paramagnetic ligands is that the effective range of the paramagnetic effect can be as far as 20 Å, ideal for detecting long-range contacts between protein and GAG. So far, GAG ligands have been successfully functionalized with the paramagnetic tag 4-amino-TEMPO using both EDC–NHS chemistry to couple it to carboxylates in uronic acids,[51] and reductive amination, which couples it to the reducing end of the ligand.[34,47,48] The reductive amination approach is especially attractive because of its specificity for the reducing end of the ligand (Scheme 11.1). Deshauer *et al.* used this approach to functionalize CS hexasaccharide ligands and showed that changes in the sulfation position of one sugar were sufficient to change the binding orientation of the CS hexasaccharide ligand on the chemokine RANTES. Figure 11.5 shows the changes in the HSQC spectra of RANTES in the presence of oxidized and reduced CS hexasaccharides labeled with TEMPO. Several signals from residues close to the GAG-binding site disappeared in the presence of oxidized TEMPO-labeled GAG ligand, but reappeared after reduction of the TEMPO, indicating they are close to the reducing end of the ligand. Quantitative measurement of the ^{1}H PRE for the amide protons revealed that substitution of a 6-*O*-sulfate for a single 4-*O*-sulfate was sufficient to change the binding orientation of the molecule. The GAG ligand orientation information obtained from these experiments was used to create the model of RANTES–CS complex.[19] Similar paramagnetically labeled heparin ligands have been used to confirm the GAG-binding sites of the bacterial adhesin DBPA and DBPB.[34,48,52] It should be noted that the paramagnetic tag is usually not attached to the ligand rigidly. This flexibility can reduce the resolution of the structural information attainable.

Besides the use of paramagnetic ligands, it may also be possible to obtain long-range contacts between protein and GAGs by extending the effective range of NOE through specific protonation of deuterated proteins. The effective range of NOE in fully protonated proteins is 5 Å. This is the result of spin diffusion due to a dipole–dipole interaction network in proteins and rapid longitudinal relaxation of proton signals. However, specifically protonating selective functional groups while deuterating the surrounding protons can greatly reduce spin diffusion and longitudinal relaxation, resulting in an extended range for the NOE. Such an approach has been used to obtain long-range contacts within large proteins,[53] and should also be applicable to protein–GAG interactions. One form of specific protonation can particularly benefit the study of protein–GAG interactions. Specifically, ε-amino group of lysine can be easily methylated through reductive amination with formaldehyde. This reaction is specific to amino groups and preserves the charges on

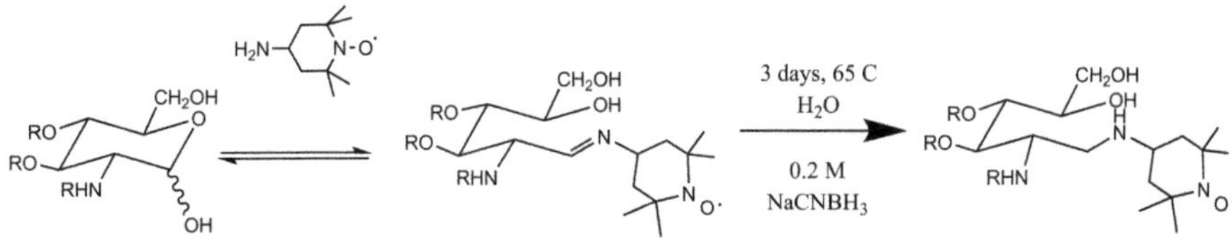

Scheme 11.1 Reductive amination of the reducing end of glycans using 4-amino-TEMPO.

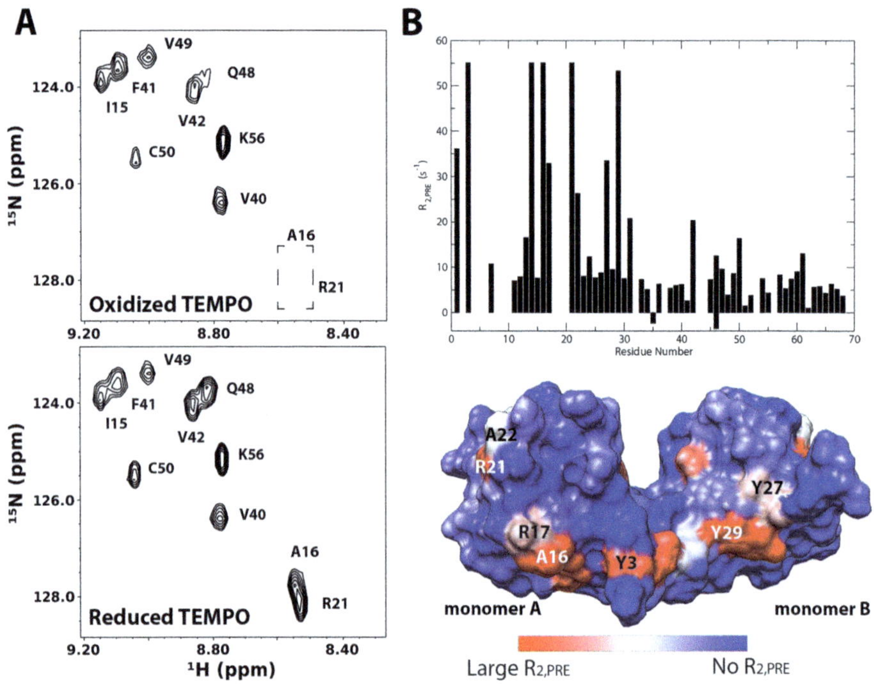

Figure 11.5 (A) ^{15}N–HSQCs of RANTES in the presence of oxidized (upper panel) and reduced (lower panel) TEMPO-labeled CS hexasaccharide. (B) Quantitative measurement of paramagnetic TEMPO-induced transverse relaxation ($R_{2,\text{pre}}$). Graph and surface plot of amide proton $R_{2,\text{pre}}$ produced by TEMPO-labeled CS hexasaccharide. Residues Y14, A16 and R21 disappeared upon addition of the ligand, but reappeared upon its reduction. They are given the same magnitude of $R_{2,\text{pre}}$ as residue Y3, which showed the biggest change. Adapted from Structure, 23, Deshauer *et al.*, Interactions of the Chemokine CCL5/RANTES with medium-sized chondroitin sulfate ligands, 1066–1077, Copyright 2015 with permission from Elsevier.[19]

lysines. One study utilizing this approach showed that even complete lysine methylation did not change the affinity of the chemokine IL-8 for GAGs.[44] This means specific methylation of a single lysine may have minimal effect on protein–GAG interactions. Such specific methylation of a single lysine is possible using amber codon and the unnatural amino acid *N*ε-methyllysine. *N*ε-methyllysine has been successfully incorporated into histone using pyrollysyl tRNA synthetase, a naturally occurring amber codon suppressor that can incorporate a large number of lysine derivatives.[54] Similar methodologies can potentially be used to specifically label lysine residues, allowing more structural information to be obtained. Besides *N*ε-methyllysine, pyrollysyl tRNA synthetase can also incorporate lysine-based paramagnetic tags;[55] this further increases the possibility of detecting these weak interactions through chemical biology.

11.6 Computational Modeling of Protein–GAG Complexes Using NMR Data

Because of the small amount of information available on intermolecular interactions between protein and GAGs, the conventional protocol for constructing biomolecular complexes using NMR data, which requires a large amount of experimental information, usually cannot be applied to the construction of protein–GAG complexes. Therefore, procedures capable of creating reasonable models using sparse information are more desirable. Using simulated data, it can be shown that approximately two unambiguous constraints per monosaccharide are sufficient in reproducing the correct pose of the glycan in the complex, even with large errors in these distance constraints. However, even this information may not be available in most situations. So far, several programs have been used by the NMR community to construct protein–GAG models. One of the most popular programs is HADDOCK.[56] A program designed to construct biological macromolecular complexes using ambiguous NMR information. HADDOCK has been used to study the interactions of heparin with both chemokine CXCL1[36] and β-defensin 6.[57] Another program that has been used to model protein–GAG complexes with sparse constraints is AMBER. Although AMBER is not able to utilize ambiguous constraints, it does handle distance constraints, but its main advantage is the use of Glycam,[58] one of the most accurate glycan force fields. AMBER is also faster than HADDOCK in performing explicit solvent simulations, which may enhance the accuracy of the models. Deshauer *et al.* used such an approach to create models of the chemokine CCL5 bound to CS hexasaccharides using only four constraints between the glycan and the protein.[19] However, both programs also have shortcomings. In particular, neither program takes into consideration the existence of multiple binding conformations. This can lead to misinterpretation of the NMR data, producing structures that are not realistic. This problem can be partially circumvented in HADDOCK through its multibody docking feature and removing interactions between the two docked ligands. However, the approach only works for conformations that coexist in equal proportions. Automated interpretation of the NOE distance constraints and ambiguous restraints deduced from chemical shift perturbation in these situations are difficult, often requiring manual intervention to produce the correct result.

There is yet no consensus on the best protocol for determining structures of protein–GAG complexes using NMR. Figure 11.6 shows one procedure that may be adopted. In the procedure, the structures of protein and ligand are first studied separately using techniques mentioned in this chapter. The two structures can then be docked, with the help of HADDOCK, using minimal amount of information obtained from either isotopically filtered–edited NOESY experiments or PRE data through the use of TEMPO labeled GAG ligand. Finally, the models can be refined with AMBER in a more realistic environment.

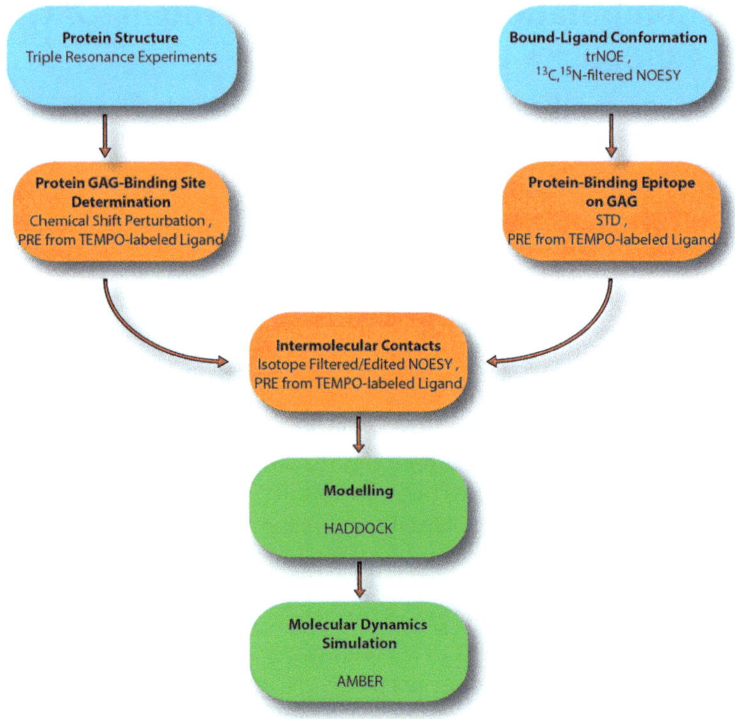

Figure 11.6 Procedure for creating models of protein–GAG complexes using NMR data.

References

1. P. N. Sanderson, T. N. Huckerby and I. A. Nieduszynski, *Biochem. J.*, 1987, **243**, 175–181.
2. U. Lindahl, G. Bäckström and L. Thunberg, *J. Biol. Chem.*, 1983, **258**, 9826–9830.
3. H. E. Conrad, *Heparin-Binding Proteins*, Academic Press, San Diego, 1998.
4. U. Lindahl and L. Kjellen, *J. Intern. Med.*, 2013, **273**, 555–571.
5. D. Xu and J. D. Esko, *Annu. Rev. Biochem.*, 2014, **83**, 129–157.
6. B. Richard, R. Swanson and S. T. Olson, *J. Biol. Chem.*, 2009, **284**, 27054–27064.
7. K. Tan, M. Duquette, J. H. Liu, K. Shanmugasundaram, A. Joachimiak, J. T. Gallagher, A. C. Rigby, J. H. Wang and J. Lawler, *J. Biol. Chem.*, 2008, **283**, 3932–3941.
8. W. J. Carter, E. Cama and J. A. Huntington, *J. Biol. Chem.*, 2005, **280**, 2745–2749.
9. D. B. Fenske and R. J. Cushley, *Chem. Phys. Lipids*, 1988, **49**, 15–29.
10. T. J. Boltje, T. Buskas and G. J. Boons, *Nat. Chem.*, 2009, **1**, 611–622.

11. M. M. Zulueta, S. Y. Lin, Y. P. Hu and S. C. Hung, *Curr. Opin. Chem. Biol.*, 2013, **17**, 1023–1029.
12. J. Liu and R. J. Linhardt, *Nat. Prod. Rep.*, 2014, **31**, 1676–1685.
13. J. Xiong, U. Bhaskar, G. Li, L. Fu, L. Li, F. Zhang, J. S. Dordick and R. J. Linhardt, *J. Biotechnol.*, 2013, **167**, 241–247.
14. W. He, L. Fu, G. Li, J. Andrew Jones, R. J. Linhardt and M. Koffas, *Metab. Eng.*, 2015, **27**, 92–100.
15. V. H. Pomin, J. S. Sharp, X. Li, L. Wang and J. H. Prestegard, *Anal. Chem.*, 2010, **82**, 4078–4088.
16. R. U. Lemieux, R. K. Kullnig and R. Y. Moir, *J. Am. Chem. Soc.*, 1958, **80**, 2237–2242.
17. M. Mayer and B. Meyer, *J. Am. Chem. Soc.*, 2001, **123**, 6108–6117.
18. G. M. Clore, A. M. Gronenborn and R. W. Davies, *J. Mol. Biol.*, 1982, **155**, 447–466.
19. C. Deshauer, A. M. Morgan, E. O. Ryan, T. M. Handel, J. H. Prestegard and X. Wang, *Structure*, 2015, **23**, 1066–1077.
20. G. Kunze, J. P. Gehrcke, M. T. Pisabarro and D. Huster, *Glycobiology*, 2014, **24**, 1036–1049.
21. J. C. Munoz-Garcia, E. Chabrol, R. R. Vives, A. Thomas, J. L. de Paz, J. Rojo, A. Imberty, F. Fieschi, P. M. Nieto and J. Angulo, *J. Am. Chem. Soc.*, 2015, **137**, 4100–4110.
22. F. Yu, S. Roy, E. Arevalo, J. Schaeck, J. Wang, K. Holte, J. Duffner, N. S. Gunay, I. Capila and G. V. Kaundinya, *Anal. Bioanal. Chem.*, 2014, **406**, 3079–3089.
23. V. Jayalakshmi and N. R. Krishna, *J. Magn. Reson.*, 2002, **155**, 106–118.
24. S. C. Tjong, T. S. Chen, W. N. Huang and W. G. Wu, *Biochemistry*, 2007, **46**, 9941–9952.
25. M. Hricovini, M. Guerrini, A. Bisio, G. Torri, M. Petitou and B. Casu, *Biochem. J.*, 2001, **359**, 265–272.
26. A. Canales, R. Lozano, B. Lopez-Mendez, J. Angulo, R. Ojeda, P. M. Nieto, M. Martin-Lomas, G. Gimenez-Gallego and J. Jimenez-Barbero, *FEBS J.*, 2006, **273**, 4716–4727.
27. C. Laguri, N. Sapay, J. P. Simorre, B. Brutscher, A. Imberty, P. Gans and H. Lortat-Jacob, *J. Am. Chem. Soc.*, 2011, **133**, 9642–9645.
28. X. Wang, *Biochemistry*, 2012, **51**, 8353–8362.
29. M. Sakakura, S. Oo-Puthinan, C. Moriyama, T. Kimura, J. Moriya, T. Irimura and I. Shimada, *J. Biol. Chem.*, 2008, **283**, 33665–33673.
30. L. E. Kay, D. A. Torchia and A. Bax, *Biochemistry*, 1989, **28**, 8972–8979.
31. G. Lipari and A. Szabo, *J. Am. Chem. Soc.*, 1982, **104**, 4546–4559.
32. G. Lipari and A. Szabo, *J. Am. Chem. Soc.*, 1982, **104**, 4559–4570.
33. F. A. Mulder, N. R. Skrynnikov, B. Hon, F. W. Dahlquist and L. E. Kay, *J. Am. Chem. Soc.*, 2001, **123**, 967–975.
34. W. Feng and X. Wang, *Biochim. Biophys. Acta*, 2015, **1854**(12), 1823–1832.
35. L. Duma, D. Haussinger, M. Rogowski, P. Lusso and S. Grzesiek, *J. Mol. Biol.*, 2007, **365**, 1063–1075.

36. K. M. Poluri, P. R. Joseph, K. V. Sawant and K. Rajarathnam, *J. Biol. Chem.*, 2013, **288**, 25143–25153.
37. P. R. Joseph, P. D. Mosier, U. R. Desai and K. Rajarathnam, *Biochem. J.*, 2015, **472**(1), 121–133.
38. X. Wang, C. Watson, J. S. Sharp, T. M. Handel and J. H. Prestegard, *Structure*, 2011, **19**, 1138–1148.
39. D. Mikhailov, H. C. Young, R. J. Linhardt and K. H. Mayo, *J. Biol. Chem.*, 1999, **274**, 25317–25329.
40. M. F. Garcia-Mayoral, A. Canales, D. Diaz, J. Lopez-Prados, M. Moussaoui, J. L. de Paz, J. Angulo, P. M. Nieto, J. Jimenez-Barbero, E. Boix and M. Bruix, *ACS Chem. Biol.*, 2013, **8**, 144–151.
41. L. Nieto, A. Canales, G. Gimenez-Gallego, P. M. Nieto and J. Jimenez-Barbero, *Chemistry*, 2011, **17**, 11204–11209.
42. A. Esadze, D. W. Li, T. Wang, R. Bruschweiler and J. Iwahara, *J. Am. Chem. Soc.*, 2011, **133**, 909–919.
43. L. Zandarashvili and J. Iwahara, *Biochemistry*, 2015, **54**, 538–545.
44. K. Mobius, K. Nordsieck, A. Pichert, S. A. Samsonov, L. Thomas, J. Schiller, S. Kalkhof, M. Teresa Pisabarro, A. G. Beck-Sickinger and D. Huster, *Glycobiology*, 2013, **23**, 1260–1269.
45. K. Singh, A. G. Gittis, P. Nguyen, D. C. Gowda, L. H. Miller and D. N. Garboczi, *Nat. Struct. Mol. Biol.*, 2008, **15**, 932–938.
46. D. Xu, J. H. Young, J. M. Krahn, D. Song, K. D. Corbett, W. J. Chazin, L. C. Pedersen and J. D. Esko, *ACS Chem. Biol.*, 2013, **8**, 1611–1620.
47. A. Morgan, K. M. Sepuru, W. Feng, K. Rajarathnam and X. Wang, *Biochemistry*, 2015, **54**, 5113–5119.
48. A. M. Morgan and X. Wang, *Biochem. J.*, 2015, **467**, 439–451.
49. J. Koehler and J. Meiler, *Prog. Nucl. Magn. Reson. Spectrosc.*, 2011, **59**, 360–389.
50. J. Iwahara, C. Tang and G. M. Clore, *J. Magn. Reson.*, 2007, **184**, 185–195.
51. E. Gemma, A. N. Hulme, A. Jahnke, L. Jin, M. Lyon, R. M. Muller and D. Uhrin, *Chem. Commun. (Camb.)*, 2007, 2686–2688.
52. A. Morgan and X. Wang, *Biochemistry*, 2013, **52**, 8237–8245.
53. R. Kerfah, M. J. Plevin, O. Pessey, O. Hamelin, P. Gans and J. Boisbouvier, *J. Biomol. NMR*, 2015, **61**, 73–82.
54. T. Yanagisawa, M. Takahashi, T. Mukai, S. Sato, M. Wakamori, M. Shirouzu, K. Sakamoto, T. Umehara and S. Yokoyama, *ChemBioChem*, 2014, **15**, 1830–1838.
55. M. J. Schmidt, J. Borbas, M. Drescher and D. Summerer, *J. Am. Chem. Soc.*, 2014, **136**, 1238–1241.
56. C. Dominguez, R. Boelens and A. M. Bonvin, *J. Am. Chem. Soc.*, 2003, **125**, 1731–1737.
57. V. S. De Paula, V. H. Pomin and A. P. Valente, *J. Biol. Chem.*, 2014, **289**, 22969–22979.
58. K. N. Kirschner, A. B. Yongye, S. M. Tschampel, J. Gonzalez-Outeirino, C. R. Daniels, B. L. Foley and R. J. Woods, *J. Comput. Chem.*, 2008, **29**, 622–655.

CHAPTER 12

Solid-State NMR Analysis of Mannose Recognition by Pradimicin A

YU NAKAGAWA*[a,b] AND YUKISHIGE ITO[b]

[a]Department of Applied Molecular Biosciences, Graduate School of Bioagricultural Sciences, Nagoya University, Furo-cho, Chikusa-ku, Nagoya 464-8601, Japan; [b]Synthetic Cellular Chemistry Laboratory, RIKEN, 2-1 Hirosawa, Wako, Saitama 351-0198, Japan
*E-mail: yu@agr.nagoya-u.ac.jp

12.1 Introduction

There is a growing body of evidence indicating that carbohydrates mediate a variety of biological processes such as protein folding, cell signaling, fertilization, embryogenesis, neuronal development, hormone activity, and cell proliferation.[1–4] Moreover, overwhelming data support the involvement of carbohydrates in diseases such as pathogen infection, inflammation, innate immune responses, and cancer.[5–7] Coupled with these significant advances in knowledge about the functions of carbohydrates in physiological and pathological processes, interest has been rapidly growing in biotechnological and therapeutic applications of lectins, proteins of non-immune origin that bind specific carbohydrates.[8–11] However, their practical applications are severely limited by several disadvantages due to nature as proteins, including difficulties of large-scale production, long storage, and chemical modification.

New Developments in NMR No.10
NMR in Glycoscience and Glycotechnology
Edited by Koichi Kato and Thomas Peters

Published by the Royal Society of Chemistry, www.rsc.org

From a therapeutic viewpoint, concerns are also raised about their immunogenicity, inflammatory activity, cellular toxicity, mitogenic stimulation of human peripheral lymphocyte cells, and hemagglutination of human red blood cells. Consequently, increasing research attention has been directed at the development of carbohydrate-binding small molecules, which can be prepared in reliable quantities and are easily tuned for structural optimization.

During the last two decades, intensive work in the area of supramolecular chemistry has generated a variety of carbohydrate receptors with non-protein structures.[12–16] However, most of them exhibit efficient binding for carbohydrates only in organic solvents. The main problem is the competition between carbohydrates and water molecules. Arrays of the hydroxyl groups on carbohydrates are almost indistinguishable from clusters of water molecules, resulting in a drastic decrease in binding affinity of carbohydrate receptors in aqueous solutions. Selectivity among carbohydrates is an additional problem that is not easily overcome. Carbohydrates are quite similar to each other in terms of chemical structure. For example, D-glucose (D-Glc) differs from D-mannose (D-Man) and D-galactose only in the stereochemistry at one chiral center. Recognition of one specific carbohydrate is thus a particular challenge in the design of carbohydrate receptors. Although several design strategies have been recently developed to overcome these problems,[17–19] the primary targets of the existing carbohydrate receptors are monosaccharides or disaccharides, and there has been a paucity of synthetic compounds that bind to glycans of biological and pathological importance.[20]

On the other hand, carbohydrate-binding small molecules are also found in nature. Pradimicins (PRMs) and benanomicins are a unique family of antibiotics with a lectin-like ability to bind D-Man in the presence of Ca^{2+} ion (Figure 12.1).[21–23] These natural products capture D-Man under

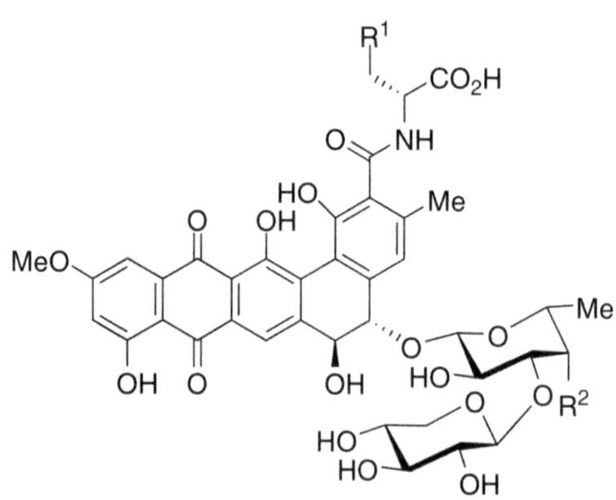

Pradimicin A (PRM-A) : R^1 = H, R^2 = NHMe
Benanomicin A : R^1 = H, R^2 = OH
BMY-28864 : R^1 = OH, R^2 = NMe_2

Figure 12.1 Structures of pradimicin A (PRM-A), benanomicin A, and BMY-28864.

physiologically relevant conditions but not L-mannose, 2-, 3-, and 4-epimers of D-Man (D-Glc, D-altrose, and D-talose, respectively), D-mannosamine, and *N*-acetyl-D-mannosamine.[24] Of further significance, PRMs have been shown to exhibit antifungal and anti-HIV activities by binding to cell wall mannans of microorganisms and high mannose-type glycans on the viral envelope of HIV, respectively.[25–28] At present, there exists no chemotherapeutics that directly target glycans of pathogens, underscoring their potential as promising drug candidates for a first-in-class approach to the infectious diseases.[29]

Notwithstanding the great potential of PRMs as the lectin mimics and therapeutic leads, their molecular basis of D-Man recognition has been scarcely investigated. In this chapter, we first outline the fundamental problems of interaction analysis of PRMs and D-Man, and then describe our strategy, methodology, and results of investigation of D-Man recognition by PRMs.

12.2 Underlying Problems in Interaction Analysis of Pradimicins and D-Mannose

Since the discovery that PRMs serve as mimics of C-type lectins, much attention has been focused on the molecular basis of their D-Man recognition. As a first step toward elucidating this issue, initial efforts have been made to examine complex-forming process of PRMs with Ca^{2+} ion and D-Man. Earlier studies found that PRMs co-precipitate with D-Man and Ca^{2+} ion to form water-insoluble aggregates, which consist of the ternary complex with a PRM : Ca^{2+} : D-Man ratio of 2 : 1 : 4.[24] Subsequent UV-visible spectrophotometric analysis using BMY-28864 (Figure 12.1),[30] an artificial PRM derivative with improved water solubility, suggested a possible process of the complex formation.[31] As shown in Scheme 12.1, two molecules of PRMs are initially bridged by Ca^{2+} ion to give the binary [PRM_2/Ca^{2+}] complex, which binds two molecules of D-Man (the primary binding). The resulting ternary [PRM_2/Ca^{2+}/D-Man_2] complex accommodates another two molecules of D-Man

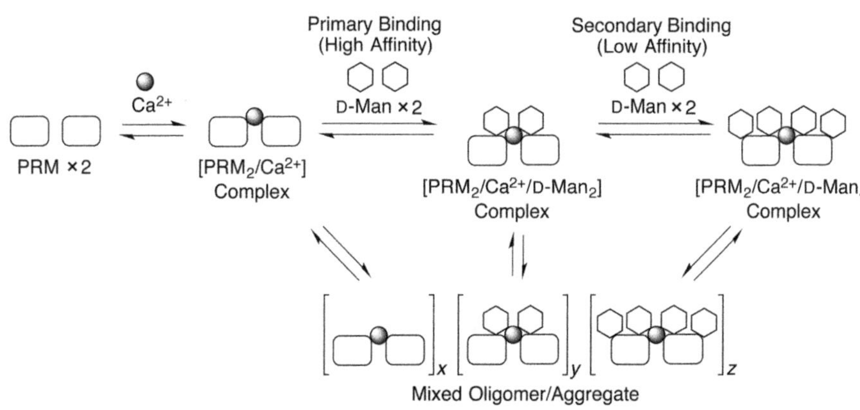

Scheme 12.1 Complex-forming process of PRM with Ca^{2+} ion and D-Man.

(the secondary binding) to form the ultimate [PRM_2/Ca^{2+}/D-Man_4] complex. Formation of the binary [PRM_2/Ca^{2+}] complex was confirmed by several solution NMR studies.[32–34]

Although the complex-forming process of PRMs with Ca^{2+} ion and D-Man has been largely understood, little is known about the molecular structures of the ternary [PRM_2/Ca^{2+}/D-Man_2] and [PRM_2/Ca^{2+}/D-Man_4] complexes. This is due to several reasons. First, these ternary complexes, including those of water-soluble BMY-28864, have a high propensity to aggregate, and are hardly crystallized. In the case of PRM-A (Figure 12.1),[22] an original member of PRMs, its binary [PRM_2/Ca^{2+}] complex easily aggregates even in the absence of D-Man. Second, under dilute conditions that prevent aggregation, the [PRM_2/Ca^{2+}], [PRM_2/Ca^{2+}/D-Man_2], and [PRM_2/Ca^{2+}/D-Man_4] complexes coexist in solution, and assemble with each other to afford the mixed oligomers.[35] This complicated equilibrium severely exacerbates the analytical difficulty. Third, PRMs possess the disaccharide moiety, whose ^{1}H- and ^{13}C-NMR signals overlap with those of D-Man. These issues have collectively frustrated X-ray crystallographic and conventional solution NMR approaches, leaving the question open as to how PRMs recognize D-Man.

12.3 Analytical Strategy Using Solid-State NMR Spectroscopy

In order to circumvent the above issues, we explored a novel analytical strategy using solid-state NMR spectroscopy. In contrast to solution NMR approach, which suffers from aggregate-forming propensity of PRMs, the solid-state NMR approach can utilize the aggregate itself for structural analysis. Additionally, there is no equilibrium in the solid sample, which eliminates the problem related to the complicated complex formation of PRMs and D-Man. However, a critical issue to be addressed is how to prepare the solid aggregate consisting of a single ternary complex. Thus, as a prerequisite for solid-state NMR analysis, we examined in more detail the aggregate formation of PRM-A.[36,37]

In accord with literature reports,[23,24] when PRM-A was mixed with 10 equivalents of $CaCl_2$ and 25 equivalents of methyl α-D-mannopyranoside (α-D-Man-OMe), a significant amount of aggregate was produced. The molar ratio of α-D-Man-OMe relative to PRM-A in the aggregate was calculated to be 2.35 by ^{1}H–NMR on the basis of integration values of PRM-A and α-D-Man-OMe after dissociation of the aggregates by acid treatment. This molar ratio indicates that the aggregate consists of the [PRM-A_2/Ca^{2+}/α-D-Man-OMe_4] complex with a small amount of non-specifically absorbed α-D-Man-OMe. The amount of non-specific binding was confirmed by a similar experiment using methyl α-D-glucopyranoside (α-D-Glc-OMe), which is insensitive to PRM-A (α-D-Glc-OMe/PRM-A = 0.34).

Following the demonstration that the main component of the aggregate is not a simple mixture but the complex of PRM-A and α-D-Man-OMe, the effect of washing on α-D-Man-OMe incorporation in the aggregate was examined. Table 12.1 summarizes α-D-Man-OMe : PRM-A ratio in the aggregate after 0–4 times washing with an aqueous $CaCl_2$ solution. After the first and second washings, the α-D-Man-OMe : PRM-A ratio stepwise decreased to the equimolar amount. However, intriguingly, the ratio was maintained after the third and forth washings. The realistic implication of these results is that the secondary D-Man binding is significantly weaker than the primary binding, and α-D-Man-OMe was released from the secondary binding site during the washing process of the aggregate to produce the 1 : 1 complex, that is, the [PRM-A_2/Ca^{2+}/α-D-Man-OMe_2] complex.

This hypothesis was supported by isothermal titration calorimetry (ITC) experiments. ITC provides values for thermodynamic parameters for ligand–receptor interaction by detecting heat changes associated with binding. Although the heat resulting from aggregation is a formidable obstacle to the estimation of interaction between PRM-A and D-Man in ITC experiments, we found that the aggregation was minimized when 1 mM PRM-A and 100 mM α-D-Man-OMe in 50 mM 3-morpholinopropane-1-sulfonic acid (MOPS) buffer (pH 7.0) containing 10 mM $CaCl_2$ were used as the titrate and titrant solutions, respectively. The obtained isothermal curve was fitted well to a two sets of sites binding model (Figure 12.2A). On the other hand, ITC experiments with α-D-Glc-OMe instead of α-D-Man-OMe provided negligible heat change (Figure 12.2B), confirming that the heat effects obtained in the titration with α-D-Man-OMe truly originated from its specific interaction with PRM-A. The calculated K_a values for two-step D-Man binding (K_1 = 10 400 and K_2 = 263 M^{-1}) indicate that the secondary D-Man binding is about 40 times weaker than the primary one. This result reinforces our hypothesis that washing of the aggregate removes α-D-Man-OMe selectively from the secondary binding site of PRM-A as well as non-specifically absorbed one.

Having confirmed that the aggregate solely composed of the [PRM-A_2/Ca^{2+}/α-D-Man-OMe_2] complex can be prepared by the extensive washing, we conducted solid-state cross polarization–magic angle spinning (CP–MAS) ^{13}C-NMR spectroscopic experiments using the aggregate. The solid-state ^{13}C-NMR spectrum of a freeze-dried sample of PRM-A was also recorded for a comparison. The aggregate exhibited clearly distinguishable signals, which were significantly sharper than those of PRM-A (Figure 12.3). The narrower line width of the signals reflects the higher structural homogeneity of the

Table 12.1 Molar ratio of α-D-Man-OMe and α-D-Glc-OMe relative to PRM-A in aggregates.

	Number of washing times				
	0	1	2	3	4
α-D-Man-OMe	2.35 (0.10)[a]	1.45 (0.05)	1.08 (0.09)	1.10 (0.04)	1.07 (0.09)
α-D-Glc-OMe	0.34 (0.03)	<0.1	<0.1	<0.1	<0.1

[a]Standard deviation of at least three separate experiments.

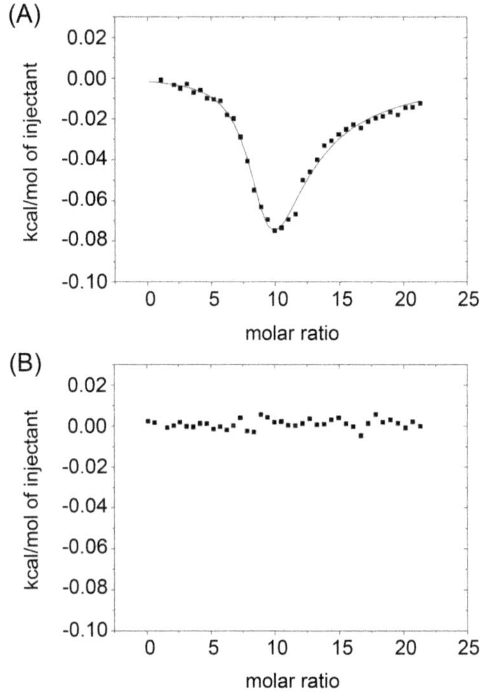

Figure 12.2 ITC profiles for PRM-A titrated with (A) α-D-Man-OMe and (B) α-D-Glc-OMe in the presence of $CaCl_2$. The experiments were performed using 1 mM PRM-A and 100 mM α-D-Man-OMe or α-D-Glc-OMe in 50 mM MOPS buffer (pH 7.0) with 10 mM $CaCl_2$ at 30 °C. The solid line in (A) represents the least-squares fit of the data to a two sets of sites binding model.

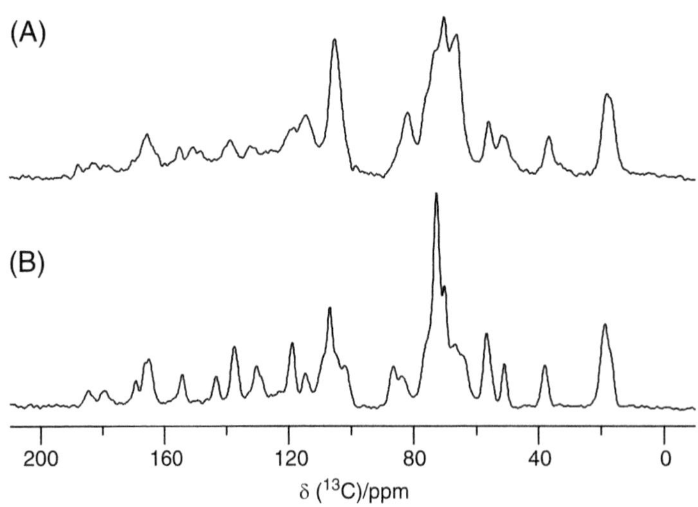

Figure 12.3 Solid-state CP–MAS ^{13}C–NMR spectra of (A) PRM-A and (B) the [PRM-A_2/Ca^{2+}/α-D-Man-OMe_2] complex.

aggregate. From this preliminary result of the solid-state NMR experiments, it was expected that our strategy using solid-state NMR spectroscopy would allow for structural analysis of the [PRM_2/Ca^{2+}/D-Man_2] complex without the problems related to aggregation and the complicated complex formation of PRMs.

12.4 Role of Ca^{2+} Ion in the Primary D-Mannose Binding of Pradimicin A

Our solid-state NMR analysis began with an examination of the role of Ca^{2+} ion in the primary D-Man binding of PRM-A.[36] Although several lines of evidence have indicated that Ca^{2+} ion might bind to the carboxylate group of PRM-A,[38,39] clear experimental support has yet to be provided. Additionally, it remains unclear whether Ca^{2+} ion simply bridges two molecules of PRM-A, or whether it also participates in D-Man binding as generally observed in C-type lectins.[40] To approach these issues, we planned solid-state CP/MAS ^{113}Cd-NMR experiments using $^{113}Cd^{2+}$ ion as a surrogate probe for Ca^{2+} ion. $^{113}Cd^{2+}$ ion has been shown to behave similarly to Ca^{2+} ion in several biological systems because its ionic radius (1.09 Å for 6-coordination)[41] is quite similar to that of Ca^{2+} ion (1.14 Å for 6-coordination).[42,43] In addition, ^{113}Cd nucleus carries a nuclear spin of 1/2, and shows a broad range of chemical shifts (>800 ppm), which is sensitive to the nature, number, and geometric arrangement of atoms coordinated with $^{113}Cd^{2+}$ ion.[44] By taking advantage of these beneficial characters of $^{113}Cd^{2+}$ ion, ^{113}Cd-NMR spectroscopy has been widely used to examine the Ca^{2+} environments present in biological systems.

As a prerequisite for the solid-state ^{113}Cd-NMR experiment, we verified that PRM-A could bind D-Man in the presence of Cd^{2+} ion. As in the case of $CaCl_2$, addition of $CdCl_2$ to a mixture of PRM-A and α-D-Man-OMe effectively produced the aggregate. The α-D-Man-OMe : PRM-A ratio in the aggregate after 2 times washing was calculated to be 0.88, which is slightly lower than that obtained using $CaCl_2$ (α-D-Man-OMe/PRM-A = 1.08). Although somewhat weak, this D-Man binding in the presence of Cd^{2+} ion was confirmed to be specific because only a negligible amount of α-D-Glc-OMe was detected when the aggregate was formed using $CdCl_2$ and α-D-Glc-OMe (α-D-Glc-OMe/PRM-A = <0.1).

After confirming that Cd^{2+} ion satisfactorily acts as a surrogate for Ca^{2+} ion in the primary D-Man binding of PRM-A, we prepared the aggregate of the ternary [PRM-A_2/$^{113}Cd^{2+}$/α-D-Man-OMe_2] complex according to our standardized procedure. The solid sample of the binary [PRM-A_2/$^{113}Cd^{2+}$] complex was also similarly prepared as aggregate to examine the Ca^{2+} coordination of PRM-A in the [PRM-A_2/Ca^{2+}] complex. The solid-state ^{113}Cd-NMR spectrum of the binary complex showed a broad signal around $\delta = -50$ ppm (Figure 12.4A). This chemical shift is close to those reported for $Cd(OAc)_2 \cdot H_2O$ ($\delta = -46$ ppm) and $Cd(O_2CCH_2CH_2CO_2) \cdot H_2O$ ($\delta = -52$ ppm),

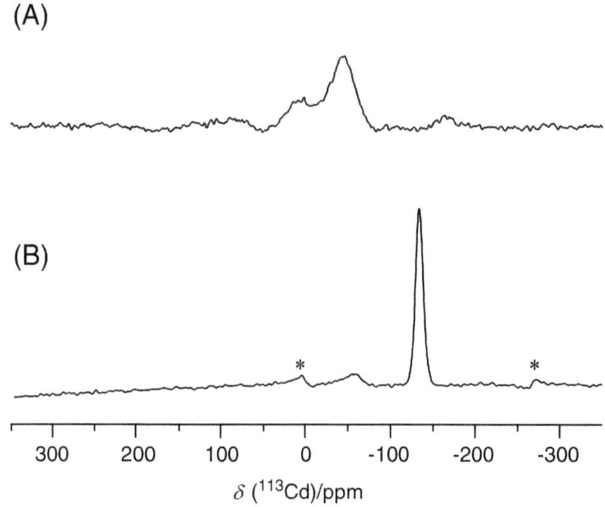

Figure 12.4 Solid-state CP–MAS ^{113}Cd–NMR spectra of (A) the [PRM-A_2/$^{113}Cd^{2+}$] and (B) the [PRM-A_2/$^{113}Cd^{2+}$/α-D-Man-OMe_2] complexes. The signals with an asterisk are the spinning side bands of the ^{113}Cd signal at $\delta = -135$ ppm.

but significantly different from those reported for solid cadmium compounds having ligands other than carboxylates, such as $Cd(OH)_2$ (δ = 158 ppm), [$Cd(en)_3Cl_2 \cdot H_2O$] (δ = 380 ppm; en = ethylenediamine), [$Na_2Cd(EDTA)$] (δ = 102 ppm; EDTA = ethylenediaminetetraacetate), $Cd(NH_2CH_2CO_2)_2 \cdot H_2O$ (δ = 112 ppm), and [Cd(glycylglycine)$_2 \cdot 2H_2O$] (δ = 169 ppm).[44–46] This comparison of the chemical shifts suggests that $^{113}Cd^{2+}$ ion binds to the carboxylate group of PRM-A in the binary [PRM-A_2/$^{113}Cd^{2+}$] complex, supporting the putative role of Ca^{2+} ion to bridge the carboxylate groups of two PRM-A molecules.

On the other hand, the aggregate of the ternary [PRM-A_2/$^{113}Cd^{2+}$/α-D-Man-OMe_2] complex exhibited a sharp signal at $\delta = -135$ ppm (Figure 12.4B). Mirroring the solid-state ^{13}C–NMR spectrum of the ternary complex (Figure 12.3B), the line width of the ^{113}Cd signal was apparently narrow, again confirming the high structural homogeneity of the complex. It is also noteworthy that the ^{113}Cd signal of the ternary [PRM-A_2/$^{113}Cd^{2+}$/α-D-Man-OMe_2] complex was markedly shifted upfield (>80 ppm) in comparison with that of the binary [PRM-A_2/$^{113}Cd^{2+}$] complex. When considering that ^{113}Cd signals upfield of $\delta = -100$ ppm are observed only for $^{113}Cd^{2+}$ ions coordinated with more than six oxygen ligands,[44] it is reasonable to assume that D-Man-OMe binds to $^{113}Cd^{2+}$ ion as an additional oxygen ligand in the [PRM-A_2/$^{113}Cd^{2+}$/α-D-Man-OMe_2] complex. These combined observations suggest a two-fold role of Ca^{2+} ion in the primary D-Man binding of PRM-A: it acts as a core to bridge two PRM-A molecules and also directly participates in D-Man binding.

12.5 D-Mannose Binding Geometry of Pradimicin A

The solid-state ^{113}Cd-NMR analysis revealed that PRM-A would bind D-Man in a Ca^{2+}-mediated manner through the carboxylate group of its D-alanine moiety. This implicated the D-alanine moiety as a part of the primary D-Man binding site of PRM-A. To test this hypothesis, we conducted two-dimensional dipolar-assisted rotational resonance (2D-DARR) experiments for detection of close contacts between α-D-Man-OMe and the D-alanine moiety of PRM-A in the [PRM-A_2/Ca^{2+}/α-D-Man-OMe_2] complex.[37] The DARR,[47,48] also known as RF assisted diffusion (RAD),[49] is a methodology of solid-state NMR spectroscopy to realize ^{13}C–^{13}C polarization transfer under MAS conditions. In this method, radiofrequency irradiation with the intensity ν_1 satisfying the rotary-resonance condition $\nu_1 = n\nu_R$ ($n = 1$ or 2; ν_R = MAS frequency) is applied to ^{1}H spins to recouple the ^{13}C–^{1}H dipolar interactions, and spectral overlap for a ^{13}C–^{13}C pair is achieved between the ^{13}C–^{1}H dipolar pattern of one ^{13}C spin and the main peak–spinning sideband ^{13}C–^{1}H dipolar pattern of the other ^{13}C spin. These spectral overlaps introduce the polarization transfer between a pair of spins with particular internuclear orientation even under MAS. This orientation-selective recoupling avoids simultaneous recovery of all ^{13}C–^{13}C couplings, resulting in isolation of a weak coupling. DARR thus can suppress the dipolar truncation effect that prevents the measurement of the weak dipolar couplings of interest in the presence of strong couplings derived from directly bound carbons. This unique feature enables us to obtain multiple ^{13}C–^{13}C distance restrains (~6 Å) simultaneously from cross-peak intensities in a 2D DARR spectrum.

One requisite for the DARR analysis is selective ^{13}C-enrichment of the D-alanine moiety of PRM-A. However, chemical synthesis of PRM-A requires a large number of synthetic steps,[50] and thus cannot supply ^{13}C-enriched PRM-A in quantities needed for DARR experiments. One possible approach is semi-synthesis *via* detachment of the D-alanine moiety of PRM-A followed by introduction of the ^{13}C-enriched counterpart. However, earlier studies reported that hydrolysis of the D-alanine moiety is laborious and material-consuming.[51] Recognizing these issues, we planned to utilize the biosynthesis of PRM-A for the preparation of ^{13}C-enriched PRM-A. Earlier biosynthetic studies showed that D-alanine is directly incorporated into the D-alanine moiety of PRMs,[52] indicating that feeding of ^{13}C-labeled D-alanine to a PRM-A-producing actinomycete could realize selective ^{13}C-enrichment of the D-alanine moiety. Thus, we conducted a feeding experiment using *Actinomadura* sp. TP-A0019 for the preparation of ^{13}C-enriched PRM-A (Figure 12.5). As expected, feeding of exogenous D-[$^{13}C_3$]alanine successfully provided [$^{13}C_3$] PRM-A with the ^{13}C-enriched D-alanine moiety (*ca.* 70 atom% ^{13}C). Similarly, on the basis of the assumption that 11-*O*- and 4′-*N*-methyl groups of PRM-A are derived from the *S*-methyl group of L-methionine,[53,54] we obtained [$^{13}C_2$] PRM-A with the ^{13}C-enriched methyl groups (*ca.* 65 atom% ^{13}C) by feeding of exogenous L-[5-$^{13}CH_3$]methionine (Figure 12.5).

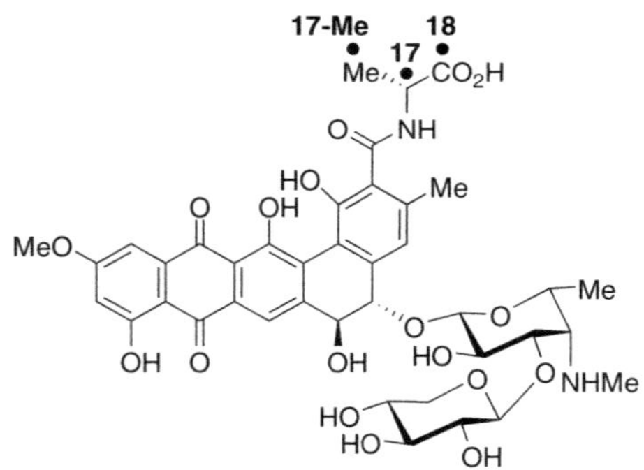

[$^{13}C_3$]PRM-A
(D-[$^{13}C_3$]Alanine feeding)
● = ca. 70 atom % ^{13}C

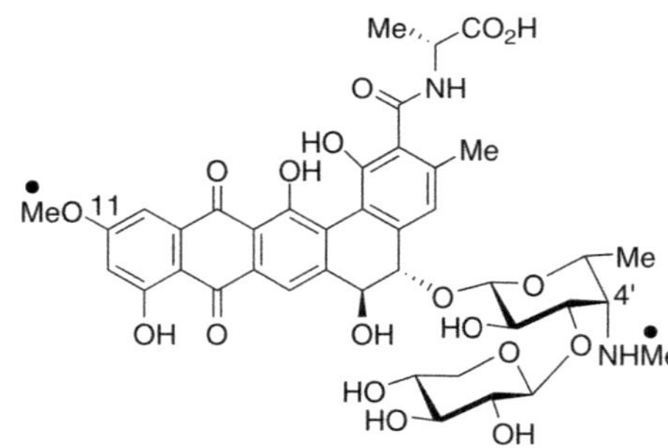

[$^{13}C_2$]PRM-A
(L-[5-^{13}C]Methionine feeding)
● = ca. 65 atom % ^{13}C

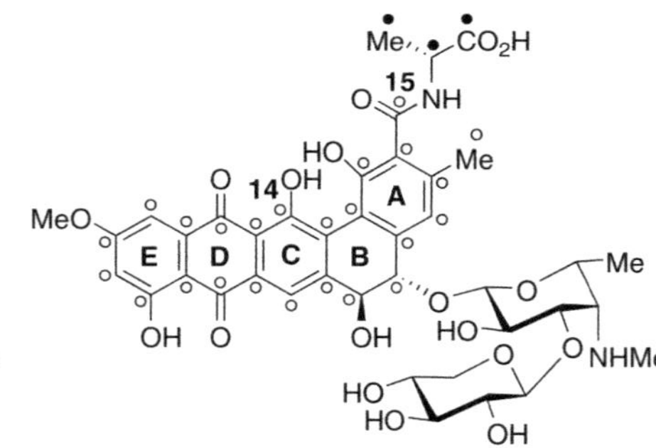

[$^{13}C_{27}$]PRM-A
(DL-[$^{13}C_3$]Alanine feeding)
● = ca. 70 atom % ^{13}C
○ = ca. 15 atom % ^{13}C

Figure 12.5 Biosynthetically ^{13}C-enriched PRM-As. Black and open circles indicate ^{13}C-enriched positions. The ^{13}C population was calculated from solution ^{1}H–NMR data on the basis of integration values of proton signals spit with ^{1}H–^{13}C couplings.

With the ^{13}C-enriched PRM-As in hand, we prepared the solid aggregates composed of the [PRM-A_2/Ca^{2+}/α-D-Man-OMe_2] complexes using ^{13}C-enriched PRM-As and α-D-[$^{13}C_6$]Man-OMe. Figure 12.6 shows the pulse sequence for DARR experiments (A) and 2D-DARR spectra of these samples (B–E), in which cross peaks represent dipolar interactions between ^{13}C nuclei that are located within approximately 6 Å. In the spectrum of the complex of [$^{13}C_3$]PRM-A at the mixing time of 500 ms (Figure 12.6B), symmetric cross peaks were clearly observed between carbon signals for the D-alanine moiety of PRM-A (δ = 20.0 ppm for 17-Me, δ = 50.8 ppm for C17, δ = 179.8 ppm for C18) and those for α-D-Man-OMe (δ = 63.0, 68.2, 71–76 ppm for C2–C6, δ = 101.4 ppm for C1). The absence of intermolecular cross peaks at the mixing time of 20 ms (Figure 12.6C) indicates that those observed at the mixing time of 500 ms originated from ^{13}C–^{13}C dipolar interaction reintroduced by DARR during mixing time. In contrast, the complex of [$^{13}C_2$]PRM-A showed no intermolecular cross peak even at the mixing time of 500 ms (Figure 12.6D), eliminating the possibility that intermolecular cross peaks observed in the complex of [$^{13}C_3$]PRM-A are derived from non-specific contacts between PRM-A and α-D-Man-OMe in the solid aggregate. These combined results indicate that α-D-Man-OMe is located within 6 Å of the D-alanine moiety of PRM-A in the ternary [PRM-A_2/Ca^{2+}/α-D-Man-OMe_2] complex, supporting our hypothesis that the D-alanine moiety constitutes the primary D-Man binding site of PRM-A.

Given the encouraging results of the initial DARR experiments, attention was next turned to the investigation of the D-Man binding geometry of PRM-A.[55] During our attempts to achieve ^{13}C-enrichment of the aromatic region of PRM-A, we found that DL-[$^{13}C_3$]alanine feeding increased ^{13}C-population of the aromatic region (*ca.* 15 atom% ^{13}C) of PRM-A as well as the D-alanine moiety (Figure 12.5). This is probably due to the metabolic conversion of L-[$^{13}C_3$]alanine to [$^{13}C_2$]acetyl-CoA, which constitutes the benzo[*a*]naphthacenequinone moiety of PRM-A.[52] This additional ^{13}C enrichment, although low, was expected to provide some information regarding interatomic distances between D-Man and the aromatic region of PRM-A. It is also practically important that this DL-[$^{13}C_3$]alanine feeding did not affect ^{13}C-population of the disaccharide moiety of PRM-A, which would remove the problem related to the signal overlapping between PRM-A and D-Man. Thus, we performed systematic 2D-DARR experiments using the [PRM-A_2/Ca^{2+}/α-D-Man-OMe_2] complexes of [$^{13}C_{27}$]PRM-A (Figure 12.5) and singly ^{13}C-labeled α-D-Man-OMe.

As shown in Figure 12.7, the complexes with α-D-[2-^{13}C]-, [3-^{13}C]-, and [4-^{13}C]Man-OMe gave distinct intermolecular cross peaks with carbon signals for the D-alanine moiety (C17, 17-Me, C18) of PRM-A, and the complex with α-D-[5-^{13}C]Man-OMe gave attenuated ones. On the other hand, only negligible or no cross peaks were observed in the complexes with α-D-[1-^{13}C]- and [6-^{13}C]Man-OMe, indicating that the C2–C4 region of D-Man is proximal to the D-alanine moiety of PRM-A. Regarding the benzo[*a*]naphthacenequinone moiety

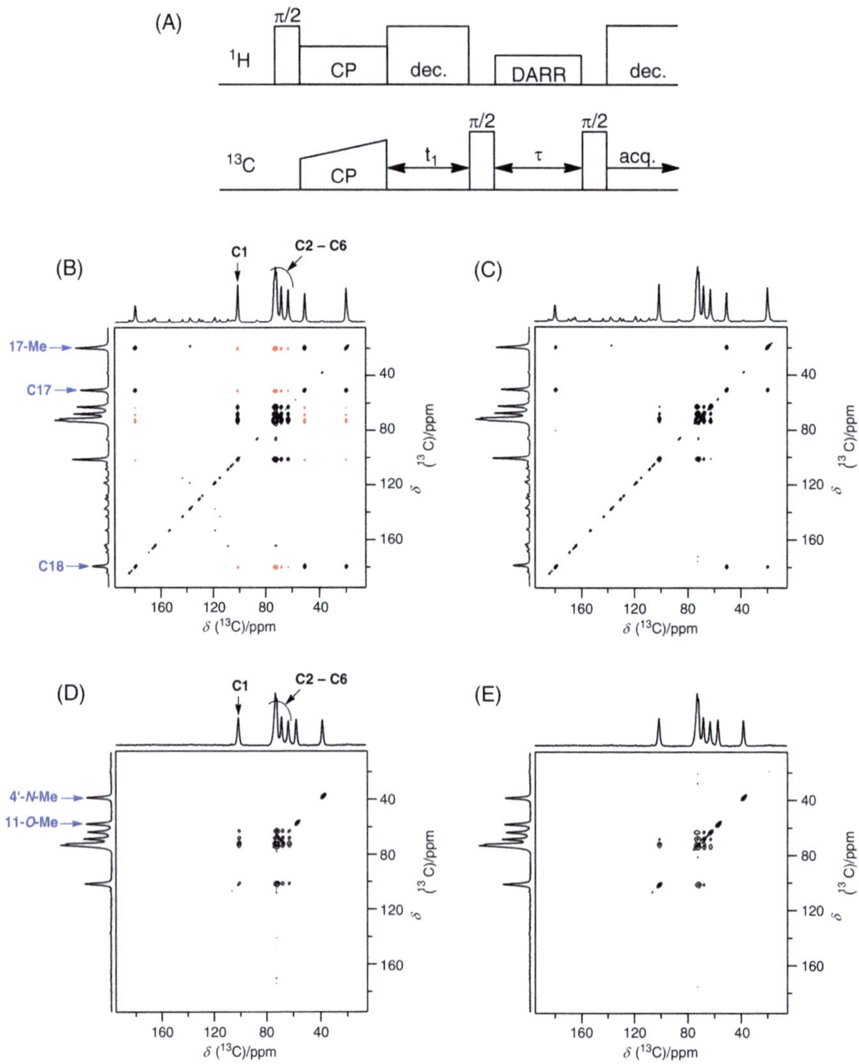

Figure 12.6 (A) Pulse sequence for DARR experiments (CP; cross polarization, t_1; evolution time, dec.; ^{1}H decoupling, τ: mixing time, acq.; acquisition). The sequence begins with variable-amplitude CP (VACP). During acquisition and evolution, two-pulse phase-modulated (TPPM) decoupling is used. (B–E) 2D-DARR spectra of the [PRM-A$_2$/Ca^{2+}/α-D-Man-OMe$_2$] complexes of [^{13}C$_3$]PRM-A with α-D-[^{13}C$_6$]Man-OMe at the mixing times of (B) 500 ms and (C) 20 ms and of [^{13}C$_2$] PRM-A with α-D-[^{13}C$_6$]Man-OMe at the mixing times of (D) 500 ms and (E) 20 ms. Blue and black arrows represent ^{13}C signals for ^{13}C-enriched PRM-A and α-D-[^{13}C$_6$]Man-OMe, respectively. Intermolecular cross peaks are shown in red.

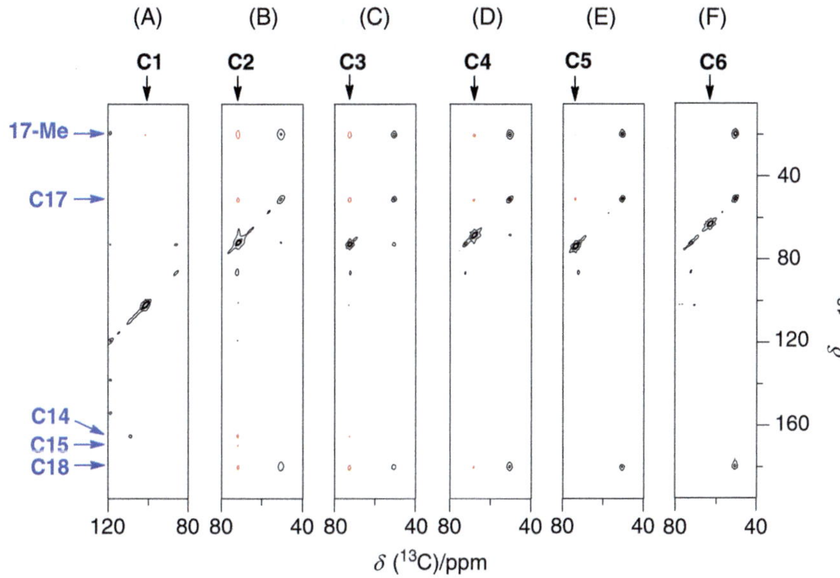

Figure 12.7 2D-DARR spectra of the [PRM-A_2/Ca^{2+}/α-D-Man-OMe_2] complexes of [$^{13}C_{27}$]PRM-A with (A) [1-^{13}C]-, (B) [2-^{13}C]-, (C) [3-^{13}C]-, (D) [4-^{13}C]-, (E) [5-^{13}C]-, and (F) [6-^{13}C]Man-OMe at the mixing time of 500 ms. Blue and black arrows represent positions of ^{13}C signals for ^{13}C-enriched PRM-A and singly ^{13}C-labeled α-D-Man-OMe, respectively. Intermolecular cross peaks are shown in red.

of PRM-A, the C2 signal of α-D-[2-^{13}C]Man-OMe gave weak cross peaks with the C14 and C15 signals of PRM-A (Figure 12.7B). These cross peaks were not detected in the corresponding spectrum at the mixing time of 20 ms (data not shown), indicating that they truly arose from ^{13}C–^{13}C dipolar interaction reinforced by DARR during mixing time. The mixing time-dependent cross peak with the C14 signal of PRM-A was also observed in the complex with α-D-[3-^{13}C]Man-OMe (Figure 12.7C). In contrast, cross peaks with aromatic carbon signals of PRM-A were hardly detectable in the complexes with α-D-[1-^{13}C]-, [4-^{13}C]-, [5-^{13}C]-, and [6-^{13}C]Man-OMe (Figure 12.7A and D–F). Although the low ^{13}C-enrichment of the benzo[*a*]naphthacenequinone moiety hampered full detection of close contacts between D-Man and the aromatic region of PRM-A, it is quite likely that the C2 and C3 positions of D-Man are in close contact with the AC rings of PRM-A.

These results collectively indicate that the D-alanine moiety and the AC rings of PRM-A constitute the primary D-Man binding site. The obtained distance information also suggests that the C2–C4 region of D-Man is located close to the binding site, while the C1 and C6 positions are relatively distant. Guided by these findings, we estimated a possible binding geometry of PRM-A with α-D-Man-OMe by molecular modeling (Figure 12.8).[55] In our model, PRM-A forms hydrogen bonds with the 2-, 3-, and 4-hydroxyl groups of α-D-Man-OMe. Moreover, the CH–π interaction[56] between the H2 proton of α-D-Man-OMe and the A ring of PRM-A possibly contributes to the stabilization of the complex as observed in many carbohydrate–lectin complexes.[57–59] Although we were not able to incorporate Ca^{2+} ion into the calculations due to lack of information regarding the geometry of Ca^{2+} coordination in the [PRM-A_2/Ca^{2+}/α-D-Man-OMe_2] complex, it can be assumed that Ca^{2+} ion coordinates with the 4-hydroxyl group of D-Man (Figure 12.8B), which is located closest to the carboxylate group of PRM-A in our model.

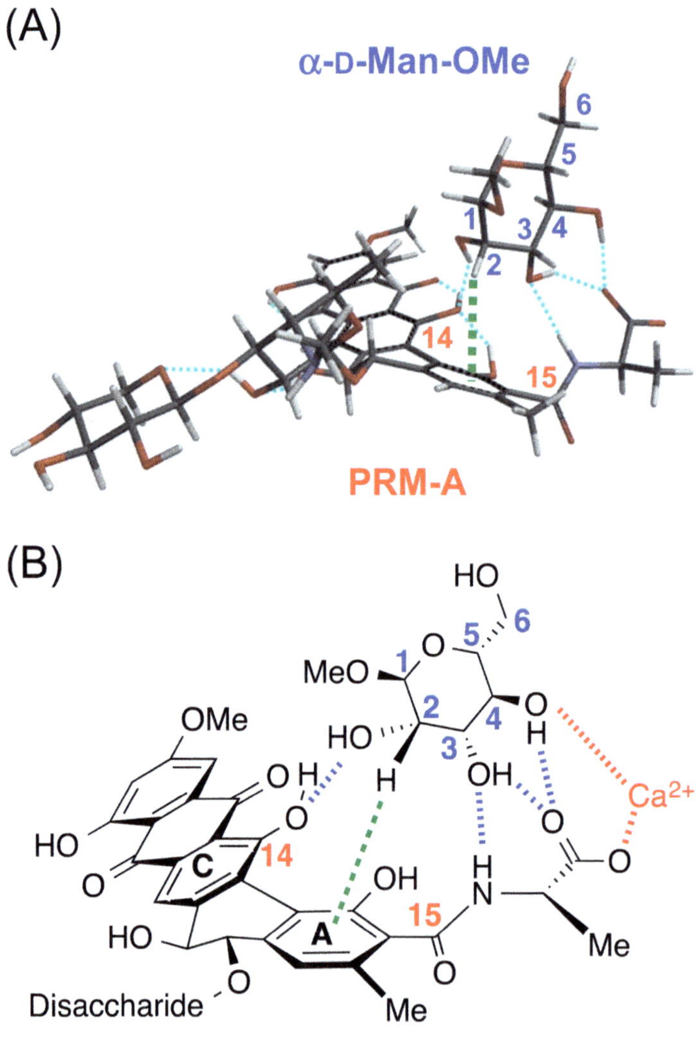

Figure 12.8 (A) Binding model of PRM-A with α-D-Man-OMe. Carbon, oxygen, nitrogen, and hydrogen atoms are shown in gray, red, blue, and white, respectively. Hydrogen bonds appear as cyan dotted lines. Hypothetical CH–π interaction is also depicted as a green dotted line. (B) Schematic representation of the binding model. Hypothetical Ca^{2+} coordination is shown as a red dotted line.

12.6 Validation of D-Mannose Binding Model of Pradimicin A

Our binding model (Figure 12.8) proposed that PRM-A recognizes the 2-, 3-, and 4-hydroxyl groups of D-Man but not the 6-hydroxyl group. To test this hypothetical recognition mode, we evaluated co-precipitation of PRM-A with deoxy derivatives of α-D-Man-OMe in the presence of Ca^{2+} ion.[55] Table 12.2 summarizes the molar ratio of 2-, 3-, 4-, and 6-deoxy-α-D-Man-OMe relative to PRM-A in the aggregate after 0, 2, and 4 times washing with an aqueous $CaCl_2$ solution. The results using α-D-Man-OMe and α-D-Glc-OMe are again listed for comparison. As in the case of α-D-Glc-OMe, specific incorporation into the aggregates was hardly observed in 2-, 3-, and 4-deoxy-α-D-Man-OMe, supporting the crucial roles of the 2-, 3-, and 4-hydroxyl groups of D-Man in its binding to PRM-A. In sharp contrast, 6-deoxy-α-D-Man-OMe was nearly indistinguishable from α-D-Man-OMe. PRM-A initially formed the 1 : 2 complex with 6-deoxy-α-D-Man-OMe, and washing of the aggregate released 6-deoxy-α-D-Man-OMe from the secondary binding site to produce the 1 : 1 complex. This observation clearly indicates the marginal contribution of the 6-hydroxyl group of D-Man to binding with PRM-A, which is also in good agreement with our binding model.

On the other hand, regarding the anomeric position of D-Man, BMY-28864 was reported to accommodate both α and β anomers of methyl, *p*-nitrophenyl, and *p*-aminophenyl D-mannopyranosides,[24] indicating that the configuration and substituents at the C1 position of D-Man are not essential for binding to PRMs. This tolerance at the anomeric position is also explained by our binding model, in which the C1 position of D-Man faces outward from the binding site of PRM-A. Taken together, the results of the solid-state NMR analysis and the co-precipitation experiments indicate that PRMs recognize the spatial array of the 2-, 3-, and 4-hydroxyl groups of D-Man. This molecular basis of recognition provides a clear explanation as to why PRM-A also binds methyl pyranosides of D-arabinose, L-fucose, and L-galactose,[24,60] all of which share the characteristic array of the three hydroxyl groups of D-Man (Figure 12.9).

After having validated the binding model, our next focus was glycan binding of PRMs. Although it has been assumed that PRMs bind only to D-Man

Table 12.2 Molar ratio of deoxy derivatives of α-D-Man-OMe relative to PRM-A in aggregates.

	Number of washing times		
	0	2	4
α-D-Man-OMe	2.35	1.08	1.07
α-D-Glc-OMe	0.34	<0.1	<0.1
2-deoxy-α-D-Man-OMe	0.30 (0.05)[a]	<0.1	<0.1
3-deoxy-α-D-Man-OMe	0.34 (0.06)	<0.1	<0.1
4-deoxy-α-D-Man-OMe	0.35 (0.05)	<0.1	<0.1
6-deoxy-α-D-Man-OMe	2.25 (0.03)	1.07 (0.03)	1.05 (0.02)

[a]Standard deviation of at least three separate experiments.

residues at non-reducing ends,[23] our model (Figure 12.8) raises a possibility that they have an ability to bind also to 6-substituted D-Man residues with free 2-, 3-, and 4-hydroxyl groups. As an initial test of this possibility, we examined whether PRM-A could accept methyl 2-deoxy-α-D-arabinohexopyranosyl(1 → 6)-α-D-mannopyranoside (**1**), in which PRM-A-insensitive 2-deoxy-D-Man residue is attached to the 6-hydroxyl group of α-D-Man-OMe (Figure 12.10).[55] As shown in Table 12.3, co-precipitation experiments revealed that **1** forms unusual complexes with PRM-A. PRM-A and **1** initially produced the aggregate consisted of the 1:1 (2:2) complex, which was converted to the 1:0.5 (2:1) complex after washing of the aggregate, indicating that **1** occupied one of the two binding sites in both the primary and secondary binding. In contrast, methyl α-D-mannopyranosyl(1 → 2)-α-D-mannopyranoside (**2**), which possesses D-Man residue at the non-reducing end, behaved similarly to α-D-Man-OMe.

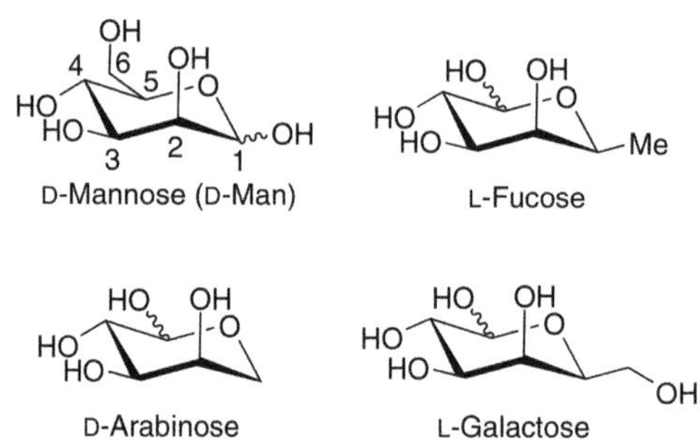

Figure 12.9 Structural comparison of D-mannose (D-Man), D-arabinose, L-fucose, and L-galactose.

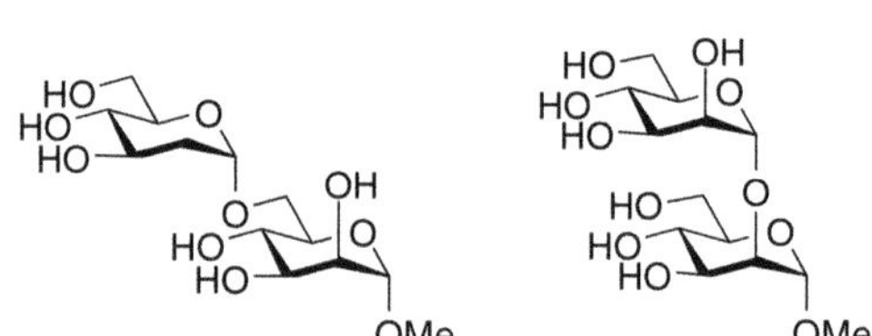

1 **2**

Figure 12.10 Structures of **1** and **2**.

Table 12.3 Molar ratio of **1** and **2** relative to PRM-A in aggregates.

	Number of washing times		
	0	2	4
1	1.23 (0.02)[a]	0.55 (0.02)	0.52 (0.05)
2	2.22 (0.05)	1.13 (0.04)	1.02 (0.01)

[a]Standard deviation of at least three separate experiments.

The origin of the difference of complex formation between **1** and **2** remains to be fully elucidated, but our binding model provides a possible explanation for the primary binding (Figure 12.11). In our model (Figure 12.8), the C6 carbon atom of D-Man is located above the carboxylate group of PRM-A. When considering that Ca^{2+} ion would bridge the carboxylate groups of two

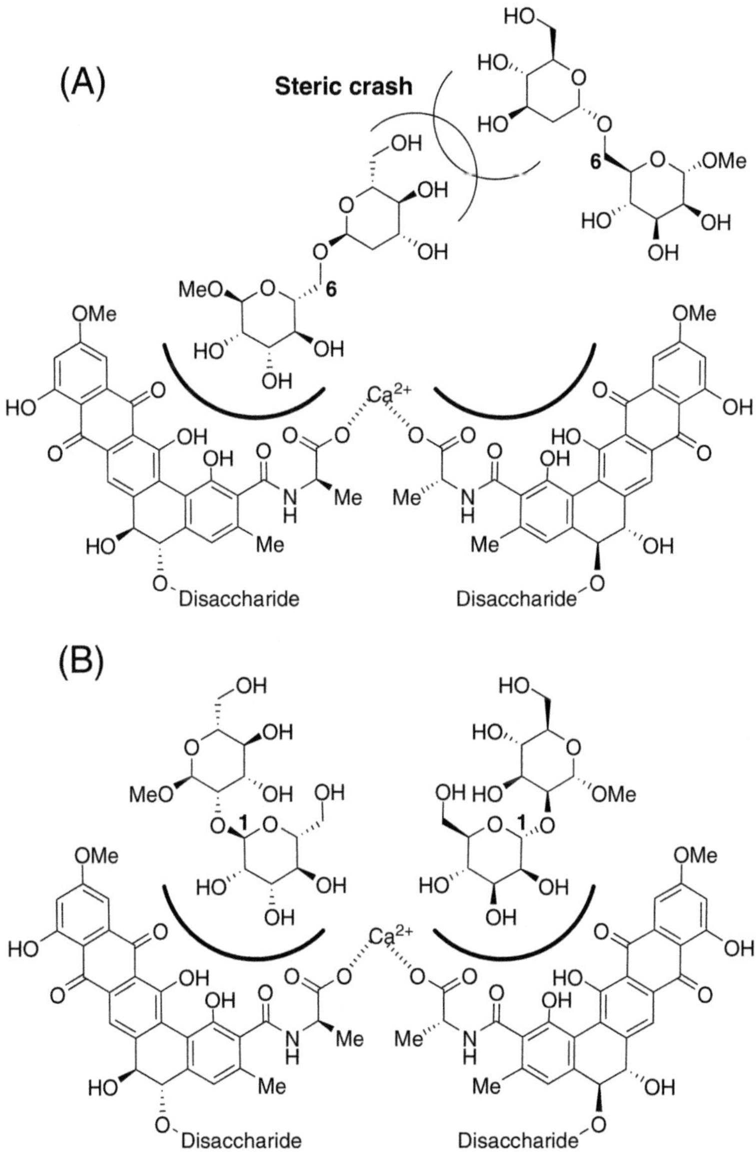

Figure 12.11 Schematic representation of the primary binding of **1** and **2** to the [PRM-A_2/Ca^{2+}] complex. (A) The 2-deoxy-D-Man residue of **1** bound in one binding site blocks occupation of the other binding site to form the 2:1 (1:0.5) complex of PRM-A and **1**. (B) The [PRM-A_2/Ca^{2+}] complex accommodates two molecules of **2** without steric hindrance to produce the 2:2 (1:1) complex of PRM-A and **2**.

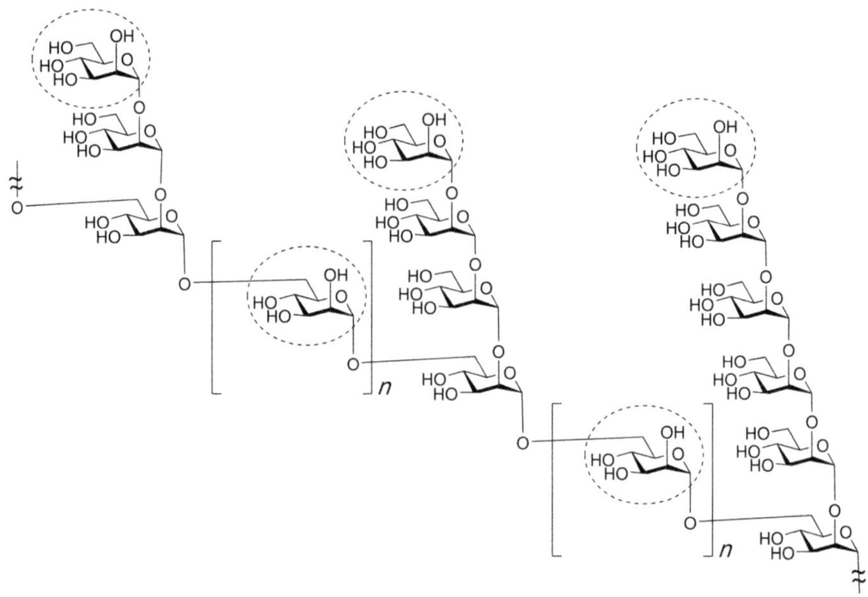

Figure 12.12 Representative structure of cell wall mannans of *Candida albican*.[61] Dotted circles represent potential binding sites for PRMs.

molecules of PRM-A, it can be assumed that these carboxylate groups are in close proximity to each other in the [PRM-A_2/Ca^{2+}] complex. Therefore, simultaneous occupation of the primary binding sites by two molecules of **1** would lead to steric clash between their 2-deoxy-D-Man residues, resulting in the formation of the 2:1 complex of PRM-A and **1** (Figure 12.11A). On the other hand, the C1 carbon atom of D-Man is oriented opposite to the carboxylate group of PRM-A, enabling accommodation of two molecules of **2** without any steric clashes to form the 2:2 complex (Figure 12.11B).

These results strongly support the possibility that PRM-A is capable of binding to 6-substituted D-Man residues in glycans. This is of particular significance for understanding of antifungal mechanism of PRMs. Earlier studies suggested that antifungal activity of PRMs is highly correlated with adsorption to cell wall mannans of pathogenic organisms.[25] However, little information is available as to how PRMs bind to the mannans, which have a comb-like structure with an α-1,6-linked polymannose backbone (Figure 12.12).[61] The present study indicates that, in addition to terminal D-Man residues in oligomannose branches of mannans, internal D-Man residues in the α-1,6-linked polymannose backbone are also potential binding targets for PRMs. Although further validation is still required, this emerging glycan specificity of PRMs could provide a new insight into their mode of antifungal action.

12.7 Conclusion and Future Perspectives

PRMs and benanomicins are a family of unique small molecules with D-Man binding ability, and thus hold great promise as structurally simper and clinically superior mimics of lectins. Recently, PRMs have figured prominently in studies directed at developing a novel therapy for HIV treatment.[8,26,27,29] They have been found to block entry of HIV into host cells by binding to glycans of envelope glycoprotein gp120, and also cause mutation of its glycosylation sites to expose cryptic immunogenic epitopes of the virus surface. Therefore, exposure of HIV to PRMs puts the virus in the dilemma of either becoming eliminated from its host by being kept suppressed by PRMs or becoming susceptible to neutralization by escaping the pressure from PRMs. In contrast to conventional anti-HIV strategies aiming to delay the development of drug resistance, this approach using PRMs utilizes viral viability and the inherent viral replication to trigger the immune system. Although this therapeutic concept is entirely unique and promising, clinical development of PRMs has been impeded by limited knowledge of how they recognize D-Man and glycans. Our solid-state NMR analysis revealed the geometry of the primary D-Man binding of PRM-A, which further led to the finding that PRM-A has the capacity to bind D-Man residues other than at non-reducing ends of glycans. The information obtained in this study could provide an opportunity to advance clinical development of PRMs. The present study also demonstrates that ligand–receptor interactions can be analyzed at an atomic level by solid-state NMR spectroscopy. Given that a number of lectins precipitate with carbohydrates or glycoproteins,[62] solid-state NMR spectroscopy is likely to become a powerful technique for investigating molecular interactions of the lectins with their carbohydrate ligands.

Acknowledgements

This research was conducted in collaboration with Professor Yasuhiro Igarashi of Toyama Prefectural University, Professor Kiyonori Takegoshi, Dr Yuichi Masuda, Mr Takashi Doi, and Mr Keita Yamada of Kyoto University, and Mr Takara Taketani of RIKEN. The work was partly supported by a MEXT Grant-in-Aid for Young Scientists (B), Scientific Research (B), Scientific Research on Innovative Areas "Chemical Biology of Natural Products", and SUNBOR GRANT from the Suntory Foundation of Life Sciences.

References

1. R. A. Dwek, *Chem. Rev.*, 1996, **96**, 683.
2. H. E. Myrrey and L. C. Hsieh-Wilson, *Chem. Rev.*, 2008, **108**, 1708.
3. Y. C. Lee, *Trends Glycosci. Glycotechnol.*, 2010, **22**, 95.
4. A. Larkin and B. Imperiali, *Biochemistry*, 2011, **50**, 4411.
5. K. Ohtsubo and J. D. Marth, *Cell*, 2006, **126**, 855.
6. K. S. Lau and J. W. Dennis, *Glycobiology*, 2008, **18**, 750.
7. G. W. Hart and R. J. Copeland, *Cell*, 2010, **143**, 672.
8. J. Balzarini, *Nat. Rev. Microbiol.*, 2007, **5**, 583.
9. J. Hirabayashi, *Methods Mol. Biol.*, 2014, **1200**, 225.
10. L. M. Koharudin and A. M. Gronenborn, *Curr. Opin. Virol.*, 2014, **7**, 95.
11. T. Yau, X. Dan, C. C. Ng and T. B. Ng, *Molecules*, 2015, **26**, 3791.
12. S. Jin, Y. Cheng, S. Reid, M. Li and B. Wang, *Med. Res. Rev.*, 2010, **30**, 171.

13. M. Mazik, *RSC Adv.*, 2012, **2**, 2630.
14. D. B. Walker, G. Joshi and A. P. Davis, *Cell. Mol. Life Sci.*, 2009, **66**, 3177.
15. Y. Nakagawa and Y. Ito, *Adv. Carbohydr. Chem. Biochem.*, 2012, **68**, 1.
16. C. E. Miron and A. Petitjean, *ChemBioChem*, 2015, **16**, 365.
17. Y. Ferrand, M. P. Crump and A. P. Davis, *Science*, 2007, **318**, 619.
18. C. Ke, H. Destecroix, M. P. Crump and A. P. Davis, *Nat. Chem.*, 2012, **4**, 718.
19. A. Pal, M. Bérubé and D. G. Hall, *Angew. Chem., Int. Ed.*, 2010, **49**, 1492.
20. O. Francesconi, C. Nativi, G. Gabrielli, I. D. Simone, S. Noppen, J. Balzarini, S. Liekens and S. Roelens, *Chem.–Eur. J.*, 2015, **21**, 1.
21. T. Takeuchi, H. Hara, H. Naganawa, M. Okada, M. Hamada, H. Umezawa, S. Gomi, M. Sezaki and S. Kondo, *J. Antibiot.*, 1988, **41**, 807.
22. T. Oki, M. Konishi, K. Tomatsu, K. Tomita, K. Saitoh, M. Tsunakawa, M. Nishio, T. Miyaki and H. Kawaguchi, *J. Antibiot.*, 1988, **41**, 1701.
23. Y. Fukagawa, T. Ueki, K. Numata and T. Oki, *Actinomycetologia*, 1993, 7, 1.
24. T. Ueki, M. Oka, Y. Fukagawa and T. Oki, *J. Antibiot.*, 1993, **46**, 465.
25. Y. Sawada, K. Numata, T. Murakami, H. Tanimichi, S. Yamamoto and T. Oki, *J. Antibiot.*, 1990, **43**, 715.
26. J. Balzarini, K. O. François, K. Van Laethem, B. Hoorelbeke, M. Renders, J. Auwerx, S. Liekens, T. Oki, Y. Igarashi and D. Schols, *Antimicrob. Agents Chemother.*, 2010, **54**, 1425.
27. J. Balzarini, K. Van Laethem, D. Daelemans, S. Hatse, A. Bugatti, M. Rusnati, Y. Igarashi, T. Oki and D. Schols, *J. Virol.*, 2007, **81**, 362.
28. Y. Igarashi and T. Oki, *Adv. Appl. Microbiol.*, 2004, **54**, 147.
29. K. O. François and J. Balzarini, *Med. Res. Rev.*, 2012, **32**, 349.
30. T. Oki, M. Kakushima, M. Nishio, H. Kamei, M. Hirano, Y. Sawada and M. Konishi, *J. Antibiot.*, 1990, **43**, 1230.
31. K. Fujikawa, Y. Tsukamoto, T. Oki and Y. C. Lee, *Glycobiology*, 1998, **8**, 407.
32. M. Hu, Y. Ishizuka, Y. Igarashi, T. Oki and H. Nakanishi, *Spectrochim. Acta, Part A*, 1999, **55**, 2547.
33. M. Hu, Y. Ishizuka, Y. Igarashi, T. Oki and H. Nakanishi, *Spectrochim. Acta, Part A*, 1999, **56**, 181.
34. M. Hu, Y. Ishizuka, Y. Igarashi, T. Oki and H. Nakanishi, *Spectrochim. Acta, Part A*, 2000, **56**, 1233.
35. S. Shahzad-ul-Hussan, R. Ghirlando, C. I. Dogo-Isonagie, Y. Igarashi, J. Balzarini and C. A. Bewley, *J. Am. Chem. Soc.*, 2012, **134**, 12346.
36. Y. Nakagawa, Y. Masuda, K. Yamada, T. Doi, K. Takegoshi, Y. Igarashi and Y. Ito, *Angew. Chem., Int. Ed.*, 2011, **50**, 6084.
37. Y. Nakagawa, T. Doi, Y. Masuda, K. Takegoshi, Y. Igarashi and Y. Ito, *J. Am. Chem. Soc.*, 2011, **133**, 17485.
38. T. Ueki, K. Numata, Y. Sawada, T. Nakajima, Y. Fukagawa and T. Oki, *J. Antibiot.*, 1993, **46**, 149.
39. T. Ueki, K. Numata, Y. Sawada, M. Nishio, H. Ohkuma, S. Toda, H. Kamachi, Y. Fukagawa and T. Oki, *J. Antibiot.*, 1993, **46**, 455.
40. K. Drickamer, *Curr. Opin. Struct. Biol.*, 1993, **3**, 393.

41. R. D. Shannon, *Acta Crystallogr.*, 1976, **A32**, 751.
42. K. McAteer, A. S. Lipton and P. D. Ellis, in *Encyclopedia of Nuclear Magnetic Resonance*, ed. D. M. Grant and R. K. Harris, John Wiley & Sons, Chichester, 1996, p. 1085.
43. I. M. Armitage, T. Drakenberg and B. Reilly, *Met. Ions Life Sci.*, 2013, **11**, 117.
44. M. F. Summers, *Coord. Chem. Rev.*, 1988, **86**, 43.
45. H. Xia and G. D. Rayson, *Adv. Environ. Res.*, 2002, **7**, 157.
46. T. Takayama, S. Ohuchida, Y. Koike, M. Watanabe, D. Hashizume and Y. Ohashi, *Bull. Chem. Soc. Jpn.*, 1996, **69**, 1579.
47. K. Takegoshi, S. Nakamura and T. Terao, *Chem. Phys. Lett.*, 2001, **344**, 631.
48. K. Takegoshi, S. Nakamura and T. Terao, *J. Chem. Phys.*, 2003, **118**, 2325.
49. C. R. Morcombe, V. Gaponenko, R. A. Byrd and K. W. Zilm, *J. Am. Chem. Soc.*, 2004, **126**, 7196.
50. M. Tamiya, K. Ohmori, M. Kitamura, H. Kato, T. Arai, M. Oorui and K. Suzuki, *Chem.–Eur. J.*, 2007, **13**, 9791.
51. S. Okuyama, H. Kamachi and T. Oki, *Chem. Pharm. Bull.*, 1993, **41**, 223.
52. M. Kakushima, Y. Sawada, M. Nishio, T. Tsuno and T. Oki, *J. Org. Chem.*, 1989, **54**, 2536.
53. S. Gomi, M. Sezaki, M. Hamada, S. Kondo and T. Takeuchi, *J. Antibiot.*, 1989, **42**, 1145.
54. S. Kakinuma, T. Furumai and T. Oki, *Actinomycetologia*, 1994, **8**, 134.
55. Y. Nakagawa, T. Doi, T. Taketani, K. Takegoshi, Y. Igarashi and Y. Ito, *Chem.–Eur. J.*, 2013, **19**, 10516.
56. M. Nishio, *Phys. Chem. Chem. Phys.*, 2011, **13**, 13873.
57. J. Asensio, A. Ardá, F. J. Cañada and J. Jiménez-Barbero, *Acc. Chem. Res.*, 2013, **46**, 946.
58. Z. R. Laughrey, S. E. Kiehna, A. J. Rierrien and M. L. Waters, *J. Am. Chem. Soc.*, 2008, **130**, 14625.
59. M. C. Fernandez-Alonso, F. J. Cañada, J. Jiménez-Barbero and G. Cuevas, *J. Am. Chem. Soc.*, 2005, **127**, 7379.
60. Y. Nakagawa, Y. Watanabe, Y. Igarashi, Y. Ito and M. Ojika, *Bioorg. Med. Chem. Lett.*, 2015, **25**, 2963.
61. N. Shibata, K. Ikuta, T. Imai, Y. Satoh, R. Satoh, A. Suzuki, C. Kojima, H. Kobayashi, K. Hisamichi and S. Suzuki, *J. Biol. Chem.*, 1995, **270**, 1113.
62. N. Sharon and H. Lis, *Glycobiology*, 2004, **14**, 53R.

CHAPTER 13

Structure and Dynamics of Polysaccharides in Plant Cell Walls from Solid-State NMR

TUO WANG[a] AND MEI HONG*[a]

[a]Department of Chemistry, Massachusetts Institute of Technology, 170 Albany Street, Cambridge, MA 02139, USA
*E-mail: meihong@mit.edu

Plant cell walls are an insoluble material formed by polysaccharides, proteins and lignin.[1] They shape plant cell morphology, maintain ionic balance, protect plant cells from environmental stress, and allow cell expansion. Until recently, information about the molecular structures of cell wall biopolymers that underlie these biochemical and mechanical functions was limited due to the difficulty of studying these insoluble and non-crystalline materials in their native states. In the last few years, two- and three-dimensional high-resolution magic-angle-spinning (MAS) ^{13}C solid-state NMR spectroscopy has been employed to investigate the intermolecular packing, conformation and dynamics of polysaccharides and proteins in native plant cell walls. In this chapter, we summarize the SSNMR methods used in these studies and key biological findings, with a special emphasis on how wall polysaccharides interact with each other and with proteins and water.

New Developments in NMR No.10
NMR in Glycoscience and Glycotechnology
Edited by Koichi Kato and Thomas Peters

Published by the Royal Society of Chemistry, www.rsc.org

13.1 2D and 3D Correlation MAS NMR Techniques for Studying Plant Cell Wall Biopolymers

While 1D ^{13}C MAS NMR has been used for decades to study unlabeled cellulose in their pure forms and much insight was obtained,[2–4] only recently has uniform ^{13}C enrichment of entire plants been conducted[5,6] to provide the crucial sensitivity for 2D and 3D ^{13}C correlation NMR. This correlation approach is necessary for resolving the signals of the mixtures of polysaccharides and proteins in the intact cell wall. Depending on the NMR pulse sequences, the cross peaks in these 2D and 3D ^{13}C correlation spectra can result from covalently bonded ^{13}C spins or from non-covalent but spatially proximal ^{13}C spins. The through-bond cross peaks allow the assignment of the chemical shifts to specific sugars in polysaccharides, amino acid residues in proteins or aromatic functional groups in lignin, while the through-space cross peaks give information on the spatial proximity of two sugars. The availability of two or three ^{13}C chemical shift dimensions sufficiently disperses the signals to allow resolution of multiple sugars with similar chemical structures and to distinguish different conformations of the same sugar. Three complementary correlation experiments have proven to be particularly useful for plant cell wall structural studies. First, a ^{13}C–^{13}C J-coupling-mediated double-quantum and single-quantum correlation experiment called INADEQUATE[6,7] is used to measure through-bond correlation spectra of dynamic matrix polysaccharides. Second, double- or zero-quantum dipolar recoupling experiments have been used to measure 2D spectra that exhibit mostly one-bond cross peaks of rigid polysaccharides.[8–10] Third, a proton-driven ^{13}C spin diffusion (PDSD) experiment based on distance-dependent ^{13}C–^{13}C dipolar coupling has been used to measure intermolecular cross peaks between carbons separated by many angstroms.[11] The PDSD experiment can be conducted in a 2D or 3D fashion to obtain site resolution.[12]

In addition to chemical structure and molecular conformation, one can also investigate molecular dynamics to gain insights into cell wall structure and function. Two SSNMR approaches have been applied to measure wall polysaccharide mobility. The first experiment, dipolar-chemical shift correlation experiment (DIPSHIFT), separates ^{13}C–^{1}H dipolar couplings by one or two ^{13}C chemical shift dimensions.[13–15] The reorientation angle of C–H bonds is manifested by the magnitude of the dipolar couplings: the larger the motional amplitude, the smaller the coupling. The order parameter, S_{CH}, is the ratio of the motionally averaged C–H dipolar coupling to the rigid-limit coupling. The second approach uses spin-lattice relaxation times in the laboratory frame (T_1) and the rotating frame ($T_{1\rho}$) to obtain information about nanosecond and microsecond motions. For example, ^{13}C T_1 and ^{1}H $T_{1\rho}$ relaxation times have been measured and found to differ significantly for different wall polysaccharides and between different plants.

Water interaction with wall polysaccharides can be investigated using ^{1}H spin diffusion approaches.[16–20] The water ^{1}H chemical shift can be encoded in the indirect dimension of a 2D experiment before the water magnetization is transferred to the polysaccharides and detected through ^{13}C signals. Alternatively, the water ^{1}H magnetization can be selected by a relaxation filter without chemical shift encoding, resulting in a simpler 1D experiment. The distance-dependent ^{1}H spin diffusion from water to polysaccharides gives information about the site-specific contact of various polysaccharides with water and about the mobility of water in the porous space between the polymers, which generally should differ from the mobility of bulk water.

13.2 Dynamics of Polysaccharides in Primary Cell Walls

13.2.1 Dicot Primary Cell Walls

Solid-state NMR is a powerful technique for investigating molecular motions over a large range of timescales of picoseconds to seconds.[21–24] Figure 13.1 shows the C–H bond order parameters and ^{13}C spin-lattice relaxation times (T_1) of several plant primary cell walls. The order parameters reflect motional amplitudes while NMR relaxation times reflect motional rates. In all plant primary walls studied so far, cellulose exhibits the largest order parameters, 0.8–0.9, indicating small motional amplitudes, while matrix polysaccharides, namely hemicellulose and pectins, show lower-order parameters of 0.4–0.6 (Figure 13.1a). The former is consistent with the extensively hydrogen-bonded and tightly packed nature of glucan chains in cellulose microfibrils,[25] while the latter indicates that matrix polysaccharides undergo large-amplitude motions. Consistent with this result, cellulose has much slower T_1 relaxation than matrix polysaccharides. For wild-type *Arabidopsis* primary walls, cellulose exhibits a single exponential relaxation with a time constant of ~5 s, while pectins show double-exponential relaxation with near-equal contributions of a short T_1 component (~0.3 s) and a long T_1 component (~4.5 s) (Figure 13.1b and c). This double-exponential behavior strongly suggests a bimodal distribution of pectins: a cellulose-bound, immobilized, fraction, and an inter-fibrillar fraction that is highly mobile. When most xyloglucan is depleted in an *Arabidopsis* mutant, *xxt1xxt2xxt5*,[5] the ^{13}C T_1 relaxation times of the remaining polysaccharides decreased substantially (Figure 13.1b), indicating faster motions on the nanosecond timescale. ^{1}H $T_{1\rho}$ values, which reflect collective motions of polymer chains on the microsecond timescale, also decreased in the xyloglucan-deficient mutant. Thus, hemicellulose depletion causes a more dynamic cell wall. However, while relaxation times are qualitatively different between wild-type and mutant *Arabidopsis* cell walls, the order parameters are comparable, suggesting that chain-packing density is not affected by xyloglucan depletion. Thus, the faster motional rates of the mutant wall polysaccharides cannot be attributed to lower packing densities but is likely due to weakened hydrogen bonds and altered three-dimensional arrangement of the wall polysaccharides.

Pectin extraction caused the opposite effect to the cell wall compared to xyloglucan depletion: ^{13}C T_1 and ^{1}H $T_{1\rho}$ relaxation times both increased, indicating that the remaining polysaccharides became less mobile.[26] We attribute

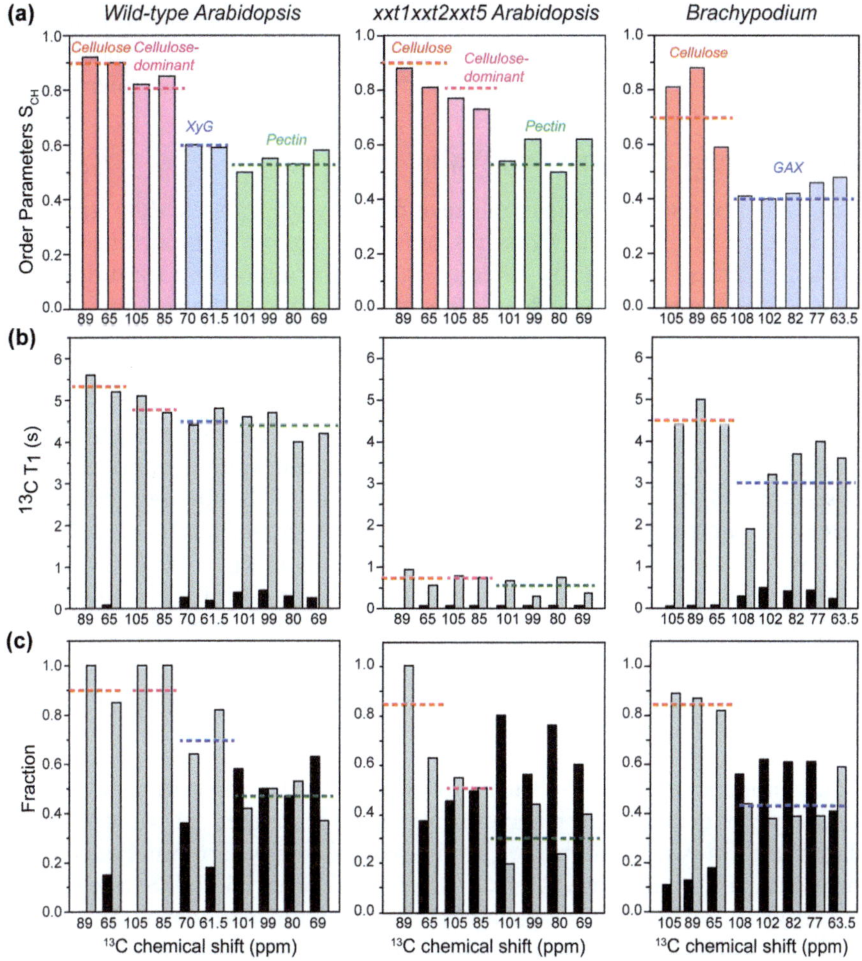

Figure 13.1 Comparison of polysaccharide mobilities in three hydrated primary cell walls based on C–H dipolar order parameters (a) and ^{13}C T_1 relaxation times (b, c). Left column: wild-type *Arabidopsis* cell wall; Middle column: xyloglucan-deficient *xxt1xxt2xxt5 Arabidopsis* cell wall; Right column: *Brachypodium* cell wall. (a) C–H order parameters obtained from quantitative ^{13}C–^{1}H DIPSHIFT experiments. (b) ^{13}C T_1 relaxation times. For polysaccharides that exhibit double-exponential relaxation, the short and long T_1 values are shown in black and gray, respectively. (c) Fractions of the fast-relaxing (black) and slow-relaxing (gray) components in (b). All data were measured at ambient temperature on a 600 MHz NMR spectrometer. Adapted with permission from M. Dick-Perez, Y. A. Zhang, J. Hayes, A. Salazar, O. A. Zabotina and M. Hong, *Biochemistry*, 2011, **50**, 989–1000, and T. Wang, A. Salazar, O. A. Zabotina and M. Hong, *Biochemistry*, 2014, **53**, 2840–2854. Copyright (2011 and 2014) American Chemical Society and ref. 61 with permission from the American Society of Plant Biologists.

this result to the higher chain-packing density of the remaining polymers in the extracted cell walls. The pectin signals show the largest increase in the relaxation times, which supports the bimodal distribution of pectins and indicates that the unextracted pectins are significantly immobilized by cellulose microfibrils.

13.2.2 Grass Primary Cell Walls

Most food crops such as wheat, maize and barley, are grasses, thus structural investigation of grass cell walls has practical economic importance. Grass cell walls have a significantly different chemical composition from dicot cell walls.[27] For example, two-week-old *Brachypodium*, a model grass, mainly contains cellulose and the hemicellulose glucuronoarabinoxylan (GAX), while pectins, xyloglucan and structural proteins exist in negligible concentrations. Interestingly, the S_{CH}'s of GAX are only ~0.4, which are significantly smaller than the S_{CH}'s of XyG and pectins in dicot cell walls (Figure 13.1a). This larger-amplitude motion of GAX suggests that grass cell wall polymers are more loosely packed compared to *Arabidopsis* primary walls. Double-exponential ^{13}C T_1 relaxation suggests that ~40% of GAX is tightly associated with cellulose. However, the two T_1 values of GAX are not as distinct as pectins in *Arabidopsis* primary walls, suggesting that cellulose interactions with matrix polysaccharides are weaker in *Brachypodium* than in the dicot cell wall of *Arabidopsis*.

13.3 Cellulose Structure in Primary Cell Walls

The structure of cellulose, the most abundant natural polymer on Earth, has been studied since the 1980's using SSNMR.[6,28–32] 1D ^{13}C MAS spectra show that extracted cellulose from bacteria, algae and secondary cell walls of higher plants contain two allomorphs, I_α and I_β, whose relative amounts depend on the organism.[2,3,33] Recently, 2D ^{13}C–^{13}C and ^{13}C–^{1}H correlation SSNMR experiments were conducted on algal and tunicate cellulose to resolve and assign the ^{13}C and ^{1}H chemical shifts of these two allomorphs.[34] It was found that each allomorph contains two magnetically inequivalent anhydroglucose residues. In I_α cellulose the inequivalent A and A′ residues alternate within each chain while in I_β cellulose the inequivalent B and B′ residues are separated into neighboring chains.[35] These SSNMR results, together with X-ray crystal structures,[36,37] provided atomic-resolution structural information of highly crystalline cellulose in algae and tunicate.

Compared to bacterial, algae and animal cellulose, higher plants' primary wall cellulose has smaller lateral dimensions and is now known to contact matrix polysaccharides on the sub-nanometer length scale.[38,39] These properties make extraction-based structural studies of primary wall cellulose suspect. Earlier SSNMR studies of cellulose in intact plant primary cell walls relied on relaxation filters to selectively detect the ^{13}C signals of rigid cellulose while suppressing the signals of the dynamic matrix polysaccharides.[40] These relaxation-filtered 1D ^{13}C MAS spectra resolved the chemical shifts of the surface amorphous glucan chains *versus* the interior crystalline glucan chains, thus allowing their relative amounts to be estimated to give information on the lateral dimension of the cellulose microfibrils.[29,40,41] In particular, the C4 and C6 ^{13}C chemical shifts differ by 3–5 ppm between the surface and interior glucan chains. This difference is commonly attributed to distinct hydrogen bonding and torsion angles between the surface and interior chains. For example, the C6 carbon adopts a *trans-gauche* (*tg*) conformation in interior chains but *gauche-gauche* (*gg*) or *gauche-trans* (*gt*) conformation in the surface chains.[29,42] With ^{13}C-labeling of intact primary cell walls, the use of 2D ^{13}C–^{13}C correlation spectroscopy, and the availability of high magnetic fields, the number of resolvable cellulose conformations have significantly increased: our most recent data reveal at least seven sets of cellulose C1–C4 and C4–C6 cross peaks (Figure 13.2). Two conformations are found for surface chains and five for interior chains. While the cellulose peaks in 1D ^{13}C spectra show inhomogeneous linewidths of 2–3 ppm, when resolved in 2D spectra, narrower linewidths of 0.7–1.0 ppm are found. Among the different conformations, type-d cellulose (Figure 13.2b), with (C1, C4, C6) chemical shifts of (105, 87.5, 65) ppm, corresponds to the cellulose bound to expansin,[43] a wall-loosening protein.[44–47] A separate study showed that this expansin-bound cellulose is also enriched in xyloglucan, as seen in 2D correlation spectra.[43] Consistently, the type-d cellulose cross peaks disappear in the xyloglucan-depleted *xxt1xxt2xxt5* mutant (Figure 13.2b), suggesting that xyloglucan is responsible for causing this cellulose conformation. Type-d cellulose cross peaks are also detected in the 2D spectra of *Brachypodium* cell walls (Figure 13.2c), suggesting that GAX similarly induces this conformation. The three main C4 chemical shifts of cellulose, 88, 89 and 90 ppm, are common

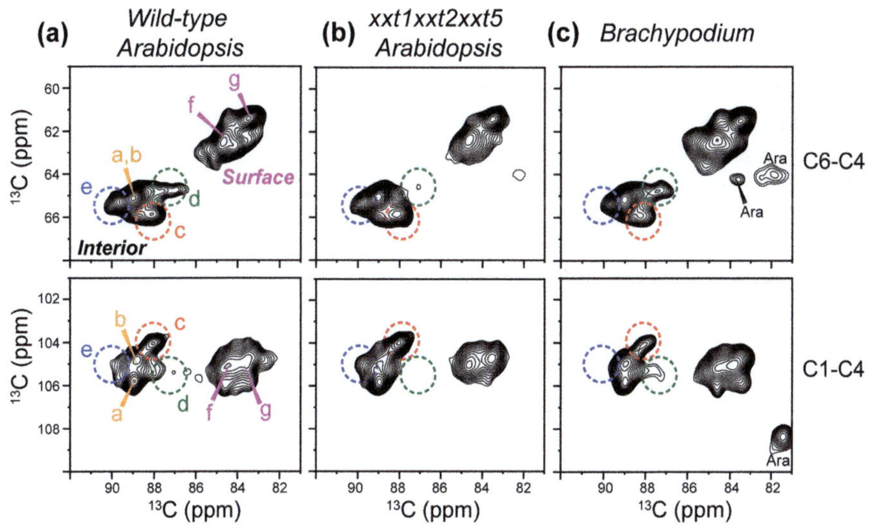

Figure 13.2 Comparison of cellulose ^{13}C chemical shifts from 2D ^{13}C–^{13}C correlation spectra of three primary cell walls. (a) Wild-type *Arabidopsis* cell wall. (b) *xxt1xxt2xxt5* mutant of the *Arabidopsis* cell wall. (c) *Brachypodium* cell wall. Multiple sets of cellulose chemical shifts are observed in each sample, indicating multiple cellulose conformations and hydrogen bonding. The peak intensities vary for different cell walls. The 2D spectra were measured on an 800 MHz spectrometer at ambient temperature.

to both primary and secondary cell walls, but the chemical shifts of the other carbons of glucose differ, indicating that cellulose structure and packing are not identical between primary and secondary cell walls. Mutations in cellulose-synthesis enzymes (CESA) have been shown to reduce cellulose crystallinity, as manifested by lower intensities of the interior cellulose C4 and C6 peaks.[48,49] Further experimental studies are needed to obtain direct conformational constraints such as torsion angles and distances for cellulose in primary cell walls and to relate these structures to function. These studies will benefit from input from quantum chemical calculations that relate ^{13}C chemical shifts to molecular conformation and packing.[50–52]

The interior and surface cellulose signals do not equilibrate to the same intensities in 2D ^{13}C–^{13}C PDSD spectra even after very long mixing times (1.5–3.0 s).[11] The lack of equilibration qualitatively indicates that the primary wall cellulose microfibrils have sufficiently large diameters that some of the interior chains do not contact the surface. These "core" chains have different conformations from interior chains that do come into molecular contact with the surface chains, since they exhibit different chemical shifts.[11] The presence of a population of deeply embedded glucan chains places a lower bound to the lateral dimension of the cellulose microfibril. Combined with the surface-to-interior intensity ratios obtained from quantitative 1D ^{13}C NMR spectra, these data indicate that cellulose microfibrils in *Arabidopsis* and *Brachypodium* primary walls have at least 24 chains.[11] Recent analyses of wide-angle X-ray scattering (WAXS) data, non-quantitative SSNMR spectra, together with emerging structural evidence that the cellulose synthesis complex (CSC) contains 18 subunits,[53,54] favored small, 18-chain, microfibrils for the plant primary cell wall.[41,55] However, the relation between the CSC composition and the cellulose dimensions is indirect, non-quantitative ^{13}C NMR spectra bias against the detection of crystalline interior chains and hence underestimate microfibril cross section, and interpretation of the WAXS data requires assumptions of the type of chain disorder. Thus, results from quantitative ^{13}C SSNMR spectra provide a more direct and objective readout of glucan-chain conformations and their relative concentrations in primary wall cellulose.

13.4 Polysaccharide Interactions with Other Molecules in Plant Primary Cell Walls

13.4.1 Polysaccharide–Polysaccharide Interactions

Intermolecular cross peaks in 2D and 3D correlation MAS spectra indicate that cellulose microfibrils in plant primary cell walls interact extensively with other biopolymers and water. The signals of different biomolecules can be resolved in the NMR spectra through ^{13}C and ^{1}H chemical shifts, relaxation filters[40,56] or differential isotope labeling. We take advantage of the unique C4 (88–90 ppm) and C6 (65 ppm) chemical shifts of interior cellulose and the unique chemical shifts of pectins at 101, 80 and 69 ppm to report the spatial proximities between cellulose and matrix polysaccharides. The 2D and 3D correlation spectra can be measured as a function of spin diffusion mixing time, and the intensity buildup of cross peaks can be extracted and compared with cross peak buildup between carbons with known distances to obtain semi-quantitative distance information.[57,58] Many intermolecular cellulose-pectin cross peaks were observed, indicating significant intermolecular contacts on the nanometer scale (Figure 13.3a). The fraction of surface cellulose in contact with pectins in *Arabidopsis* primary walls was estimated to be 25–50% based on the equilibrium intensity of the cross peaks.[59,60] These cellulose-pectin contacts are independent of the hydration history of the cell wall and are retained in partially depectinated walls, suggesting that the interactions are site-specific rather than non-specific due to molecular crowding. These cross peaks suggest that pectins may be entrapped within a microfibril or between multiple microfibrils.[61–63] Compared to pectins, xyloglucan, the main hemicellulose in dicot primary walls, shows relatively few cross peaks with cellulose, indicating that xyloglucan cannot extensively coat

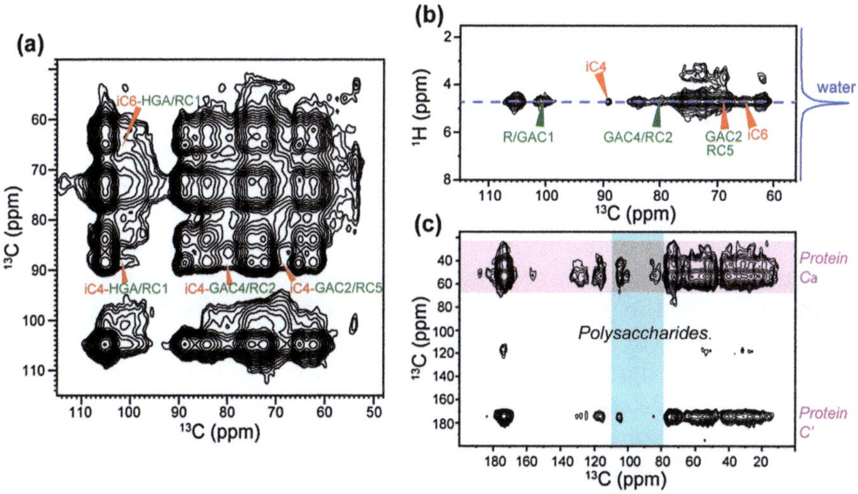

Figure 13.3 Intermolecular contacts of polysaccharides in *Arabidopsis* cell walls from 2D correlation NMR spectra. (a) Cellulose-pectin cross peaks (assigned in red and green) in a 2D ^{13}C–^{13}C PDSD spectrum measured with a 1.5 s spin diffusion mixing time at 253 K on a 600 MHz spectrometer. (b) Polysaccharide-water cross peaks from a ^{1}H–^{13}C 2D correlation spectrum measured with 1 ms ^{1}H spin diffusion at 293 K on a 900 MHz spectrometer. The spectra in (a-b) are adapted from ref. 61 with permission from the American Society of Plant Biologist. (c) Expansin-polysaccharide cross peaks from a ^{15}N–^{13}C dipolar filtered 2D ^{13}C–^{13}C correlation spectrum. The spectrum was measured with a 3 s spin diffusion mixing time at 102 K on a 600 MHz NMR spectrometer equipped with a 395 GHz gyrotron to increase the sensitivity by DNP. The spectrum in (c) is adapted from ref. 43 with permission from the National Academy of Sciences, USA.

the microfibril surface[5,12] as depicted in the tethered network model. These spectral observations indicate that all three types of polysaccharides interact with each other and form a single network in *Arabidopsis* primary walls.

This single-network structural model of primary cell walls has found strong independent biochemical support. Enzymatic digestion and biomechanical measurements showed that the majority of xyloglucan has no load-bearing function, and only a small fraction of xyloglucan tethers the cellulose microfibrils.[64,65] MD simulations suggested that xyloglucan binds multiple microfibrils to form sandwiched complexes, which may provide mechanical strength to the cell walls.[66] It was also observed that xyloglucan-depleted cell walls partially retain their mechanical strength,[67] suggesting that other interactions such as cellulose–cellulose and cellulose–pectin interactions make significant contributions to cell wall strength.

The *Brachypodium* cell walls exhibit numerous cross peaks between cellulose and the backbone and sidechains of GAX. For example, cellulose-xylose, cellulose-arabinose, and cellulose-ferulic acid cross peaks were observed.[68] Similar to pectins in dicot cell walls, GAX in *Brachypodium* primary wall exhibits two dynamically distinct domains (Figure 13.1b and c), with the more mobile domain filling the interfibrillar space while the more rigid domain in contact with cellulose. Since two-week old *Brachypodium* cell wall has a simple polysaccharide composition, other grass cell walls should be investigated in the future to understand how mixed-linkage glucan, another important hemicellulose, participates in the grass wall network.[69,70] Future solid-state NMR studies will also be useful for determining how GAX binds cellulose in the cell wall.

In addition to primary cell wall, 2D and 3D correlation SSNMR have been used to investigate secondary cell walls.[71–73] Secondary walls are located between the primary wall and the plasma membrane and account for the majority of plant biomass. Secondary walls mainly consist of cellulose, xylan, lignin and structural proteins. An early study of 10% ^{13}C-labeled wood detected only intra-sugar cross peaks but no intermolecular cross peaks for spin diffusion mixing times up to 8 s, indicating that cellulose, xylan and lignin are separated by more than a few nanometers in wood.[71] This result was somewhat unexpected because cellulose microfibrils are thought to be coated by a xylan–lignin complex[74] and may also associate with glucomannan.[75] It is possible that the lack of cellulose cross peaks with matrix polysaccharides in this early study was partly due to the low ^{13}C labeling level of 10%, which reduces the ^{13}C–^{13}C polarization transfer efficiency, lowers the statistical probability of ^{13}C–^{13}C spin pairs, and decreases the experimental sensitivity. On the other hand, if cellulose microfibrils aggregate sufficiently in secondary cell walls,[76] then limited contact with matrix polysaccharides should be expected. Very recently, uniformly ^{13}C-labeled secondary cell walls of *Arabidopsis* stems were studied using 2D and 3D correlation NMR experiments, and clear cross peaks between cellulose and non-carbohydrate molecules, including lignin and proteins, were reported,[72] but still no cellulose-xylan and cellulose-glucomannan cross peaks were detected. Thus, this *Arabidopsis* study partly supports the wood study and suggests that there is limited cellulose–hemicellulose contact in the secondary cell wall. This study also detected multiple sets of xylan chemical shifts, many of which differ from the chemical shifts of xylan in solution and in primary cell walls, suggesting that secondary wall xylan chains adopt distinct conformations for function. Future solid-state NMR investigation will be useful for elucidating the differences in the three-dimensional packing, conformations, and mobilities of polysaccharides and other biopolymers in primary and secondary cell walls.

13.4.2 Polysaccharide–Water Interactions

Plant cell walls are highly hydrated materials. How water interacts with wall polysaccharides and proteins gives insights into the structure and dynamics of cell walls. For studying polysaccharide hydration, a useful approach is to select the mobile water ^{1}H magnetization with a relaxation filter or dipolar filter, then transferring the selected magnetization to the more rigid biopolymers. Figure 13.3b shows a 2D ^{13}C–^{1}H correlation spectrum of *Arabidopsis* primary walls, where the ^{1}H dimension mainly exhibits the water ^{1}H signal while the polymer ^{1}H signals[18,77] were suppressed by a ^{13}C–^{1}H dipolar filter. The 2D spectrum shows many polysaccharide-water cross peaks, indicating that the wall polymers are well hydrated. By monitoring the intensity buildup from water, we can compare the relative hydration of various wall polysaccharides.[78] Cellulose shows weaker water cross peaks and slower water transfer than pectins, indicating, not surprisingly, that the glucan chains in the microfibrils are not well hydrated and further the microfibril surface may have limited water exposure due to blockage by matrix polysaccharides.[61] Unexpectedly, partial extraction of the cell wall exerted profound influences on the water-polysaccharide polarization transfer. Removal of Ca^{2+} ions, which chelate homogalacturonan, dramatically slowed down polarization transfer, even though the reduced number of polymer chains is expected to *increase* the water-exposed surface area of the remaining polysaccharides.[78] This result, together with the observed water ^{1}H linewidths, was explained by the fact that the chelating Ca^{2+} ions immobilize water molecules around the polysaccharides, which increase the polarization transfer rates compared to bulk water. Further digestion of hemicellulose restored the water-polysaccharide polarization transfer rates, indicating that the increased water-exposed surface area improves hydration. Thus, the water-polysaccharide polarization transfer experiments give insights into both the amount and the mobility of water in the cell wall network, which are not easily available from bulk physical chemical measurements. The fact that the water-cellulose polarization transfer consistently lags behind water-pectin polarization transfer also supports the single-network model of the plant cell wall, in which pectins and cellulose have direct molecular contact with each other.

13.4.3 Polysaccharide–Protein Interactions

Protein ^{13}C signals can be readily distinguished from polysaccharides ^{13}C signals by chemical shifts and by the presence of ^{15}N labels in the backbone and sidechains. One solid-state NMR strategy for detecting protein–polysaccharide interactions is a ^{15}N–^{13}C dipolar filtered spin diffusion experiment, which selects the protein ^{13}C magnetization by virtue of their dipolar coupling with the amide ^{15}N, then allowing this protein ^{13}C magnetization to transfer to the neighboring polysaccharides *via* distance-dependent ^{13}C–^{13}C spin diffusion.[43] The experiment can be designed to show the protein Cα and CO signals in the indirect dimension and all ^{13}C signals in the direct dimension to assign protein–polysaccharide cross peaks unambiguously. Since all observed ^{13}C signals originate from the protein magnetization, and most carbohydrate-binding proteins in cell walls have low concentrations, such an experiment has very low sensitivity. Even for major structural proteins, their interactions with carbohydrates may be limited, thus evading detection from conventional SSNMR. For example, the arabinogalactan proteins (AGPs) can contain less than ~5% protein and more than ~95% carbohydrate.[79] We overcame this challenge by using dynamic nuclear polarization (DNP), a technique that enhances the NMR sensitivity by one to two orders of magnitude by transferring the electron spin polarization to nuclei.[80–83] Figure 13.3c shows a DNP-enhanced ^{15}N–^{13}C filtered 2D ^{13}C–^{13}C correlation spectrum, obtained on a sample that contains 0.14 mg of expansin and ~14 mg of *Arabidopsis* primary cell wall. The sensitivity enhancement allowed us to show that expansin transfers its magnetization most rapidly to cellulose and this cellulose binding correlates with the wall-loosening activity,[43] while expansin binding to pectins is unrelated to activity. Moreover, the expansin-bound cellulose has a slightly different conformation from bulk cellulose and is enriched in xyloglucan. The latter is in good agreement with biomechanical assays[64,66] showing that only a small fraction of xyloglucan, entrapped by cellulose microfibrils, is load bearing. This DNP–NMR method is generally applicable to studying the binding target of trace amounts of proteins in biomolecular mixtures.[84] This expansin binding study for *Arabidopsis* cell walls used bacterial expansin,[85] which can be expressed and ^{13}C,^{15}N-labeled. In comparison, the binding target of plant expansins in grass cell walls has not been investigated[45,86,87] and will be of interest in the future.

13.5 Perspectives

Multidimensional ^{13}C MAS NMR has provided some of the most significant molecular-level structural and dynamical information of cellulose and matrix polysaccharides in near-native, intact, plant cell walls. We envision that future SSNMR investigations of plant cell walls will broaden the ^{13}C-based experiments to ^{1}H, by taking advantage of the availability of ultrafast MAS probes and increasing magnetic field strengths.[88,89] ^{17}O NMR may also become useful, since DNP can be applied to increase the sensitivity and reduce the requirement of ^{17}O enrichment.[90–93] DNP–NMR can also be used to measure ^{13}C MAS spectra of *unlabeled* plant cell walls as a function of developmental stage to screen chemical compositions and extract conformational information. The ever-enlarging toolbox of SSNMR spectroscopy makes this an indispensable and powerful technique for characterizing the structure of native plant cell walls.

Acknowledgements

This work is supported by the Center for Lignocellulose Structure and Formation, an Energy Frontier Research Center funded by the U.S. Department of Energy, Office of Science, Basic Energy Sciences under Award # DE-SC0001090.

References

1. P. Albersheim, A. Darvill, K. Roberts, R. Sederoff and A. Staehelin, *Plant Cell Walls*, Garland Science, Taylor & Francis Group, LLC, New York, 2011.
2. R. H. Atalla and D. L. Vanderhart, *Science*, 1984, **223**, 283–285.
3. R. H. Atalla and D. L. Vanderhart, *Solid State Nucl. Magn. Reson.*, 1999, **15**, 1–19.
4. R. H. Newman, M. A. Ha and L. D. Melton, *J. Agric. Food Chem.*, 1994, **42**, 1402–1406.
5. M. Dick-Perez, Y. A. Zhang, J. Hayes, A. Salazar, O. A. Zabotina and M. Hong, *Biochemistry*, 2011, **50**, 989–1000.
6. A. Lesage, C. Auger, S. Caldarelli and L. Emsley, *J. Am. Chem. Soc.*, 1997, **119**, 7867–7868.
7. A. Bax, R. Freeman and S. P. Kempsell, *J. Am. Chem. Soc.*, 1980, **102**, 4849–4851.
8. A. E. Bennett, J. H. Ok, R. G. Griffin and S. Vega, *J. Chem. Phys.*, 1992, **96**, 8624–8627.
9. M. Hohwy, C. M. Rienstra, C. P. Jaroniec and R. G. Griffin, *J. Chem. Phys.*, 1999, **110**, 7983–7992.
10. M. Hong, *J. Magn. Reson.*, 1999, **136**, 86–91.
11. T. Wang and M. Hong, *J. Exp. Bot.*, 2016, **67**, 503–514.
12. S. Li, Y. Zhang and M. Hong, *J. Magn. Reson.*, 2010, **202**, 203–210.
13. M. G. Munowitz, R. G. Griffin, G. Bodenhausen and T. H. Huang, *J. Am. Chem. Soc.*, 1981, **103**, 2529–2533.
14. M. Hong, J. D. Gross, C. M. Rienstra, R. G. Griffin, K. K. Kumashiro and K. Schmidt-Rohr, *J. Magn. Reson.*, 1997, **129**, 85–92.
15. E. R. deAzevedo, K. Saalwachter, O. Pascui, A. A. de Souza, T. J. Bonagamba and D. Reichert, *J. Chem. Phys.*, 2008, **128**, 104505.
16. J. K. Williams and M. Hong, *J. Magn. Reson.*, 2014, **247**, 118–127.
17. D. Huster, X. L. Yao and M. Hong, *J. Am. Chem. Soc.*, 2002, **124**, 874–883.
18. S. Li, Y. Su, W. Luo and M. Hong, *J. Phys. Chem. B*, 2010, **114**, 4063–4069.

19. X. L. Yao, K. Schmidt-Rohr and M. Hong, *J. Magn. Reson.*, 2001, **149**, 139–143.
20. W. Luo and M. Hong, *J. Am. Chem. Soc.*, 2010, **132**, 2378–2384.
21. A. Krushelnitsky, D. Reichert and K. Saalwächter, *Acc. Chem. Res.*, 2013, **46**, 2028–2036.
22. J. R. Lewandowski, *Acc. Chem. Res.*, 2013, **46**, 2018–2027.
23. A. G. Palmer, J. Williams and A. McDermott, *J. Phys. Chem.*, 1996, **100**, 13293–13310.
24. M. Hong, Y. Zhang and F. Hu, *Annu. Rev. Phys. Chem.*, 2012, **63**, 1–24.
25. M. Jarvis, *Nature*, 2003, **426**, 611–612.
26. M. Dick-Perez, T. Wang, A. Salazar, O. A. Zabotina and M. Hong, *Magn. Reson. Chem.*, 2012, **50**, 539–550.
27. J. Vogel, *Curr. Opin. Plant Biol.*, 2008, **11**, 301–307.
28. P. T. Larsson, E. L. Hult, K. Wickholm, E. Pettersson and T. Iversen, *Solid State Nucl. Magn. Reson.*, 1999, **15**, 31–40.
29. L. H. Thomas, V. T. Forsyth, A. Sturcova, C. J. Kennedy, R. P. May, C. M. Altaner, D. C. Apperley, T. J. Wess and M. C. Jarvis, *Plant Physiol.*, 2013, **161**, 465–476.
30. A. N. Fernandes, L. H. Thomas, C. M. Altaner, P. Callow, V. T. Forsyth, D. C. Apperley, C. J. Kennedy and M. C. Jarvis, *Proc. Natl. Acad. Sci. U. S. A.*, 2011, **108**, E1195–E1203.
31. P. T. Larsson and P. O. Westlund, *Spectrochim. Acta, Part A*, 2005, **62**, 539–546.
32. H. Kono, S. Yunoki, T. Shikano, M. Fujiwara, T. Erata and M. Takai, *J. Am. Chem. Soc.*, 2002, **124**, 7506–7511.
33. D. L. Vanderhart and R. H. Atalla, *ACS Symp. Ser.*, 1987, **340**, 88–118.
34. H. Kono, T. Erata and M. Takai, *Macromolecules*, 2003, **36**, 5131–5138.
35. H. Kono and Y. Numata, *Cellulose*, 2006, **13**, 317–326.
36. Y. Nishiyama, U. J. Kim, D. Y. Kim, K. S. Katsumata, R. P. May and P. Langan, *Biomacromolecules*, 2003, **4**, 1013–1017.
37. Y. Nishiyama, P. Langan and H. Chanzy, *J. Am. Chem. Soc.*, 2002, **124**, 9074–9082.
38. D. J. Cosgrove, *Curr. Opin. Plant Biol.*, 2014, **22C**, 122–131.
39. G. Guerriero, J. Fugelstad and V. Bulone, *J. Integr. Plant Biol.*, 2010, **52**, 161–175.
40. R. H. Newman, L. M. Davies and P. J. Harris, *Plant Physiol.*, 1996, **111**, 475–485.
41. R. H. Newman, S. J. Hill and P. J. Harris, *Plant Physiol.*, 2013, **163**, 1558–1567.
42. R. J. Vietor, R. H. Newman, M. A. Ha, D. C. Apperley and M. C. Jarvis, *Plant J.*, 2002, **30**, 721.
43. T. Wang, Y. B. Park, M. A. Caporini, M. Rosay, L. H. Zhong, D. J. Cosgrove and M. Hong, *Proc. Natl. Acad. Sci. U. S. A.*, 2013, **110**, 16444–16449.
44. S. McQueen-Mason, D. M. Durachko and D. J. Cosgrove, *Plant Cell*, 1992, **4**, 1425–1433.
45. D. J. Cosgrove, *Nature*, 2000, **407**, 321–326.
46. D. J. Cosgrove, *Plant Physiol.*, 2001, **125**, 131–134.

47. D. J. Cosgrove, *Nat. Rev. Mol. Cell Biol.*, 2005, **6**, 850–861.
48. D. M. Harris, K. Corbin, T. Wang, R. Gutierrez, A. L. Bertolo, C. Petti, D. M. Smilgies, J. M. Estevez, D. Bonetta, B. R. Urbanowicz, D. W. Ehrhardt, C. R. Somerville, J. K. C. Rose, M. Hong and S. DeBolt, *Proc. Natl. Acad. Sci. U. S. A.*, 2012, **109**, 4098–4103.
49. N. Sorek, H. Sorek, A. Kijac, H. J. Szemenyei, S. Bauer, K. Hematy, D. E. Wemmer and C. R. Somerville, *J. Biol. Chem.*, 2014, **289**, 34911–34920.
50. T. Mori, E. Chikayama, Y. Tsuboi, N. Ishida, N. Shisa, Y. Noritake, S. Moriya and J. Kikuchi, *Carbohydr. Polym.*, 2012, **90**, 1197–1203.
51. J. D. Kubicki, M. N. A. Mohamed and H. D. Watts, *Cellulose*, 2013, **20**, 9–23.
52. J. D. Kubicki, H. D. Watts, Z. Zhao and L. H. Zhong, *Cellulose*, 2014, **21**, 909–926.
53. M. Gonneau, T. Desprez, A. Guillot, S. Vernhettes and H. Hofte, *Plant Physiol.*, 2014, **166**, 1709–1712.
54. J. L. Hill, M. B. Hammudi and M. Tien, *Plant Cell*, 2014, **26**, 4834–4842.
55. M. C. Jarvis, *Plant Physiol.*, 2013, **163**, 1485–1486.
56. T. Komatsu and J. Kikuchi, *J. Phys. Chem. Lett.*, 2013, **4**, 2279–2283.
57. B. H. Meier, *Adv. Magn. Opt. Reson.*, 1994, **18**, 1–116.
58. J. N. Dumez and L. Emsley, *Phys. Chem. Chem. Phys.*, 2011, **13**, 7363–7370.
59. T. Wang, O. Zabotina and M. Hong, *Biochemistry*, 2012, **51**, 9846–9856.
60. T. Wang, J. K. Williams, K. Schmidt-Rohr and M. Hong, *J. Biomol. NMR*, 2015, **61**, 97–107.
61. T. Wang, Y. B. Park, D. J. Cosgrove and M. Hong, *Plant Physiol.*, 2015, **168**, 871.
62. A. Zykwinska, J. F. Thibault and M. C. Ralet, *J. Exp. Bot.*, 2007, **58**, 1795–1802.
63. A. Peaucelle, S. Braybrook and H. Hofte, *Front. Plant Sci.*, 2012, **3**, 121.
64. Y. B. Park and D. J. Cosgrove, *Plant Physiol.*, 2012, **158**, 1933–1943.
65. Y. B. Park and D. J. Cosgrove, *Plant Physiol.*, 2012, **158**, 465–475.
66. Z. Zhao, V. H. Crespi, J. D. Kubicki, D. J. Cosgrove and L. H. Zhong, *Cellulose*, 2014, **21**, 1025–1039.
67. D. M. Cavalier, O. Lerouxel, L. Neumetzler, K. Yamauchi, A. Reinecke, G. Freshour, O. A. Zabotina, M. G. Hahn, I. Burgert, M. Pauly, N. V. Raikhel and K. Keegstra, *Plant Cell*, 2008, **20**, 1519–1537.
68. T. Wang, A. Salazar, O. A. Zabotina and M. Hong, *Biochemistry*, 2014, **53**, 2840–2854.
69. N. C. Carpita, *Phytochemistry*, 1984, **23**, 1089–1093.
70. N. C. Carpita, *Annu. Rev. Plant Physiol.*, 1996, **47**, 445–476.
71. M. Bardet, L. Emsley and M. Vincendon, *Solid State Nucl. Magn. Reson.*, 1997, **8**, 25–32.
72. R. Dupree, T. J. Simmons, J. C. Mortimer, D. Patel, D. Iuga, S. P. Brown and P. Dupree, *Biochemistry*, 2015, **54**, 2335–2345.
73. H. L. Wheeler, R. Soong, D. Courtier-Murias, A. Botana, B. Fortier-Mcgill, W. E. Maas, M. Fey, H. Hutchins, S. Krishnamurthy, R. Kumar, M. Monette, H. J. Stronks, M. M. Campbell and A. Simpson, *Magn. Reson. Chem.*, 2015, **53**, 735–744.

74. D. J. Cosgrove and M. C. Jarvis, *Front. Plant Sci.*, 2012, **3**, 204.
75. L. Salmen and E. Bergstrom, *Cellulose*, 2009, **16**, 975–982.
76. L. Donaldson, *Wood Sci. Technol.*, 2007, **41**, 443–460.
77. X. L. Yao and M. Hong, *J. Biomol. NMR*, 2001, **20**, 263–274.
78. P. B. White, T. Wang, Y. B. Park, D. J. Cosgrove and M. Hong, *J. Am. Chem. Soc.*, 2014, **136**, 10399–10409.
79. B. B. Buchanan, W. Gruissem and R. L. Jones, *Biochemistry and Molecular Biology of Plants*, American Society of Plant Physiologists, Rockville, MD, 2000.
80. Y. Su, L. Andreas and R. G. Griffin, *Annu. Rev. Biochem.*, 2015, **84**, 465–497.
81. D. A. Hall, D. C. Maus, G. J. Gerfen, S. J. Inati, L. R. Becerra, F. W. Dahlquist and R. G. Griffin, *Science*, 1997, **276**, 930–932.
82. Q. Z. Ni, E. Daviso, T. V. Can, E. Markhasin, S. K. Jawla, T. M. Swager, R. J. Temkin, J. Herzfeld and R. G. Griffin, *Acc. Chem. Res.*, 2013, **46**, 1933.
83. T. V. Can, Q. Z. Ni and R. G. Griffin, *J. Magn. Reson.*, 2015, **253**, 23–35.
84. H. Ivanir and A. Goldbourt, *J. Biomol. NMR*, 2014, **59**, 185–197.
85. N. Georgelis, A. Tabuchi, N. Nikolaidis and D. J. Cosgrove, *J. Biol. Chem.*, 2011, **286**, 16814–16823.
86. N. H. Yennawar, L. C. Li, D. M. Dudzinski, A. Tabuchi and D. J. Cosgrove, *Proc. Natl. Acad. Sci. U. S. A.*, 2006, **103**, 14664–14671.
87. A. Tabuchi, L. C. Li and D. J. Cosgrove, *Plant J.*, 2011, **68**, 546–559.
88. S. Asami, P. Schmieder and B. Reif, *J. Am. Chem. Soc.*, 2010, **132**, 15133–15135.
89. I. Bertini, L. Emsley, M. Lelli, C. Luchinat, J. F. Mao and G. Pintacuda, *J. Am. Chem. Soc.*, 2010, **132**, 5558–5559.
90. F. A. Perras, T. Kobayashi and M. Pruski, *J. Am. Chem. Soc.*, 2015, **137**, 8336–8339.
91. F. Blanc, L. Sperrin, D. A. Jefferson, S. Pawsey, M. Rosay and C. P. Grey, *J. Am. Chem. Soc.*, 2013, **135**, 2975–2978.
92. T. H. Sefzik, J. B. Houseknecht, T. M. Clark, S. Prasad, T. L. Lowary, Z. Gan and P. J. Grandinetti, *Chem. Phys. Lett.*, 2007, **434**, 312–315.
93. V. K. Michaelis, E. G. Keeler, T. C. Ong, K. N. Craigen, S. Penzel, J. E. C. Wren, S. Kroeker and R. G. Griffin, *J. Phys. Chem. B*, 2015, **119**, 8024–8036.

CHAPTER 14

New Methods for the Analysis of Heterogeneous Polysaccharides – Lessons Learned from the Heparin Crisis

TIMOTHY R. RUDD*[a], EDWIN A. YATES*[b] AND MARCO GUERRINI*[c]

[a]National Institute for Biological Standards and Control, Blanche Lane, South Mimms, Potters Bar, Hertfordshire, EN6 3QG UK; [b]Department of Biochemistry, Institute of Integrative Biology, University of Liverpool, Crown Street, Liverpool, L69 7ZB UK; [c]Istituto di Ricerche Chimiche e Biochimiche "*G. Ronzoni*", *Via* Giuseppe Colombo 81, Milano, 20133 Italy
*E-mail: tim.rudd@nibsc.org, eayates@liv.ac.uk, guerrini@ronzoni.it

14.1 Heparin

14.1.1 What Is Heparin?

The anionic linear glycosaminoglycan (GAG) polysaccharide, heparin, is one of the principal pharmaceutical agents used in the treatment and prevention of blood clots or thrombi, and consequently finds widespread application in surgery, the treatment of heart attacks, the prevention of surface contact-induced clotting during blood dialysis and as a coating material for a range of medical devices. Heparin exerts antithrombotic and anticoagulant activities

New Developments in NMR No.10
NMR in Glycoscience and Glycotechnology
Edited by Koichi Kato and Thomas Peters

Published by the Royal Society of Chemistry, www.rsc.org

principally *via* two coagulation enzymes: factor-Xa and thrombin (factor IIa), working through the activation of antithrombin (AT), which is considered as one of the major physiological coagulation inhibitors. While the acceleration of inhibition of factor Xa (anti-Xa activity) requires only a specific pentasaccharide sequence, the potentiation of thrombin inhibition (anti-IIa activity) requires a minimum total chain length of 18 saccharides.[1] Heparin, however, is not a single defined product. Pharmaceutical heparin is obtained almost exclusively from porcine intestinal mucosa, one pig gut yielding in the order of one dose of heparin. Although there is a predominant repeating unit and many similarities between batches, each sample of heparin is, nevertheless, composed of numerous polysaccharide chains that have distinct sequences and lengths. Despite this variability, porcine intestinal mucosa heparin remains a reliable and very widely used anticoagulant. The history of heparin has been discussed by Barrowcliffe and Casu,[2,3] while the general properties of the polysaccharide are reviewed by Rabenstein.[4]

14.1.2 The Structure of Heparin

The polysaccharide backbone of heparin and HS is composed of alternating α-D-glucosamine (**GlcN**) residues 1-4 linked to β-D-glucuronic acid (**G**) or α-L-iduronic acid (**I**) residues. Glucosamine can be *N*-sulfated (**GlcNS**) or *N*-acetylated (**GlcNAc**) at position 2, *O*-sulfated at position 6 (**Glc6S**) and more rarely at position 3 (**GlcNS,3S**). The uronic acid residues, **I** and more rarely **G**, can be 2-*O*-sulfated (**I2S/G2S**). Even though in heparin the most prevalent disaccharide is **-I2S-GlcNS,6S-**, short, under-sulfated domains and more complex sequences are present along the heparin chains. The most significant sequence is the pentasaccharide -GlcNAc,6S-G-GlcNS,3S,6S-I2S-ANS,6S- (**AGA*IA**), that exhibits high affinity for antithrombin,[5] the serine protease inhibitor of several proteases of the coagulation system (mainly thrombin and factor-Xa). Recent work by the authors though has shown that extension of this sequence can produce binding sites that have a greater affinity for antithrombin than the canonical pentasaccharide.[6] Another unsulfated sequence is the linkage region, located at the reducing end of the polymer, the site of attachment through a serine residue to the peptide core of the original proteoglycan.[7] A schematic representation of the structure of heparin and heparan sulfate can be found in Figure 14.1.

Further minor features, strictly arising from the purification process and not related to the biosynthesis of heparin, can be present in commercial heparin. Oxidation under alkaline conditions, which is used to produce a material without colour that does not cause fever on administration, may also induce oxidation at the reducing end of the carbohydrate, particularly the serine and **GlcNAc** residues with the formation of glycoserine derivatives[9] and 2-*N*-acetyl-2-deoxy-glucono-1,5-lactone,[10] respectively. The alkaline conditions may also contribute to the selective epoxidation between C2 and C3 of the 2-*O*-sulfated iduronic acid residue with formation of 2,3-anhydro-α-L-guluronic acid.[11]

A number of subtle structural features distinguish heparin from different animal sources or different organs of the same animal. Notably, bovine

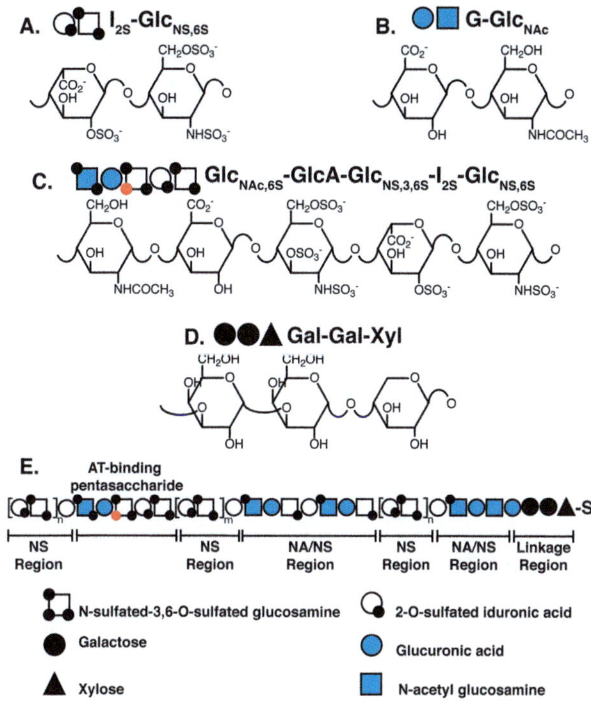

Figure 14.1 *The structure of heparan sulfate and heparin.* **A.** The major repeating disaccharide of heparin. **B.** One of the rarest disaccharides found in heparin and a prominent component of the NA regions of heparan sulfate. **C.** The **AGA*IA** antithrombin binding pentasaccharide sequence. **D.** The linkage region that attaches the polysaccharide to its protein core.[7] **E.** The domain structure of heparan sulfate, illustrating the three different regions proposed to exist in the chain, NA, NA–NS and NS regions.[8]

lung heparin, historically commonly used in therapy, contains much more of the trisulfated disaccharide (**-I2S-ANS,6S-**) than porcine mucosal heparin, while bovine mucosal heparin exhibits lower sulfation at the 6-*O* position of glucosamine residues than either bovine lung or porcine mucosal heparin,[12–14] as well as more sulfation at the 2-*O* position of glucuronic acid residue.[15] Finally, recent work has demonstrated the existence of different AT-binding pentasaccharide variants with diverse non-reducing and reducing end extensions.[6,16–19] Their molecular ratios provide further fingerprints of the structural diversity among heparins of different origins.[14,15,20]

14.1.3 Biosynthesis

The biosynthesis of heparin occurs in the Golgi compartment of mast cells. Following production heparin is stored in the cytoplasmic granules of the cell as serglycin, along with cytokines, proteases and histamine. The biosynthesis starts with the formation of the tetrasaccharide (D-GlcA-β(1 → 3)-D-Gal-β(1 → 3)-D-Gal-β(1 → 3)-D-Xyl-β(1→), which is linked to a serine residue on the core protein. The chain is then elongated by addition of alternating 1-4 linked β D-glucuronate and α *N*-acetyl-D-glucosamine residues from their respective UDP-sugars. The elongation of the chain leads to the formation of a polysaccharide with a molecular weight of 60–100 kDa, whereas, the molecular weight range of commercial heparin lies between 15–20 kDa. During the elongation of the polysaccharide the juvenile heparin chains are modified by a series of enzymes, resulting in a highly sulfated polysaccharide. The maturation of the chain involves a large number of enzymes:

- De-*N*-acetylation and *N*-sulfation of *N*-acetyl-glucosamine residues by de-*N*-acetyl-sulfotransferases (NDST).
- C5-epimerisation of D-GlcA to L-IdoA by C5-epimerase.
- 2-*O*-sulfation of the resulting L-IdoA residues and, to a lesser extent, GlcA units by 2-*O*-sulfotransferase.
- 6-*O*-sulfation of the D-glucosamine residues by 6-*O*-sulfotransferases.
- Finally, the 3-*O*-sulfation of D-glucosamine residue by 3-*O*-sulfotransferase in the biosynthesis of the antithrombin-binding pentasaccharide of heparin and elsewhere.

None of these reactions are template driven or reach completion, providing structural differences in the final chains. The substrate specificity of the various enzymes, often present in several isoforms, varies according to the organ and the animal species, but the causes of these differences are not yet known.[21] The enzyme pathway illustrated in Figure 14.2 is the orthodox representation of the biosynthesis of HS and heparin, with the enzymes acting on the chain in a linear fashion. The authors proposed a more detailed scheme routing, in which the synthesis of the common and rare disaccharides occur by two distinct branched pathways, allowing the full diversity of HS–heparin disaccharides to be produced.[22]

14.1.4 Source of Heparin

The original tissue from which heparin was extracted was dog liver but, this source was deemed unsuitable for large-scale production. The use, first of bovine liver, then of bovine lung, providing higher yields, allowed more of the drug to be produced, which was marketed in 1939 as a pharmaceutical product in the USA. Bovine lung continued as a source of heparin until the mid-1950s, when porcine and bovine intestinal mucosa were introduced as alternative sources. Porcine mucosal heparin was subsequently found to be a cleaner and less expense source tissue, requiring fewer steps in the extraction and purification processes than bovine lung heparin.[2,3] The appearance of bovine spongiform encephalopathy (BSE) in Europe in the 1990s and the concerns of prion contamination decreased the worldwide production and use of bovine heparin drastically, owing to fears of 'mad cow disease', and of ovine heparin, following concerns over scrapie.[23] The production of bovine

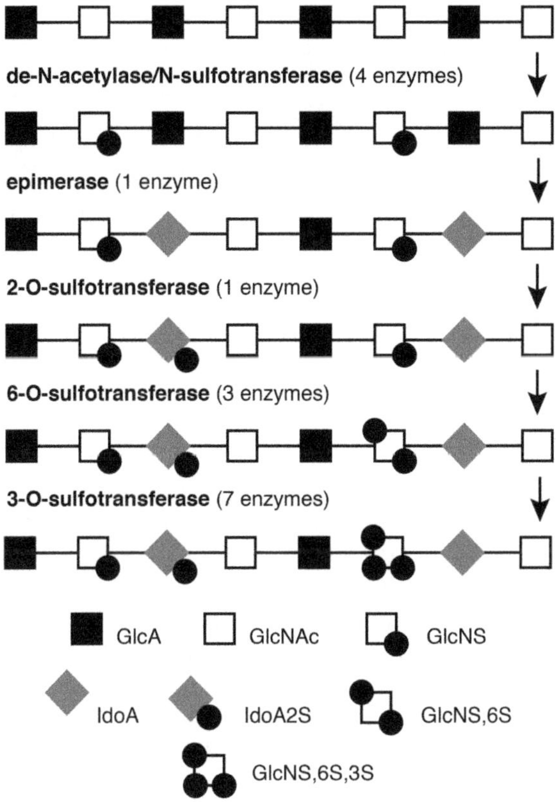

Figure 14.2 *Schematic of the complex biosynthesis of heparan sulfate and heparin.* This representation shows the enzymatic modifications that occur after the chain has been initiated; this occurs at a serine through the attachment of a xylose group and the subsequent addition of two galactose resides and finally the linkage region is finished by the addition of glucuronic acid. Then elongated; after the tetrasaccharide linker the chain is extended in a stepwise manner, with the addition of glucuronic acid and *N*-acetyl glucosamine, which is carried out by a complex that is composed of the proteins exostosin-1 and -2 (EXT1 and EXT2).

heparin is currently restricted to South America and some Islamic countries, while ovine heparin use is confined almost exclusively to laboratory studies (Figure 14.3).

At the time of writing (2015/2016) the Food and Drug Administration (FDA) is considering the reintroduction of bovine heparin as a drug product to the US market, to help diversify the supply chain and minimise the possibility of potential drug shortages that are associated with utilising a single animal species as the primary source of this major life saving drug.[24]

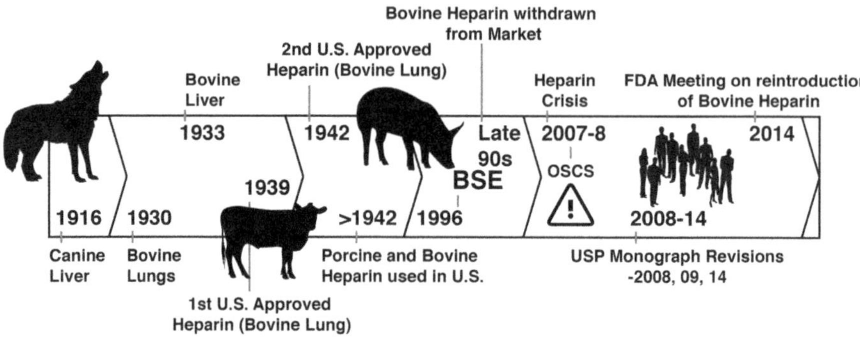

Figure 14.3 *History of pharmaceutical heparin in the United States of America.*

14.1.5 Manufacturing Process

The main biological source for pharmaceutical-grade heparin is porcine or bovine intestinal mucosa. The extraction of heparin usually occurs at the pig slaughtering facility from intestinal mucosa or the whole intestine ("hashed pork guts"), yielding so-called "crude heparin". The material often contains other polysaccharides, including heparan sulfate (HS), chondroitin sulfate (CS), dermatan sulfate (DS), as well as residual nucleic acids. Heparin is then purified using various methods that take advantage of its high anionic charge density (2.2–2.7 sulfate groups per disaccharide unit), aqueous solubility and generally good stability to heat, acids, alkalis and solvents. The purification process includes several physical and chemical treatments that are necessary to remove heavy metals, residual proteins or nucleotides, solvents, other GAG polysaccharides (DS, CS, HS), bacterial endotoxins, and viruses.[25] According to the process used to extract and purify heparin, the final API product can vary in its molecular weight distribution, sulfation degree or potency.[25] It is worth noting that the average mean molecular weight (Mw) of heparin ranges from 15 to 19 kDa, with a polydispersity (Mw/Mn) of 1.1–1.3, whereas the sulfation degree for disaccharide can vary between 2.2 for bovine mucosal heparin to 2.3–2.5 for porcine and ovine mucosal heparin, and can reach 2.7 for bovine lung heparin.[20,26]

14.2 Low Molecular-Weight Heparins

The low molecular weight heparins (LMWHs), derived from heparin through different depolymerisation processes, possess several therapeutic advantages and reduced side effects compared to heparin, maintaining substantially similar monosaccharide composition and oligosaccharide sequences to heparin.[27] Owing to their lower molecular weight compared to the parent heparins, LMWHs can claim a more predictable pharmacological profile, better bioavailability and longer half-life.[28] The structural heterogeneity that is biosynthetically imprinted, as well as the introduction of additional chemical diversity induced by the depolymerisation process, make their structure even more complex than that of heparin.[29]

14.2.1 Production of LMWHs

Three major strategies are used to prepare LMWHs. These are depolymerisation by chemical deamination and β-elimination, either enzymatically or chemically (Figure 14.4).

In the first method, heparin undergoes nitrosylation at the amino group of *N*-sulfo-glucosamine. The unstable *N*-nitroso-sulfonamide residues rearrange to generate a carbocation at C2. Subsequent ring-contraction and hydrolysis of the adjacent glycosidic bond generate an anhydromannose residue that is stabilised by reduction with sodium borohydride to form a terminal anhydromannitol residue[30] (Figure 14.4).

The chemical β-eliminative method to produce enoxaparin is based on cleavage, in alkaline medium, of the benzyl ester of heparin, which is formed by treatment of the heparin benzethonium salt with benzyl chloride.[31] Under these conditions, β-elimination generates an unsaturated uronate residue at the non-reducing end and an *N*-sulfo-glucosamine at the reducing end. The alkaline hydrolysis of the benzyl ester of heparin generates two unique bicyclic structures, 2-sulfo-amino-1,6-anhydro-2-deoxy-β-D-glucopyranose and 2-sulfo-amino-1,6-anhydro-2-deoxy-β-D-mannopyranose at the reducing end, which can be used as markers for enoxaparin[32] (Figure 14.4).

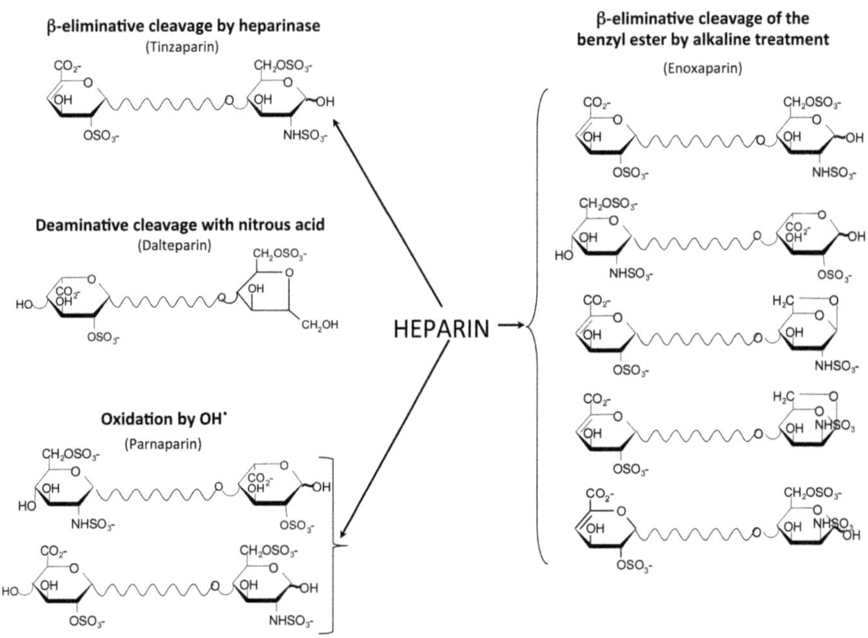

Figure 14.4 *A summary of the various methods used to produce low-molecular weight heparins.* The different methods produce distinct non-reducing and reducing end modifications that can assist in the identification of the various products.

The enzymatic depolymerisation, through β-elimination, used to prepare tinzaparin, also generates a unique non-reducing residue, in this circumstance a 4,5 unsaturated uronate. However, this residue is mostly 2-*O*-sulfated owing to the selectivity of the enzyme (heparinase I) for the sequence[33] **-GlcNS-I2S-** (Figure 14.4).

14.3 Heparin Crisis—2008 *OSCS Contamination*

The urgent need for improved quality control and analysis of pharmaceutical heparin was emphasised by the heparin contamination incidents in 2007–2008. In November 2007, the Children's Hospital in St. Louis, began noticing serious allergic reactions and low blood pressure in their dialysis patients. This led to the creation of a multidisciplinary team assembled from the FDA, academia and industry, including international experts who, by early March 2008, had identified the major contaminant as a modified polysaccharide, over-sulfated chondroitin sulfate (OSCS),[34,35] that had been added to batches of heparin from Baxter at some point in the supply chain. This contamination officially resulted in 94 associated deaths and 574 adverse reactions in the US with broader adverse events recorded worldwide. One of the main factors that made this adulteration possible was the complexity of the international heparin supply chain. Most of the workshops for crude heparin production are remote from the location of subsequent drug production, being located in China, and many cannot guarantee complete traceability from the animal to the intermediate product (crude heparin).

In the case of OSCS contamination, crude heparin was sold to Scientific Protein Laboratories (SPL), at that time an American company, who manufactured the purified heparin product and had one plant in China and another in Wisconsin, USA. This heparin, Active Pharmaceutical Ingredient (API) was then sold to Baxter, who manufactured the finished heparin products. It should be recalled that, before the crisis, the heparin monograph employed the potency assay as the main method of heparin testing. The addition of OSCS, which also possesses anticoagulant properties, did not alter this characteristic, allowing the contaminated product to pass the control procedures imposed by the then existing pharmacopoeia rules. Following the crisis, both the United States and the European Pharmacopoeias proposed defined analytical and bio-analytical methods to test the properties and purity of heparin sodium. The key tests within the revised monographs were 1-dimensional ^{1}H NMR spectroscopy, which establishes regions within the spectrum of a heparin sample that must be free of any unidentified signal above a defined threshold, and capillary electrophoresis, later substituted by HPLC analysis, which verifies that no peaks corresponding to over-sulfated chondroitin sulfate are detected as eluting later than the peak corresponding to heparin. These tests, which were augmented recently by the determination of molecular weight, together with the requirement for increased potency, collectively reduce the possibility of heparin contamination, making heparin control more secure than in the past. Figure 14.5 contains an example comparing

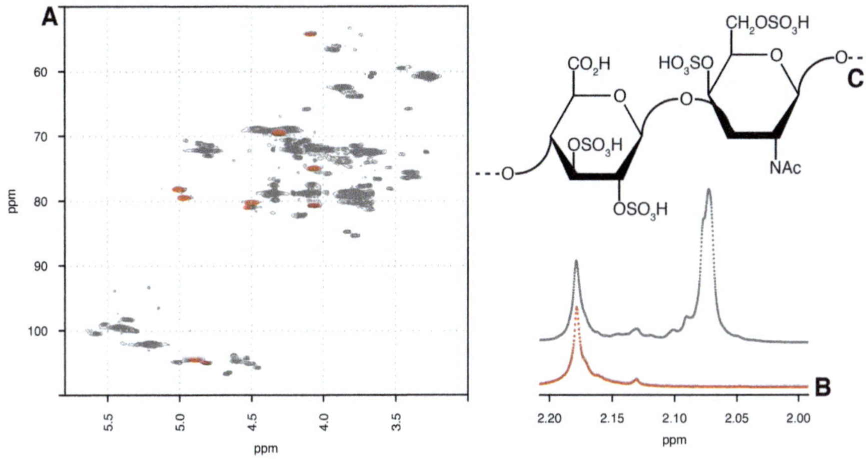

Figure 14.5 Panel **A.** shows the superimposition of the HSQC spectra of a contaminated heparin sample (grey) and the isolated contaminant OSCS (red). The *N*-acetyl region of 1H NMR clearly shows that this region of the spectrum can be used to identify whether heparin is adulterated with OSCS or not (**B**). The major repeating disaccharide unit of OSCS is shown in panel **C**.

the HSQC and 1H NMR of heparin contaminated with OSCS and the pure adulterant itself.

Following the challenge posed by the contamination of heparin with OSCS, which had been met initially by modification of the regulatory requirements, there also arose a desire to refine those methods to allow much more subtle, or challenging (*vide infra*) contaminants, to be detected as well. While OSCS is, relatively speaking, easy to detect by ^{1}H NMR spectroscopy once its presence is suspected, since it presents ^{1}H NMR signals that lie outside the regions in which the signals of heparin fall, the same cannot be said of all other potential contaminants. It is even possible to produce sulfated polysaccharides, whose ^{1}H NMR signals lie entirely under those of heparin, as a test of the analysis. Indeed, methods capable of meeting these challenges have also been developed,[36,37] and will be described in Sections 14.4.1 and 14.4.2. An even more challenging problem concerns the possibility of detecting the presence of heparin from a different animal origin within a pharmaceutical heparin preparation. This problem, the most demanding of all, has also been met[38] and is described.

14.3.1 Conventional NMR Solutions to Heparin Safety

14.3.1.1 Current Regulations

The main regulatory agencies, EDQM and USP, defined the new heparin monographs, which include new requirements for porcine heparin. Even though the number of required tests is similar, slightly different approaches are used from the two regulatory agencies. The tests, at present required from the two pharmacopoeias, are summarised in Table 14.1 in comparison with those present in the USP monograph before the heparin crisis. Although the new tests introduced in the monographs enhance the safety and control of the heparin supply chain, a harmonisation of specifications and standards worldwide would be strongly recommended. Moreover, the possible reintroduction of bovine heparin in the US market will require two distinct monographs (*e.g.* bovine and porcine heparin), which should include a test for tissue and species identification.

14.3.1.2 *Quantitative 2D Heteronuclear Single Quantum Correlation NMR*

The heteronuclear single quantum correlation spectroscopy (HSQC) NMR experiment allows overlapping signals in the directly measured ^{1}H domain to be dispersed in an additional dimension, carbon or nitrogen[42] (Figure 14.6). This experiment is used widely for carbohydrate characterisation since their mono-dimensional spectra exhibit many overlapping proton signals, especially in the non-anomeric region. Recently, HSQC experiments were applied to the quantitative analysis by volume integration of the corresponding cross-peaks, selecting those in which the effect of $^{1}J_{CH}$ couplings are minimal and differences in transverse relaxation times (T_2) are negligible.[43] The monosaccharide and disaccharide content (as percentages) of heparin and LMWHs were calculated by normalising volumes with reference to the sum of volumes of signals corresponding to the same monosaccharide type (glucosamine or uronic acid).[39,40] The advantage of this method compared to HPLC disaccharide analysis after enzymatic digestion is that it is quicker (from 2 up to 12 hours, depending on the probe sensitivity), furthermore, it does not require sample manipulation, *i.e.*, it is non-destructive. The FDA has recently accepted HSQC NMR as a physico-chemical comparison of the innovator and the generic version of enoxaparin.[41] The results from this method showed that the precision of detection can be lower than 5% for the most abundant fragments and less than 10% for all fragments more intense than the limit of quantification (LOQ), defined as the percentage of residue which corresponding signals have a signal to noise ratio of 10.[67]

For completeness, other multidimensional NMR experiments have been applied to the problem of heparin safety, notably NOESY[44] and DOSY (Diffusion-ordered spectroscopy),[45,46] as heparin is heterogeneous in length as well as composition DOSY, where proton signals are separated in a second dimension which indicates the diffusion coefficients of the proton, maybe of interest when monitoring heparin from different sources and the chemical modification–depolymerisation of the material.

Table 14.1 US and European Pharmacopoeia requirements for porcine heparin before and after the heparin crisis.

Monograph section	Test	USP before crisis	USP	EDQM
Identification	1H-NMR	No present	Magnetic field NLT 500 MHz No unidentified signals greater than 4% of the mean of signal height of 1 and 2 are present in the following ranges: 0.10–2.00, 2.10–3.20, and 5.73 0–8.00 ppm. No signals greater than 200% signal height of the mean of the signal height of 1 and 2 are present in the 3.75–4.55 ppm for porcine heparin.	Magnetic field NLT 300 MHz No unidentified signals greater than 4% compared to the heparin signal at 5.42 ppm are present in the following ranges: 0.10–2.00, 2.10–3.20, and 5.70–8.00 ppm.
	Chromatographic identity by SAX–HPLC	No present	The retention time of the major peak from the sample solution corresponds to that of the standard solution.	The retention time of the major peak from the sample solution corresponds to that of the standard solution.
	Molecular weight	No present	M_{24000} is NMT 20%; Mw is between 15 kDa and 19 kDa; ratio of $M_{8000-16000}$ to $M_{16000-24000}$ is NLT 1	In preparation
	Sodium	Requirements of flame test for sodium	Requirements of flame test for sodium	Requirements of flame test for sodium
Assay	Assay of heparin potency	NLT 140 USP heparin units/mg (clotting)	NLT 180 USP heparin units/mg anti-IIa assay	NLT 180 EU heparin units/mg anti-IIa assay
	Activity	89 to 120 (anti-Xa)	Anti-Xa/anti-IIa ratio (0.9–1.1)	The estimated potency is NLT 90% and NMT 111% of the stated potency.
Impurities	Organic impurities	No present	Galactosamine in total hexosamine (NMT 1%)	Not more than the area of the corresponding peak in the chromatogram obtained with reference solution (2%)
	Absence of OSCS	No present	References Identification	No identification
	Nucleotidic and protein impurities	No turbidity observed	Protein: NMT 0.1% Nucleotidic impurities: NMT 0.1%	Protein: NMT 0.5% Nucleotidic impurities: the absorbance at 260 nm not greater than 0.15
	Bacterial Endotoxins	NML 0.03 USP Unit	NML 0.01 USP International Unit	NML 0.01 International Unit
Other components	Nitrogen Determination	No present	1.3%–2.5%	1.5%–2.5%

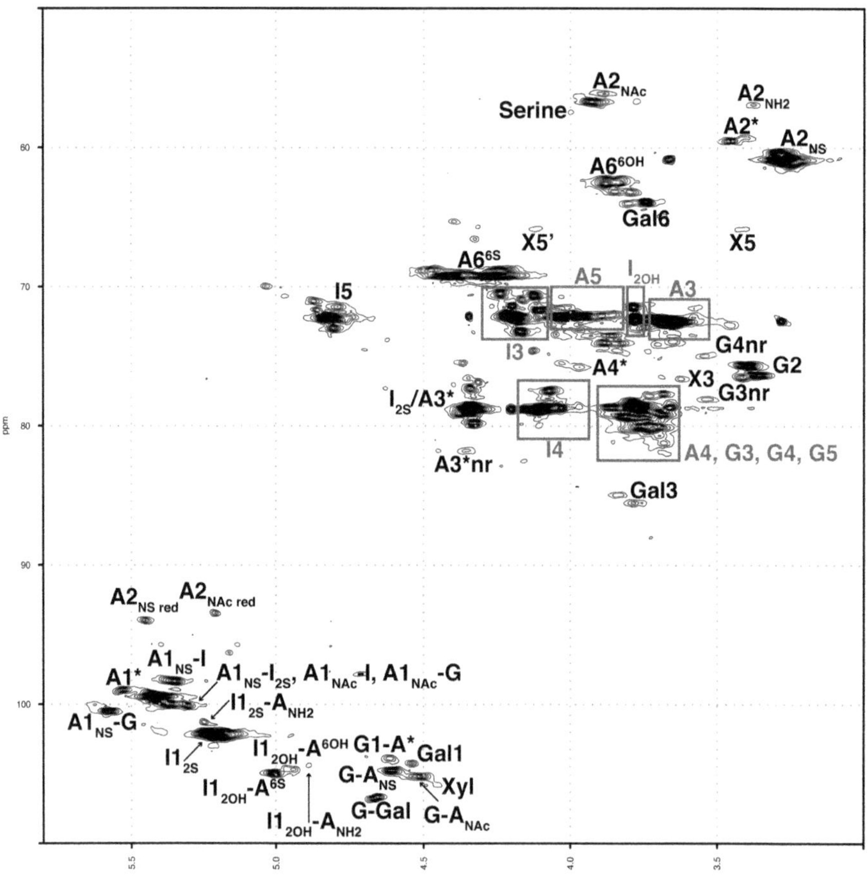

Figure 14.6 *HSQC NMR spectrum of heparin*; many of these signals are involved in the HSQC analysis of heparin, for further details see ref. 39–40.

14.3.2 Conventional Chemometric–NMR Solutions to Heparin Safety

14.3.2.1 The Problem: Heterogeneous Material

How to Define an Innately Variable Material? A paper by Rudd *et al.*,[37] introduced the key change in perception and definition of the problem, which follows the realisation that there is no unique, accepted definition of "heparin", in a way that the structural definition of a conventional small molecule defines a typical pharmaceutical agent (*e.g.* paracetamol). Seeking such a definition, or continuing to apply such approaches, as though the polydisperse and heterogeneous product, heparin, can be considered in this way is, therefore, not only destined to be fruitless, it is nonsensical.

Heparin is not only structurally heterogeneous within molecular populations of a single sample, but is also variable between samples. Its analysis requires that any sample of heparin that needs to be quality controlled or analysed (*e.g.* for contaminants or to assure provenance) must be compared to the population of heparins that are (at that moment) considered to be *bona fide*, within whatever field of activity is relevant. These may consist, for example, of samples produced during a manufacturing run, or commercial heparin samples available in a particular market or country. The group of acceptable heparins can be termed a library of *bona fide* heparins and collectively constitute what has been termed "*The Current Statistical Definition of Heparin*" (*CSDH*) for that particular application. The decision as to what constitutes the library of *bona fide* heparin samples in any particular situation rests ultimately with manufacturers, regulators and analysts. A further key point is that this library will change regularly, and any *CSDH* for a particular application must be continually updated as more samples are added to it.

The definition of heparin should be composed of experimental data that encompasses the heterogeneity of the material. The method used should require a reasonable amount [following the pharmacopeia methods this is 20 mg of heparin, depending on the field strength of the spectrometer and the probe (room temperature or cryoprobe) being utilised; this amount could be decreased, but if multidimensional experiments are of interest the amount of heparin maybe increased to decrease experimental time] of material and should be relatively quick to perform. The technique chosen by the authors and others is ^{1}H NMR (Figure 14.7). This experiment reports the chemical environment that the protons are in and is applied extensively to the analysis of metabolites in biological fluids, tissue extracts and whole tissues, allowing the detection of spectral features of hundreds of different endogenous metabolites present.[47] The chemical shifts for protons in heparin are well known and the various disaccharide building blocks of heparin have distinct ^{1}H chemical shifts that can be used to differentiate them.[48,49]

14.3.2.2 Principal Component Analysis

If a large dataset is going to be the definition of an inherently heterogeneous material, then tools are required that are able to both explore this dataset and compare a test sample with it. The most common tool for this task is principal component analysis (PCA). This technique decomposes a dataset into a set of orthogonal components. The weighted linear summation of these components can reconstruct any member of the dataset. Principal component analysis is an unsupervised method that does not require prior knowledge of the sample being investigated, whose variables are separated by the observables that have the largest variance. The authors have used PCA to explore heparin ^{1}H and ^{13}C NMR spectra, differentiating heparin from porcine and bovine sources and investigating heparin contaminated with over-sulfated carbohydrates.[37,50] A research team led by Alban and Holzgrabe has also used PCA to interrogate a large number of batches of German heparin and heparin contaminated with OSCS; that was employed in order to establish the limit of quantification.[51,52]

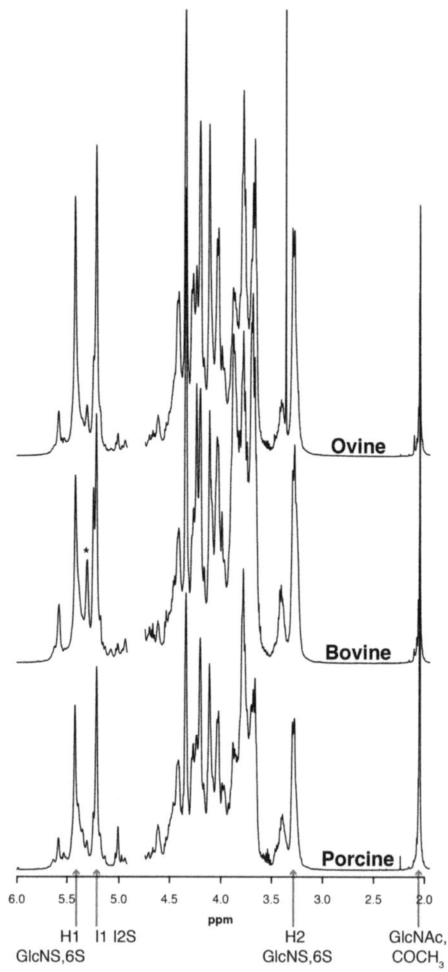

Figure 14.7 The ^{1}H NMR spectra of heparin from the same organ, intestinal mucosa, but different animal sources. The complexity of the ^{1}H NMR spectra makes it difficult to easily differentiate the samples by eye, one difference that is relative easy to distinguish is the different amounts of 6-*O*-sulfation in the sample that originates from beef (*).

14.3.2.3 Taught Analyses, Known Contaminants

If sufficient information is known about a material than a library of standard samples can be used to make predictions about a test sample. Such an approach has been used by a number of groups to predict the properties of heparin, Ruiz-Calero *et al.*,[53–55] Keire *et al.*[56–59] and most recently Monakhova and Diehl.[60] The former cases concentrated primarily on predicting the quantity of non-heparin GAGs in the heparin test sample, while the later used regression techniques to predict the amount of bovine heparin in porcine heparin.

14.4 Unknown Contaminants?

One of the major problems with guaranteeing the safety of heparin, or any other highly heterogeneous biological medicine, is the detection of unknown contaminants, whether added deliberately or not and, naturally, it is not possible to construct a reference library of heparin standards contaminated with all potential contaminants, so another approach is required.

14.4.1 Other Solutions—1D NMR Correlation Spectroscopy-Filtering

Instead of relying on knowing what every possible contaminant of heparin is, it is more efficient to concentrate on defining heparin thoroughly. Once a comprehensive library of ^{1}H NMR spectra of porcine intestinal mucosal heparin samples has been formed, then this can be used to uncover alien features found in the ^{1}H NMR spectrum of test samples. This technique is known as spectral filtering.[36]

14.4.1.1 Discussion of Two-Dimensional Correlation Spectroscopy

Spectral filtering is based on a spectroscopic approach devised by Isao Noda, know as Two-Dimension Correlation Spectroscopy (*2D-COS*).[61] This technique was invented so that one-dimensional techniques, such as infra-red spectroscopy, could be transformed mathematically into bi-dimensional spectroscopy, providing connectivity information analogous to that provided by COSY–TOCSY NMR experiments. Traditionally, the approach is applied to a set of linearly perturbed spectra. The perturbation can either be mechanical or chemical. The *2D-COS* spectrum of the perturbed dataset is produced by determining the inner-product of the dataset to generate what is known as the synchronous spectrum, which highlights transitions that occur in-phase with the external perturbation. The laboratory of Nicholson subsequently applied *2D-COS* to the ^{1}H NMR spectra of bodily fluids,[62] using metabolic profiling to investigate toxicology and the signatures of disease states in model animal systems and humans. The *2D-COS* spectrum of a set of ^{1}H NMR spectra takes the form of a pseudo-TOCSY spectrum.

14.4.1.2 Two-Dimensional Correlation Spectroscopy-Filtering (2D-COS-f)

The principle behind the formation and use of the library of accepted heparins is straightforward and was first described by Rudd *et al.*[36] First, a group of n heparin samples, considered to be genuine using whatever criteria have been defined by the party concerned (*e.g.* manufacturer or regulator) are taken. ^{1}H NMR spectra of each of the n samples are then recorded individually, in each case providing a series of m data points for each sample. These spectra are then arranged into a matrix with n rows and m columns, and a covariance matrix (**CM**) is formed by performing the simple matrix inner product $1/(n-1) \times X{\cdot}X^T$, which has dimensions $m \times m$. This covariance matrix contains cross-peaks only where NMR spectral features reside and hence, structural features, correlate between all members of the library. Next, one additional set of data (*i.e.* its ^{1}H NMR spectrum) from the individual sample under test is added to those of the library and a second covariance matrix is formed. This second covariance matrix will also be of dimension $m \times m$ but, now also contains information regarding the similarity between the test sample and all the other samples in the library. Then, a simple subtraction of the first covariance matrix from the second, producing a difference covariance matrix (**ΔCM**), provides features in the test sample that are not present in the library thus:

$$\Delta\mathbf{CM}\ (\text{2D-COS-F spectrum}) = \mathbf{CM}_{\text{library+test}} - \mathbf{CM}_{\text{library}} = ((1/n_{\text{library+test}} - 1) \times \mathbf{X}_{\text{library+test}}\mathbf{X}_{\text{library+test}}{}^{\text{T}}) - ((1/n_{\text{library}} - 1) \times \mathbf{X}_{\text{library}}\mathbf{X}_{\text{library}}{}^{\text{T}}) \quad (14.1)$$

where $n_{\text{library+test}}$ and n_{library} ($n_{\text{library+test}} = n_{\text{library}} + 1$) are the number of spectra in the test and reference data sets. The matrices $\mathbf{X}_{\text{library+test}}$ and $\mathbf{X}_{\text{library}}$, contain the test and library datasets, respectively. Each column of the matrices represents one spectrum.

Before the difference covariance matrix is formed, each dataset ($\mathbf{X}_{\text{test}}$ and $\mathbf{X}_{\text{ref}}$) is pre-processed independently. This involves binning the data (averaging 5 data points to form one), normalising for area, mean centring and finally, Pareto scaling (where the mean-centered dataset is divided by the square root of the standard deviation). The final step involves normalising the data by the square root of the standard deviation of the data, making each point equally as important as the other points.

In effect, this process "filters" the covariance matrix containing the test sample from that without the sample to reveal any aberrant structures originating in the test sample, whether chemically attached to the sample molecules or not, but which do not correlate with, hence are not present in, the library samples. Using this procedure, a perfectly acceptable heparin test sample would give rise to a filtered spectrum practically devoid of features, whereas one containing modified structures, foreign compounds or contaminants, would reveal the NMR signals originating from these. The filtered spectrum can also be used directly as a means of identifying such compounds. Figure 14.8 shows an example of filtering with *2D-COS* for a heparin sample contaminated with a small amount of OSCS, 3% w/w. This technique is able to detect a few percent of a polysaccharide contaminant whose ^{1}H NMR signals lie entirely under those of heparin. The method, in effect, identifies any signals that do not lie directly on the correlation matrix of the library. Having now derived a method capable of identifying compounds whose signals lay under those of heparin, the challenge was to refine this to improve sensitivity. For this, we now turn to 2-dimensional correlation spectroscopic filtering with iterative random sampling (*2D-COS-firs*).

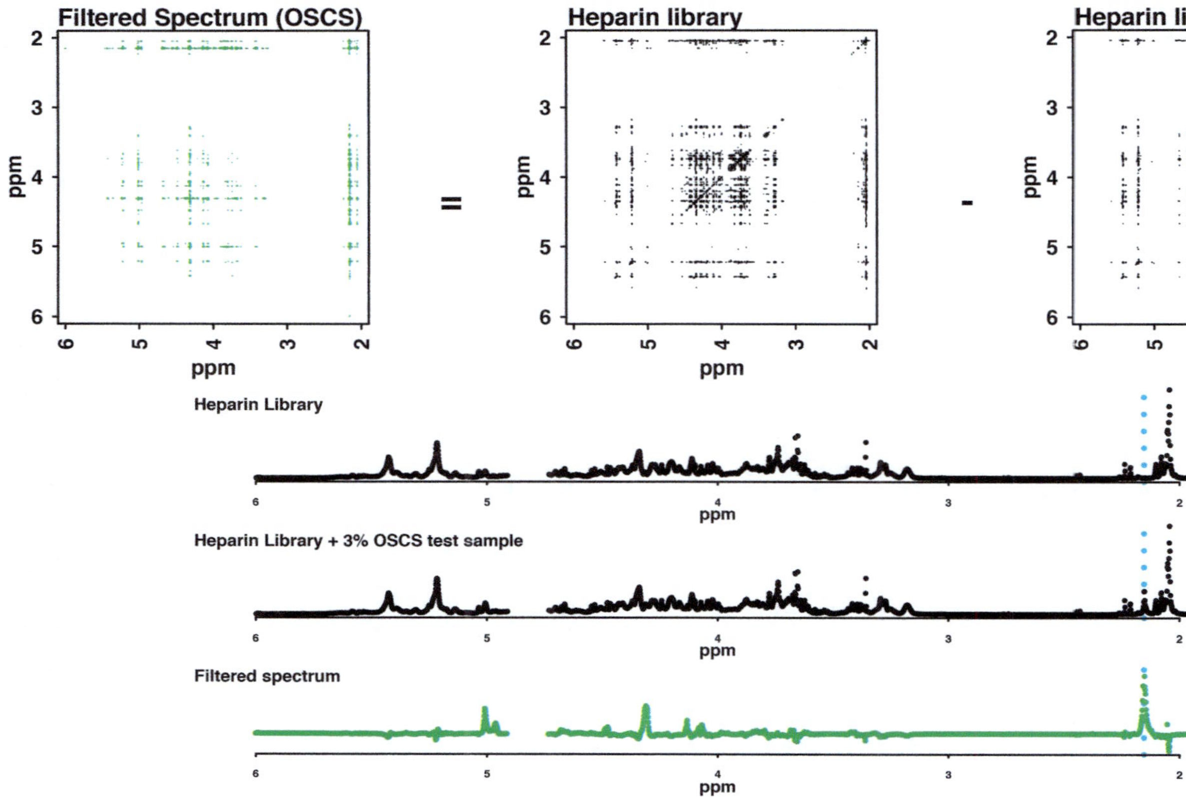

Figure 14.8 *2D-COS-f of a heparin sample contaminated with 3% OSCS w/w.* The panels at the top of the figure show the covariance matrices formed by the ^{1}H NMR spectra that comprise the heparin library and the heparin library plus the test sample, subtracting these produces the 'filtered' spectrum, the difference covariance matrix. The 1D-spectra below are the diagonals from the respective covariance matrices.

14.4.1.3 Two-Dimensional Correlation Spectroscopy-Filtering with Iterative Random Sampling (2D-COS-Firs)

This technique[38] employs a difference covariance matrix acting on a first library of accepted heparins - as above (termed library A, which includes the test sample) except that now, a second library (library B) of *bona fide* heparin samples is also utilised. Library B forms the basis of the natural degree of variation within the heparin population. Filtering of the test sample against library A identifies any alien features within the test sample, whereas, filtering library A (the definition of heparin) against library B (a smaller library also consisting of accepted heparin samples) identifies the degree of natural variation within the heparin population. Both procedures can be iterated for, say, 1500 cycles, to provide a test of whether the features extracted from the test sample (using library A) lie within those comprising the natural variation of heparin (using library B). If they do, then the test sample can be considered to be consistent with the other library members, while if not, the sample contains some alien features and can be rejected as a *bona fide* sample of heparin. This process is explained in schematic form in Figure 14.9.

An illustration of the process in operation is provided in Figure 14.10. Here, a library of porcine mucosal heparin samples forms the two libraries (both A and B) and the test sample comprises a further sample of porcine heparin that contains, as an illustrative example, 1% ovine lung heparin [Figure 14.10]. The *2D-COS-firs* technique can identify this level of contamination

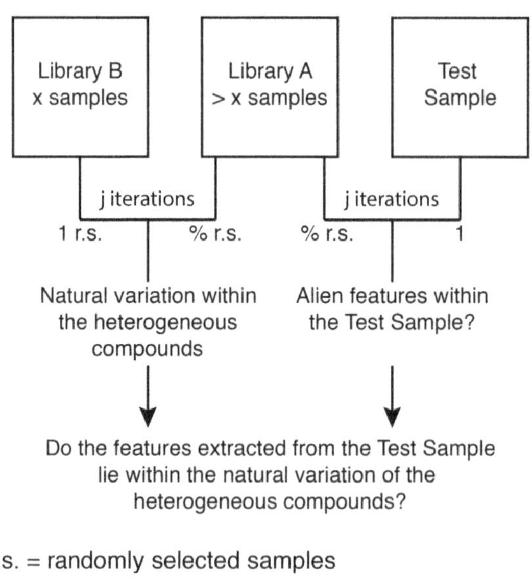

Figure 14.9 *Schematic of the process underlying 2D-COS-f.* The analysis requires two libraries of *bona fide* reference material, the first to filter the test sample (Library A) and the second as a measure of the natural variation of the material being investigated (Library B).

that, it should be stressed, contains no signals that are not already present in the library, but only variations in their relative intensities, with ease.

The technique is suitable to any situation in which a comparison of a single entity needs to be made to a family of heterogeneous entities, particularly natural products and biosimilars. In keeping with the techniques developed above, *2D-COS-firs* does not require any prior knowledge of the nature of the contamination. It is therefore also ideally suited to process monitoring and quality control of any heterogeneous products.

14.4.2 Correlation Spectroscopy Applied to HSQC NMR Spectra

As can be seen in Figure 14.7 the ^{1}H NMR spectrum of heparin has many overlapping signals. This is overcome by using an HSQC experiment (Figure 14.6), in which the proton signals are dispersed by virtue of the signal dispersion available for the carbon atoms with which the protons are associated in the ^{13}C NMR spectrum. The HSQC NMR spectrum of a heparin sample

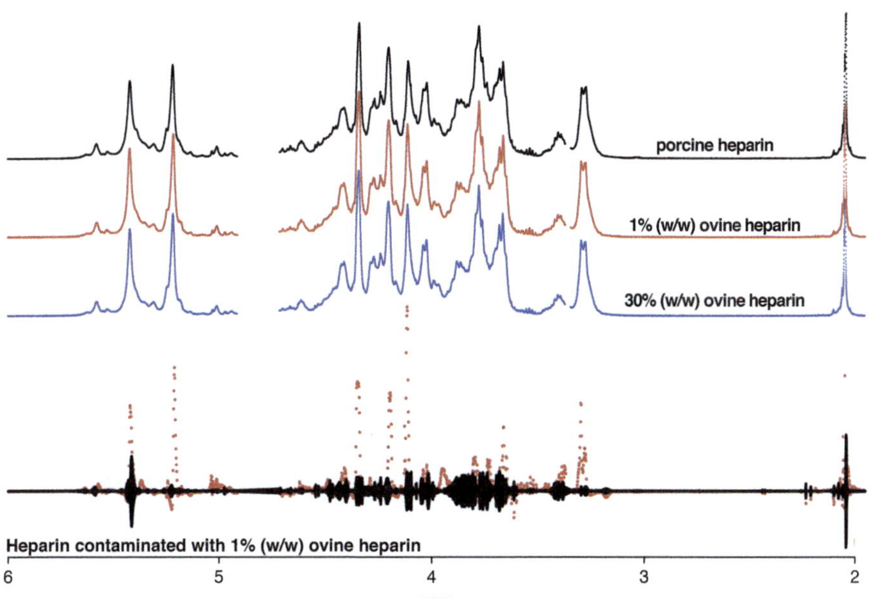

Figure 14.10 *2D-COS-firs* analysis of a porcine heparin contaminated with 1% w/w ovine heparin. The top three ^{1}H NMR spectra illustrate the challenge of differentiating heparin and heparin that is contaminated with heparin from a different source. The bottom spectra reports the result of the *2D-COS-firs* analysis, the red *2D-COS-firs* spectrum shows the features within the test sample that are not similar to those found in the *bona fide* heparin library, while the black spectrum represents the natural variation of heparin, *i.e.*, the result from the comparison of *Library A* and *Library B*.

provides a chemical signature for the compound, but lacks the connectivity information contained in other NMR experiments, such as COSY, TOCSY and NOESY experiments. Applying correlation spectroscopy to HSQC spectra allows statistical sequence information to be recovered and this approach has been applied to heparin[63] as well as to biofluids.[64]

As with the 1D variant a library of HSQC spectra are required to perform HSQC correlation spectroscopy (*HSQCcos*). Unlike the 1D variant where the dataset can be acted upon without modification, if desired, the 2D datasets have to be converted from a matrix to a vector. This makes determining the inner product of the dataset very difficult as the resultant covariance matrix would have the dimensions of at least 1048576 × 1048576 (for HSQC spectra of the dimensions 1024 × 1024), in this case each element of the matrices are handled independently, with the correlation between the elements of interest in the HSQC spectra (υ_1, υ_2 (1H ppm,13C ppm)) being correlated with every other element in the HSQC spectra (Figure 14.11). This produces a correlation map, highlighting which signal in the HSQC spectrum correlates strongly with the other signals present, allowing statistical information to be found for a heterogeneous material, indicating chemical features that occur concurrently in the heterogeneous mixture of polymer.

14.4.2.1 *Statistical Sequence Information—HSQCcos of I12S-GlcNH2 Signal in Heparin*

The use of *HSQCcos*[63] allows statistical sequence information to be extracted for a population of samples, where signals relate to those features that appear concurrently in a material, but not necessarily adjacent to each other in the molecule. The example shown here uses a library of heparin samples and investigates the statistical environment of the disaccharide **-I12S-GlcNH2-** (Anomeric Signal (**I1**) of **-I12S-GlcNH2-** (101.3/5.25 ppm)) a very minor signal in heparin. An example of the ability of correlation

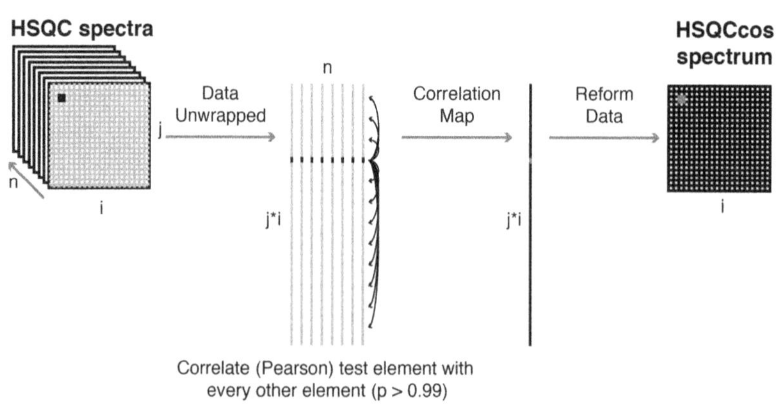

Figure 14.11 *Schematic of the process underlying HSQCcos.*

spectroscopy applied to HSQC spectra to extract signals from related structures is shown in Figure 14.12. Evaluation of this single feature extracts signals which are due to the disaccharide **-I12S-GlcNH2-** (Figure 14.12 bold assignment), isolating the entire disaccharide **I1** to 5 and **A1** to 6 and the heparin linkage-region,[63] the assignments of the linkage-region can be found in.[7] Prior to this analysis it was not known whether the disaccharide **-I12S-GlcNH2-** was sulfated at position 6 or not. This analysis clearly indicates that it is 6-*O*-sulfated (Figure 14.12 - feature labelled **A6**). The other signals found are due specifically to the heparin linkage-region, the region of the heparin chain that connects the carbohydrate to its protein core. The analyses extract positions-3, -4, -5, and -5′ of xylose, positions-1, -2, -3, and -4 of galactose (**Gal-Gal** and **Gal-G**) (Figure 14.12 *italic assignments*).[63] This correlation pattern indicates that **-I12S-GlcNH2-** is adjacent

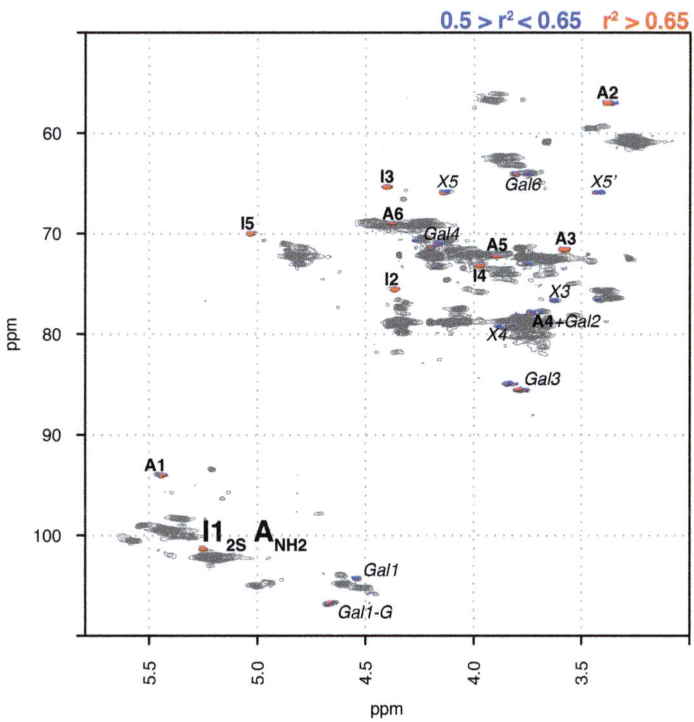

Figure 14.12 *HSQCcos analysis of* ***I12S-ANH2*** *(whose signals fall at 101.3, 5.25 ppm).* By examining a single signal, which belongs to **I12S-ANH2**, *HSQCcos* analysis extracts other signals that are associated with that disaccharide (assignments in bold). Here, this confirms the disaccharide as **I2S-ANH2**, which is *O*-sulfated at position A6, while signals arising from the heparin linkage-region (*italic* assignments) show that the disaccharide **I12S-A6SNH2** is located near the linkage-region. The underlying spectrum (in grey) represents the HSQC spectrum of porcine intestinal mucosal heparin. This also appears in subsequent correlation spectra.

to the non-reducing end of the heparin samples tested, thus, **-I12S-GlcNH2-**...-G-Gal-Gal-Xyl-Ser, (Ser represents serine and Xyl stands for xylose), is a possible statistical sequence. The disaccharide is not represented as being attached to the linkage tetrasaccharide because the analysis can only find features that arise simultaneously in the heparin samples measured. It cannot differentiate whether they are physically connected or not. The linkage region is usually assumed to be adjacent to un-sulfated region of the heparin chain and this result was surprising.

14.4.3 Correlation Spectroscopy-Filtering Applied to HSQC NMR Spectra

As with correlation spectroscopy, spectroscopic filtering can be applied to 2D-NMR spectra. Conceptually the approach is similar to the 1D approach. Practically though it is a little more complicated. Owing to the greater levels of information afforded by 2D-NMR spectra, the spectroscopic filtering of HSQC spectra can be used to produce signatures that are characteristic to a specific heparin. This could be valuable when comparing generic and branded materials.[65]

The filtering of 1D-NMR spectra was previously introduced in Section 14.4.1 and by the authors,[36] being coined two-dimensional correlation spectroscopy-filtering (*2D-COS-f*). The approach allowed a single test spectrum to be compared to a set of reference spectra, removing common features from the test spectrum that appear in the reference spectra, thereby allowing differences in the test spectrum to be observed easily. The technique is based on the application of a difference covariance matrix, which involves forming the covariance matrix (**CM**) of the library spectra ($\mathbf{CM}_{\text{library}}$) and the covariance matrix of the library spectra plus the test spectra ($\mathbf{CM}_{\text{library+test}}$) and subtracting one from the other. The result is a difference covariance matrix ($\Delta\mathbf{CM}$) the "*filtered*" spectrum, (eqn (14.1)). A **CM** is a symmetrical matrix, the diagonal of the **CM** contains the variance of the data, while the covariances of the data are contained in the off-diagonal elements. Therefore, the $\Delta\mathbf{CM}$ can be used to identify the regions of the test spectrum that increase the variance within the $\mathbf{X}_{\text{library+test}}$ dataset, the matrix formed from the library spectra plus the test spectrum.

An HSQC spectrum can be thought to be a matrix composed of many rows of ^{1}H NMR spectra (the directly measured component of an HSQC spectrum). For spectroscopic filtering to be applied to HSQC spectra, HSQC correlation spectroscopy-filtering (*HSQCcos-f*), the approach outlined above has to be applied to each row, individual proton projections that comprise the HSQC spectrum, independently (Figure 14.13). The diagonal of the $\Delta\mathbf{CM}$ formed for a specific row then becomes the row of the filtered spectrum. Before *HSQCcos-f* was performed the datasets $\mathbf{X}_{\text{test}}$ and $\mathbf{X}_{\text{ref}}$ were independently pre-processed and, in the examples shown here, are normalised for area, mean centred and Pareto scaled.

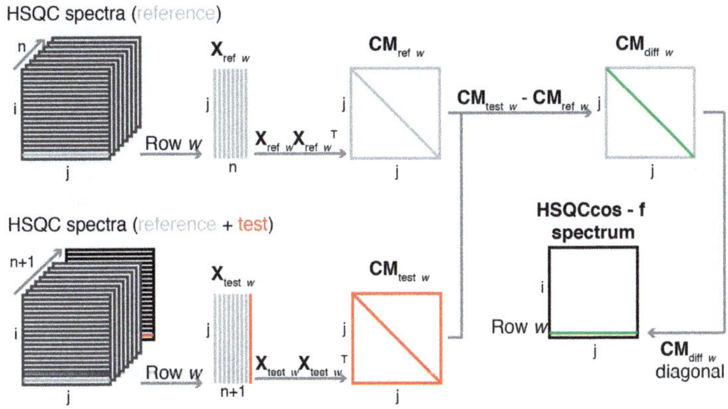

Figure 14.13 *Schematic of the process that underlies HSQCcos-f.* A ^{13}C-^{1}H HSQC spectrum can be thought of as a matrix comprised of many ^{1}H NMR spectra (rows of the HSQC spectrum matrix). Consequently, for spectral filtering to be applied to an HSQC spectrum it has to be applied to each row separately. For illustrative purposes the above schematic illustrates the spectral filtering of a single row, row *w*. For the complete filtered spectrum to be formed each row (*i* rows) of the HSQC spectra would have to undergo the same process.

14.4.3.1 *HSQCcos-f*—OSCS

In 2008 pharmaceutical porcine intestinal mucosal heparin was contaminated with a chemically modified glycosaminoglycan, chondroitin sulfate, which had been oversulfated (OSCS).[34,35] This action resulted in the widespread pollution of the pharmaceutical material and the injury and death of many people. It has been proposed that the adulterant was added to dilute the API heparin, allowing a greater profit to be earned. Oversulfated chondroitin sulfate is formed of a repeating disaccharide of D-glucuronic acid linked *via* a β(1 → 3) glycosidic linkage to β-*N*-acetyl-D-galactosamine. This disaccharide is *O*-sulfated in positions-2 and -3 of the glucuronic acid, as well as positions-4 and -6 of the D-galactosamine residue. The example shown here is the *HSQCcos-f* of a heparin sample contaminated with 1% (w/w) OSCS. Although this level of contamination is not economically viable, it does demonstrate that the analysis can detect small amounts of material within a heterogeneous environment.

In this analysis the HSQC NMR spectra of 23 heparin samples were used to filter the HSQC NMR spectrum of the test sample; heparin contaminated with 1% (w/w) OSCS. The analysis clearly reveals features due to the adulterant OSCS, the strong signals arising from position-2 (^{13}C–^{1}H δ 80.23, 4.48 ppm), -3 (^{13}C–^{1}H δ 79.53, 4.96 ppm) and -5 (^{13}C–^{1}H δ 82.34, 4.13 ppm) of 2,3 *O*-sulfated glucuronic acid and position-4 (^{13}C–^{1}H δ 78.20, 5.00 ppm), -5/-5′ (^{13}C–^{1}H δ 74.62, 4.13 and 74.85, 4.07 ppm) and -6 (^{13}C–^{1}H δ 69.47, 4.30 ppm) of 4,6 *O*-sulfated *N*-acetyl galactosamine (Figure 14.14). Knowing the contaminant allows the weaker, but distinct, signals arising from position-1

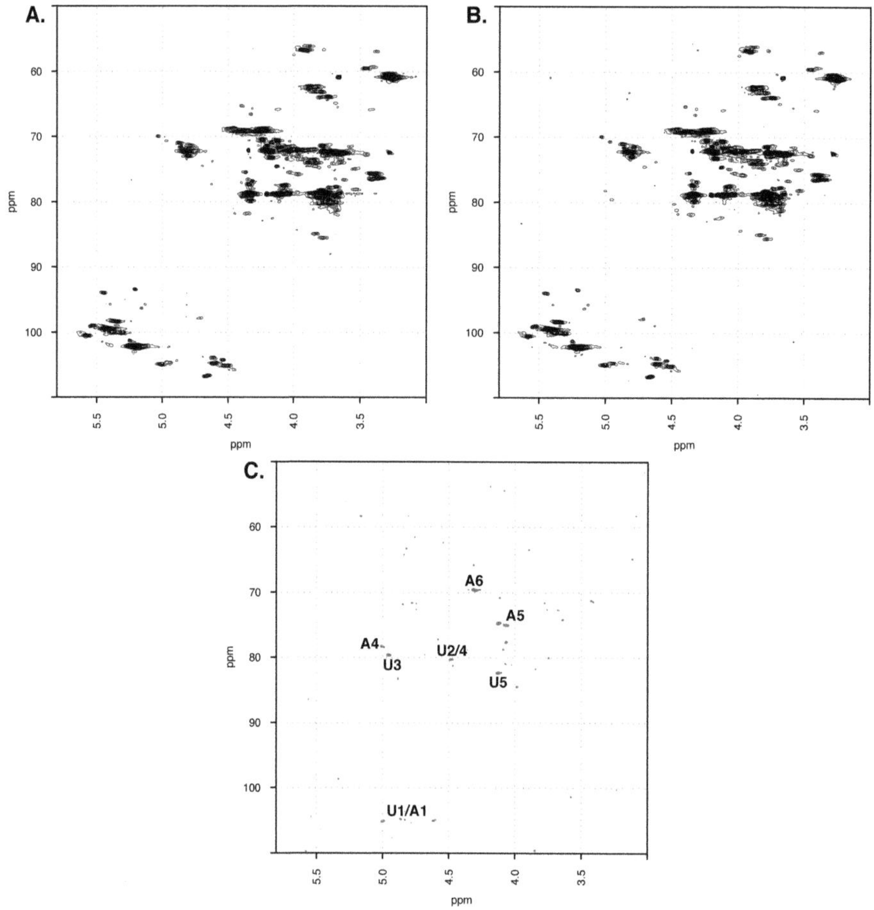

Figure 14.14 *HSQCcos-f of heparin adulterated with 1% OSCS.* **A.** HSQC NMR spectra of heparin. **B.** HSQC NMR spectra of heparin contaminated with 1% (w/w) OSCS. **C.** *HSQCcos-f* spectrum of heparin contaminated with 1% (w/w) OSCS. Twenty three heparin HSQC spectra were used to filter the test sample, before filtering occurred the datasets (library and library plus test sample) were normalised for the area under the spectra, mean centred and Pareto scaled. The figure contains the positive values of the *HSQCcos-f* spectrum, *i.e.*, signals that exhibit an increase in variance. The blue notation relates to signals arising from glucuronic acid, while green are from galactosamine.

(^{13}C–^{1}H δ 104.79, 4.83 ppm), position-2 (^{13}C–^{1}H δ 54.19, 4.09 ppm) and position-3 (^{13}C–^{1}H δ 80.78, 4.07 ppm) of 4,6 *O*-sulfated *N*-acetyl galactosamine and position-1 (^{13}C–^{1}H δ 104.63, 4.86 ppm) and position-4 (^{13}C–^{1}H δ 81.17, 4.47 ppm) of 2,3 *O*-sulfated glucuronic acid to be assigned (Figure 14.14). The other signals found are due to differences in the heparin used to make up the test sample compared to those used to construct the reference spectra. The assignments for OSCS were taken from Guerrini *et al.*[35]

14.4.3.2 HSQCcos-f—Enoxaparin

Low molecular weight heparins (LMWH) have been produced by depolymerising full-length heparin, resulting in reduced length and increased prevalence of the AT binding pentasaccharide sequence (AGA*IA). The reason for using depolymerised heparin instead of unmodified heparin is that the pharmacodynamics are more predictable and that the chance of heparin-induced thrombocytopenia (HIT), an autoimmune response to the complex of heparin and platelet factor-4, is greatly decreased.

There are a number of different LMWHs, which are differentiated by the method of depolymerisation. Here, we compare the branded Lovenox™ material produced by Sanofi US with an Enoxaparin sample, the generic version of the latter, produced by Amphastar Pharmaceuticals, Inc. The means of depolymerisation of both Lovenox™ and Enoxaparin is β-eliminative cleavage of the benzyl ester of heparin by alkaline treatment. This reaction creates a number of terminating residues, 2-*O*-sulfated uronic acid unsaturated at positions-4 and -5 at the non-reducing end and *N*-sulfated, 6-*O*-sulfated glucosamine at the reducing end are the most common terminating residues. Other reducing terminal groups are *N*-sulfated, 6-*O*-sulfated mannosamine, 1,6-anhydro-2-deoxy-β-D-glucopyranose, 2-sulfoamine-1,6-anhydro-2-deoxy-β-mannopyranose and 2-*O*-sulfated iduronic acid. The latter has *N*-sulfated, 6-*O*-sulfate glucosamine at the non-reducing end of the polysaccharide.[32] In addition to the heterogeneity of the terminal groups the use of alkaline conditions during the depolymerisation modifies the uronic acid residues, forming an epoxide ring between carbons 2 and 3, while subsequent acid treatment converts the epoxide modified uronic acid into an α-L-galacturonic acid residue.[11,66] Both signals from epoxide and α-L-galactose are present in the HSQC NMR spectra of Enoxaparin.

These products are from the US market. The Food and Drug Administration (FDA) legislated that the generic material must have the same API as the branded material, Lovenox™. The '*sameness*' of the material is compared using 5 criteria; the physicochemical characteristics of Enoxaparin, the nature of the heparin starting material and the chemical process used to depolymerise this material, the nature and arrangement of components that constitute Enoxaparin, the product's anticoagulant activity, and the *in vivo* effect of the sample. Here *HSQCcos-f* is used to examine the sameness of the chemical content of the Enoxaparin produced by Amphastar Pharmaceuticals, Inc. compared to the branded material (Figure 14.15).

The principal differences between the HSQC spectra of the 10 Lovenox samples used to filter the test sample and the HSQC spectra of the Enoxaparin sample are found to comprise the content of linkage region monosaccharides and modifications to the polysaccharide caused by the harsh conditions used for the depolymerisation of the mother heparin. The linkage region is tetrasaccharide linker (-**G**-β(1 → 3)-**Gal**-β(1 → 3)-**Gal**-β(1 → 4)-**Xyl**-) that attaches the glycosaminoglycan chain to a serine in the

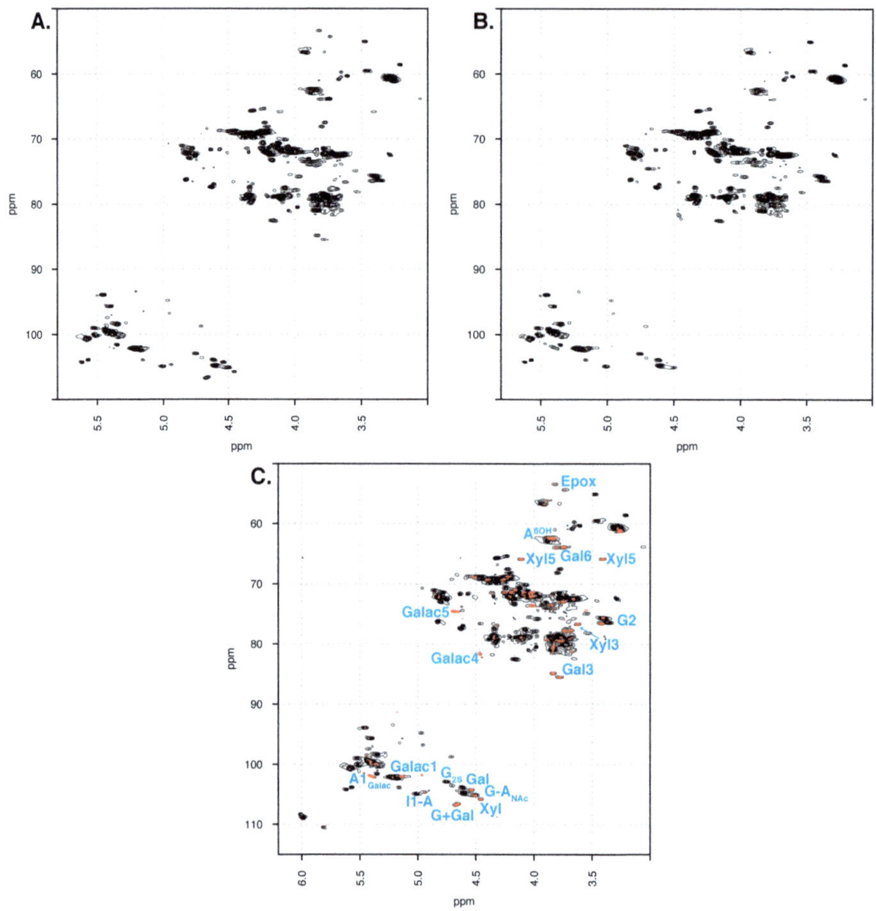

Figure 14.15 *HSQCcos-f spectrum of an Enoxaparin (generic LMWH) sample (Amphastar Pharmaceuticals, Inc.).* **A.** HSQC NMR spectrum of the Enoxaparin sample that was subjected to *HSQCcos-f*. **B.** Representative HSQC NMR spectrum of the Lovenox samples used to filter the Enoxaparin sample. **C.** *HSQCcos-f* spectrum of the Enoxaparin sample (red signals) overlaid on the average HSQC NMR spectrum of the ten Lovenox samples used to filter the Enoxaparin sample (black signals). Ten Lovenox HSQC spectra were used to filter the test sample. Before filtering occurred the datasets (library and library plus test sample) were normalised for the area under the spectra, mean centred and Pareto scaled. The *HSQCcos-f* spectrum panel contains the positive values of the *HSQCcos-f* spectrum (red signals), *i.e.*, signals that exhibit an increase in variance.

proteoglycan protein core *via* the non-reducing end xylose residue of the linkage region. In the *HSQCcos-f* spectrum of the Enoxaparin sample signals from positions-1, -3 and -5 of xylose, positions-1, -3 and -6 of galactose and position-1 of glucuronic acid linked to galactose are clearly observed (Figure 14.15C). The Enoxaparin sample has experienced strong base and subsequent acid conditions evinced by signals arising from epoxide (position-2 and -3 of the epoxide modified uronic acid), as well as signals from galacturonic acid (positions-1, -4 and -5 of galacturonic acid and position-1 of *N*-sulfated glucosamine linked to galacturonic acid). Further differences lie in the glucuronic content of the Enoxaparin test sample, with the signals from position-2 of glucuronic acid, position-1 of 2-*O*-sulfated glucuronic acid and position-1 of glucuronic acid linked to *N*-acetyl glucosamine being observed. The sulfation pattern of the Enoxaparin sample may be different to those seen in the Lovenox reference samples, with signals due to 6-de-/6-*O*-sulfated glucosamine and *N*-sulfated–*N*-acetyl glucosamine seen in the *HSQCcos-f* spectrum of the Enoxaparin sample (Figure 14.15C). Another small difference is the level of 2-de-*O*-sulfated iduronic acid present in the Enoxaparin sample, position-1 of 2-de-*O*-sulfated iduronic acid linked to 6-de-*O*-sulfated glucosamine is visible in the *HSQCcos-f* spectrum. For more information regarding the difference between Lovenox™ and the Enoxaparins available see.[65]

14.5 Conclusions

NMR spectroscopy, which does not demand any sample manipulation, provides both qualitative and quantitative structural information of heparin and LMWHs. Following the heparin crisis, all of the major Pharmacopoeias introduced NMR for the purposes of identification and characterisation of the active principle, and also allowed NMR to be used to demonstrate the degree of '*sameness*' between branded and generic LMWHs. However, a simple proton spectrum, such as that requested by Pharmacopoeias for heparin, cannot fully guarantee the quality of the sample from intentional contamination, including cross-species contamination. Only the combination of NMR and statistical analysis of an ensemble of spectra of validated "*bona fide*" samples forming a library can perform this task. Spectroscopic filtering can differentiate any signal from those of the heparin library as long as the contaminant contains some anomaly that is not common to all the samples in the heparin library, even if this anomaly is accompanied by signals that are otherwise indistinguishable from accepted heparin, or consist of the absence of some signals from accepted heparin. Furthermore, spectroscopic filtering is capable of identifying compounds that are contaminated with material that is made of the same building blocks as the heterogeneous compound being investigated. This technique can be applied to any heterogeneous mixture of compounds, including bio-drugs, food and diagnostic medical materials.

Acknowledgements

The authors would like to acknowledge the contributions of Davide Gaudesi, Eleonora Macchi, Marcelo Lima, Laura Muzi, Cesare Cosentino, Cristina Gardini and Lucio Mauri.

References

1. E. Gray, J. Hogwood and B. Mulloy, in *Heparin–A Century of Progress*, ed. R. Lever, B. Mulloy and P. C. Page, Springer Berlin Heidelberg, Berlin, Heidelberg, 2012, pp. 43–61.
2. T. W. Barrowcliffe, in *Heparin–A Century of Progress*, ed. R. Lever, B. Mulloy and P. C. Page, Springer Berlin Heidelberg, Berlin, Heidelberg, 2012, pp. 3–22.
3. B. Casu, A. Naggi and G. Torri, *Carbohydr. Res.*, 2015, **403**, 60.
4. D. L. Rabenstein, *Nat. Prod. Rep.*, 2002, **19**, 312.
5. U. Lindahl, G. Bäckström, L. Thunberg and I. G. Leder, *Proc. Natl. Acad. Sci. U. S. A.*, 1980, **77**, 6551.
6. M. Guerrini, S. Guglieri, B. Casu, G. Torri, P. Mourier, C. Boudier and C. Viskov, *J. Biol. Chem.*, 2008, **283**, 26662.
7. M. Iacomini, B. Casu, M. Guerrini, A. Naggi, A. Pirola and G. Torri, *Anal. Biochem.*, 1999, **274**, 50.
8. J. E. Turnbull and J. T. Gallagher, *Biochem. J.*, 1991, **273**, 553.
9. P. Mourier, P. Anger, C. Martinez, F. Herman and C. Viskov, *Anal. Chem. Res.*, 2015, **3**, 46.
10. E. Kellenbach, K. Sanders, P. J. Michiels and F. C. Girard, *Anal. Bioanal. Chem.*, 2011, **399**, 621.
11. S. Piani, B. Casu, E. G. Marchi, G. Torri and F. Ungarelli, *J. Carbohydr. Chem.*, 1993, **12**, 507.
12. B. Casu, M. Guerrini, A. Naggi, G. Torri, L. De-Ambrosi, G. Boveri, S. Gonella, A. Cedro, L. Ferro, E. Lanzarotti, M. Paterno, M. Attolini and M. G. Valle, *Arzneim.-Forsch.*, 1996, **46**, 472.
13. B. Casu, M. Guerrini, A. Naggi, G. Torri, L. De-Ambrosi, G. Boveri and S. Gonella, *Thromb. Haemostasis*, 1995, **74**, 1205.
14. L. Fu, G. Li, B. Yang, A. Onishi, L. Li, P. Sun, F. Zhang and R. J. Linhardt, *J. Pharm. Sci.*, 2013, **102**, 1447.
15. A. Naggi, C. Gardini, G. Pedrinola, L. Mauri, E. Urso, A. Alekseeva, B. Casu, G. Cassinelli, M. Guerrini, M. Iacomini, V. Baigorria and G. Torri, *J. Pharm. Biomed. Anal.*, 2016, **118**, 52.
16. M. Guerrini, S. Elli, P. Mourier, T. R. Rudd, D. Gaudesi, B. Casu, C. Boudier, G. Torri and C. Viskov, *Biochem. J.*, 2013, **449**, 343.
17. M. Guerrini, S. Guglieri, D. Beccati, G. Torri, C. Viskov and P. A. Mourier, *Biochem. J.*, 2006, **399**, 191.
18. M. Guerrini, P. A. Mourier, G. Torri and C. Viskov, *Glycoconjugate J.*, 2014, **31**, 409.
19. C. Viskov, S. Elli, E. Urso, D. Gaudesi, P. Mourier, F. Herman, C. Boudier, B. Casu, G. Torri and M. Guerrini, *J. Biol. Chem.*, 2013, **288**, 25895.
20. G. Li, B. Yang, F. Zhang, C. Xue and R. J. Linhardt, *Anal. Biochem.*, 2014, **455**, 3.
21. P. Carlsson and L. Kjellén, Heparin Biosynthesis, in *Heparin–A Century of Progress*, ed. R. Lever, B. Mulloy and P. C. Page, Springer Berlin Heidelberg, Berlin, Heidelberg, 2012, pp. 23–41.
22. T. R. Rudd and E. A. Yates, *Mol. BioSyst.*, 2012, **8**, 1499.

23. L. B. Schonberger, *Infect. Dis. Clin. North Am.*, 1998, **12**, 111.
24. D. A. Keire, B. Mulloy, C. Chase, A. Al-Hakim, D. Cairatti, E. Gray, J. Hogwood, T. Morris, P. Mourao, M. Da Luz, C. Soares and A. Szajek, *Pharm. Technol.*, 2015, **39**, 28.
25. R. J. Linhardt and N. S. Gunay, *Semin. Thromb. Hemostasis*, 1999, **25**(suppl. 3), 5.
26. S. Bertini, A. Bisio, G. Torri, D. Bensi and M. Terbojevich, *Biomacromolecules*, 2005, **6**, 168.
27. T. W. Barrowcliffe, in *Heparin: Chemical and Biological Properties*, ed. D. A. Lane and U. Lindahl, Edward Arnold, London, 1989, pp. 393–416.
28. J. Fareed, Q. Ma, M. Florian, J. Maddineni, O. Iqbal, D. A. Hoppensteadt and R. L. Bick, *Semin. Thromb. Hemostasis*, 2004, **30**(suppl. 1), 89.
29. M. Guerrini and A. Bisio, in *Heparin–A Century of Progress*, ed. R. Lever, B. Mulloy and P. C. Page, Springer Berlin Heidelberg, Berlin, Heidelberg, 2012, pp. 127–157.
30. J. C. Lormeau, M. Petitou and J. Choay, 1998, US Patent #4981955.
31. A. Uzan, 1998, US Patent #5849721.
32. G. Mascellani, M. Guerrini, G. Torri, L. Liverani, F. Spelta and P. Bianchini, *Carbohydr. Res.*, 2007, **342**, 35.
33. R. S. Langer, R. J. Linhardt, C. L. Cooney and G. Fitzgerald, 1983, US Patent #4396762.
34. T. K. Kishimoto, K. Viswanathan, T. Ganguly, S. Elankumaran, S. Smith and K. Pelzer, *et al.*, *N. Engl. J. Med.*, 2008, **358**, 2457.
35. M. Guerrini, D. Beccati, Z. Shriver, A. Naggi, K. Viswanathan and A. Bisio, *et al.*, *Nat. Biotechnol.*, 2008, **26**, 669.
36. T. R. Rudd, D. Gaudesi, M. A. Lima, M. A. Skidmore, B. Mulloy, G. Torri, H. B. Nader, M. Guerrini and E. A. Yates, *Analyst*, 2011, **136**, 1390.
37. T. R. Rudd, D. Gaudesi, M. A. Skidmore, M. Ferro, M. Guerrini, B. Mulloy, G. Torri and E. A. Yates, *Analyst*, 2011, **136**, 1380.
38. T. R. Rudd, E. Macchi, C. Gardini, L. Muzi, M. Guerrini, E. A. Yates and G. Torri, *Anal. Chem.*, 2012, **84**, 6841.
39. M. Guerrini, A. Naggi, S. Guglieri, R. Santarsiero and G. Torri, *Anal. Biochem.*, 2005, **337**, 35.
40. M. Guerrini, S. Guglieri, A. Naggi, R. Sasisekharan and G. Torri, *Semin. Thromb. Hemostasis*, 2007, **33**, 478.
41. S. Lee, A. Raw, L. Yu, R. Lionberger, N. Ya, D. Verthelyi, A. Rosenberg, S. Koziowski, K. Webber and J. Woodcock, *Nat. Biotechnol.*, 2013, **31**, 220.
42. G. Bodenhausen and D. J. Ruben, *Chem. Phys. Lett.*, 1980, **69**, 185.
43. P. Giraudeau, *Magn. Reson. Chem.*, 2014, **52**, 259.
44. P. Bigler and R. Brenneisen, *J. Pharm. Biomed. Anal.*, 2009, **49**, 1060.
45. E. Bednarek, J. Sitkowski, W. Bocian, B. Mulloy and L. Kozerski, *J. Pharm. Biomed. Anal.*, 2010, **53**, 302.
46. J. Sitkowski, E. Bednarek, W. Bocian and L. Kozerski, *J. Med. Chem.*, 2008, **51**, 7663.
47. J. C. Lindon, E. Holmes and J. K. Nicholson, *FEBS J.*, 2007, **274**, 1140.
48. E. A. Yates, F. Santini, M. Guerrini, A. Naggi, G. Torri and B. Casu, *Carbohydr. Res.*, 1996, **294**, 15.

49. E. A. Yates, F. Santini, B. De Cristofano, N. Payre, C. Cosentino, M. Guerrini, A. Naggi, G. Torri and M. Hricovini, *Carbohydr. Res.*, 2000, **329**, 239.
50. T. R. Rudd, M. A. Skidmore, S. E. Guimond, C. Cosentino, G. Torri, D. G. Fernig, R. M. Lauder, M. Guerrini and E. A. Yates, *Glycobiology*, 2009, **19**, 52.
51. S. Alban, S. Luhn, S. Schiemann, T. Beyer, J. Norwig, C. Schilling, O. Radler, B. Wolf, M. Matz, K. Baumann and U. Holzgrabe, *Anal. Bioanal. Chem.*, 2011, **399**, 605.
52. T. Beyer, M. Matz, D. Brinz, O. Radler, B. Wolf, J. Norwig, K. Baumann and U. Holzgrabe, *Eur. J. Pharm. Sci.*, 2010, **40**, 297.
53. V. Ruiz-Calero, J. Suarina, S. Hernandez-Cassou, M. T. Galceran and L. Puignou, *Analyst*, 2002, **127**, 407–415.
54. V. Ruiz-Calero, J. Saurina, M. T. Galceran, S. Hernandez-Cassou and L. Puignou, *Anal. Bioanal. Chem.*, 2002, **373**, 259.
55. V. Ruiz-Calero, J. Saurina, M. T. Galceran, S. Hernandez-Cassou and L. Puignou, *Analyst*, 2000, **125**, 933.
56. Q. Zang, D. A. Keire, R. D. Wood, L. F. Buhse, C. M. Moore, M. Nasr, A. Al-Hakim, M. L. Trehy and W. J. Welsh, *Anal. Bioanal. Chem.*, 2011, **399**, 635.
57. Q. Zang, D. A. Keire, R. D. Wood, L. F. Buhse, C. M. Moore, M. Nasr, A. Al-Hakim, M. L. Trehy and W. J. Welsh, *Anal. Chem.*, 2011, **83**, 1030.
58. Q. Zang, D. A. Keire, R. D. Wood, L. F. Buhse, C. M. Moore, M. Nasr, A. Al-Hakim, M. L. Trehy and W. J. Welsh, *J. Pharm. Biomed. Anal.*, 2011, **54**, 1020.
59. Q. Zang, D. A. Keire, L. F. Buhse, R. D. Wood, D. P. Mital, S. H. Shankar Srinivasan, C. M. Moore, M. Nasr, A. Al-Hakim, M. L. Trehy and W. J. Welsh, *Anal. Bioanal. Chem.*, 2011, **401**, 939.
60. Y. B. Monakhova and B. W. Diehl, *J. Pharm. Biomed. Anal*, 2015, **115**, 543.
61. I. Noda and Y. Ozaki, *Two-Dimensional Correlation Spectroscopy: Applications in Vibrational and Optical Spectroscopy*, Wiley, 2005.
62. O. Cloarec, M. E. Dumas, A. Craig, R. H. Barton, J. Trygg, J. Hudson, C. Blancher, D. Gauguier, J. C. Lindon, E. Holmes and J. Nicholson, *Anal. Chem.*, 2005, **77**, 1282.
63. T. R. Rudd, E. Macchi, L. Muzi, M. Ferro, D. Gaudesi, G. Torri, B. Casu, M. Guerrini and E. A. Yates, *Anal. Chem.*, 2013, **85**, 7487.
64. T. Oman, M. B. Tessem, T. F. Bathen, H. Bertilsson, A. Angelsen, M. Hedenstrom and T. Andreassen, *BMC Bioinf.*, 2014, **15**, 413.
65. M. Guerrini, T. R. Rudd, L. Mauri, E. Macchi, J. Fareed, E. A. Yates, A. Naggi and G. Torri, *Anal. Chem.*, 2015, **87**, 8275.
66. M. Jaseja, R. N. Rej, F. Sauriol and A. S. Perlin, *Can. J. Chem.*, 1989, **67**, 1449.
67. L. Mauri, G. Boccardi, G. Torri, M. Karfunkle, E. Macchi, L. Muzi, D. Keire and M. Guerrini, *J. Pharm. Biomed. Anal.*, 2017, **136**, 92.

CHAPTER 15

NMR Chemical Shift Predictions and Structural Elucidation of Oligo- and Polysaccharides by the Computer Program CASPER

JONAS STÅHLE[a] AND GÖRAN WIDMALM*[a]

[a]Department of Organic Chemistry, Arrhenius Laboratory, Stockholm University, S-106 91 Stockholm, Sweden
*E-mail: goran.widmalm@su.se

15.1 Introduction

Techniques for structural determination of oligosaccharides, polysaccharides and glycoconjugates[1] include chemical analysis by gas or liquid chromatography,[2] electrophoresis,[3] mass spectrometry (MS),[4,5] circular dichroism,[6] Raman optical activity[7] and nuclear magnetic resonance (NMR) spectroscopy.[8] For carbohydrates ^{1}H, ^{13}C, ^{15}N and ^{31}P are naturally occurring spin-1/2 nuclei suitable for NMR spectroscopy analysis. 1D ^{1}H or ^{31}P NMR experiments rapidly give a good overview of the glycan material to be investigated, due to their very high natural abundance, whereas due to the low occurrence of the ^{13}C and ^{15}N nuclei they are better analyzed by heteronuclear correlation experiments detected *via* ^{1}H nuclei. Higher dimensionality experiments, 3D and *n*D, may efficiently resolve spectral overlap but require longer experiments. To reduce the duration of the NMR experiments Hadamard transform

New Developments in NMR No.10
NMR in Glycoscience and Glycotechnology
Edited by Koichi Kato and Thomas Peters

Published by the Royal Society of Chemistry, www.rsc.org

spectroscopy, TILT or APSY experiments and non-uniform sampling techniques offer efficient means to this end.[9–11]

Computerized methods based on NMR chemical shift data emerged in the 1980s, in particular using ^{13}C chemical shifts and were based on increment rules, in which corrections (positive or negative) are added to the base value of an atom in a structure as a result of a substituent and/or steric interactions, if present.[12] In the carbohydrate field these developments include CASPER,[13] an acronym for Computer Assisted SPectrum Evaluation of Regular polysaccharides, a ^{13}C NMR approach to the structural analysis of carbohydrates,[14] a carbon-13 NMR methodology for their structural elucidation[15,16] as well as utilization of ^{13}C NMR for identification of manno-oligosaccharides and glycopeptides.[17] CASPER was subsequently enhanced to also include ^{1}H NMR data,[18,19] 2D NMR representations of spectra[18] and to address oligosaccharide structures.[20,21] The program GlyNest[22] estimates the ^{1}H and ^{13}C chemical shifts of glycans based on a spherical environment encoding scheme, an approach that also supplies the standard deviation of the chemical shift. Other developments including database approaches or algorithms for NMR-based determinations based on chemical shift predictions for structural elucidations are BIOPSEL for ^{13}C data,[23] SOACS using ^{1}H chemical shifts,[24] a machine learning approach for the prediction of ^{13}C NMR chemical shifts[25,26] and an improved structure generalization scheme for ^{13}C NMR simulations referred to as GODESS,[27] its further developments that also contain ^{1}H NMR data[28] and simulation of 2D NMR spectra.[29]

The CASPER program is able to (i) predict ^{1}H and ^{13}C NMR chemical shifts for a given carbohydrate structure in the module 'Calculate NMR Chemical Shifts', (ii) carry out a component analysis based on NMR data from hydrolyzed, methanolyzed or butanolized carbohydrate material in the module 'Component Analysis', and (iii) in 'Determine Glycan Structure' elucidate the primary structure based on component data from chemical analysis, NMR spectroscopic analysis or bioinformatics, the latter *via* interaction with the database ECODAB,[30,31] together with chemical shift information from *unassigned* NMR spectra, *e.g.* uploaded from a CCPN project,[32,33] producing tentative assignments of NMR resonances, downloadable into CCPN, as outlined in the flowchart of Figure 15.1. In conjunction with the program CarbBuilder[34,35] files in pdb-format can be made representing a canonical 3D structure of the oligo- or polysaccharide. We herein describe the basis for prediction of NMR chemical shifts, input to and output from the program as well as examples of structural elucidations including an alditol-containing oligosaccharide and a polysaccharide containing also an *O*-acetyl group as a part of the repeating unit.

15.2 CASPER is Based on Increment Rules

NMR spectroscopy facilitates structural information at atomic resolution since each nucleus gives rise to a resonance at a specific chemical shift being dependent on the structural environment of the atom under consideration.

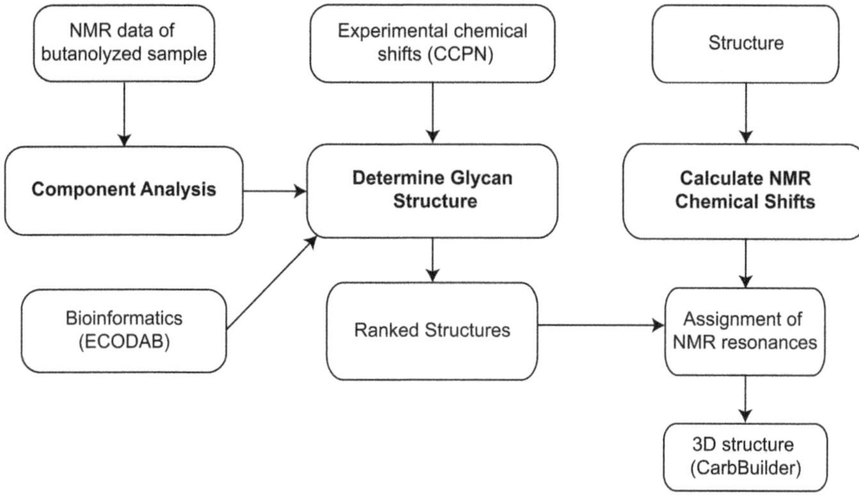

Figure 15.1 Flowchart of NMR-based structural elucidation of carbohydrates using CASPER. The three modules are shown with bold text. The inputs shown in the top row are required for the respective module, but additional information may be supplied. For example, some structural data can be given when carrying out 'Determine Glycan Structure' to reduce the number of permutations needed for the calculations. This structural data can be generated from the Component Analysis module, bioinformatics or from other component analysis. When performing 'Calculate NMR Chemical Shift', any experimental NMR chemical shifts that are also given will be assigned. CASPER is able to handle unknowns as part of the input, *i.e.*, a sugar residue that has not been determined prior to the NMR-based structural elucidation, as well as a smaller number of NMR resonances than what is anticipated from the input data based on component analysis, typically a result of spectral overlap of resonances. The output consists of structural suggestions ranked according to the best-fit to experimental data, NMR chemical shift assignments as well as 3D structural representations in the pdb-format generated by the CarbBuilder program as part of the CASPER-GUI.

Furthermore, the scalar spin–spin couplings between nuclei, when of sufficient magnitude, report on interactions between nuclei and atomic networks through their mutual correlations. The chemical shift dispersion of carbohydrates is quite limited and the majority of the ^{1}H and ^{13}C resonances in NMR spectra are present at 3–4 ppm and 60–80 ppm, respectively. Some resonances reside outside the central region and may be used to unravel spin–spin coupling networks within and between sugar residues. In formulation of the NMR chemical shift prediction procedure that CASPER utilizes it was noted that significant chemical shift displacements took place between a disaccharide and its constituent monosaccharides depending on anomeric configuration as well as substitution position at the glycosidic linkage. Notably, both large and small chemical shifts changes were present and it was decided to make use of all of them, even though some were small and not easily accounted for from a structural point of view; this is in contrast to only considering chemical shift changes just a few bonds away in the structure, although these are usually largest in magnitude. Consequently, the procedure employed by CASPER is based on chemical shift displacements also referred to as increments[36] (positive or negative) that depend on geometrical arrangements at the glycosidic linkage[37] and substituent effects. Thus, for a given structure described by its constituent monosaccharides, their absolute configuration, the ring form(s), anomeric configuration(s), substitution pattern (linkages) and substituents, when present, the monosaccharide chemical shifts (δ) are stored as the first step of the chemical shift prediction procedure (Figure 15.2). Glycosylation shifts ($\Delta\delta$), *i.e.*, chemical shift differences between a disaccharide and its constituent monosaccharides, are subsequently added. When a sugar residue is vicinally disubstituted, *i.e.* by glycosyl residues at two adjacent positions, O2 and O3 or O3 and O4 in hexopyranoses,[37,38] are substituted, deviations from pure additivity of the chemical shift increments may occur and correction terms to the three sugar residues should then be added, referred to as a correction set ($\Delta\Delta\delta$). In a similar way corrections may be needed for a residue that is substituted at O2 by a sugar residue that participates in a glycosidic linkage by substituting another sugar residue, *e.g.* in a (1 → 2), (1 → 3)-linked constellation.[39] Any chemical shift increments due to substituents[40] are subsequently added and the ^{1}H and ^{13}C NMR chemical shifts of a given structure have thus been predicted.

When a di- or trisaccharide correction is not known, *i.e.*, when there are no data for the exact linkage between sugar residues, approximations must be made. This is done by substituting either constituent in a way that minimizes the stereochemical difference around the affected linkages. An α-D-Gal*p*-(1 → 3)-β-D-Gal*p*-linkage, for example, could be approximated using the disaccharide correction from an α-D-Glc*p*-(1 → 3)-β-D-Gal*p*-linkage, since the stereochemical difference in position 4 of the terminal residue will have little impact on the chemical shifts at the glycosidic linkage of the disaccharide.[37] The database currently keeps record of around 400 di- and trisaccharide entries that are used as a basis for these approximations.

15.2.1 CASPER Input and Output

CASPER is run from a graphical-user-interface (GUI) available at http://www.casper.organ.su.se/casper/ where either ‘Calculate NMR Chemical Shifts’ for predictions, ‘Component Analysis’ for monosaccharide determinations or ‘Determine Glycan Structure’ for structural elucidation may be chosen.[41,42] For the chemical shift predictions a large number of monosaccharides and a few amino acids are available from drop-down menus including anomeric configurations and linkage positions. Substituents may be linked onto the sugar residues and are found under advanced options. The ^{1}H and ^{13}C NMR chemical shifts for an entered structure are then predicted within seconds leading to tentative resonance assignments.

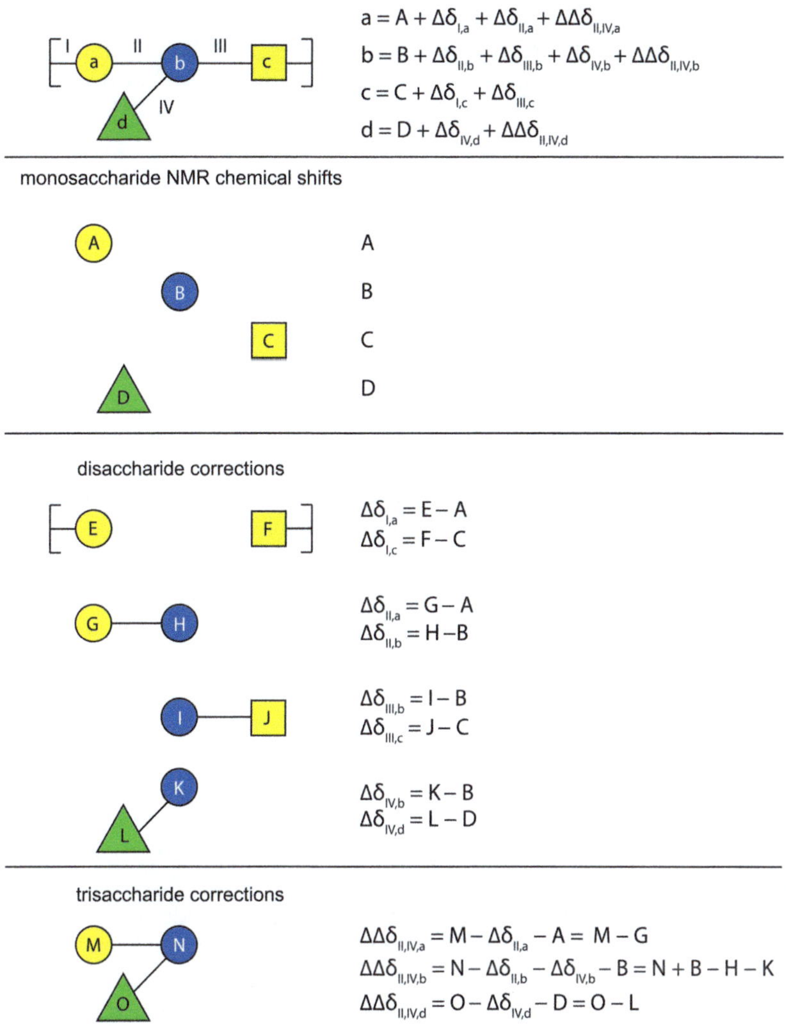

Figure 15.2 Schematic representation of the chemical shift predictions of a single polysaccharide structure (one set of anomeric configurations, ring forms, linkages and absolute configurations) where the sugar residues are represented in SNFG-format. Chemical shifts of monosaccharides (δ) are utilized to which glycosylation shifts ($\Delta\delta$) are added followed by corrections for vicinal disubstitution ($\Delta\Delta\delta$), which may occur in branched structures *e.g.* in $(1 \rightarrow 2)$, $(1 \rightarrow 3)$-linked arrangements.

The structural elucidation carried out by CASPER is based on unassigned NMR data but in order not to have unreasonably long calculation times (due to the large number of sugars and substituents available, linkage positions, anomeric configurations and ring forms) components are required as input, although all of them need not be given. To this end sugar components can be determined by other means such as GC-MS or LC-MS techniques but may also be obtained by NMR analysis of an oligo- or polysaccharide hydrolysate determining the constituent sugar residues by *e.g.* 2D ^{1}H,^{13}C-HSQC spectra, as methyl glycosides in the same fashion, or obtaining also their absolute configuration by the use of optically active 2-butyl glycosides,[43,44] noting that if the chemical shifts of monosaccharide, methyl or 2-butyl glycoside have been entered in the database, reference spectra do not have to be acquired, only those of the unknown. Information on sugar components, anomeric configuration and linkage positions may be obtained based on bioinformatics-derived data stored in the database ECODAB available at http://www.casper.organ.su.se/ECODAB/.[30,31]

For the 'Component Analysis' or 'Determine Glycan Structure' NMR spectral data can be entered directly in the GUI or uploaded from a CCPN project.[32] The unassigned NMR spectra, the chemical shifts of which have typically been obtained by automatic peak-picking, are supplied but not all resonances need to be supplied as input,[45] although the more the faster the computation will be. In order to quicken the structural elucidation the number of possibilities and permutations may be decreased by addition of $^{3}J_{H1,H2}$ and $^{1}J_{C1,H1}$ coupling constants of anomeric protons that in many cases are readily obtained from ^{1}H and coupled 2D ^{1}H,^{13}C-HSQC spectra, respectively, since chemical shifts for combinations that are not possible, based on the J values, will not be computed. Another way of limiting the structures that should be tried with respect to different permutations is to employ 'Biological rules' that make use of what is known about glycan structures and their biosynthesis. Presently those related to N-glycans, O-glycans, WecA-based repeating units, backbone of *Shigella flexneri* O-antigens, except serotype 6, and conserved parts of the core structure of *Haemophilus influenza* lipopolysaccharides that have short-chain LPS are possible to select.

The results of 'Calculate NMR Chemical Shifts' include proposed ^{1}H and ^{13}C chemical shifts of the sugar residues and substituents, if present. Schematic representations of the ^{13}C NMR spectrum and a 2D ^{1}H,^{13}C-correlated plot, the option to print a report in pdf-format and to make a 3D structure in pdb-format generated by the CarbBuilder program[34,35] run in a seamless way from within CASPER, transparent to the user. It is also possible to download the CarbBuilder program separately and to run it standalone using input in CASPER format from the command line. When the 'Component Analysis' is carried out the program proposes the constituent monosaccharides and as part of the procedure it also lists unused resonances in the procedure since the presence of several unused chemical shifts may indicate that an additional sugar component is present but not accounted for, highlighting the fact that further analysis of the NMR spectra may be required. For the 'Determine Glycan Structure' the structural propositions are ranked according to best fit, where the best fit is normalized to unity, and the chemical shift data of the different structures can be inspected from the ranking list. The structures can in many cases be plotted in SNFG-format. Furthermore, the tentative NMR assignments can be transferred back to a CCPN project for continued analysis of the NMR spectra. The output options described for 'Calculate NMR Chemical Shifts' are also available when using the 'Determine Glycan Structure' module.

15.3 Structural Elucidation of a Glucan Oligosaccharide

Lactobacillus reuteri 121 produces an exopolysaccharide by utilizing a number of glucansucrases (GS) and glucosyltransferases. Recently, a novel GS designated GTFB was purified and investigated for specificity.[46] It was shown that GTFB is inactive with the usual substrate, sucrose, but instead it acts on malto-oligosaccharides to produce linear α-(1 → 4)- and α-(1 → 6)-linked glucose-containing oligosaccharides. Both the oligosaccharide products and their corresponding alditols were characterized and provide an opportunity to investigate the predictive capabilities of CASPER. The specific structural elements of the glucans give rise to different ^{1}H and ^{13}C NMR chemical shifts being dependent of the flanking sugar residues. One of the oligosaccharides isolated and characterized in the study by Dobruchowska *et al.*[46] was the pentasaccharide α-D-Glc*p*-(1 → 4)-α-D-Glc*p*-(1 → 6)-α-D-Glc*p*-(1 → 6)-α-D-Glc*p*-(1 → 4)-D-Glc-ol (also referred to as BEDC′R-ol in the original publication). The characteristic ^{1}H and ^{13}C NMR chemical shifts of the glucan pentasaccharide together with four Glc residues, an unknown hexitol and no restrictions to

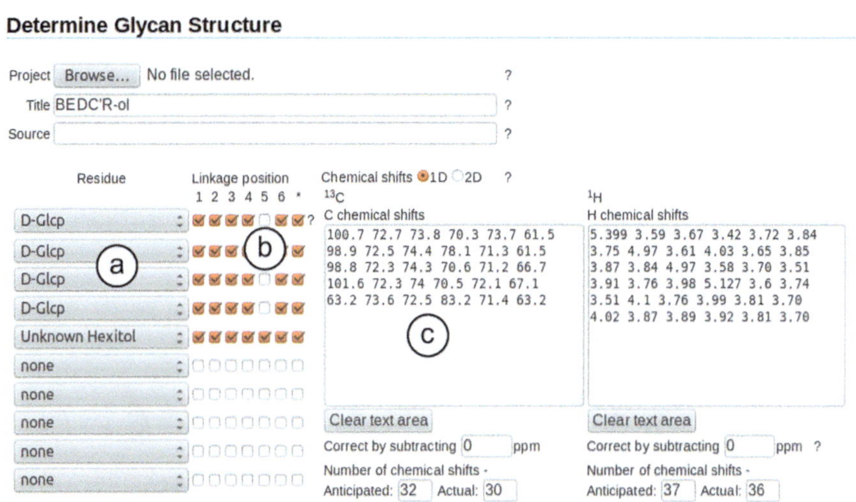

Figure 15.3 Input data to CASPER for the structural determination of a glucan pentasaccharide; (a) Sugar residues, (b) linkage positions as wildcard characters, (c) ^{13}C and ^{1}H NMR chemical shifts entered under the '1D-option'.

linkage positions using the wildcard option denoted by an asterisk (meaning that all ticked positions are tried as 'and/or') were given as input data to CASPER (Figure 15.3). The calculations carried out by CASPER take less than one minute on a personal computer allocated to this end. The output of structural suggestions rank the above pentasaccharide alditol highest (Figure 15.4), but the difference in relative ranking to the second proposal is small which means that to differentiate the two top-ranked structural suggestions a separate additional NMR experiment, such as a $^{1}H,^{1}H$-NOESY or a $^{1}H,^{13}C$-HMBC experiment, would be needed, which in conjunction with the tentative NMR assignments given by CASPER as part of the chemical shift predictions may be sufficient to resolve the two closely ranked structures. As part of the output of the determination of the primary structure of the oligosaccharide a 3D molecular model can be generated from the results pages within CASPER by the software tool CarbBuilder[35] producing a file in pdb-format of the pentasaccharide that can be viewed by a suitable graphics visualization program (Figure 15.5).

	Proposed structures for BEDC'R-ol	Ranking
1.	α-D-Glc-(1→4)-α-D-Glc-(1→6)-α-D-Glc-(1→6)-α-D-Glc-(1→4)-D-Glc-ol	1.00
2.	α-D-Glc-(1→6)-α-D-Glc-(1→4)-α-D-Glc-(1→6)-α-D-Glc-(1→4)-D-Glc-ol	1.01
3.	α-D-Glc-(1→6)-α-D-Glc-(1→6)-α-D-Glc-(1→4)-α-D-Glc-(1→4)-D-Glc-ol	1.09
4.	α-D-Glc-(1→4)[α-D-Glc-(1→6)]α-D-Glc-(1→6)-α-D-Glc-(1→4)-D-Glc-ol	1.35
5.	α-D-Glc-(1→4)[α-D-Glc-(1→6)-α-D-Glc-(1→6)]α-D-Glc-(1→4)-D-Glc-ol	1.36

Figure 15.4 Ranking of structures by CASPER for a glucan pentasaccharide alditol based on ^{13}C and ^{1}H NMR chemical shifts. The sugar residues are denoted by BEDC'R-ol as given in the original publication.[46] The relative ranking is based on a normalized value of unity for the structure with the best fit of experimental NMR chemical shifts compared to those predicted by CASPER.

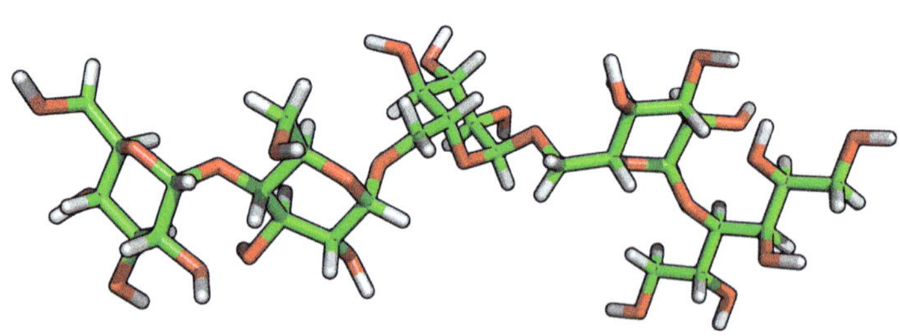

Figure 15.5 3D structure of glucan pentasaccharide alditol α-D-Glc*p*-(1 → 4)-α-D-Glc*p*-(1 → 6)-α-D-Glc*p*-(1 → 6)-α-D-Glc*p*-(1 → 4)-D-Glc-ol produced by CarbBuilder as part of the CASPER output and visualized using The PyMOL Molecular Graphics System, version 1.3 Schrödinger, LLC.

15.4 Structural Elucidation of a Lipopolysaccharide Having *O*-Acetyl Groups as Substituents

NMR spectroscopy analysis of O-antigen structures is usually performed on materials that have been treated with dilute aqueous acid to cleave the O-antigen from the lipid A part. Sometimes NMR spectra of native lipopolysaccharides (LPS) show excellent quality,[47] and highly similar NMR spectra may be observed between different types of preparations.[48] In cases when the polymer is degraded under acidic conditions, the LPS can instead be treated with dilute aqueous ammonia, which results in *O*-deacylation of lipids from lipid A as well as any *O*-acyl groups in the O-antigen polysaccharide, also referred to as LPS-OH, rendering a material useful for NMR analysis in aqueous solution.[49]

In the following analysis it will be assumed that the bacterial species (*Salmonella*) is known *a priori* by serotyping but that the serogroup is unknown. The ^{1}H,^{13}C-HSQC NMR spectrum of the LPS was of excellent quality. O-antigen biosynthesis in *Salmonella* is either initiated by WecA, forming an UndPP-GlcNAc residue which can be converted to UndPP-GalNAc by an epimerase encoded by *gnu*, or by WbaP, forming an UndPP-Gal residue.[50,51] Inspection of the ^{1}H,^{13}C-HSQC spectrum revealed a lack of nitrogen-bearing carbons of *N*-acetylhexosamine sugars in the ^{13}C spectral region ~48–58 ppm, suggesting a galactosyl-initiated repeating unit O-antigen structure. The following three sugars should then be part of the O-antigen, *viz.*, D-Man, L-Rha and D-Gal. About two-thirds of the Gal-initiated *Salmonella* O-antigens also contain 3,6-dideoxyhexoses, which will give characteristic signals from the deoxygenated positions. A peak in addition to the expected C6 from L-Rha in the methyl region and a CH_2-group resonating at ~30 ppm in the ^{13}C NMR spectrum reveals an O-antigen containing a 3,6-dideoxyhexose residue. The presence of *O*-acetyl groups as substituents on O-antigens usually make the spectral appearance much more complicated[52,53] and from a permutation point of view it will make the structural determination by CASPER more complicated and time-consuming; therefore, a two-step procedure was initially chosen based on *O*-deacetylation and determination of the carbohydrate structure of the O-polysaccharide followed by locating the position of the *O*-acetyl group on the LPS using CASPER also in the second step.[54]

Four cross-peaks in the anomeric region of the ^{1}H,^{13}C-HSQC NMR spectrum of the LPS-OH suggested a tetrasaccharide repeating unit (Figure 15.6a). As input to CASPER the three hexoses and one unknown 3,6-dideoxyhexose residue were chosen (Figure 15.7a) where all linkage positions are set as possible using the wildcard option (Figure 15.7b). When supplying NMR chemical shifts to CASPER the system can be underdetermined (*i.e.* overlapping signals can be left unresolved) and some signals from experimental data will be duplicated to satisfy the anticipated number of signals required. In conjunction with information from the ^{13}C NMR spectrum peak-picking of the ^{1}H,^{13}C-HSQC spectrum resulted in ^{13}C and ^{1}H chemical shifts that were

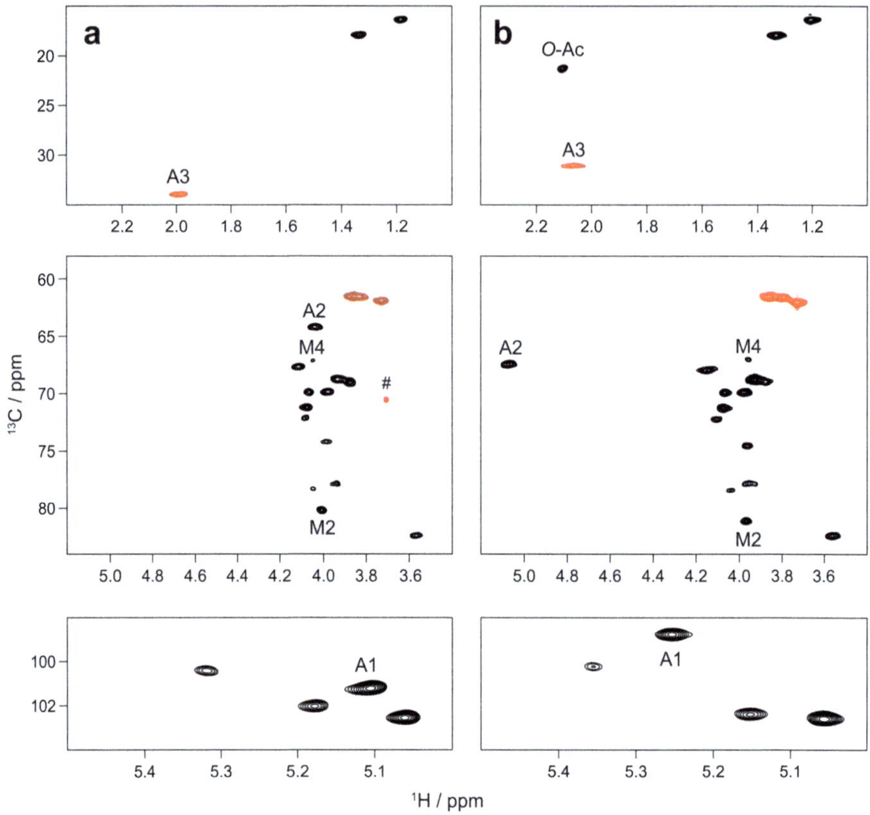

Figure 15.6 Spectral regions of ^{1}H,^{13}C-HSQC NMR spectrum of an unknown *Salmonella* LPS-OH (a) and LPS (b) showing the anomeric region (bottom), the ring atom and hydroxymethyl group region (middle), and the methylene and methyl group region (top). Cross-peaks that significantly change their chemical shifts or are removed upon treatment under basic conditions are annotated, where A refers to abequose and M to mannose. A polyethylene glycol impurity is denoted by a hash symbol.

entered under the '1D-option' (Figure 15.7c). The structural elucidation by CASPER is then ready to execute.

Once the calculation, which takes ~1 min, is finished, the user will be redirected to the results page listing up to ten proposed structures. The highest ranked structure had a backbone of three sugar residues, →2)-α-D-Man*p*-(1 → 4)-α-L-Rha*p*-(1 → 3)-α-D-Gal*p*-(1→, onto which an abequose group is α-(1 → 3)-linked to the mannosyl residue (Figure 15.8, top). This O-antigen structure corresponds to O-factor 4 in *S.* Typhimurium,[55] having a structure in common with that from *Citrobacter freundii* O22.[56] The excellent agreement between experimental (manually determined using additional NMR experiments) and CASPER-predicted ^{1}H and ^{13}C NMR chemical shifts of LPS-OH is shown in Figure 15.9a.

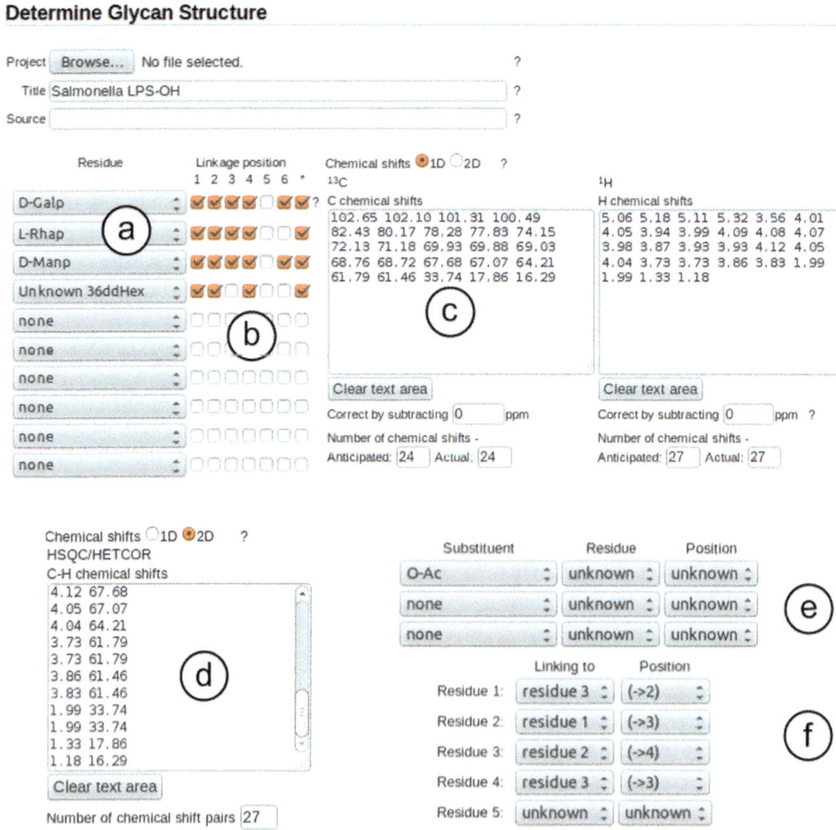

Figure 15.7 Input data to CASPER for the structural determination the LPS-OH of an unknown *Salmonella* strain; (a) sugar residues, (b) linkage positions including wildcard characters, (c) ^{13}C and ^{1}H chemical shifts entered under the '1D-option' and (d) and the chemical shifts entered as ^{1}H/^{13}C pairs. Additional input data for the structural determination the LPS; (e) *O*-acetyl substituent and (f) constraints for linkage positions of the sugar residues, *viz.*, Residue 1 = Gal, Residue 2 = Rha, Residue 3 = Man and Residue 4 = Abe. See Appendix 15.1 for NMR chemical shift data.

In the subsequent step of the O-antigen structure determination of the LPS containing an *O*-acetyl group, the structure determined for the LPS-OH is constrained. By going to the advanced options an *O*-acetyl group is added (Figure 15.7e) but the linkage position and the residue where it is located are still unknown. The now known linkages between the four different sugar residues are specified (Figure 15.7f). NMR chemical shifts from the ^{1}H,^{13}C-HSQC NMR spectrum (Figure 15.6b), including those at δ_H/δ_C 2.11, 21.21 from the methyl group of an *O*-acetyl group (but not the one from the carbonyl group at δ_C 173.98 which should only be entered when ^{13}C NMR data under the 1D-option is used) and constituent residues are entered in the same way

Proposed structures for *Salmonella* LPS-OH		Ranking
1.	→2)[α-Abe-(1→3)]α-D-Man-(1→4)-α-L-Rha-(1→3)-α-D-Gal-(1→	1.00
2.	→3)[α-Col-(1→3)]α-D-Gal-(1→2)-α-D-Man-(1→4)-α-L-Rha-(1→	1.07
3.	→3)[α-D-Gal-(1→2)[α-Abe-(1→3)]α-D-Man-(1→4)]α-L-Rha-(1→	1.15
4.	→3)[α-Abe-(1→2)]α-D-Man-(1→4)-α-L-Rha-(1→2)-α-D-Gal-(1→	1.16
5.	→2)[α-Abe-(1→3)]α-D-Man-(1→4)-α-L-Rha-(1→2)-α-D-Gal-(1→	1.16

Proposed structures for *Salmonella* LPS		
1.	→2)[α-Abe2Ac-(1→3)]α-D-Man-(1→4)-α-L-Rha-(1→3)-α-D-Gal-(1→	1.00
2.	→2)[α-Abe4Ac-(1→3)]α-D-Man-(1→4)-α-L-Rha-(1→3)-α-D-Gal-(1→	1.80
3.	→2)[α-Abe2Ac-(1→3)]α-D-Man-(1→4)-β-L-Rha-(1→3)-α-D-Gal-(1→	1.86
4.	→2)[α-Abe-(1→3)]α-D-Man-(1→4)-α-L-Rha-(1→3)-α-D-Gal2Ac-(1→	1.97
5.	→2)[α-Abe2Ac-(1→3)]β-D-Man-(1→4)-α-L-Rha-(1→3)-α-D-Gal-(1→	2.04

Proposed structures for *Salmonella* LPS		
1.	→2)[α-Abe2Ac-(1→3)]α-D-Man-(1→4)-α-L-Rha-(1→3)-α-D-Gal-(1→	1.00
2.	→3)[α-D-Gal-(1→2)[α-Abe2Ac-(1→3)]α-D-Man-(1→4)]α-L-Rha-(1→	1.10
3.	→3)[α-Col2Ac-(1→3)-α-D-Gal-(1→2)-α-D-Man-(1→4)]α-L-Rha-(1→	1.10
4.	→2)[α-Abe2Ac-(1→3)]α-D-Man-(1→4)-α-L-Rha-(1→2)-α-D-Gal-(1→	1.14
5.	→3)[α-Abe2Ac-(1→2)]α-D-Man-(1→4)-α-L-Rha-(1→2)-α-D-Gal-(1→	1.16

Figure 15.8 Ranking of structures by CASPER for the structural determination of the O-antigen polysaccharide from a *Salmonella* strain; (top) LPS-OH, (middle) LPS elucidation in which location of *O*-acetyl group a variable to the known O-antigen structure determined from the LPS-OH material and (bottom) LPS elucidation in which the hexose residues D-Man, L-Rha and D-Gal should be present in the backbone of the *Salmonella* polysaccharide and used as a constraint since the polymer lacks an *N*-acetylhexosamine sugar.

as previously described, but since the *O*-acetylation position is unknown all linkage positions are set as possible, despite the fact that the inter-residual linkage positions are known and defined under the 'advanced options'. The structural elucidation of the LPS by CASPER is then ready to execute.

The calculations, which only take a second, produce a top-ranked structure that has the *O*-acetyl group assigned to position O2 on the abequose residue (Figure 15.8, middle), *i.e.*, and the resulting O-antigen structure →2)[α-Abe*p*2Ac-(1 → 3)]-α-D-Man*p*-(1 → 4)-α-L-Rha*p*-(1 → 3)-α-D-Gal*p*-(1→, corresponds to *S. Typhimurium* O-factor 5.[53] The second ranked structure

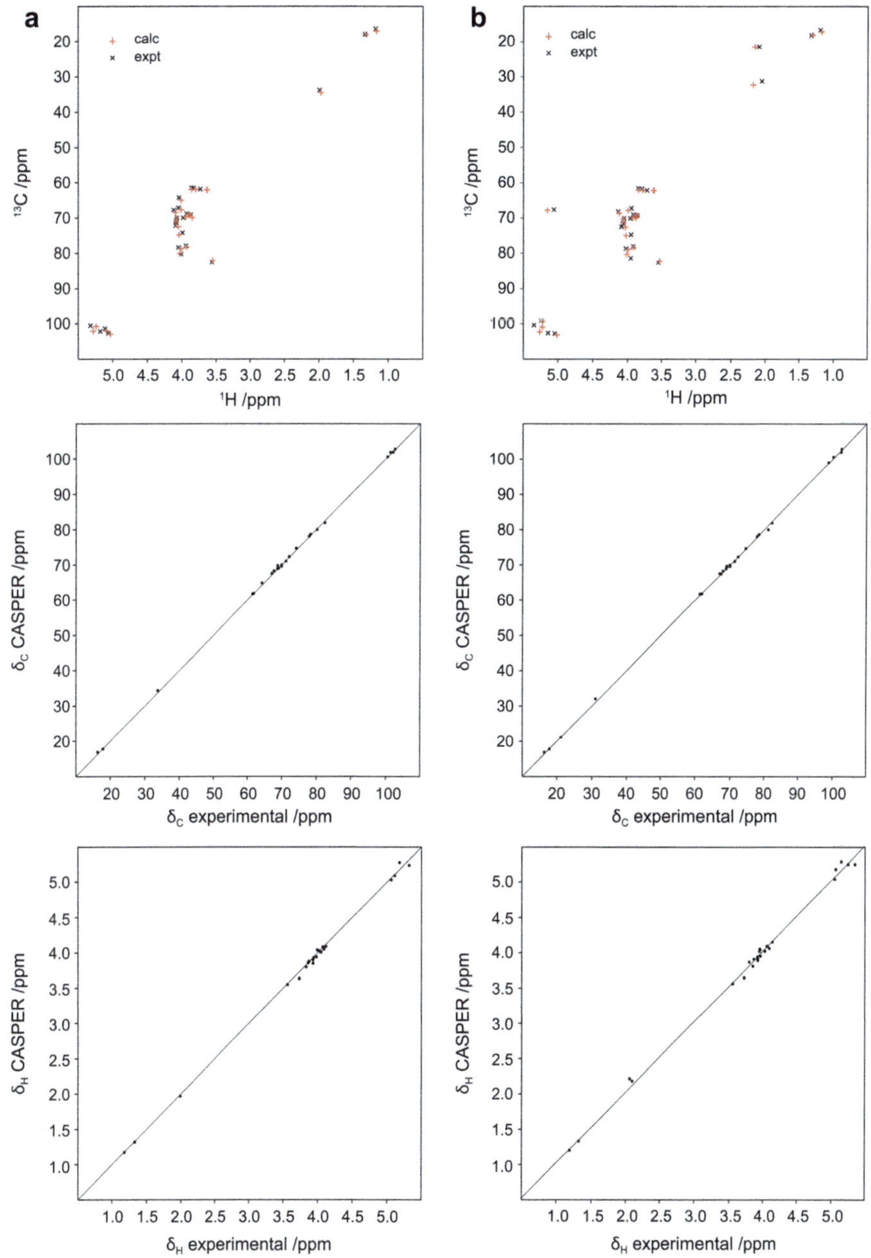

Figure 15.9 Comparison between the NMR chemical shifts of the experimental and CASPER-predicted assignments for ^{1}H (bottom), ^{13}C (middle) and in a 2D ^{1}H/^{13}C-correlation map (top) of the *O*-deacetylated LPS-OH (a) and the LPS (b) materials from a *Salmonella* strain.

had a relative ranking value of 1.86, which strongly suggests that the first proposed structure is the only possible alternative, since a difference in magnitude of ~20% usually is sufficient for distinguishing the structural proposals. The ^{1}H NMR chemical shift of H2 in Abe at the *O*-acetylated position is displaced significantly downfield[40,57] and the chemical shifts of neighboring positions are also subject to some changes (Figure 15.6b). The ^{1}H and ^{13}C NMR chemical shift differences between the experimental and calculated data were slightly larger for these neighboring positions, but the overall assignments made by CASPER were still in good agreement (Figure 15.9b) when compared to those manually determined using additional NMR experiments. Using a two-step procedure to determine the polysaccharide structure *per se* and then the *O*-acetyl substitution position, the structure could be unequivocally determined. Structural as well as NMR spectroscopical aspects of *O*-acetylation at position O2 of abequose, which results in O-factor 5, have been studied in detail by several research groups.[52,53,58,59]

In the absence of *N*-acetylhexosamine sugars in *Salmonella* O-antigens, the hexose residues D-Man, L-Rha and D-Gal should be present in the backbone. This was used as a constraint together with the ^{1}H,^{13}C–HSQC NMR data, an unknown 3,6-dideoxyhexose residue and the presence of an *O*-acetyl group but with unknown location. Thus, the above described two-step procedure is here exchanged for a single step, where information based on 'Biological rules' is utilized for decreasing the number of permutations that CASPER needs to consider. The calculations take ~6 min in this case and the output of the CASPER-based methodology shows that the *S. Typhimurium* O-factor 5 structure is again the top-ranked one with a moderate difference in the relative ranking (Figure 15.8, bottom). Notably, other 3,6-dideoxy sugars are present in the highly ranked structures.

15.5 Future Developments

In future developments CASPER will be directed toward using MS data as additional input information to expedite the process of generating structures for which chemical shifts predictions should be carried out, *i.e.*, limiting the number of permutations that will be needed to be taken into account, *cf.* the pruning of structures that are inconsistent with the number of pertinent spin–spin coupling constants of anomeric protons present. Furthermore, when the differences in ranking values are small between top-ranked structures additional experiments need to be performed, which typically would be a suitable 2D NMR experiment; the use of ^{1}H NMR data employing NMR spin-simulation approaches, such as PERCH,[38,40,60] to generate regular 1D ^{1}H-based spin–spin coupled NMR spectra that can be compared to the experimental spectrum may differentiate the top-ranked structures in a fully automatic fashion thereby giving higher credence to the proposed structure that is ranked highest in a similar way to what was noted when 2D ^{1}H,^{13}C-correlations were added to a structural determination initially based on ^{13}C NMR data.[18] The current option to read CCPN projects as input could be extended to also handle peak lists from Bruker Topspin. In addition, information on constituent sugar components should be possible to elucidate, also with respect to their absolute configuration, by bioinformatics approaches unraveling the biosynthetic pathways and enzymes carrying out these transformations. Substituents present on the sugar residues should also be identifiable employing bioinformatics information.[61]

Appendix 15.1

^{13}C NMR chemical shifts of LPS-OH

102.65 102.10 101.31 100.49 82.43 80.17 78.28 77.83 74.15 72.13 71.18 69.93 69.88 69.03 68.76 68.72 67.68 67.07 64.21 61.79 61.46 33.74 17.86 16.29

^{1}H NMR chemical shifts of LPS-OH

5.06 5.18 5.11 5.32 3.56 4.01 4.05 3.94 3.99 4.09 4.08 4.07 3.98 3.87 3.93 3.93 4.12 4.05 4.04 3.73 3.73 3.86 3.83 1.99 1.99 1.33 1.18

^{1}H,^{13}C-HSQC NMR chemical shifts of LPS-OH

5.06 102.65 5.18 102.10 5.11 101.31 5.32 100.49 3.56 82.43 4.01 80.17 4.05 78.28 3.94 77.83 3.99 74.15 4.09 72.13 4.08 71.18 4.07 69.93 3.98 69.88 3.87 69.03 3.93 68.76 3.93 68.72 4.12 67.68 4.05 67.07 4.04 64.21 3.73 61.79 3.73 61.79 3.86 61.46 3.83 61.46 1.99 33.74 1.99 33.74 1.33 17.86 1.18 16.29

^{13}C NMR chemical shifts of LPS

102.54 102.43 100.17 98.78 82.34 81.16 78.40 77.83 74.50 72.24 71.28 69.88 69.88 69.01 68.85 68.80 67.94 67.41 67.02 61.93 61.46 30.99 21.21 17.88 16.38 173.98

^{1}H NMR chemical shifts of LPS

5.06 5.16 5.36 5.26 3.57 3.97 4.04 3.93 3.96 4.11 4.08 4.07 3.98 3.87 3.93 3.93 4.15 5.07 3.96 3.73 3.73 3.80 3.86 2.07 2.07 2.11 1.34 1.21

^{1}H,^{13}C-HSQC NMR chemical shifts of LPS

5.06 102.54 5.16 102.43 5.36 100.17 5.26 98.78 3.57 82.34 3.97 81.16 4.04 78.40 3.93 77.83 3.96 74.50 4.11 72.24 4.08 71.28 4.07 69.88 3.98 69.88 3.87 69.01 3.93 68.85 3.93 68.80 4.15 67.94 5.07 67.41 3.96 67.02 3.73 61.93 3.73 61.93 3.80 61.46 3.86 61.46 2.07 30.99 2.07 30.99 2.11 21.21 1.34 17.88 1.21 16.38

Acknowledgements

This work was supported by grants from the Swedish Research Council and The Knut and Alice Wallenberg Foundation. Professor C. Whitfield is thanked for kindly providing the LPS sample.

References

1. G. Widmalm, *Carbohydr. Res.*, 2013, **378**, 123–132.
2. H. Yagi, N. Takahashi, Y. Yamaguchi, N. Kimura, K. Uchimura, R. Kannagi and K. Kato, *Glycobiology*, 2005, **15**, 1051–1060.
3. G. Jarvas, M. Szigeti and A. Guttman, *Electrophoresis*, 2015, **36**, 3094–3096.
4. S. Daikoku, G. Widmalm and O. Kanie, *Rapid Commun. Mass Spectrom.*, 2009, **23**, 3713–3719.
5. Y. Kostyukevich, A. Kononikhin, I. Popov and E. Nikolaev, *J. Mass Spectrom.*, 2015, **50**, 1150–1156.
6. T. Kurtán, G. Pescitelli, P. Salvadori, Á. Kenéz, S. Antus, L. Szilágyi, T.-Z. Illyes and I. Szabó, *Chirality*, 2008, **20**, 379–385.
7. C. Mensch, R. Pendrill, G. Widmalm and C. Johannessen, *ChemPhysChem*, 2014, **15**, 2252–2254.
8. C. Fontana, H. Kovacs and G. Widmalm, *J. Biomol. NMR*, 2014, **59**, 95–110.
9. E. Kupče, T. Nishida and R. Freeman, *Prog. Nucl. Magn. Reson. Spectrosc.*, 2003, **42**, 95–122.
10. E. Kupče, T. Nishida, G. Widmalm and R. Freeman, *Magn. Reson. Chem.*, 2005, **43**, 791–794.
11. M. R. Palmer, C. L. Suiter, G. E. Henry, J. Rovnyak, J. C. Hoch, T. Polenova and D. Rovnyak, *J. Phys. Chem. B*, 2015, **119**, 6502–6515.
12. E. Pretsch, A. Fürst, M. Badertscher, R. Bürgin and M. E. Munk, *J. Chem. Inf. Comput. Sci.*, 1992, **32**, 291–295.
13. P.-E. Jansson, L. Kenne and G. Widmalm, *Carbohydr. Res.*, 1987, **168**, 67–77.
14. N. K. Kochetkov, A. S. Shashkov, G. M. Lipkind and Y. A. Knirel, *Sov. Sci. Rev., Sect. B*, 1989, **13**, 1–73.
15. M. K. McIntyre and G. W. Small, *Anal. Chem.*, 1987, **59**, 1805–1811.
16. G. W. Small and M. K. McIntyre, *Anal. Chem.*, 1989, **61**, 666–674.
17. D. A. Cumming, C. Hellerqvist and O. Touster, *Carbohydr. Res.*, 1988, **179**, 369–380.
18. P.-E. Jansson, L. Kenne and G. Widmalm, *Pure Appl. Chem.*, 1989, **61**, 1181–1192.
19. P.-E. Jansson, L. Kenne and G. Widmalm, *J. Chem. Inf. Comput. Sci.*, 1991, **31**, 508–516.
20. P.-E. Jansson, L. Kenne and G. Widmalm, *Anal. Biochem.*, 1991, **199**, 11–17.
21. K. Hermansson, P.-E. Jansson, L. Kenne, G. Widmalm and F. Lindh, *Carbohydr. Res.*, 1992, **235**, 69–81.

22. A. Loss, R. Stenutz, E. Schwarzer and C.-W. von der Lieth, *Nucleic Acids Res.*, 2006, **34**, W733–W737.
23. F. V. Toukach and A. S. Shashkov, *Carbohydr. Res.*, 2001, **335**, 101–114.
24. E. Maes, F. Bonachera, G. Strecker and Y. Guerardel, *Carbohydr. Res.*, 2009, **344**, 322–330.
25. F. Pereira, *Carbohydr. Res.*, 2011, **346**, 960–972.
26. F. Pereira, *Molecules*, 2012, **17**, 3818–3833.
27. R. R. Kapaev, K. S. Egorova and P. V. Toukach, *J. Chem. Inf. Model.*, 2014, **54**, 2594–2611.
28. R. R. Kapaev and P. V. Toukach, *Anal. Chem.*, 2015, **87**, 7006–7010.
29. R. R. Kapaev and P. V. Toukach, *J. Chem. Inf. Model.*, 2016, **56**, 1100–1104.
30. M. Lundborg, V. Modhukur and G. Widmalm, *Glycobiology*, 2010, **20**, 366–368.
31. M. A. Rojas-Macias, J. Ståhle, T. Lütteke and G. Widmalm, *Glycobiology*, 2015, **25**, 341–347.
32. W. F. Vranken, W. Boucher, T. J. Stevens, R. H. Fogh, A. Pajon, M. Llinas, E. L. Ulrich, J. L. Markley, J. Ionides and E. D. Laue, *Proteins*, 2005, **59**, 687–696.
33. M. Lundborg and G. Widmalm, *Anal. Chem.*, 2011, **83**, 1514–1517.
34. M. Kuttel, Y. Mao, G. Widmalm and M. Lundborg, *2011 IEEE Seventh Int. Conf. eScience*, 2011, 395–402.
35. M. M. Kuttel, J. Ståhle and G. Widmalm, *J. Comput. Chem.*, 2016, **37**, 2098–2105.
36. R. Stenutz, in *Bioinformatics for Glycobiology and Glycomics: An Introduction*, ed. C.-W. von der Lieth, T. Lütteke and M. Frank, John Wiley & Sons, Ltd, 2009, pp. 311–320.
37. P. Söderman, P.-E. Jansson and G. Widmalm, *J. Chem. Soc., Perkin Trans. 2*, 1998, 639–648.
38. M. U. Roslund, E. Säwén, J. Landström, J. Rönnols, K. H. M. Jonsson, M. Lundborg, M. V. Svensson and G. Widmalm, *Carbohydr. Res.*, 2011, **346**, 1311–1319.
39. A. Adeyeye, P.-E. Jansson, L. Kenne and G. Widmalm, *J. Chem. Soc., Perkin Trans. 2*, 1991, 963–973.
40. J. Rönnols, R. Pendrill, C. Fontana, C. Hamark, T. Angles d'Ortoli, O. Engström, J. Ståhle, M. V. Zaccheus, E. Säwén, L. E. Hahn, S. Iqbal and G. Widmalm, *Carbohydr. Res.*, 2013, **380**, 156–166.
41. P.-E. Jansson, R. Stenutz and G. Widmalm, *Carbohydr. Res.*, 2006, **341**, 1003–1010.
42. M. Lundborg and G. Widmalm, in *Glycoinformatics*, ed. T. Lütteke and M. Frank, Springer New York, New York, NY, 2015, vol. 1273, pp. 29–40.
43. M. Lundborg, C. Fontana and G. Widmalm, *Biomacromolecules*, 2011, **12**, 3851–3855.
44. C. Fontana, A. Weintraub and G. Widmalm, *Chempluschem*, 2013, **78**, 1327–1329.
45. R. Stenutz, P.-E. Jansson and G. Widmalm, *Carbohydr. Res.*, 1998, **306**, 11–17.

46. J. M. Dobruchowska, G. J. Gerwig, S. Kralj, P. Grijpstra, H. Leemhuis, L. Dijkhuizen and J. P. Kamerling, *Glycobiology*, 2012, **22**, 517–528.
47. S. N. Senchenkova, Y. Zhang, A. V Perepelov, X. Guo, A. S. Shashkov, A. Weintraub, B. Liu, G. Widmalm and Y. A. Knirel, *Glycobiology*, 2016, **26**, 335–342.
48. M. Linnerborg, A. Weintraub and G. Widmalm, *Eur. J. Biochem.*, 1999, **266**, 460–466.
49. T. Ali, A. Weintraub and G. Widmalm, *Carbohydr. Res.*, 2006, **341**, 1878–1883.
50. M. M. Cunneen, B. Liu, L. Wang and P. R. Reeves, *PLoS One*, 2013, **8**, e67646.
51. B. Liu, Y. A. Knirel, L. Feng, A. V. Perepelov, S. N. Senchenkova, P. R. Reeves and L. Wang, *FEMS Microbiol. Rev.*, 2014, **38**, 56–89.
52. C. De Castro, R. Lanzetta, S. Leone, M. Parrilli and A. Molinaro, *Carbohydr. Res.*, 2013, **370**, 9–12.
53. F. Micoli, N. Ravenscroft, P. Cescutti, G. Stefanetti, S. Londero, S. Rondini and C. A. Maclennan, *Carbohydr. Res.*, 2014, **385**, 1–8.
54. C. Fontana, M. Lundborg, A. Weintraub and G. Widmalm, *Glycobiology*, 2014, **24**, 450–457.
55. K. Bock, M. Meldal, D. R. Bundle, T. Iversen, P. J. Garegg, T. Norberg, A. A. Lindberg and S. B. Svensson, *Carbohydr. Res.*, 1984, **130**, 23–34.
56. E. Katzenellenbogen, N. A. Kocharova, P. V. Toukach, S. Górska, A. Korzeniowska-Kowal, M. Bogulska, A. Gamian and Y. A. Knirel, *Carbohydr. Res.*, 2009, **344**, 1724–1728.
57. P.-E. Jansson, L. Kenne and E. Schweda, *J. Chem. Soc., Perkin Trans. 1*, 1987, 377–383.
58. G. Zandomeneghi, K. Ilg, M. Aebi and B. H. Meier, *J. Am. Chem. Soc.*, 2012, **134**, 17513–17519.
59. K. Ilg, G. Zandomeneghi, G. Rugarabamu, B. H. Meier and M. Aebi, *Carbohydr. Res.*, 2013, **382**, 58–64.
60. G. F. Pauli, S. Chen, D. C. Lankin, J. Bisson, R. J. Case, L. R. Chadwick, T. Go, T. Inui, A. Krunic, B. U. Jaki, J. B. McAlpine, S. Mo, J. G. Napolitano, J. Orjala, J. Lehtivarjo, S.-P. Korhonen and M. Niemitz, *J. Nat. Prod.*, 2014, **77**, 1473–1487.
61. B. Liu, A. V. Perepelov, M. V. Svensson, S. D. Shevelev, D. Guo, S. N. Senchenkova, A. S. Shashkov, A. Weintraub, L. Feng, G. Widmalm, Y. A. Knirel and L. Wang, *Glycobiology*, 2010, **20**, 679–688.

CHAPTER 16

NMR Databases for Plant Cell Wall Biopolymers

WILLIAM S. YORK*[a], MEHDI ALLAHYARI[b] AND JOHN D. HOGAN[c]

[a]University of Georgia, Complex Carbohydrate Research Center, 315 Riverbend Road, Athens, GA 30602, USA; [b]University of Georgia, Department of Computer Science, 415 Boyd Graduate Studies Research Center, Athens, GA 30602, USA; [c]Boston University School of Medicine, 670 Albany St., Suite 504, Boston, MA 02118, USA
*E-mail: will@ccrc.uga.edu

16.1 Introduction

NMR is an extremely powerful method for determining the structures of complex glycans.[1] *De novo* analysis of 2-dimensional (2D) NMR spectra can provide information about the identity of the monosaccharide residues that compose the glycan, the anomeric configuration of these resonances, and the identity and attachment site of their substituents, including other glycosyl residues and non-glycosyl moieties. The first step in interpreting this NMR data is to assign individual resonances in the spectra to specific nuclei in the glycan. The chemical shift of each nucleus is influenced primarily by the structure and substitution pattern of the residue in which it resides, but to a lesser extent by the overall molecular context of the residue within the glycan. Although the subtle chemical shift effects that vary according to molecular context are difficult to accurately predict from first principles,

New Developments in NMR No.10
NMR in Glycoscience and Glycotechnology
Edited by Koichi Kato and Thomas Peters

Published by the Royal Society of Chemistry, www.rsc.org

assignment of the NMR spectra of a collection of structurally similar glycans makes it possible to leverage these effects to find correlations between specific structural motifs and the chemical shifts of certain diagnostic resonances. This so-called structural reporter approach[2] has been widely used to rapidly obtain detailed information about glycan structures by analysis of one-dimensional (1D) NMR spectra.

The assignment of resonances in the NMR spectra of glycans[3] is the key prerequisite for implementing the structural reporter approach. This assignment is facilitated by the fact that each monosaccharide residue in a glycan usually comprises a single isolated spin system. That is, the (non-exchangeable) protons within each residue interact with each other *via* intra-residue 2- and 3-bond scalar coupling. Homonuclear inter-residue scalar coupling (*i.e.*, between protons in two different residues) are rarely observed, because there are invariably more than three intervening bonds between such protons. The resonances within a given isolated spin system are readily identified using various types of 2D homonuclear correlation spectroscopy, including COSY and TOCSY.[3] The geometry of the residue can be inferred by analysis of the magnitude of the scalar couplings, which vary according to the dihedral angle between coupled nuclei, as described by the Karplus equation.[4] This makes it possible to identify characteristic spin systems of isomeric residues (*e.g.*, glucose and mannose) that are difficult to discern by other techniques such as mass spectrometry. The mapping of each isolated spin system to a specific monosaccharide residue is very convenient in that it allows the analyst to think about the NMR spectra using well-established partonomic concepts.[5] That is, a glycan is composed of parts (monosaccharide residues) that are connected in ways that can be unambiguously and explicitly defined. The structural reporter approach provides a powerful method to identify these parts and their molecular context by empirical evaluation of the subtle effects that other parts of the molecule have on the magnetic environment of each residue.

Application of the structural reporter approach depends on the availability of accurate chemical shift data for collections of glycans with closely related structures. Databases that provide this information are essential,[6,7] as they provide a means to identify correlations between chemical shifts and molecular context. The most comprehensive NMR database providing experimental data for oligosaccharides and other glycans is accessible *via* the Internet (www.glycosciences.de/database/nmr/).

The most basic data required are the chemical shifts of each assigned nucleus in each glycan in the database. This allows the database user to generate lists, for example, of all of the nuclei that have a chemical shift similar to a particular resonance in the spectrum of an unknown. Without additional information, the analyst must rely on personal experience and/or spend considerable time to examine each resonance in the list to identify the structural information contained in the chemical shift. Therefore, databases that provide several different representations of structural motifs containing the resonances on the list make it much easier for the analyst to interpret the data.

As described herein, we have developed the xyloglucan oligoglycosyl alditol database (XGOA-DB) to correlate chemical shift information to molecular context for this specific class of glycans. This was facilitated by the well-established nomenclature for xyloglucan oligosaccharide structure,[8,9] which specifies oligosaccharide structures in terms of substructures containing one or more monosaccharide residues. Providing information about the substructures containing magnetically active nuclei makes it much easier to identify the structural information in a chemical shift measurement. However, this database is limited to xyloglucan oligosaccharides that are described by a particular structural nomenclature. Therefore, we are developing CeWaN, a more generally applicable database for NMR analysis of plant cell wall polymers. CeWaN, also described herein, uses the partonomic model of GLYDE-II (manuscript submitted) to describe molecular structure, making it possible to specify and evaluate the molecular context of assigned nuclei in diverse classes of biologically important glycans. CeWaN also facilitates the identification of pairs of nuclei that are correlated by scalar coupling, a task that can be very time-consuming even for the experienced analyst. This allows the structural reported approach to be extended to include crosspeaks in 2D spectra, significantly increasing the amount of structural information that is directly supplied in response to database queries.

16.2 Experimental

16.2.1 NMR Data

Oligosaccharides were prepared and their NMR spectra were recorded and assigned as described.[10–18] To obtain the data contained in XGOA-DB, oligosaccharides were prepared by endoglucanase digestion of xyloglucans from various plant sources and reduced with sodium borohydride to generate oligoglycosyl alditols, which were purified by a combination of methods, notably size exclusion chromatography and reversed-phase chromatography. Samples were dissolved in deuterium oxide and NMR spectra were recorded at a probe temperature of 300 K, using standard pulse sequences provided by the instrument manufacturers (Bruker and Varian). ^{1}H chemical shifts were measured relative to internal acetone (2.225 ppm). The same data are available through CeWaN at the time of this writing. However, CeWaN is intended to be much broader in scope than XGOA-DB, and we are continuously adding new data sets to CeWaN.

16.2.2 Software Development

XGOA-DB was implemented using MYSQL and PHP.[19] CeWaN was implemented using postgresql (http://www.postgresql.org/), java (https://www.oracle.com/java/), javaservlets (www.oracle.com/technetwork/java/javaee/), the JSME structure editor,[20] the JSmol molecular visualization tool,[21] and the National Cancer Institute's Online SMILES Translator and Structure File Generator (https://cactus.nci.nih.gov/translate/).

16.3 Results and Discussion

16.3.1 The Xyloglucan Oligoglycosyl Alditol Database (XGOA-DB)

XGOA-DB consists of a searchable table containing the ^{1}H-NMR chemical shifts of 53 xyloglucan oligoglycosyl alditols (XGOAs). Xyloglucans are highly branched plant cell wall polysaccharides with a cellulosic backbone consisting of β-(1 → 4)-linked D-Glc*p* residues.[14] Many of the backbone residues bear glycosyl side chains, most often at O6. If each glucosyl residue in the backbone, along with all of its pendant side chains, is considered to be a substructure of the polymer, then the polymer can be considered as a linear array of these substructures. This is the basis for the most commonly used nomenclature used to specify xyloglucan structures.[8,9] Due to the regular pattern of sidechain substitution in xyloglucans, well-defined oligosaccharide subunits of the polymer can be generated by treating it with a xyloglucan-specific endoglucanase.[14] Reduction of the oligosaccharides with sodium borohydride simplifies their purification and spectroscopic analysis by transforming the interconverting anomeric forms of the reducing residues into a single, stable moiety (an alditol). The ^{1}H-NMR chemical shifts of the resulting XGOAs are similar, but not identical to those of the parent reducing oligosaccharides.

The chemical shifts of the resonances in an XGOA are, to a first approximation, a function of a three well-defined parameters:

- The identity of the sugar residue containing the assigned nucleus, including its anomeric configuration and linkage
- The identity of the substructure containing the sugar residue
- The environment of the substructure containing the sugar residue. This includes end effects arising from the proximity of the side chain to the non-reducing or alditol end of the oligomer and the presence of other side chains in the immediate vicinity.

In order to concisely represent these structural features, XGOA-DB uses a nomenclature[8,9] that represents each substructure as a single letter (Figure 16.1). Each distinct substructure consists of a backbone β-D-Glc*p* residue along with its characteristic substituents. For example, an unbranched β-D-Glc*p* residues are represented by the letter G and β-D-Glc*p* residues bearing a monosaccharide sidechain at O6 are represented by the letter X. Reducing glucose residues that have been converted to alditol moieties are indicated by the code "Gol". The most commonly found xyloglucan structures represented by this nomenclature are illustrated in Figure 16.1.

XGOA-DB users can query the database using web interface described in Figure 16.2. Each row in the table corresponds to a specific monosaccharide residue in a specific oligoglycosyl alditol. Each column of the query results table contains specific information described in the header at the top of the table, as described next.

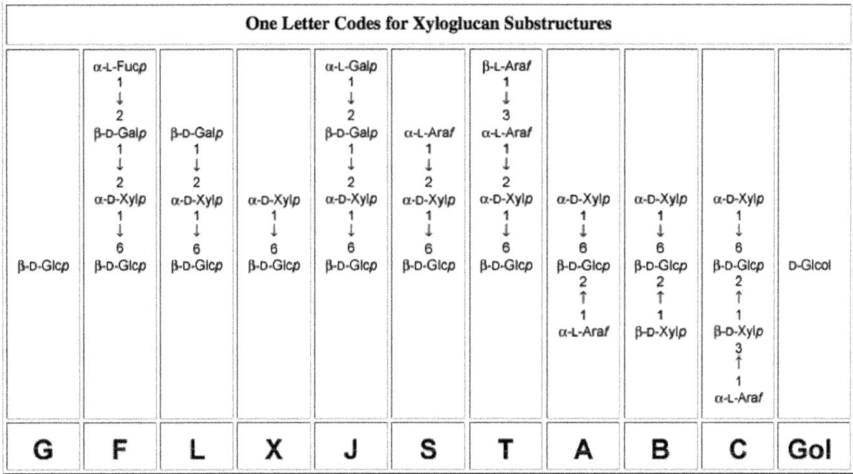

Figure 16.1 Xyloglucan structure nomenclature used by XGOA-DB. An uppercase letter is assigned to each unique sidechain substitution pattern that has been identified in xyloglucan. The structure of an XGOA is unambiguously described by listing the code letters corresponding to each β-D-Glc*p* residue in the backbone, starting with the non-reducing end. For example, the pentasaccharide β-D-Gal*p*-(1 → 2)-α-D-Xyl*p*-(1 → 6)-β-D-Glc*p*-(1 → 4)-β-D-Glc*p*-(1 → 4)-D-glucitol is completely and concisely defined by the code "LGGol".

- *Oligo*—The complete chemical structure of the XGOA, specified using the nomenclature described in Figure 16.1.
- *Env*—The local environment of the substructure containing the residue described in the row. This column contains three letters that, using the nomenclature described in Figure 16.1, specify the part of the XGOA that includes the residue listed in the row. The middle letter refers to the substructure that actually contains the residue, while the two other letters indicate adjacent substructures.
- *SS*—The substructure containing the residue described in the row.
- *Link*—The substitution pattern (linkage) of the residue described in the row. The letter "T" indicates a terminal (unsubstituted) residue and numerical entries indicate the sites that bear glycosyl residues.
- *Sugar*—The identity of the monosaccharide residue described in the row, specified using standard IUPAC nomenclature.
- *H1, H2, etc.*—The identity of the proton whose chemical shift is listed in the column. Dashes ("--") in this column indicate that a proton by this name does not exist in the residue; "n.a." indicates that the named proton exists, but was not assigned.

The query illustrated in Figure 16.2 generates a result table containing resonances in XGOA-DB having a chemical shift of 4.940 ± 0.005 ppm. Figure 16.3 shows the first 5 of the 33 rows in the results table generated by this

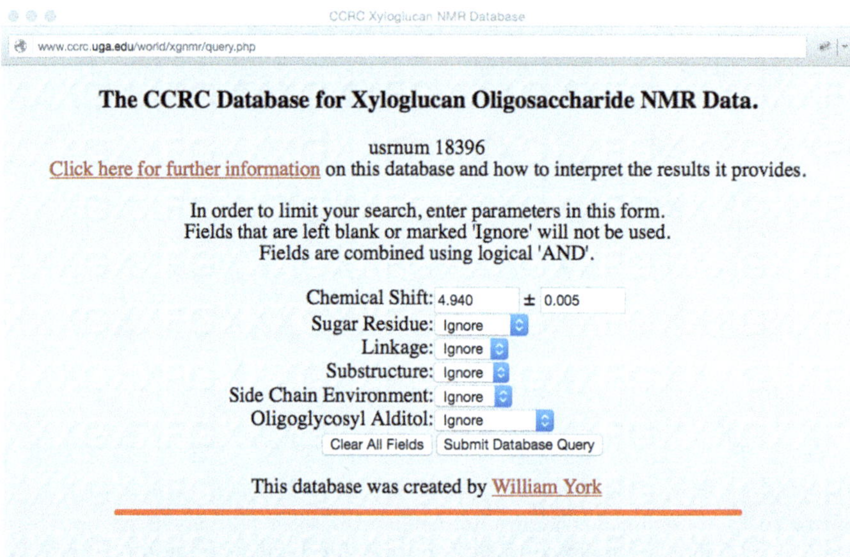

Figure 16.2 Query page for XGOA-DB. This web interface allows users to search for NMR data using several different criteria, including chemical shift range, the identity of the monosaccharide residue containing nuclei whose resonances have been assigned, the substitution pattern (linkage) of the residue, the substructure containing the residue, the molecular environment of the substructure, and the oligosaccharide containing the substructure. All query parameters are combined using logical *and* to generate a list of annotated resonances that fulfill all of the criteria entered by the user. The illustrated query will retrieve all assigned resonances in XGOA-DB having a chemical shift of 4.940 ± 0.005 ppm.

query. The table includes a concise representation of the molecular context of each of the nuclear resonances that are consistent with the query. All rows in this particular table have "α-D-Xyl*p*" in the *Sugar* column, the letter "T" in the *Link* column, the letter "X" in the SS column and a string of characters beginning with "_X" in the *Env* column. The third character in the each entry in the *Env* column and identity of the XGOA (*Oligo* column) vary from row to row. However, when all of the rows in the table are considered, the data suggest that a resonance with a chemical shift of 4.940 ppm is diagnostic for α-D-Xyl*p* residues terminating so-called "X" sidechains that are located at the non-reducing end of the XGOA.

The query illustrated in Figure 16.4 requests information about all nuclei in α-D-Xyl*p* residues that terminate X substructures. Of the 93 rows in the results table (Figure 16.5), 34 correspond to α-D-Xyl*p* residues terminating X substructures at the non-reducing end of the XGOA, as indicated by character strings starting with "_X" in the *Env* column. All 34 have H1 resonances between δ 4.934 and δ 4.943, with 32 having chemical shifts between δ 4.938

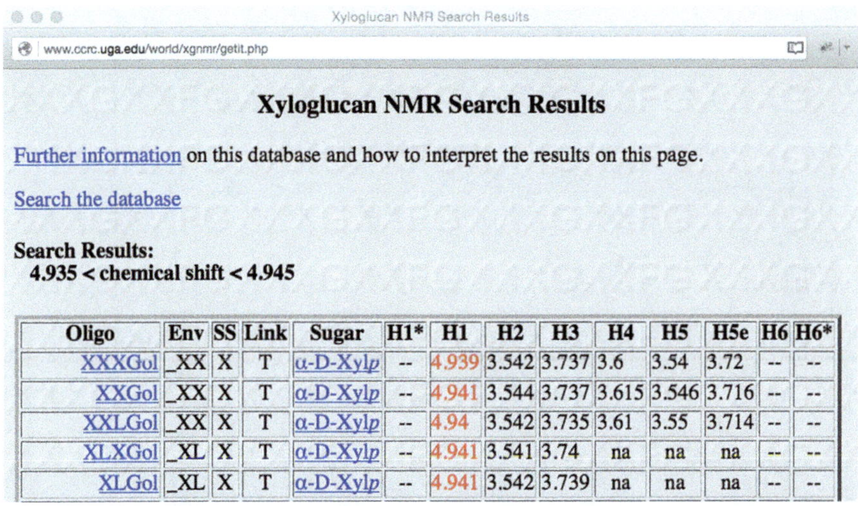
Xyloglucan NMR Search Results

www.ccrc.uga.edu/world/xgnmr/getit.php

Xyloglucan NMR Search Results

Further information on this database and how to interpret the results on this page.

Search the database

Search Results:
4.935 < chemical shift < 4.945

Oligo	Env	SS	Link	Sugar	H1*	H1	H2	H3	H4	H5	H5e	H6	H6*
XXXGol	_XX	X	T	α-D-Xyl*p*	--	4.939	3.542	3.737	3.6	3.54	3.72	--	--
XXGol	_XX	X	T	α-D-Xyl*p*	--	4.941	3.544	3.737	3.615	3.546	3.716	--	--
XXLGol	_XX	X	T	α-D-Xyl*p*	--	4.94	3.542	3.735	3.61	3.55	3.714	--	--
XLXGol	_XL	X	T	α-D-Xyl*p*	--	4.941	3.541	3.74	na	na	na	--	--
XLGol	_XL	X	T	α-D-Xyl*p*	--	4.941	3.542	3.739	na	na	na	--	--

Figure 16.3 Partial results generated by the query illustrated in Figure 16.2. As described in the text, the query results table provides information about the chemical environment of the nuclei giving rise to resonances that fulfill the search criteria along with the chemical shifts of other resonances in the residues that contain these nuclei. In this case, the results indicate that several ^{1}H resonances at δ 4.940 have been assigned to H1 of α-D-Xyl*p* residues that terminate X substructures at the non-reducing end of the oligosaccharide.

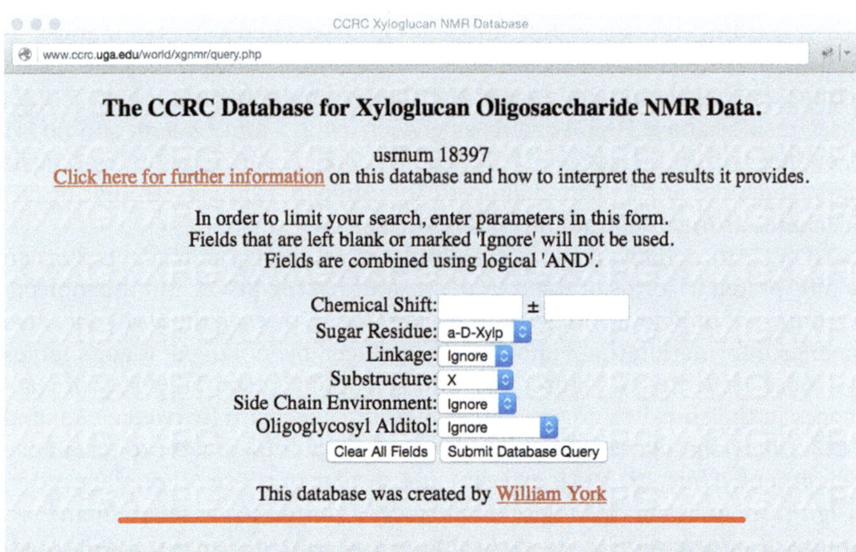
CCRC Xyloglucan NMR Database

www.ccrc.uga.edu/world/xgnmr/query.php

The CCRC Database for Xyloglucan Oligosaccharide NMR Data.

usrnum 18397
Click here for further information on this database and how to interpret the results it provides.

In order to limit your search, enter parameters in this form.
Fields that are left blank or marked 'Ignore' will not be used.
Fields are combined using logical 'AND'.

Chemical Shift: ±
Sugar Residue: a-D-Xylp
Linkage: Ignore
Substructure: X
Side Chain Environment: Ignore
Oligoglycosyl Alditol: Ignore
Clear All Fields | Submit Database Query

This database was created by William York

Figure 16.4 XGOA-DB query requesting all resonances assigned to nuclei in α-D-Xyl*p* residues that terminate a so-called "X" substructure.

Xyloglucan NMR Search Results

www.ccrc.uga.edu/world/xgnmr/getit.php

Xyloglucan NMR Search Results

Further information on this database and how to interpret the results on this page.

Search the database

Search Results:
Sugar Residue is α-D-Xylp
and Substructure is X

Oligo	Env	SS	Link	Sugar	H1*	H1	H2	H3	H4	H5	H5e	H6	H6*
XGol	_XGol	X	T	α-D-Xylp	--	4.934	3.541	3.677	3.623	3.544	3.711	--	--
XXXGol	_XX	X	T	α-D-Xylp	--	4.939	3.542	3.737	3.6	3.54	3.72	--	--
XXXGol	XXX	X	T	α-D-Xylp	--	4.957	3.547	3.732	3.62	3.58	3.73	--	--
XXXGol	XXGol	X	T	α-D-Xylp	--	4.952	3.547	3.671	3.63	3.57	3.73	--	--
XXGol	_XX	X	T	α-D-Xylp	--	4.941	3.544	3.737	3.615	3.546	3.716	--	--
XXGol	XXGol	X	T	α-D-Xylp	--	4.955	3.546	3.67	3.63	3.578	3.735	--	--
XXLGol	_XX	X	T	α-D-Xylp	--	4.94	3.542	3.735	3.61	3.55	3.714	--	--
XXLGol	XXL	X	T	α-D-Xylp	--	4.956	3.545	3.732	3.63	3.588	3.73	--	--

Figure 16.5 Partial results (8 of 93 rows) generated by the query shown in Figure 16.4. Examination of this table (see text) indicates that the precise chemical shift of H1 of an α-D-Xyl*p* residue terminating an X substructure depends on the overall molecular context of the sidechain.

and δ 4.942. This supports the hypothesis (based on data shown in Figure 16.3) regarding the structural information associated with resonances at δ 4.940. The other 53 rows in the table correspond to α-D-Xyl*p* residues terminating X substructures that are *not* at the non-reducing end of the XGOA. These residues have H1 resonances between δ 4.915 and δ 4.980, and no H1 resonance within the range δ 4.938 and δ 4.942, confirming the hypothesis that resonances at 4.940 are diagnostic for α-D-Xyl*p* residues terminating "X" sidechains at the non-reducing end of the XGOA.

XGOA-DB thus allows users to rapidly identify chemical shift effects that can be interpreted in terms of the overall structure of the XGOA and the molecular context of individual nuclei within the XGOA. However, the NMR data also encodes information that is not readily deduced by looking at results tables like those presented in Figures 16.3 and 16.5. This is especially true for resonances in the so-called envelope region of the spectrum (between δ 3.0 and δ 4.1), where signal overlap can be a considerable impediment to resonance assignment. Thus, 2D NMR experiments similar to those originally used to assign resonances in this region are usually required to associate them with structural information when analyzing oligosaccharide mixtures. Although recording the required 2D COSY and TOCSY spectra is quite straightforward, assigning crosspeaks in these spectra just using information from the database XGOA-DB can be very time consuming. Of course, if the sample is dominated by one XGOA structure, the skilled analyst could perform a complete *de novo* analysis of the spectrum to assign the resonances. However, this could be extremely challenging if the sample contains a mixture of XGOAs. In such cases, the analyst could generate lists of resonances with chemical shifts corresponding to the two-dimensional locations of crosspeaks in the spectrum, and then identify pairs of nuclei in these lists that are likely to have an observable mutual scalar coupling. This process becomes more difficult as the lists become longer. Furthermore, the XGOA-DB is limited to the analysis of xyloglucan oligosaccharides, as it is based on a nomenclature specifically designed to describe xyloglucan structures. We therefore decided to develop a new NMR database called CeWaN, which is described next. CeWaN is not limited by the structural nomenclature that it uses and is capable of rapidly identifying crosspeaks in 2D NMR spectra that correspond to pairs of nuclei that exhibit mutual scalar coupling.

16.3.2 CeWaN—The Cell Wall NMR Database

The CeWaN database is currently under development to address the challenges described in the last paragraph. At the time of this writing, it contains approximately 35 datasets, although this number is rapidly increasing as the graphical user interface becomes more mature. These data were recorded and assigned as described.[10–18] Additional functionality, not available in the current version is also being added. This includes, for example, a more robust support for provenance, including explicit literature references supporting the chemical shift assignments in the DB.

CeWaN uses the GLYDE-II format (http://glycomics.ccrc.uga.edu/GLYDE-II/GLYDE-description.pdf manuscript submitted) to describe molecular structure. The molecular partonomy model implicit in this format and its ability to explicitly describe molecular topology provide a robust basis for extracting structural information from NMR spectra of many different kinds of glycans. Although CeWaN focuses on plant cell wall polymers, the code base used for its implementation could be used to create NMR databases for structurally diverse glycans from different sources. The GLYDE-II format that it utilizes is not limited to glycans, but has the ability to represent all sorts of glycoconjugate structures, including lignin-carbohydrate complexes (found in plant in cell walls) as well as glycoproteins and glycolipids that are found in many different biological samples.

Notably, CeWaN leverages the topological information contained in GLYDE-II representation of a molecular structure to facilitate the assignment of crosspeaks in 2D correlation spectra. The pairs of scalar coupled resonances that are identified using CeWaN can be considered structural reporters. This extension of the structural reporter approach has considerable potential facilitate the identification of structures and structural motifs that are not readily inferred by analysis of 1D spectra.

Several challenges had to be met in order to develop CeWaN. The first was to make a user interface that facilitates creation of GLYDE-II representations of structurally diverse building blocks, including monosaccharides, lignin monomers, amino acids, and lipids. GLYDE-II uses GlycoCT[22] as one method to concisely describe the structural properties of a monosaccharide, including the number of carbons it contains along with its ring form, anomeric configuration, and the stereochemistry of each of its asymmetric atoms. GlycoCT is thus extremely powerful and convenient, but it is specifically designed to represent monosaccharides and glycans. Although GLYDE-II provides a way to name (monosaccharide, monolignol, lipid or amino acid) residues using commonly accepted semantics, it was designed to implement a multilevel partonomic model with or without any abstraction of these structural features. That is, it can also represent stereochemical features by explicitly specifying the Cartesian coordinates of the atoms in the residue. The flexibility of the GLYDE-II representation can thus circumvent issues associated with interpreting the different conventions that have evolved to represent these features in different contexts. Furthermore, GLYDE-II provides a straightforward atom connection table describing the molecular topology of each monomer. The information in this table can be used to infer the existence of scalar coupling of two atoms in the monomer, based on the number of intervening molecular bonds.

It is noteworthy that the general structural representation problem outlined here has been elegantly solved for small molecules by development of the IUPAC International Chemical Identifier (InChI) standard.[23] However, specification of highly complex biopolymers composed of many smaller residues using InChI is far from trivial. For example, extracting the InChI representation of a monosaccharide residue from the InChI representation of a glycan containing the monosaccharide residue can be quite demanding, as this process often requires the parameters that specify local stereochemistry to be modified. In contrast, GLYDE-II implements a very simple and general partonomic model that can be used to describe the complex structures of virtually any kind of biopolymer at several levels of granularity. For example, GLYDE-II is well suited to generate unambiguous representations of glycoconjugates that contain both glycan and non-glycan moieties.

GLYDE-II is based in part on the notion that each distinct molecular building block should be defined once as an archetypal structure that is subsequently referenced each time it occurs in the biopolymer. Figure 16.6 illustrates the CeWaN user interface for defining such archetypal structures. The residue structure is drawn using the JSME structure editor[20] in the upper left corner of the figure, which generates a SMILES representation of the molecular topology and stereochemistry. The SMILES representation is automatically submitted to the National Cancer Institute's Online SMILES Translator and Structure File Generator (https://cactus.nci.nih.gov/translate/), which calculates an energetically accessible conformation that is returned to the application as a Protein Data Bank (PDB) file.[24] This PDB file is shown in the text area at the bottom of the figure and used by JSmol[21] to draw a three-dimensional structure of the residue (upper right). JSmol is configured to allow the user to systematically rename each atom according to a scheme that is deemed appropriate for the residue type. As they are changed,

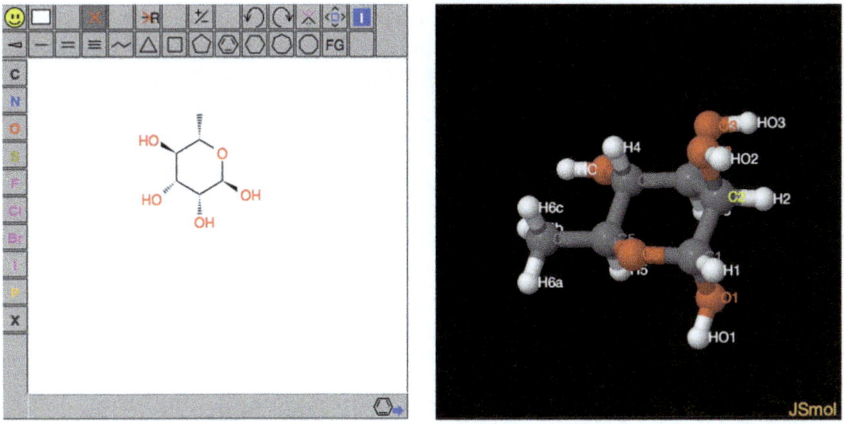

PDB Content

```
HEADER    NONAME 29-Feb-16                                              NONE   1
TITLE                                                                   NONE   2
AUTHOR    Apache                                                        NONE   3
REVDAT   1  29-Feb-16     0                                             NONE   4
ATOM      1  C1           0       0.910  -1.481  -0.063  0.00  0.00           C
ATOM      2  H1           0       1.541  -2.226   0.422  0.00  0.00           H
ATOM      3  O1           0       0.807  -1.778  -1.457  0.00  0.00           O
ATOM      4  O5           0      -0.391  -1.502   0.526  0.00  0.00           O
ATOM      5  C5           0      -1.311  -0.576  -0.056  0.00  0.00           C
ATOM      6  H5           0      -1.418  -0.791  -1.120  0.00  0.00           H
ATOM      7  C6           0      -2.671  -0.707   0.631  0.00  0.00           C
ATOM      8  C4           0      -0.781   0.849   0.126  0.00  0.00           C
ATOM      9  H4           0      -0.712   1.079   1.189  0.00  0.00           H
ATOM     10  O4           0      -1.671   1.773  -0.505  0.00  0.00           O
ATOM     11  C3           0       0.607   0.955  -0.513  0.00  0.00           C
```

Figure 16.6 CeWaN web interface for defining archetypal residues that are used to build biopolymers. This figure illustrates the process of defining the monosaccharide α-L-Rha*p*, which is referenced any time this structure appears as a residue in a biopolymer in the CeWaN database.

the updated names of the atoms automatically appear in the PDB code in the text box, so the user can verify them for completeness and accuracy. When the user is satisfied with the names of the atoms in the representation, it is submitted to the database along with an automatically generated GLYDE-II representation.

Upon receipt of a new residue structure from the user, the CeWaN system populates the database with several new records that include the GLYDE-II and PDB representations. Other new records that are generated comprise lists of magnetically active nuclei (including ^{1}H, ^{13}C and ^{15}N) in the molecule and lists of pairs of these nuclei that are likely to exhibit mutual scalar coupling by virtue of the fact that they separated by three or fewer molecular bonds. At this point, the user can define default values for the scalar coupling constants that are characteristic for the residue. For example, the value of $^{3}J_{1,2}$ (the three-bond scalar coupling constant for the interaction of H1 with H2) of a β-D-Glc*p* residue is typically about 7.8 Hz, and this can be entered as the default value for this parameter. These values are automatically copied for each instance of the residue that is generated when a molecule containing the residue is created and submitted by the user. This facilitates population of CeWaN with reasonable scalar coupling constants, but allows the user to override these values when they have been measured for specific molecules. These explicit measurements can be important for pairs of nuclei (such as H5 and H6R in a β-D-Glc*p* residue) whose scalar coupling constant depends on the overall molecular context of the residue, which significantly affects the average dihedral angle for the two nuclei.

Having created archetypal monomeric residues, the user can then use these to create representations of polymers (molecules) that contain them. Figure 16.7 illustrates the creation of a new molecule. A new user interface was created to implement this function, as most existing interfaces, such as the one implemented in GlycoWorkbench,[25] are limited to certain classes of molecules (such as glycans).

After a molecule has been generated and submitted to CeWaN, NMR data sets can be created and associated with it. Each data set is annotated with information about the NMR experiments used to generate the chemical shift and coupling constant data it contains. These include the solvent, the chemical shift standard, the chemical shifts of the standard for each type of nucleus (*e.g.*, ^{1}H or ^{13}C), the probe temperature, and the ^{1}H Larmor frequency of the spectrometer (MHz).

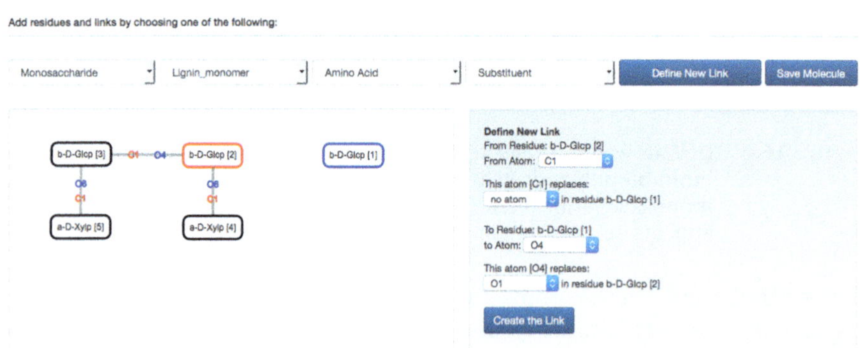

Figure 16.7 CeWaN interface for creating new polymeric molecules. This shows the creation of the pentasaccharide named XXG using the xyloglucan structure nomenclature described in Figure 16.1. The user selects residues comprising the molecule from the pull-down lists at the top of the figure. Each selected residue is added to the molecule drawing panel at the lower left, and the user can then drag it to an appropriate position. Directed chemical bonds (such as glycosidic linkages) from one residue to another are added by selecting one residue as the "from" residue (b-D-Glcp[2] in this case) and another as the "to" residue (b-D-Glcp[1]). The dialog box at the lower right allows the user to specify the replacement of a specific atom in one residue with an atom in the other, which often occurs during bond formation (*e.g.*, in the dehydrating condensation mechanism that forms glycosidic bonds). In this case, O1 of b-D-Glcp[2] is replaced with O4 of b-D-Glcp[1] upon formation of a glycosidic bond between the two residues.

The user can then use the CeWaN data input dialogs to manually enter chemical shift and coupling constant information that comprises the data set. Alternatively, the user can upload a text file that contains the chemical shift data for any or all of the magnetically active nuclei in the molecule. Several different data sets can be generated for each molecule, for example, using spectra that were recorded at different temperatures or using different solvents.

CeWaN is intended to have a much broader scope than XGOA-DB and thereby contain data recorded under various conditions and assigned using various chemical shift standards, as appropriate for each sample being analyzed. Thus, each data set must be associated with the name and chemical shift of the standard used to assign the data. The database contains a list of common standard compounds and a "recommended" chemical shift value that is associated with each. For example, acetone in D_2O is assigned a recommended chemical shift of 2.225 ppm. However, the analyst who makes chemical shift assignments may not have used this recommended standard chemical shift value. CeWaN normalizes all chemical shift values to be consistent with values that are expected if the analyst had used the recommended shift value for the standard. For example, is the analyst entering a data set specifies a value of 2.200 ppm for the standard chemical shift of acetone in D_2O, all of the chemical shift values are automatically adjusted by adding 0.025 ppm to the entered values and the result is saved. This is not apparent to the user, who sees the chemical shift values in the data set as entered. A similar process occurs when the user queries CeWaN. The query must include the name and chemical shift value of the standard used to measure chemical shifts in the spectrum of the "unknown". Upon submission of the query, the target chemical shifts are adjusted to conform with the recommended chemical shift of the standard. For example, if the chemical shifts of unknown were assigned relative to acetone in D_2O with a chemical shift of 2.200 ppm, then 0.025 ppm is added to the value of the target chemical shift such that the query is compatible with the recommended value for the chemical shift of acetone, and thus easily comparable to normalized chemical shifts in the CeWaN database.

It is important to note that chemical shift queries return data that are defined *relative the standard chemical shift values specified in the query itself.* That is, the chemical shift values returned by a query may be numerically distinct from the initially entered and the normalized chemical shifts saved in CeWaN. This maintains consistency between the query results and the query itself, thereby facilitating evaluation of the spectral data that the user is attempting to interpret. Thus, the user does not have to do any arithmetic to compare an observed resonance to one returned by querying the CeWaN database.

The NMR data in CeWaN is queried using the interface shown in Figure 16.8, which illustrates how a user might request information that is required to assign a crosspeak in the COSY spectrum of a mixture of xyloglucan oligosaccharides. Assigning such a crosspeak by *de novo* analysis could be very challenging if the mixture contains several different oligosaccharides. The information provided by CeWaN allows such crosspeaks to be interpreted as structural reporters. That is, the pairs of interacting nuclei returned upon submission of a 2D query can be examined in order to find common

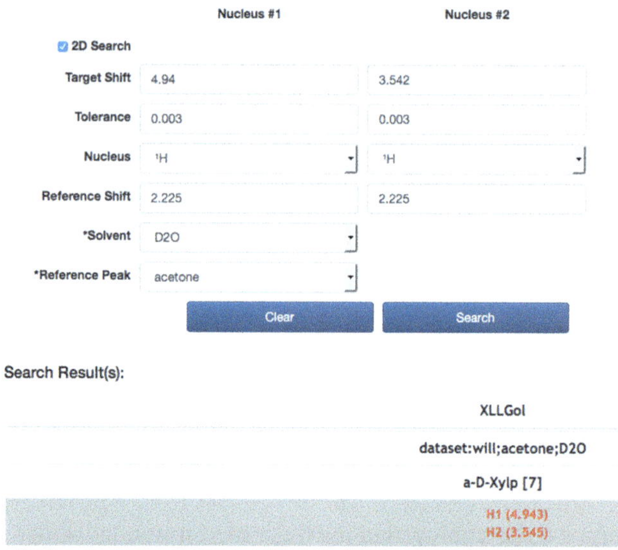

Figure 16.8 Two-dimensional query of CeWaN. The chemical shifts of two interacting nuclei are entered in the dialog box along with information about the experimental conditions used to measure these shifts. Pairs of nuclei that match these criteria and that are separated by three or fewer intervening bonds are retrieved. The chemical shifts of each pair are listed below in a hierarchical format organized according to molecule, data set, and residue. Only the first of the nine pairs returned for this query is shown.

structural motifs that can be associated with crosspeaks at the specified spectral position. In the example shown in Figure 16.8, all nine pairs of nuclei that are returned correspond to H1 and H2 of a terminal xylose residue at the non-reducing end of a xyloglucan oligosaccharide. This allows the analyst to conclude that, for NMR spectra of xyloglucan oligosaccharides, COSY crosspeaks at δ 4.940, δ 3.542 are diagnostic for this structure.

The user can also retrieve chemical shift and coupling constant data for specific structures that are in the database. This is done using the same interface that is used to enter chemical shift data into the system. However, unless the dataset was originally entered by the active user, this data can only be viewed, not modified.

16.4 Conclusions

Two databases (XGOA-DB and CeWaN) providing information about the NMR spectra of plant cell wall polymers were developed. XGOA-DB is specifically designed to provide information about the ^{1}H spectra of oligoglycosyl alditols generated by chemical reduction of the products formed upon treatment of xyloglucans with an endoglucanase. This highly focused database leverages the concise nomenclature often used to represent xyloglucan structures to generate queries and responses that reveal details about the chemical context of nuclei that are observed in their NMR spectra. CeWaN is designed to be much more flexible, handling NMR data from structurally diverse biopolymers that are found in the plant cell wall. However, the database schema and software used to implement CeWaN can be used to implement databases that contain NMR data for other types of biopolymers, such as human glycoproteins that have biomedical relevance.

Due to its limited scope, we do not foresee any further development of XGOA-DB, although it has proven quite useful in its present form. In contrast, at the time of this writing, only an alpha version of the CeWaN database exists. We plan to implement the following extensions of CeWaN by the time this chapter becomes available to the scientific community.

- Extensive population of the database with information for diverse plant cell wall biopolymers.
- Improvement of the user interface to streamline the entry and editing of structural information.
- Improvement of the user interface to facilitate the interpretation of query results, for example, to identify 2D structural reporters. This includes methods to visualize each molecule returned upon querying the database along with the chemical shift data that are provided.
- Import and export of structural information to external software and databases, such as the glycan structure registry GlyTouCan,[26] that support uniform semantic annotation of data.
- Implementation of web services and/or command line interfaces to facilitate automated assignment of crosspeaks in 2D correlation spectra.
- Support of motif structures to complement information for explicitly defined molecules.

Acknowledgements

This research was funded by the U.S. Department of Energy Bioenergy Research Center supported by the Office of Biological and Environmental Research in the DOE Office of Science. We also acknowledge the U.S. Department of Energy-funded Center for Plant and Microbial Complex Carbohydrates (Grant DE-FG02-93ER20097) for supporting the instrumentation and computer infrastructure required to complete this research.

References

1. M. Lundborg and G. Widmalm, *Anal. Chem.*, 2011, **83**, 1514–1517.
2. B. R. Leeflang, E. J. Faber, P. Erbel and J. F. G. Vliegenthart, *J. Biotechnol.*, 2000, **77**, 115–122.
3. J. Ø. Duus, C. H. Gotfredsen and K. Bock, *Chem. Rev.*, 2000, **100**, 4589–4614.
4. C. A. G. Haasnoot, F. A. A. M. De Leeuw and C. Altona, *Tetrahedron*, 1980, **36**, 2783–2792.

5. J. Rogers and A. Rector, *Proc. Annu. Symp. Am. Med. Inf. Assoc.*, 2000, 714–718.
6. J. A. van Kuik, K. Hard and J. F. Vliegenthart, *Carbohydr. Res.*, 1992, **235**, 53–68.
7. T. Lutteke, A. Bohne-Lang, A. Loss, T. Goetz, M. Frank and C. W. von der Lieth, *Glycobiology*, 2006, **16**, 71r–81r.
8. S. T. Tuomivaara, K. Yaoi, M. A. O'Neill and W. S. York, *Carbohydr. Res.*, 2015, **402**, 56–66.
9. S. C. Fry, W. S. York, P. Albersheim, A. G. Darvill, T. Hayashi, J.-P. Joseleau, Y. Kato, E. P. Lorences, G. A. Maclachlan, M. McNeil, A. J. Mort, J. S. G. Reid, H. U. Seitz, R. R. Selvendran, A. G. J. Voragen and A. R. White, *Physiol. Plant.*, 1993, **89**, 1–3.
10. S. Hantus, M. Pauly, A. G. Darvill, P. Albersheim and W. S. York, *Carbohydr. Res.*, 1997, **304**, 11–20.
11. Z. H. Jia, Q. Qin, A. G. Darvill and W. S. York, *Carbohydr. Res.*, 2003, **338**, 1197–1208.
12. W. S. York, L. K. Harvey, R. Guillen, P. Albersheim and A. G. Darvill, *Carbohydr. Res.*, 1993, **248**, 285–301.
13. W. S. York, V. S. K. Kolli, R. Orlando, P. Albersheim and A. G. Darvill, *Carbohydr. Res.*, 1996, **285**, 99–128.
14. W. S. York, H. van Halbeek, A. G. Darvill and P. Albersheim, *Carbohydr. Res.*, 1990, **200**, 9–31.
15. M. J. Pena, A. G. Darvill, S. Eberhard, W. S. York and M. A. O'Neill, *Glycobiology*, 2008, **18**, 891–904.
16. M. J. Pena, Y. Kong, W. S. York and M. A. O'Neill, *Plant Cell*, 2012, **24**, 4511–4524.
17. M. J. Pena, R. Zhong, G. K. Zhou, E. A. Richardson, M. A. O'Neill, A. G. Darvill, W. S. York and Z. H. Ye, *Plant Cell*, 2007, **19**, 549–563.
18. M. Hoffman, Z. Jia, M. J. Peña, M. Cash, A. Harper, A. R. Blackburn Ii, A. Darvill and W. S. York, *Carbohydr. Res.*, 2005, **340**, 1826–1840.
19. H. E. Williams and D. Lane, *Web Database Applications with PHP and MySQL*, O'Reilly Media, Inc., 2004.
20. B. Bienfait and P. Ertl, *J. Cheminf.*, 2013, **5**, 24.
21. R. M. Hanson, J. Prilusky, Z. Renjian, T. Nakane and J. L. Sussman, *Isr. J. Chem.*, 2013, **53**, 207–216.
22. S. Herget, R. Ranzinger, K. Maass and C.-W. von der Lieth, *Carbohydr. Res.*, 2008, **343**, 2162–2171.
23. S. R. Heller, A. McNaught, S. Stein, D. Tchekhovskoi and I. V. Pletnev, *J. Cheminf.*, 2013, **5**, 7.
24. N. Guex and M. C. Peitsch, *Electrophoresis*, 1997, **18**, 2714–2723.
25. D. Damerell, A. Ceroni, K. Maass, R. Ranzinger, A. Dell and S. M. Haslam, *Biol. Chem.*, 2012, **393**, 1357–1362.
26. K. Aoki-Kinoshita, S. Agravat, N. P. Aoki, S. Arpinar, R. D. Cummings, A. Fujita, N. Fujita, G. M. Hart, S. M. Haslam, T. Kawasaki, M. Matsubara, K. W. Moreman, S. Okuda, M. Pierce, R. Ranzinger, T. Shikanai, D. Shinmachi, E. Solovieva, Y. Suzuki, S. Tsuchiya, I. Yamada, W. S. York, J. Zaia and H. Narimatsu, *Nucleic Acids Res.*, 2016, **44**, D1237–D1242.

CHAPTER 17

Polysaccharides as Major Carbon Sources in Environmental Biodiversity

JUN KIKUCHI*[a,b,c] AND TAKANORI KOMATSU[a,b]

[a]RIKEN Center for Sustainable Resource Science, 1-7-22 Suehiro-cho, Tsurumi-ku, Yokohama, Kanagawa 230-0045, Japan; [b]Graduate School of Medical Life Science, Yokohama City University, 1-7-29 Suehiro-cho, Tsurumi-ku, Yokohama, Kanagawa 230-0045, Japan; [c]Graduate School of Bioagricultural Sciences, Nagoya University, 1 Furo-cho, Chikusa-ku, Nagoya, Aichi 464-0810, Japan
*E-mail: jun.kikuchi@riken.jp

17.1 Introduction

17.1.1 Carbon Cycle in Environmental Biodiversity

Natural ecosystems can be conceptually thought of as interconnected environmental and metabolic systems; *e.g.* biologically driven carbon flux through ecosystems at a global scale, but also through biochemical pathways at an organismal or population scale (Figure 17.1). Although global climatic change is controversial, a significant increase in carbon emission from soil over the last 40 years has been reported to have contributed slightly to global climate change.[1] Another more direct aspect of these issues can be the result

New Developments in NMR No.10
NMR in Glycoscience and Glycotechnology
Edited by Koichi Kato and Thomas Peters

Published by the Royal Society of Chemistry, www.rsc.org

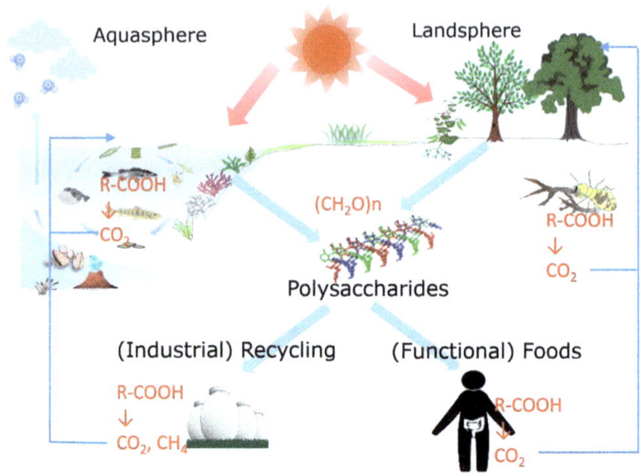

Figure 17.1 Material cycling originated from polysaccharides in nature. Most polysaccharides are primarily produced *via* photosynthesis in land and marine plants, and consumed by animals or microflora. Polysaccharides also contribute to energy production as feedstock for methane fermentation and ethanol production.

of direct resource extraction by humans. For example, overfishing can completely restructure patterns across trophic levels in coastal ecosystems.[2]

Promoting human society and preserving natural ecosystems are both issues of great importance. Advances in agricultural/fishery science are providing approaches to overcome food shortages arising as a result of increasing population in developing countries.[3] However, agricultural/fishery research in industrialized countries today emphasizes food processing and added-value products rather than alleviating food shortages in developing countries, which have greater food demands.[4,5] Considering the remarkable diversity of food culture over the world, it is important to gain an understanding of natural ecology in various environment.[6]

On the basis of these reports on the direct and indirect effects of agriculture, forestry and fisheries on natural ecosystems, which produce biomass resource, we proposed the application of a recent engineering paradigm to ecosystem research. There currently is considerable effort to transition from petrochemical-based energy and manufacturing to a bio-based model; *i.e.* from oil refineries to bio-refineries.[7] Efforts to replace these underground resources as raw materials for industry by compounds and minerals obtained from renewable plant biomass are underway.[8] Similarly, identification of renewable enzymes to be used as reactive catalysts for chemical reactions leading to biomass production is a major focus.[9,10] For example, it is important to monitor reactions and yields of intermediates as raw materials are converted to biomass products in a quantitative manner in the chemical engineering field.[11,12]

17.1.2 Polysaccharides in Landsphere

Plant biomass is the most abundant and important material in the terrestrial biosphere. Its major components are cellulose, hemicellulose, and lignin. These components are all complex molecules and are abundantly produced in plant cell walls. Cellulose is a linear condensation polymer that consists of β-1, 4-linked D-glucose units with a degree of polymerization from 100 to 20 000. Cellulose has strong interchain hydrogen bonding between adjacent cellulose by hydroxyl groups and this strong interchain hydrogen-bonding network makes crystalline cellulose resistant to enzymatic hydrolysis. Hemicelluloses are branched and the molecular masses are lower than that of cellulose. The main constituents are glucose, mannose, galactose, xylose, and arabinose. Lignins are reticulated, cross-linked macromolecules composed of phenylpropanoid units, which include *p*-hydroxyphenyl, guaiacyl, and syringyl. Lignin is the second most available biological polymer on earth although the structures are one of the most recalcitrant for biodegradation of all natural polymers. These persistent substances have a highly formed supramolecular structure which includes cellulose and hemicellulose cemented by lignin, called lignocellulose. Lignocellulosic cell walls in plants play important roles in providing plant strength, protecting plants from microbial attack, and surviving in severe environmental conditions.

17.1.3 Polysaccharides in Aquasphere

Approximately 8 million tons of wet seaweeds are harvested annually worldwide; stranded seaweeds on the beach constitute a considerable part of this. Stranded seaweeds are harvested from beaches and utilized for a variety of purposes such as feed, fertilizer, and a source of raw material for the industrial production of phytochemicals of commercial importance. In addition, seaweeds have a biosorption function, and metal-sorbing potential of their constituent polysaccharides has been indicated. To measure chemical components such as macromolecules, metabolites, and elements, a variety of analytical methods and their integrated data analysis are required for comprehensive characterization and evaluation of the seaweed biomass. Therefore, NMR in combination with other analytical techniques are useful to investigate characterization of algal polysaccharides.[13,14]

Microalgae (planktons) play important roles in element cycles in hydrosphere as primary producers.[15–17] After they perish, they also play roles as source materials for sedimentary organic matter (SOM),[18,19] particulate organic matter (POM),[20] and dissolved organic matter (DOM).[21–34] The Redfield ratio is a widely accepted empirical rule of elemental ratios (C:N:P = 106:16:1) in marine plankton, and it has contributed to the study of element cycles in the hydrosphere. Elemental ratios closely relate to the metabolism of organisms, therefore chemical speciation analyses can provide detailed information regarding the underlying mechanism of material cycles.[35] Therefore, we developed cellular solid-state NMR techniques for comprehensive analyses of biological macromolecules.[36] The developed method was applied *Euglena gracilis*, a flagellated protist, cultured in a different nitrogen concentration to demonstrate the power of this method for profiling and comparing cellular components, as described Section 3.

17.1.4 Contribution of NMR and Computing Techniques Toward Environmental Biodiversity

NMR is one of the most powerful tools for characterization of molecules. Solution NMR has been used to determine primary and stereochemical structures of synthesized[37,38] and extracted small molecules in organic and natural chemistry,[27,30,39] respectively. Methodological advances in multidimensional NMR analysis over the past four decades have extended its use to the biological field, in which it has made significant contributions, such as the signal assignment of biomacromolecular complexities.[40,41] In particular, the sequential assignment and structural analysis of high molecular weight proteins have become possible by combining information from chemical fragments, such as H–N, N–C, and C–C, using stable isotope labeling of samples and three-dimensional NMR.[42,43] After the sequential assignment, the secondary structure of proteins can be determined using conformational-dependent information from ^{1}H and ^{13}C chemical shifts.[44,45] Furthermore, interaction analysis of huge macromolecular complexes, such as GroES/EL and 20S proteasome, is also possible based on the assigned chemical shift perturbation analysis.[46,47] Recently, solution NMR has also been used for metabolic profiling in which NMR spectra of a mixture of biological small molecules in multiple-analyte are recorded and these spectra are submitted to further multivariate analyses for the purposes of identifying a biomarker of disease,[48,49] evaluate nutrients in foods[50–53] and so on. Stable isotope labeling with multidimensional NMR brought a technical breakthrough for analyzing complicated biological macromolecules including proteins[54–58] and nucleic acids, as well as sugars, and enabled determination of their 3D structures. Comprehensive analysis of biological macromolecules performed without purification is the next frontier in the advancement of insights into living systems. Breakthrough technologies in basic biological research toward polysaccharide complexity can be provided by profiling the approach to biomass synthesis and degradation as an intermediate-monitoring system. We have reported on methodologies for profiling metabolites in plants,[59–62] animals[50,52,63–65] and microbial systems,[66–68] and annotating metabolites.[69–72] Innovation in "non-targeted" approaches for a wide range of biological systems in ecosystems is important for better biomass production and the sustainability of such systems.

17.2 Methodology

17.2.1 Sample Preparation

Both solution and solid-state NMR has been developed in NMR analyses for polysaccharides, therefore sample preparation procedures depend on which you use. In general, solution NMR is higher-resolution than solid-state NMR where solubilization processes are required. Solid-state NMR, is powerful for structure and dynamics analyses in native state of polysaccharide, and requires less sample preparation procedures than solution state. The resultant NMR spectra strongly depend on sample preparation, but unfortunately sample preparation procedures are not normalized yet. In this section, several common sample preparation procedures of polysaccharide for vascular plants are introduced. Note that it is recommended to select sample preparation procedures carefully according to the analyte, especially in algae which have quite different physical and chemical properties (Figure 17.2).

17.2.1.1 Stable Isotope Labeling

Stable isotope labeling, especially with ^{13}C, tremendously enhances sensitivity, and versatility in both solution[73–76] and solid-state NMR[36,77–89] of polysaccharides. For the plant polysaccharides, there are two different strategies for stable isotope labeling with ^{13}C, one is *via* photosynthesis using ^{13}C-labeled carbon dioxide, and another is *via* uptake ^{13}C-labeled organic compounds from roots.[90]

A cautionary notice in letter strategy is that ^{13}C-enrichments decrease by incorporation of $^{12}CO_2$ *via* photosynthesis.[91] And also note that there are biases in incorporation of ^{13}C (and ^{12}C) due to different metabolic activities.

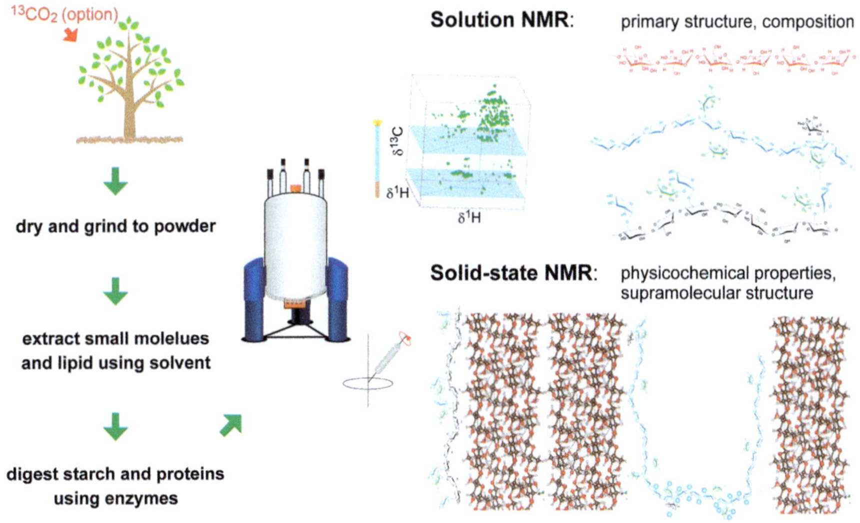

Figure 17.2 Analyzing cell wall polysaccharides in plants by NMR. Isotope labeling improves the versatility of NMR. Water, metabolite, lipid, protein, and starch are typically removed before NMR analyses. For solution NMR analyses, cell wall materials are dissolve into solvent based on DMSO-*d6* after fine milling with a ball mill machine. Solution NMR spectra provide information about the primary structure of polysaccharides and their compositions. While signals are generally broader than those in solution NMR spectra, solid-state NMR spectra provide information about the physicochemical properties originating from the supramolecular structure of the cell wall.

Metabolic flux analyses indicated nitrogen-rich amino acids such as arginine and asparagine are major metabolite synthesized in root and translocated to other tissues.[91,92] This might result in heterogenic cell wall isotope labeling. Almost uniform (>97 atom% ^{13}C) labeled plants with ^{13}C can be obtained using the former strategy using $^{13}CO_2$ when $^{12}CO_2$ is almost completely removed. However this strategy requires special equipment such as a hermetically-sealed plant chamber, and generally costs more than latter strategy.[90] Isotope labeling services of plants are also offered by private companies.[93]

17.2.1.2 *Obtaining Polysaccharide Fractions*

One of the most common procedures to obtain polysaccharide fractions are called alcohol insoluble residue (AIR)-based methods, in which ethanol solution and a mixture of methanol and chloroform are used to remove small molecules and lipids. At first, the dried plant material is ground to powder using a milling machine, and then 70 or 80% (v/v) ethanol solution added to remove polar metabolites. In one study, this procedure was performed under boiling conditions to denature enzymes, and in another study, it was performed at 4 °C to keep enzymes inactive.[94] In some study, a suspension with phenol–acetic acid–water mixture (2: 1:1) was used to deactivate enzymes completely.[95] Subsequently, the residue was suspended in a mixture of methanol and chloroform (1:1) and extracted cellular lipids. Then, enzymatic digestion of starch was performed with α-amylase. In addition, enzymatic digestion of proteins was performed with protease. As a result of these procedures, cell wall fractions were obtained.

Cell wall fractions from vascular plant contain cellulose, hemicellulose, pectin, and lignin. Additional extractions are required to fractionate into each component. A solution of chelating agents such as CDTA (cyclohexane-1,2-diaminetetraacetic acid) is used for pectin extraction because chelating agents remove pectin linkage *via* calcium ion with galacturonate. Alkaline solution is used to extract the hemicellulose fraction.

17.2.1.3 *Solubilization*

Cell wall fractions barely dissolve in any organic solvent, therefore grinding to a fine powder is essential to be solubilized. Ball mill instruments are frequently used for cell wall solubilization.[96] Cell wall acetylation is performed to enhance the solubility toward chloroform and dimethylsulfoxide (DMSO).[96] Gel-state cell wall NMR techniques have been developed using DMSO,[97] DMSO with 1-methylimidazole,[98] and DMSO with pyridine.[99] Several ionic liquids and molten salt were used as co-solvent including 1-butyl-3-methylimidazolium chloride,[100] 1-ethyl-3-methylimidazolium acetate ([Emim] OAc),[101] pyridinium chloride[102] with DMSO. Several solvent systems using ionic liquids can dissolve cell wall materials including cellulose microfibrils without ball milling procedure. The dissolving mechanisms were investigated using ^{1}H–^{13}C HSQC–NOESY spectra to assess proximity between cellulose with [Emin]OAc using intermolecular nuclear Overhauser effect. There was the strongest cross peak between H2 in [Emim]OAc with H2 in cellulose, indicating imidazolium cation associated with the collapse of hydrogen bonds in cellulose.[103]

17.2.2 NMR Measurements and Data Processing

17.2.2.1 Solution NMR

Signals in ^{1}H NMR spectra of polysaccharide such as cell wall polysaccharides in vascular plants are often broad line in shape due to high viscosity of sample, therefore ^{1}H–^{13}C HSQC is used for cell wall composition analyses.[73,97–99,101,102] Intensity, area, and volume of the peaks in ^{1}H–^{13}C HSQC spectra are not quantitative because they strongly depend on $^{1}J_{CH}$ and relaxation rate of the nuclei. Heikkinen *et al.* reported an optimized ^{1}H–^{13}C HSQC experiment using variable polarization transfer delay in INEPT building blocks, and it was applied for an analysis of milled wood lignin.[104] Hu *et al.* developed $HSQC_0$ experiment in which quantitative intensity, area, or volume of each signals were estimated using a series of multiple HSQC building block with regression extrapolation.[105] And it was applied quantitatively to whole cell wall analysis of *Miscanthus*.[101]

Multidimensional NMR with ^{13}C-labeled plants provided a lot of detailed resonance assignments of plant cell wall especially polysaccharides. Komatsu and Kikuchi reported resonance assignment of plant cell wall from *Zea mays* (corn), *Solanum tuberosum* L. (potato), and *Cichorium intybus* (chicory) with 119 resonance assignment of ^{1}H–^{13}C correlations in ^{1}H–^{13}C HSQC spectra.[73] 3D HCCH–COSY experiment was powerful for resonance assignment of polysaccharides because crowded peaks from polysaccharides were spread into 3D space.[73,74] ^{1}H–^{13}C HSQC–NOESY spectra provided inter-sugar correlations in polysaccharide. Inter-sugar correlations cannot be obtained in ^{1}H–^{1}H COSY, ^{1}H–^{1}H TOCSY, ^{1}H–^{13}C HSQC, and also 3D HCCH–COSY spectra. Long range correction spectroscopy such as HMBC also provide inter-sugar correlations in polysaccharide, however it is unstable for polysaccharide because sometimes polysaccharide solutions have high viscosity leading coherence lifetime too short to evolve long range J_{CH} in the HMBC experiment. These analyses provided detailed resonance assignment such as arabinofuranose in glucuronoarabinoxylan in corn and arabinoxyloglucan in potato. These hemicelluloses are specially in some plant families and results agree with previous knowledge in plant science.

17.2.2.2 Solid-State NMR

Solid-state NMR is useful for detecting structural and dynamical features of analyte that are lost on dissolution or cannot be solubilized. Solid-state NMR of polysaccharides, especially in plant cell wall, provides plenty of information involving structure, composition, their dynamics, and proximities of each component in native state. Relaxation phenomena in NMR are strongly related to molecular motions, therefore relaxation time analyses are employed to assess dynamics of polysaccharides.[106] There are several relaxation time constant including longitudinal relaxation of ^{1}H (^{1}H T_1), rotating-frame ^{1}H (^{1}H $T_{1\rho}$), ^{13}C (^{13}C T_1), transverse relation of ^{1}H (^{1}H T_2), ^{13}C (^{13}C T_2), and sensitive frequencies of molecular motion in each relaxation process are different.[106] Note that relaxation time constant in ^{1}H represent means of molecular motion within domains due to rapid ^{1}H–^{1}H spin diffusion. On the other hand, ^{13}C relation time constants represent local molecular motions of each nucleus.

In general, ^{1}H and ^{13}C T_1 and ^{1}H $T_{1\rho}$ is longer, and ^{1}H and ^{13}C T_2 is shorter in cellulose than matrix polysaccharides in hydrated plant cell wall. Hong and her co-workers investigated differences in cell wall physicochemical properties *via* ^{13}C T_1 and ^{1}H local $T_{1\rho}$ (without spin diffusion using ^{1}H–^{1}H homonuclear decoupling) using various primary cell wall of Arabidopsis including wild-type, chemically pectin-depleted, homozygous xxt1xxt2xxt5 mutant (xyloglucan-depleted mutant). These time constants in the primary cell wall of the xyloglucan-depleted mutant were far smaller than those in wild-type.[78,79] This indicated xyloglucan had roles to enhance the rigidity of primary cell wall in dicot plants.

Different relaxation time between each component can be used as spectral filtering techniques. Tang *et al.* developed proton relaxation-induced spectral editing techniques (PRISE) using differences of ^{1}H T_1, $T_{1\rho}$, and T_2 for plant cell wall analysis by high-resolution solid state NMR.[107] Renard and Jarvis applied spectral editing based on ^{1}H $T_{1\rho}$-delay, CP kinetics (T_{CH}), short recycling delay with direct polarization (^{13}C T_1) for analysis of polysaccharides in sugar beet cell wall.[108] Komatsu and Kikuchi developed ^{13}C–^{13}C correlation spectroscopy coupled with spectral filtering techniques to analyze matrix polysaccharides by removing signals from cellulose.[77] Dipolar dephasing was introduced as a spectral filter in which coherence was dephased by ^{1}H–^{13}C dipolar interaction and amplitude strongly depend on molecular motion.

^{13}C–^{13}C spin-diffusion-based correlation spectroscopy, such as Proton-Driven Spin Diffusion (PDSD),[109] Dipolar Assisted Rotational Resonance (DARR),[110] Phase Alternated Recoupling Irradiation Schemes (PARIS),[111] Second-order Hamiltonian among Analogous Nuclei Generated by Heteronuclear Assistance Irradiation (SHANGHAI),[112] SHA+[113], provided information about proximities among each cell wall component. Foston *et al.* applied DARR experiment with selective excitation of carbon in lignin using SELDOM[114] (selectivity by destruction of magnetization) to assess the proximity between lignin and cellulose and hemicelluloses in the plant cell wall.[89] As a result, the average distance between lignin and cellulose and hemicelluloses was estimated 0.7–2.0 nm from build-up curves of ^{13}C spin diffusion.[89] Wang *et al.* investigated proximities of cellulose, matrix polysaccharides, and cell wall proteins using a series of mixing time of PDSD spectra of ^{13}C uniform labeled primary cell wall of Arabidopsis.[85] It was estimated that approximately 25–50% of cellulose contacted with pectin whereas far less cell wall protein content with any polysaccharide was enough to cause $^{13}C–^{13}C$ spin diffusion.[85] Water and polysaccharides interactions determine physical properties of the material, therefore those interactions in the plant cell wall are also considered to be important to understand cell wall material. White *et al.* assessed water and polysaccharides interactions in the hydrated primary cell wall of Arabidopsis using a combination of ^{13}C cross polarization, $^{1}H–^{1}H$ water to polysaccharide spin diffusion, and initial ^{1}H T_2 filter.[86] In the primary cell wall at 263K, $^{1}H–^{1}H$ water to polysaccharide in pectin build up faster than that in cellulose as functions of mixing time. This indicated pectin interacted with water stronger than cellulose. $^{1}H–^{1}H$ water to polysaccharide build up shower in Ca^{2+} extracted primary cell wall than intact cell wall, indicating there are bound water near the linkage structure *via* Ca^{2+} in homogalacturonan and it might determine property in primary cell wall in dicot plant.[86]

17.2.2.3 Multivariate Analysis

Principal component analysis (PCA) is a powerful tool to summarize multivariate in a small number of components. Kikuchi *et al.* performed PCA to characterize the cell wall in various plants, as an application of Ecomics web tool, using solution state $^{1}H–^{13}C$ HSQC spectra of 11 plant cell walls.[115] Bm-char tool in Ecomics enables you to assign query chemical shifts to 88 known chemical signals of lignin and hemicellulose (Table 17.1), including

Table 17.1 List of useful databases and web tools for the NMR analysis of polysaccharides and their metabolism.

Name	Remarks
CASPER	Web-based database for predicting the oligo/polysaccharide structure[162]
ECODAB	Relational DB for *E.coli* O-antigens[163]
ECOMICS	$^{1}H/^{13}C$ chemical shift DB of lignocellulose and correlation analysis[10]
BMRB	Biomolecular NMR database including lignocellulose and other metabolites[164]
HMDB	Human metabolome database including NMR parameters for sugars and other metabolites[165]
COLMAR	^{1}H -total correlation spectroscopy DB[166]
TOCCATA	^{13}C -total correlation spectroscopy DB[167]
SpinCouple	^{1}H chemical shift and spin-coupling DB of sugars and other metabolites[168]
MetaboAnalyst	Web-tool for processing, data normalization, statistical analysis[169]
*i*coshift	Signal alignment tool for 1D-NMR spectra[170]
MVAPACK	Source platform for complete NMR metabolomics data handling[171]
KODAMA	Unsupervised and semisupervised learning algorithm[172]
BATMAN	Automated metabolite deconvolution and quantification from complex NMR spectra[173]
rNMR	R-based HSQC analytical free software[174]

signals of 42 aromatic and 17 aliphatic sites in lignin, 26 hemicellulosic sites, and 3 uncategorized sites.[10,115] A score plot in PCA indicated first principal component (PC) 1 and PC 2 in vectors represented tree to grass specific polysaccharides and guaiacyl to syringyl lignin predominantly existed in the cell wall, respectively. HetMap tool in Ecomics web tool provide 2D correlation matrix as heat map using Pearson, Spearman correlation coefficient, or cosine similarity. Correlation analysis was performed cell wall ^{1}H–^{13}C HSQC spectra of 23 plant cell walls,[10] providing a statistical TOCSY[116] type self-correlation map useful for structural elucidation of cell wall components.

Solid-state NMR spectra were generally broader than solution NMR, and spectral overlapping hampered our interpretations from the spectra such as how many hemicelluloses are there. Therefore, peak fitting is key techniques for the later multivariate analyses. Okushita *et al.* performed PCA of bacterial cellulose using CP-MAS spectra with a series of contact times to extract information from the peak top of different forms of cellulose because each of them has different relaxation rate of T_{CH} and ^{1}H $T_{1\sigma}$.[117] Ito *et al.* investigated relationship between polysaccharide with ions in seasonal corrected *Sargassum fusiforme* (macro alga) by integration analysis using multiple data measurement including solid-state NMR of macromolecules, solution NMR for small molecules, inductively coupled plasma-optical emission spectrometry (ICP-OES) for ions, Fourier transform infrared (FT-IR) spectroscopy for molecular vibration, and thermogravimetry-differential thermal analysis (TG-DTA).[13] In that study, PCA and self-organizing maps were used as comprehensive feature extraction methods, then performed correlation analysis to explore important chemical relationships between chemicals, and finally the relationships were verified using structural equation modelling. ^{13}C CP-MAS, FI-IR spectra were pre-processed to isolate pure peaks by removing noise and separating overlapping signals using the multivariate curve resolution alternating least-squares (MCR-ALS) method before integrated analyses.[13] Wei *et al.* also performed integrative analysis of multiple measurements and employed pre-processing ^{13}C CP-MAS and FT-IR spectra using non-negative matrix factorization (NMF) method with MCR-ALS.[14]

17.2.2.4 Theoretical Calculations

Chemical shift is closely related with conformation of molecules, therefore chemical shift is a good indicator of conformation of molecules. Conformations of pyranosyl polysaccharides largely depend on few dihedral angles, ϕ, ψ (and ω). Theoretical NMR calculation of cellobiose using DFT methods agreed with empirical conformation and chemical shift correlation of pyranosyl polysaccharides including cellulose.[118] Mori *et al.* explored the conformation of amorphous cellulose using experimental and theoretical ^{13}C chemical shift calculated by *ab* initio shielding constant calculation toward structural ensemble of cellobiose generated molecular dynamics (MD) simulations.[119] MD simulations provided three stable conformers, and the comparison between experimental and theoretical ^{13}C chemical shift indicated one of them was a quasi-stable structure of amorphous cellulose. Komatsu and Kikuchi reported ^{13}C chemical shift differences between solution and solid state in cellulose were far larger than those in xyloglucan, and glucuronoarabinoxylan, indicating the differences were sensitive with structural and dynamical features of analyte that are lost on dissolution.[73]

Recently, we also demonstrated that theoretical chemical shift calculations, such as the DFT method, can be useful to verify the "annotated" metabolites as "identified" structures in metabolic mixture analysis.[120–122] However, caution should be used to ensure that the conformational dependence of theoretical chemical shifts is of higher significance than the selection of the basis set or calculation theory levels, such as B3LYP and Hartree-Fock, as well as second order Møller–Plesset perturbation (Figure 17.3).[123]

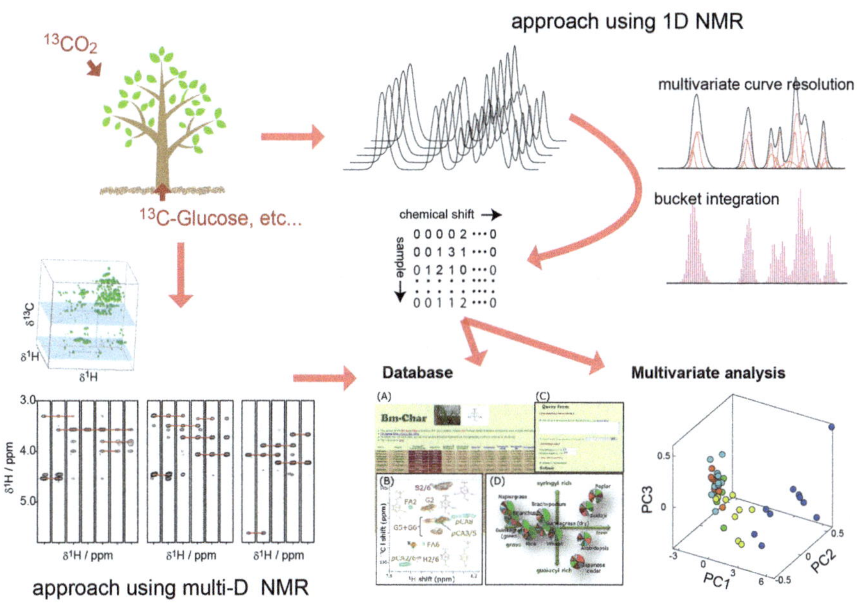

Figure 17.3 NMR data processing of cell wall polysaccharides in plants. Multidimensional NMR spectra support the resonance assignment of cell wall polysaccharides to construct a chemical shift database. ^{13}C labeling enhances multidimensional NMR analyses. Multiple NMR spectra are recorded for multiple cell wall materials of interest in the study, which are then converted into a numerical matrix by bucket integration or deconvolution techniques. Multivariate analyses are performed using that numeric matrix to explore discriminative molecules.

17.3 NMR Analyses of Polysaccharides Biosynthesized by Producers

17.3.1 NMR Approach for Land Plant Polysaccharide

Plant biomass is one of the most abundant biomaterial in nature, and they consist predominantly of polysaccharides. Polysaccharides in plants are divided into two categories, one is the storage polysaccharides such as starch, and another is structural polysaccharides such as cellulose. Vascular plants have complex cell wall systems, including primary and secondary cell walls, and they provide plants with mechanical strength, resistance to environmental stress, and a high level of tolerance toward microbial degradation.

Cell walls consist predominantly of cellulose, matrix polysaccharides, and lignin. Cellulose is crystalline microfibril of the homopolymer of β-(1,4)-linked D-glucopyranose with a high degree of polymerization. In the natural cellulose microfibril, D-glucopyranose chains locate anti-parallel. Matrix polysaccharides composed with hemicelluloses and pectin. Pectin consists can be divided several structural parts, rhamnogalacturonan I, rhamnogalacturonan II, homogalacturonan and so on. They exist in primary cell wall and less in secondary cell wall, because they are degraded accompanied with the processes of secondary cell wall formation.[124]

In 1984, Atalla and Vanderhart proposed that there are two distinct crystalline forms in the native cellulose from ^{13}C solid-state NMR spectra of cellulose in which there are sum of 2 components.[125] Nowadays, they are considered as cellulose Iα (triclinic crystal) and Iβ (monoclinic crystal), and their crystalline coordinate are reported using X-ray and neutron fiber diffractions.[126,127] Solid-state NMR has also contributed an analysis of less periodic cellulose such as para-crystal and fibril surface.[128] Wickholm *et al.* applied PCA using ^{13}C CP-MAS spectra of a series of hydrolysis and peak fitting techniques toward C4 carbon signals, resulting in the establishment of a valid model of peak fitting of various forms of cellulose including accessible fibril surfaces, cellulose at inaccessible fibril surfaces, and para-crystalline cellulose.[129] There are many other application of solid-state NMR in cellulose, however they are outlined in the other review article.[128]

In dicot plants, the major hemicellulose is xyloglucan in which D-xylopyranose constituted with the main chain of β-(1,4)-linked D-glucopyranose and various sugars such as D-galactopyranose, L-fucopyranose, L-arabinofuranose constituted to D-xylopyranose. Constitutive sugars depend on the taxonomy of the plant and they are summarized in previous report.[130] Solution NMR has contributed to analysis of the primary structure of xyloglucan, in other words which kinds of sugar constitute xyloglucan in each family of plant.[73,130]

Recently, solid-state NMR with stable isotope labeling techniques has opened a promising avenue to understand structures, interactions, and physicochemical properties of the cell wall.[80] Hong and her co-workers have studied structures, interactions, and physicochemical properties of the primary cell wall system using *Arabidopsis* and *Brachypodium*, (they are model dicot and grass plants), by relaxation and dynamics analysis, ^{13}C–^{13}C correlation multidimensional NMR, spin diffusion experiment,[78,79,82,83,85,86] resulting in successively introducing a structural model of the primary cell wall based on physicochemical properties.

17.3.2 NMR Approach for Algae Polysaccharides

There are plenty of polysaccharides in hydrosphere as structural and storage materials in algae. For example, alginic acid, fucoidan, and agarose are polysaccharides from algae for industrial use. These are called viscous polysaccharides which located in intercellular space as well as the cell wall. *Phaeophyceae* (brown algae) has laminaran which is β-1,3 glucan (with β-1,6 side chain of glucopyranose) as a storage polysaccharide. *Rhodophyceae* (red algae) has starch as a storage polysaccharide which is located in cytoplasm whereas in vascular plants it is located in the plastid.

Date *et al.* developed chemical profiling methods of macroalgae using 4 *Chlorophyceae* (green algae), 3 red algae, and 9 brown algae, by solution NMR as well as FT-IR.[131] Solution NMR using DMSO–pyridine solvent system can identify neutral polysaccharides for example agar and laminaran in *Rhodophyceae* and *Phaeophyceae*, respectively. Wei *et al.* applied integrated analysis toward 107 algal samples using multi-instrumental data including solid-state NMR, solution NMR, FT-IR, and ICP-OES to characterize them based on their chemical diversity.[14] The correlation analysis using multi-instrumental data provided chemotaxonomic clusters based on their chemical diversity and these clusters agreed with genetical linkages. The integrated analysis also indicated the relationship between alginate and ions including cadmium and arsenicum which is toxic for mammalians.

Recent genetical studies have opened avenues to understand the evolutional history of chemical diversities of polysaccharides in algae. In 2010, genome sequence analysis of *Ectocarpus siliculosus*, which was multi-cellular brown alga, was reported.[132] Alginate is unique in brown algae. Molecular phylogenetics indicated an alginate biosynthetic pathway acquired by horizontal gene transfer from an Actinobacterium.[133] In brown algae has unique carbon metabolism in which photo-assimilate is converted into D-mannitol, and stored as laminaran, β-1,3-glucan. Approaches using bioinformatics and molecular phylogenetics indicated central and storage carbon metabolism in the brown algae in which there were enzymes for mannitol, trehalose, and laminarin biosynthesis, and the mannitol pathway was acquired by lateral gene transfer from Actinobacteria.[134]

Euglena gracilis, a flagellated protist, accumulates paramylon which is crystallized and triple-helical β-1,3-glucan as storage polysaccharide like brown algae.[135] Conformational changes induced by hydration in β-1,3-glucan including paramylon and curdran, single-helical glucan, were investigated by solid-state NMR.[136] Komatsu *et al.* developed a cellular NMR approach to profile cellular biomacromolecules by solid-state NMR using cells of ^{13}C and ^{15}N-labeled *Euglena gracilis* without any purification and fractionation.[36]

Solid-state HSQC and double-CP experiments, which are ^{1}H-detected ^{1}H–^{13}C Inverse correlation spectroscopy using INEPT and cross polarization as magnetization transfer respectively, were employed. Proteins, paramylon, and lipids were detected in double-CP spectra, on the other hand, only mobile lipids were detected in solid-state HSQC spectra. This approach was then applied to *E. gracilis* cultured in high and low ammonium media to demonstrate the power of this method for profiling cellular components. Cellular NMR spectra indicated that ammonium-induced both paramylon degradation and ammonium amination. Arginine was stored as a nitrogen reserve and ammonium replaced by arginine catabolism *via* the arginine dihydrolase pathway. ^{15}N and ^{31}P cellular ssNMR indicated arginine and polyphosphate accumulation in *E. gracilis*, respectively.

17.4 Polysaccharides Degradation by Decomposers

Lignocellulose has a strong and stable structure and is resistant to biodegradation. However, recent studies have revealed how microbial communities mediate plant biomass deconstruction in animals, including detritivores, ruminants, termites, and omnivores. Primordial microbial communities that perform anaerobic digestion are considered the most ancient metabolic communities. These communities became segregated from aerobic environments by the emergence of photosynthetic organisms that initially increased oxygen levels on Earth. Many species in the microbial communities of anaerobic environments, such as soil and aquatic bottom sediments, can degrade the biomass produced by photosynthetic organisms. These species are currently used in material recycling, such as in the degradation of animal or plant remains and as nitrogen-fixing agents for agricultural use. Thus, anaerobic digestion remains essential to the biogeochemical cycling of organic and inorganic matter.

Among these, anaerobic microbial communities that are responsible for the digestion of plant saccharides into short-chain fatty acids (SCFAs) have been investigated in several systems, including the animal gut,[137,138] waste management,[139–141] and soil from paddy fields.[142] Because animals and plants have evolved to utilize the metabolic activity of microbial symbionts, major symbiont-derived substances, such as SCFAs, may be essential for the survival of the symbiont's host such as animals and plants. Moreover, SCFAs act as electron donors and play a role in reducing heavy metals such as Fe(III) and Mn(IV) under different environmental conditions. In addition, SCFA-producing bacteria such as *Lactobacillus* spp. stimulate plant germination and the growth of shoots and roots in soil ecosystems. Thus, SCFAs and the bacteria that produce them are key factors that affect the composition of soil environments.

Plant-derived saccharides are degraded by soil microbial communities and eventually return to the soil.[143] Recycling in an ecosystem occurs through symbiosis between termite gut microbial communities, specific protists and fungi, and soil microbial communities. However, little information is available on the biomass degradation process with respect to complex soil microbial communities, especially the dynamics of microbial communities and SCFA metabolism.

Metabolic profiling is a powerful tool for the analysis of complex reactions in microbial ecosystems. This technique can be used to characterize total metabolic activity and identify biomarkers. NMR is one of the most common approaches to identify molecules, and is used in a wide variety of systems.[144–146] Although ^{1}H-NMR is easy to perform and versatile, its limited spectral resolution is insufficient to tackle the overlapping signals of saccharides. The introduction of ^{13}C-nuclei to NMR analysis was a useful technological advance for the resolution of overlapping signals.[147] Advantages of the ^{13}C-labeling method also include the characterization of compounds in complex components, the determination of compound structures, and the tracking of microbial metabolic pathways using NMR or other analytical techniques.[147] We have previously successfully performed ^{13}C-labeling in plants[90,148] and microbial systems,[68] as well as metabolic profiling using two- and three-dimensional NMR techniques (Figure 17.4).[61–149]

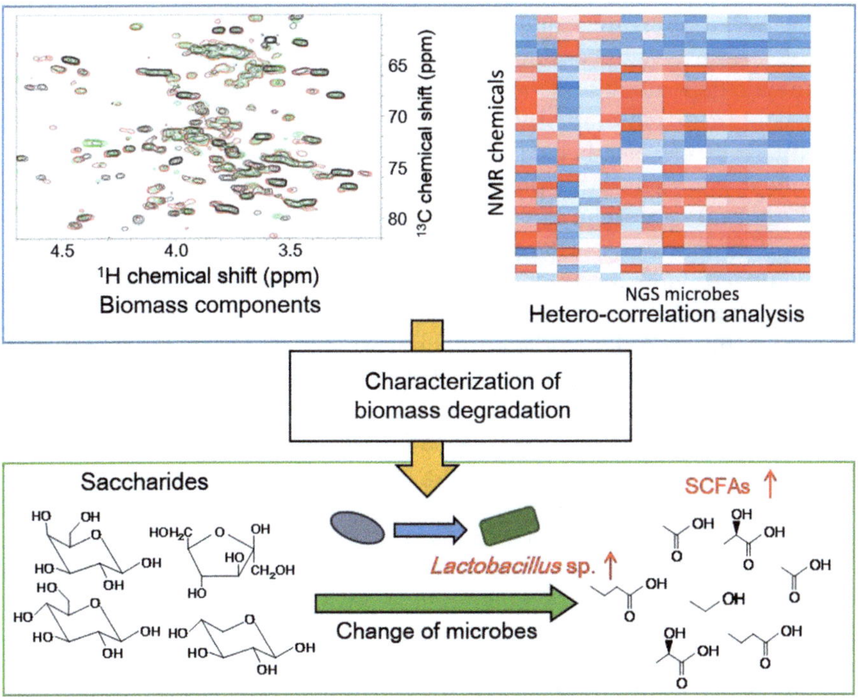

Figure 17.4 A method to reveal relationships between molecules and microbes involved in polysaccharide degradation. NMR spectra for metabolites and 16S or 18S rDNA sequences from the next generation are recorded for multiple samples of interest in the study. The correlation matrix of heterogeneous data, *i.e.*, NMR spectra and the taxonomic composition of microbes by rRNA gene, is generated. Positive correlation coefficients show that the microbe produces the molecule or, conversely, negative correlation coefficients show that the molecule is consumed or degraded by the microbe.

17.5 Polysaccharides as Major Energy Source in Animals and Human Beings

Nowadays it is more demanded than ever before for a sustainable and non-harmful energy and chemical resources from inedible plant biomass that is replaceable with fossil oils, gasses, and nuclear power.[150–153] Plant-biomass eaters, such as termites are among the most attractive living organisms for this purpose because of their considerable global impact on the decay of dead plant materials. The primary constituent of the plant biomass is cellulose, consisting of glucose linked by β-1,4-glucosidic bonds. Despite that most animals cannot use this polysaccharide, termites are thought to be capable of hydrolyzing 74–99% of cellulose in woody materials. Therefore, this efficient cellulolytic system in termites has attracted many researchers and has been extensively studied for these decades.

An emerging technique for clarifying complex metabolic processes such as cellulose digestion involves the generation of comprehensive metabolite profiles *via* 2-dimensional (2D) nuclear magnetic resonance (NMR) analyses.[90,154] We have recently demonstrated the utility of this technique in animal systems by feeding ^{13}C-glucose and amino acids to *Bombyx mori* and monitoring the time-dependent metabolic dynamics during insect morphogenesis.[70] In the context of symbiotic interactions, we have also successfully demonstrated using ^{13}C-labeled compounds the critical role of acetate production by resident gut bacteria in protection from enteropathogenic infection in mice.[137] Advantages of the use of ^{13}C-labeling methodologies in combination with multidimensional NMR analyses include the separation of crowded signals from metabolite mixtures, and the opportunity to perform quantitative, multivariate analyses of time-series of acquired data at an atomic level.[66–68] We have performed the first comprehensive metabolomic investigation of cellulose digestion in a termite.[52] We fed ^{13}C-labeled cellulose to the termite *Hodotermopsis sjostedti* and detected a total of 256 ^{13}C-signals using 2D heteronuclear single quantum coherence (HSQC)–NMR. ^{13}C-labeled cellulose hydrolysates were detected in the foregut soon after feeding, providing the first direct *in situ* evidence for the activity of endogenous cellulases. We obtained the first evidence that hindgut bacteria contribute to cellulolysis, *via* phosphorolysis of cellodextrins or cellobiose to glucose-1-phosphate. Temporal dynamics of most non-essential amino acids in the midgut and hindgut were synchronized with cellulose catabolism. In contrast, biosynthesis of essential amino acids was independent of cellulose hydrolysis, indicating that termites obtain essential amino acids *via* proctodeal trophallaxis (Figure 17.5).

The co-evolution of mammalian-microbial symbiosis accompanies extensive interactive modulations of metabolism and physiology, facilitated by the crosstalk between the host and symbiotic community (Table 17.2). Microbial symbionts often provide traits that their hosts have not evolved on their own,[155] and may synthesize essential amino acids and vitamins or process otherwise indigestible components in the diet, such as plant polysaccharides.

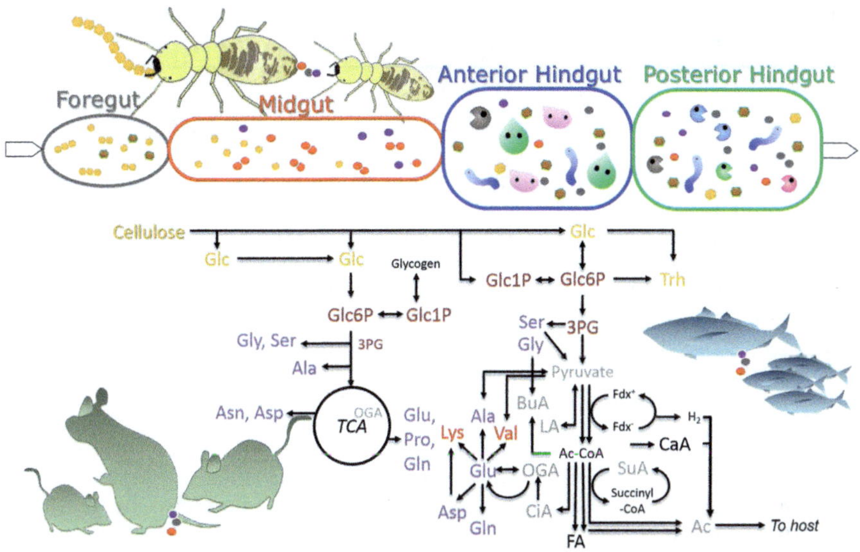

Figure 17.5 Isotope-assisted NMR metabolic profiling revealed metabolic system-specialized cellulose degradation in termites. Uniformly ^{13}C labeled cellulose was fed to termites, and the products of catabolism were traced in ^{1}H–^{13}C HSQC spectra. The results confirmed that the secretion of endogenous cellulases by the host may be significant to the digestive process and indicated a major contribution by the hindgut bacteria to the phosphorolysis of cellodextrins or cellobiose.

The composition of intestinal microbial communities is highly variable, and can be significantly affected by alterations in diet. Interactive modulations of individuals with variations in their microbial symbionts are likely to affect human health and disease.

The interactive modulations affecting human health are considerably engaged by beneficial microbial symbionts such as *Lactobacilli* and *Bifidobacteria*, which are currently the most marketed probiotic bacteria worldwide. The beneficial microbial symbionts are responsible for preventing infection, enhancing the immune system, and providing increased nutritional value to food.[156] The growth and activity of these beneficial microbial symbionts are enhanced by functional foods such as fructo-oligosaccharide (FOS) and galacto-oligosaccharides in the human gastrointestinal tract.[157] Therefore, evaluation of the effects of functional food on dietary interactive modulations of the host and the beneficial microbial symbionts is important for human health. The evaluations at an initial stage of validation are usually performed using animal experiments such as mouse and rat. However, the animal experiments are time-consuming and have ethical issues, thus simple and rapid *in vitro* method without animal experiments is aspired as a better way to evaluate the effects of functional food.[158]

Table 17.2 List of published reports on metabolic profiling related to health and medical concerns.

Targets	Technical remarks
Heart disease	Early proposal for diagnosis use[175]
Schistosoma infection	Biomarker identification[176]
Fatty liver phenotype	Gut microbial contribution[177]
Personalized drug treatment	Prediction of personalized metabolic fate[159]
Diabetic rat model	Metabolome-QTL for rat[178]
Silent phenotype	Intact worm analysis by HR-MAS[179]
Human diet and blood pressure	Large-scale exploratory study of human[160]
Human gut microbiome	Microbiome *vs.* host correlation[98]
Trypanosoma infection	Related with gut microbial metabolism[180]
Human drug metabolism	Related with gut microbial metabolism[181]
Regulation of behavior and development	Differential COSY/TOCSY analysis[182]
Red blood cell	^{13}C metabolic flux analysis in blood cells[183]
Intact tissue	General protocols using HR-MAS[184]
Protection of 157 infection by microbial acetate	Multiple-omics approach for biomarker finding[137]
Detoxification capacity of urine	Metabolome-QTL for human[185]
Drug screening	High-throughput cell-line analysis[186]
Cardiometabolic disorder	Metabolome-QTL for human[187]
Gut inflammatory suppression by microbial butyrate	Multiple-omics approach for biomarker finding[138]
Diet/health for aged peoples	Diet, gut microbiome and metabolome analysis[188]
Malawian kwashiorkor	Gut microbiome–metabolome correlation[189]
Menopause and aging	High-throughput epidemiology analysis[190]
Microbial acetate reduces appetite	^{13}C metabolic tracing in mice[191]
Shigella infection	^{13}C metabolic tracing in human epithelial cells[192]
Cancer cell metabolism	Real-time monitoring of cellular metabolism[193]

In contrast, the NMR-based metabonomic approach is widely used for investigating the metabolic characterization in mammals and their intestinal symbiotic systems.[51,159–161] Particularly in human studies, metabolic characterizations from non-invasive samples, such as urine, saliva, and feces are well known to have a large quantity of vital and beneficial information for human health and disease.

We have described an advanced analytical approach for evaluating the metabolic, mineral, and microbiota fluctuations of all fecal, urinary, and salivary samples from humans together and synchronously related with variations in dietary intakes under unlimited and uncontrolled diets.[50] We used non-invasive human sampling in combination with an NMR-based metabonomic approach, inductively coupled plasma-optical emission spectrometry (ICP-OES)-based ionomic approach, and microbiota profiling, to focus on an advanced analytical approach for evaluating metabolic responses associated with input–output variations and for characterizing metabolic profiles modulated by microbiota in the symbiotic systems of humans using this advanced data-driven approach.

References

1. E. A. Schuur, A. D. McGuire, C. Schadel, G. Grosse, J. W. Harden, D. J. Hayes, G. Hugelius, C. D. Koven, P. Kuhry, D. M. Lawrence, S. M. Natali, D. Olefeldt, V. E. Romanovsky, K. Schaefer, M. R. Turetsky, C. C. Treat and J. E. Vonk, *Nature*, 2015, **520**, 171.
2. E. K. Pikitch, *Science*, 2012, **338**, 474.
3. D. Butler, *Nature*, 2008, **453**, 8.
4. J. Jacquet, D. Pauly, D. Ainley, S. Holt, P. Dayton and J. Jackson, *Nature*, 2010, **467**, 28.
5. M. P. Squicciarini, A. Guariso and J. Swinnen, *Nature*, 2013, **501**, 492.
6. K. S. Peh, *Nature*, 2011, **473**, 284.
7. A. J. Ragauskas, C. K. Williams, B. H. Davison, G. Britovsek, J. Cairney, C. A. Eckert, W. J. Frederick Jr, J. P. Hallett, D. J. Leak, C. L. Liotta, J. R. Mielenz, R. Murphy, R. Templer and T. Tschaplinski, *Science*, 2006, **311**, 484.
8. C. Somerville, H. Youngs, C. Taylor, S. C. Davis and S. P. Long, *Science*, 2010, **329**, 790.
9. E. M. Rubin, *Nature*, 2008, **454**, 841.
10. Y. Ogata, E. Chikayama, Y. Morioka, R. C. Everroad, A. Shino, A. Matsushima, H. Haruna, S. Moriya, T. Toyoda and J. Kikuchi, *PLoS One*, 2012, **7**, e30263.
11. D. M. Harris, K. Corbin, T. Wang, R. Gutierrez, A. L. Bertolo, C. Petti, D. M. Smilgies, J. M. Estevez, D. Bonetta, B. R. Urbanowicz, D. W. Ehrhardt, C. R. Somerville, J. K. Rose, M. Hong and S. Debolt, *Proc. Natl. Acad. Sci. U. S. A.*, 2012, **109**, 4098.
12. R. Vanholme, I. Cesarino, K. Rataj, Y. Xiao, L. Sundin, G. Goeminne, H. Kim, J. Cross, K. Morreel, P. Araujo, L. Welsh, J. Haustraete, C. McClellan, B. Vanholme, J. Ralph, G. G. Simpson, C. Halpin and W. Boerjan, *Science*, 2013, **341**, 1103.
13. K. Ito, K. Sakata, Y. Date and J. Kikuchi, *Anal. Chem.*, 2014, **86**, 1098.
14. F. Wei, K. Ito, K. Sakata, Y. Date and J. Kikuchi, *Anal. Chem.*, 2015, **87**, 2819.
15. R. C. Everroad, S. Yoshida, Y. Tsuboi, Y. Date, J. Kikuchi and S. Moriya, *J. Vis. Exp.*, 2012, e3163.
16. D. M. Ogawa, S. Moriya, Y. Tsuboi, Y. Date, A. R. Prieto-da-Silva, G. Radis-Baptista, T. Yamane and J. Kikuchi, *PLoS One*, 2014, **9**, e110723.
17. K. L. Poulson-Ellestad, C. M. Jones, J. Roy, M. R. Viant, F. M. Fernandez, J. Kubanek and B. L. Nunn, *Proc. Natl. Acad. Sci. U. S. A.*, 2014, **111**, 9009.
18. Y. Gelinas, J. A. Baldock and J. I. Hedges, *Science*, 2001, **294**, 145.
19. T. Asakura, Y. Date and J. Kikuchi, *Anal. Chem.*, 2014, **86**, 5425.
20. J. I. Hedges, J. A. Baldock, Y. Gelinas, C. Lee, M. Peterson and S. G. Wakeham, *Nature*, 2001, **409**, 801.
21. E. Kaiser, A. J. Simpson, K. J. Dria, B. Sulzberger and P. G. Hatcher, *Environ. Sci. Technol.*, 2003, **37**, 2929.
22. J. Peuravuori, *Environ. Sci. Technol.*, 2005, **39**, 5541.

23. X. Feng, A. J. Simpson and M. J. Simpson, *Environ. Sci. Technol.*, 2006, **40**, 3260.
24. B. Lam and A. J. Simpson, *Anal. Chem.*, 2006, **78**, 8194.
25. B. Lam and A. J. Simpson, *Analyst*, 2008, **133**, 263.
26. H. A. Abdulla, E. C. Minor and P. G. Hatcher, *Environ. Sci. Technol.*, 2010, **44**, 8044.
27. G. C. Woods, M. J. Simpson, P. J. Koerner, A. Napoli and A. J. Simpson, *Environ. Sci. Technol.*, 2011, **45**, 3880.
28. J. Mao, X. Kong, K. Schmidt-Rohr, J. J. Pignatello and E. M. Perdue, *Environ. Sci. Technol.*, 2012, **46**, 5806.
29. B. G. Pautler, G. C. Woods, A. Dubnick, A. J. Simpson, M. J. Sharp, S. J. Fitzsimons and M. J. Simpson, *Environ. Sci. Technol.*, 2012, **46**, 3753.
30. A. Nebbioso and A. Piccolo, *Anal. Bioanal. Chem.*, 2013, **405**, 109.
31. S. J. Goldberg, G. I. Ball, B. C. Allen, S. G. Schladow, A. J. Simpson, H. Masoom, R. Soong, H. D. Graven and L. I. Aluwihare, *Nat. Commun.*, 2015, **6**, 6347.
32. K. Ikeya and A. Watanabe, *Anal. Bioanal. Chem.*, 2016, **408**, 651.
33. N. Mladenov, Y. Zheng, B. Simone, T. M. Bilinski, D. M. McKnight, D. Nemergut, K. A. Radloff, M. M. Rahman and K. M. Ahmed, *Environ. Sci. Technol.*, 2015, **49**, 10815.
34. C. Groger, K. Lutz and E. Brunner, *Prog. Nucl. Magn. Reson. Spectrosc.*, 2009, **54**, 54.
35. A. Rivas-Ubach, J. Sardans, M. Perez-Trujillo, M. Estiarte and J. Penuelas, *Proc. Natl. Acad. Sci. U. S. A.*, 2012, **109**, 4181.
36. T. Komatsu, T. Kobayashi, M. Hatanaka and J. Kikuchi, *Environ. Sci. Technol.*, 2015, **49**, 7056.
37. I. Hirao, M. Kimoto, S. Yamakage, M. Ishikawa, J. Kikuchi and S. Yokoyama, *Bioorg. Med. Chem. Lett.*, 2002, **12**, 1391.
38. K. Nakatani, S. Sando, H. Kumasawa, J. Kikuchi and I. Saito, *J. Am. Chem. Soc.*, 2001, **123**, 12650.
39. P. H. Willoughby, M. J. Jansma and T. R. Hoye, *Nat. Protoc.*, 2014, **9**, 643.
40. R. R. Ernst, *Angew. Chem., Int. Ed.*, 2010, **49**, 8310.
41. K. Wuthrich, *Angew. Chem., Int. Ed.*, 2003, **42**, 3340.
42. M. Ikura, L. E. Kay and A. Bax, *Biochemistry*, 1990, **29**, 4659.
43. A. Bax and S. Grzesiek, *Acc. Chem. Res.*, 1993, **26**, 131.
44. D. S. Wishart, B. D. Sykes and F. M. Richards, *Biochemistry*, 1992, **31**, 1647.
45. M. P. Williamson, J. Kikuchi and T. Asakura, *J. Mol. Biol.*, 1995, **247**, 541.
46. J. Fiaux, E. B. Bertelsen, A. L. Horwich and K. Wuthrich, *Nature*, 2002, **418**, 207.
47. M. P. Latham, A. Sekhar and L. E. Kay, *Proc. Natl. Acad. Sci. U. S. A.*, 2014, **111**, 5532.
48. J. K. Nicholson, E. Holmes, J. M. Kinross, A. W. Darzi, Z. Takats and J. C. Lindon, *Nature*, 2012, **491**, 384.
49. J. K. Nicholson and J. C. Lindon, *Nature*, 2008, **455**, 1054.
50. T. Misawa, Y. Date and J. Kikuchi, *J. Proteome Res.*, 2015, **14**, 1526.

51. T. Asakura, K. Sakata, S. Yoshida, Y. Date and J. Kikuchi, *PeerJ*, 2014, **2**, e550.
52. G. Tokuda, Y. Tsuboi, K. Kihara, S. Saitou, S. Moriya, N. Lo and J. Kikuchi, *Proc. Biol. Sci.*, 2014, **281**, 20140990.
53. Y. Shiokawa, T. Misawa, Y. Date and J. Kikuchi, *Anal. Chem.*, 2016, **88**, 2714.
54. J. Kikuchi and T. Asakura, *J. Biochem. Biophys. Methods*, 1999, **38**, 203.
55. J. Kikuchi, T. Asakura, K. Hasuda, T. Ito, K. Ohwaku, H. Araki and M. P. Williamson, *J. Biochem. Biophys. Methods*, 2000, **42**, 35.
56. K. Fukushima, J. Kikuchi, S. Koshiba, T. Kigawa, Y. Kuroda and S. Yokoyama, *J. Mol. Biol.*, 2002, **321**, 317.
57. E. Sakata, Y. Yamaguchi, E. Kurimoto, J. Kikuchi, S. Yokoyama, S. Yamada, H. Kawahara, H. Yokosawa, N. Hattori, Y. Mizuno, K. Tanaka and K. Kato, *EMBO Rep.*, 2003, **4**, 301.
58. O. Serve, Y. Kamiya, A. Maeno, M. Nakano, C. Murakami, H. Sasakawa, Y. Yamaguchi, T. Harada, E. Kurimoto, M. Yagi-Utsumi, T. Iguchi, K. Inaba, J. Kikuchi, O. Asami, T. Kajino, T. Oka, M. Nakasako and K. Kato, *J. Mol. Biol.*, 2010, **396**, 361.
59. K. Mochida, T. Furuta, K. Ebana, K. Shinozaki and J. Kikuchi, *BMC Genomics*, 2009, **10**, 568.
60. M. Okamoto, Y. Tsuboi, E. Chikayama, J. Kikuchi and T. Hirayama, *Plant Biotechnol.*, 2009, **26**, 551.
61. C. Tian, E. Chikayama, Y. Tsuboi, T. Kuromori, K. Shinozaki, J. Kikuchi and T. Hirayama, *J. Biol. Chem.*, 2007, **282**, 18532.
62. M. Watanabe, Y. Ohta, S. Licang, N. Motoyama and J. Kikuchi, *Food Chem.*, 2015, **169**, 387.
63. H. Morita, H. Toh, S. Fukuda, H. Horikawa, K. Oshima, T. Suzuki, M. Murakami, S. Hisamatsu, Y. Kato, T. Takizawa, H. Fukuoka, T. Yoshimura, K. Itoh, D. J. O'Sullivan, L. L. McKay, H. Ohno, J. Kikuchi, T. Masaoka and M. Hattori, *DNA Res.*, 2008, **15**, 151.
64. H. Motegi, Y. Tsuboi, A. Saga, T. Kagami, M. Inoue, H. Toki, O. Minowa, T. Noda and J. Kikuchi, *Sci. Rep.*, 2015, **5**, 15710.
65. C. Y. Hsieh, T. Osaka, E. Moriyama, Y. Date, J. Kikuchi and S. Tsuneda, *Physiol. Rep.*, 2015, **3**, e12327.
66. S. Fukuda, Y. Nakanishi, E. Chikayama, H. Ohno, T. Hino and J. Kikuchi, *PLoS One*, 2009, **4**, e4893.
67. Y. Date, Y. Nakanishi, S. Fukuda, T. Kato, S. Tsuneda, H. Ohno and J. Kikuchi, *J. Biosci. Bioeng.*, 2010, **110**, 87.
68. Y. Nakanishi, S. Fukuda, E. Chikayama, Y. Kimura, H. Ohno and J. Kikuchi, *J. Proteome Res.*, 2011, **10**, 824.
69. E. Chikayama, Y. Sekiyama, Y. Okamoto, Y. Nakanishi, Y. Tsuboi, K. Akiyama, K. Saito, K. Shinozaki and J. Kikuchi, *Anal. Chem.*, 2010, **82**, 1653.
70. E. Chikayama, M. Suto, T. Nishihara, K. Shinozaki and J. Kikuchi, *PLoS One*, 2008, **3**, e3805.
71. Y. Sekiyama, E. Chikayama and J. Kikuchi, *Anal. Chem.*, 2010, **82**, 1643.

72. Z. Yang, R. Nakabayashi, Y. Okazaki, T. Mori, S. Takamatsu, S. Kitanaka, J. Kikuchi and K. Saito, *Metabolomics*, 2014, **10**, 543.
73. T. Komatsu and J. Kikuchi, *Anal. Chem.*, 2013, **85**, 8857.
74. T. Mori, Y. Tsuboi, N. Ishida, N. Nishikubo, T. Demura and J. Kikuchi, *Sci. Rep.*, 2015, **5**, 11848.
75. M. Foston, R. Samuel and A. J. Ragauskas, *Analyst*, 2012, **137**, 3904.
76. L. Lam, R. Soong, A. Sutrisno, R. de Visser, M. J. Simpson, H. L. Wheeler, M. Campbell, W. E. Maas, M. Fey, A. Gorissen, H. Hutchins, B. Andrew, J. Struppe, S. Krishnamurthy, R. Kumar, M. Monette, H. J. Stronks, A. Hume and A. J. Simpson, *J. Agric. Food Chem.*, 2014, **62**, 107.
77. T. Komatsu and J. Kikuchi, *J. Phys. Chem. Lett.*, 2013, **4**, 2279.
78. M. Dick-Perez, T. Wang, A. Salazar, O. A. Zabotina and M. Hong, *Magn. Reson. Chem.*, 2012, **50**, 539.
79. M. Dick-Perez, Y. Zhang, J. Hayes, A. Salazar, O. A. Zabotina and M. Hong, *Biochemistry*, 2011, **50**, 989.
80. T. Wang and M. Hong, *J. Exp. Bot.*, 2016, **67**, 503.
81. T. Wang, Y. B. Park, M. A. Caporini, M. Rosay, L. Zhong, D. J. Cosgrove and M. Hong, *Proc. Natl. Acad. Sci. U. S. A.*, 2013, **110**, 16444.
82. T. Wang, Y. B. Park, D. J. Cosgrove and M. Hong, *Plant Physiol.*, 2015, **168**, 871.
83. T. Wang, A. Salazar, O. A. Zabotina and M. Hong, *Biochemistry*, 2014, **53**, 2840.
84. T. Wang, J. Williams, K. Schmidt-Rohr and M. Hong, *J. Biomol. NMR*, 2015, **61**, 97.
85. T. Wang, O. Zabotina and M. Hong, *Biochemistry*, 2012, **51**, 9846.
86. P. B. White, T. Wang, Y. B. Park, D. J. Cosgrove and M. Hong, *J. Am. Chem. Soc.*, 2014, **136**, 10399.
87. M. Bardet, L. Emsley and M. Vincendon, *Solid State Nucl. Magn. Reson.*, 1997, **8**, 25.
88. R. Dupree, T. J. Simmons, J. C. Mortimer, D. Patel, D. Iuga, S. P. Brown and P. Dupree, *Biochemistry*, 2015, **54**, 2335.
89. M. Foston, R. Katahira, E. Gjersing, M. F. Davis and A. J. Ragauskas, *J. Agric. Food Chem.*, 2012, **60**, 1419.
90. J. Kikuchi and T. Hirayama, *Methods Mol. Biol.*, 2007, **358**, 273.
91. Y. Sekiyama and J. Kikuchi, *Phytochemistry*, 2007, **68**, 2320.
92. T. Komatsu, R. Ohishi, A. Shino, K. Akashi and J. Kikuchi, *Metabolites*, 2014, **4**, 1018.
93. A. Gorissen, P. J. Kuikman, J. H. van Ginkel, H. van de Beek and A. G. Jansen, *Plant Soil*, 1996, **179**, 81.
94. F. A. Pettolino, C. Walsh, G. B. Fincher and A. Bacic, *Nat. Protoc.*, 2012, **7**, 1590.
95. R. R. Selvendran, *Phytochemistry*, 1975, **14**, 1011.
96. S. D. Mansfield, H. Kim, F. Lu and J. Ralph, *Nat. Protoc.*, 2012, **7**, 1579.
97. H. Kim, J. Ralph and T. Akiyama, *BioEnergy Res.*, 2008, **1**, 56.
98. D. J. Yelle, J. Ralph and C. R. Frihart, *Magn. Reson. Chem.*, 2008, **46**, 508.
99. H. Kim and J. Ralph, *Org. Biomol. Chem.*, 2010, **8**, 576.
100. J. S. Moulthrop, R. P. Swatloski, G. Moyna and R. D. Rogers, *Chem. Commun.*, 2005, 1557.

101. K. Cheng, H. Sorek, H. Zimmermann, D. E. Wemmer and M. Pauly, *Anal. Chem.*, 2013, **85**, 3213.
102. R. Samuel, M. Foston, N. Jaing, S. Cao, L. Allison, M. Studer, C. Wyman and A. J. Ragauskas, *Fuel*, 2011, **90**, 2836.
103. K. Okushita, E. Chikayama and J. Kikuchi, *Biomacromolecules*, 2012, **13**, 1323.
104. S. Heikkinen, M. M. Toikka, P. T. Karhunen and I. A. Kilpeläinen, *J. Am. Chem. Soc.*, 2003, **125**, 4362.
105. K. Hu, W. M. Westler and J. L. Markley, *J. Am. Chem. Soc.*, 2011, **133**, 1662.
106. M. C. Jarvis and M. C. McCann, *Plant Physiol. Biochem.*, 2000, **38**, 1.
107. H.-R. Tang, Y.-L. Wang and P. S. Belton, *Solid State Nucl. Magn. Reson.*, 2000, **15**, 239.
108. C. M. G. C. Renard and M. C. Jarvis, *Plant Physiol.*, 1999, **119**, 1315.
109. N. M. Szeverenyi, M. J. Sullivan and G. E. Maciel, *J. Magn. Reson.*, 1982, **47**, 462.
110. K. Takegoshi, S. Nakamura and T. Terao, *Chem. Phys. Lett.*, 2001, **344**, 631.
111. M. Weingarth, D. E. Demco, G. Bodenhausen and P. Tekely, *Chem. Phys. Lett.*, 2009, **469**, 342.
112. B. Hu, O. Lafon, J. Trébosc, Q. Chen and J.-P. Amoureux, *J. Magn. Reson.*, 2011, **212**, 320.
113. B. Hu, J. Trébosc, O. Lafon, Q. Chen, Y. Masuda, K. Takegoshi and J.-P. Amoureux, *ChemPhysChem*, 2012, **13**, 3585.
114. P. Tekely, J. Brondeau, K. Elbayed, A. Retournard and D. Canet, *J. Magn. Reson. (1969)*, 1988, **80**, 509.
115. J. Kikuchi, Y. Ogata and K. Shinozaki, *J. Ecosyst. Ecography*, 2011, **S2**, 001.
116. O. Cloarec, M.-E. Dumas, A. Craig, R. H. Barton, J. Trygg, J. Hudson, C. Blancher, D. Gauguier, J. C. Lindon, E. Holmes and J. Nicholson, *Anal. Chem.*, 2005, 77, 1282.
117. K. Okushita, T. Komatsu, E. Chikayama and J. Kikuchi, *Polym. J.*, 2012, **44**, 895.
118. S. Suzuki, F. Horii and H. Kurosu, *J. Mol. Struct.*, 2009, **921**, 219.
119. T. Mori, E. Chikayama, Y. Tsuboi, N. Ishida, N. Shisa, Y. Noritake, S. Moriya and J. Kikuchi, *Carbohydr. Polym.*, 2012, **90**, 1197.
120. T. Misawa, T. Komatsu, Y. Date and J. Kikuchi, *Chem. Commun. (Camb.)*, 2016, **52**, 2964.
121. K. Ito, Y. Tsutsumi, Y. Date and J. Kikuchi, *ACS Chem. Biol.*, 2016, **11**, 1030.
122. T. Komatsu, R. Ohishi, A. Shino and J. Kikuchi, *Angew. Chem., Int. Ed.*, 2016, **44**, 6000.
123. E. Chikayama, Y. Shimbo, K. Komatsu and J. Kikuchi, *J. Phys. Chem. B.*, 2016, **120**, 3479.
124. Y.-H. Xu, R. Liu, L. Yan, Z.-Q. Liu, S.-C. Jiang, Y.-Y. Shen, X.-F. Wang and D.-P. Zhang, *J. Exp. Bot.*, 2012, **63**, 1095.
125. R. H. Atalla and D. L. Vanderhart, *Science*, 1984, **223**, 283.
126. Y. Nishiyama, P. Langan and H. Chanzy, *J. Am. Chem. Soc.*, 2002, **124**, 9074.

127. Y. Nishiyama, J. Sugiyama, H. Chanzy and P. Langan, *J. Am. Chem. Soc.*, 2003, **125**, 14300.
128. M. Foston, *Curr. Opin. Biotechnol.*, 2014, **27**, 176.
129. K. Wickholm, P. T. Larsson and T. Iversen, *Carbohydr. Res.*, 1998, **312**, 123.
130. M. Hoffman, Z. Jia, M. J. Peña, M. Cash, A. Harper, A. R. Blackburn Ii, A. Darvill and W. S. York, *Carbohydr. Res.*, 2005, **340**, 1826.
131. Y. Date, K. Sakata and J. Kikuchi, *Polym. J.*, 2012, **44**, 888.
132. J. M. Cock, L. Sterck, P. Rouze, D. Scornet, A. E. Allen, G. Amoutzias, V. Anthouard, F. Artiguenave, J. M. Aury, J. H. Badger, B. Beszteri, K. Billiau, E. Bonnet, J. H. Bothwell, C. Bowler, C. Boyen, C. Brownlee, C. J. Carrano, B. Charrier, G. Y. Cho, S. M. Coelho, J. Collen, E. Corre, C. Da Silva, L. Delage, N. Delaroque, S. M. Dittami, S. Doulbeau, M. Elias, G. Farnham, C. M. Gachon, B. Gschloessl, S. Heesch, K. Jabbari, C. Jubin, H. Kawai, K. Kimura, B. Kloareg, F. C. Kupper, D. Lang, A. Le Bail, C. Leblanc, P. Lerouge, M. Lohr, P. J. Lopez, C. Martens, F. Maumus, G. Michel, D. Miranda-Saavedra, J. Morales, H. Moreau, T. Motomura, C. Nagasato, C. A. Napoli, D. R. Nelson, P. Nyvall-Collen, A. F. Peters, C. Pommier, P. Potin, J. Poulain, H. Quesneville, B. Read, S. A. Rensing, A. Ritter, S. Rousvoal, M. Samanta, G. Samson, D. C. Schroeder, B. Segurens, M. Strittmatter, T. Tonon, J. W. Tregear, K. Valentin, P. von Dassow, T. Yamagishi, Y. Van de Peer and P. Wincker, *Nature*, 2010, **465**, 617.
133. G. Michel, T. Tonon, D. Scornet, J. M. Cock and B. Kloareg, *New Phytol.*, 2010, **188**, 82.
134. G. Michel, T. Tonon, D. Scornet, J. M. Cock and B. Kloareg, *New Phytol.*, 2010, **188**, 67.
135. A. K. Monfils, R. E. Triemer and E. F. Bellairs, *Phycologia*, 2011, **50**, 156.
136. H. Saito, M. Yokoi and Y. Yoshioka, *Macromolecules*, 1989, **22**, 3892.
137. S. Fukuda, H. Toh, K. Hase, K. Oshima, Y. Nakanishi, K. Yoshimura, T. Tobe, J. M. Clarke, D. L. Topping, T. Suzuki, T. D. Taylor, K. Itoh, J. Kikuchi, H. Morita, M. Hattori and H. Ohno, *Nature*, 2011, **469**, 543.
138. Y. Furusawa, Y. Obata, S. Fukuda, T. A. Endo, G. Nakato, D. Takahashi, Y. Nakanishi, C. Uetake, K. Kato, T. Kato, M. Takahashi, N. N. Fukuda, S. Murakami, E. Miyauchi, S. Hino, K. Atarashi, S. Onawa, Y. Fujimura, T. Lockett, J. M. Clarke, D. L. Topping, M. Tomita, S. Hori, O. Ohara, T. Morita, H. Koseki, J. Kikuchi, K. Honda, K. Hase and H. Ohno, *Nature*, 2013, **504**, 446.
139. Y. Date, T. Iikura, A. Yamazawa, S. Moriya and J. Kikuchi, *J. Proteome Res.*, 2012, **11**, 5602.
140. A. Yamazawa, T. Iikura, Y. Morioka, A. Shino, Y. Ogata, Y. Date and J. Kikuchi, *Metabolites*, 2013, **4**, 36.
141. A. Yamazawa, T. Iikura, A. Shino, Y. Date and J. Kikuchi, *Molecules*, 2013, **18**, 9021.
142. T. Ogura, Y. Date and J. Kikuchi, *PLoS One*, 2013, **8**, e66919.
143. T. Ogura, Y. Date, Y. Tsuboi and J. Kikuchi, *ACS Chem. Biol.*, 2015, **10**, 1908.

144. D. I. Schlipalius, N. Valmas, A. G. Tuck, R. Jagadeesan, L. Ma, R. Kaur, A. Goldinger, C. Anderson, J. Kuang, S. Zuryn, Y. S. Mau, Q. Cheng, P. J. Collins, M. K. Nayak, H. J. Schirra, M. A. Hilliard and P. R. Ebert, *Science*, 2012, **338**, 807.
145. H. K. Kim, Y. H. Choi and R. Verpoorte, *Nat. Protoc.*, 2010, **5**, 536.
146. C. A. Sellick, R. Hansen, G. M. Stephens, R. Goodacre and A. J. Dickson, *Nat. Protoc.*, 2011, **6**, 1241.
147. K. Bingol, F. Zhang, L. Bruschweiler-Li and R. Bruschweiler, *Anal. Chem.*, 2013, **85**, 6414.
148. J. Kikuchi, K. Shinozaki and T. Hirayama, *Plant Cell Physiol.*, 2004, **45**, 1099.
149. Y. Sekiyama, E. Chikayama and J. Kikuchi, *Anal. Chem.*, 2011, **83**, 719.
150. T. Watanabe, A. Shino, K. Akashi and J. Kikuchi, *Plant Biotechnol.*, 2012, **29**, 163.
151. T. Watanabe, A. Shino, K. Akashi and J. Kikuchi, *PLoS One*, 2014, **9**, e106893.
152. K. Sasaki, M. Okamoto, T. Shirai, Y. Tsuge, H. Teramura, D. Sasaki, H. Kawaguchi, T. Hasunuma, C. Ogino, F. Matsuda, J. Kikuchi and A. Kondo, *Biotechnol. Biofuels*, 2015, **8**, 88.
153. H. Teramura, K. Sasaki, T. Oshima, S. Aikawa, F. Matsuda, M. Okamoto, T. Shirai, H. Kawaguchi, C. Ogino, M. Yamasaki, J. Kikuchi and A. Kondo, *PLoS One*, 2015, **10**, e0128417.
154. A. Choe, T. Chuman, S. H. von Reuss, A. T. Dossey, J. J. Yim, R. Ajredini, A. A. Kolawa, F. Kaplan, H. T. Alborn, P. E. Teal, F. C. Schroeder, P. W. Sternberg and A. S. Edison, *Proc. Natl. Acad. Sci. U. S. A.*, 2012, **109**, 20949.
155. J. K. Nicholson, E. Holmes, J. Kinross, R. Burcelin, G. Gibson, W. Jia and S. Pettersson, *Science*, 2012, **336**, 1262.
156. H. Sugahara, T. Odamaki, S. Fukuda, T. Kato, J. Z. Xiao, F. Abe, J. Kikuchi and H. Ohno, *Sci. Rep.*, 2015, **5**, 13548.
157. T. Kato, S. Fukuda, A. Fujiwara, W. Suda, M. Hattori, J. Kikuchi and H. Ohno, *DNA Res.*, 2014, **21**, 469.
158. Y. Date, Y. Nakanishi, S. Fukuda, Y. Nuijima, T. Kato, M. Umehara, H. Ohno and J. Kikuchi, *Food Chem.*, 2014, **152**, 251.
159. T. A. Clayton, J. C. Lindon, O. Cloarec, H. Antti, C. Charuel, G. Hanton, J. P. Provost, J. L. Le Net, D. Baker, R. J. Walley, J. R. Everett and J. K. Nicholson, *Nature*, 2006, **440**, 1073.
160. E. Holmes, R. L. Loo, J. Stamler, M. Bictash, I. K. Yap, Q. Chan, T. Ebbels, M. De Iorio, I. J. Brown, K. A. Veselkov, M. L. Daviglus, H. Kesteloot, H. Ueshima, L. Zhao, J. K. Nicholson and P. Elliott, *Nature*, 2008, **453**, 396.
161. S. Yoshida, Y. Date, M. Akama and J. Kikuchi, *Sci. Rep.*, 2014, **4**, 7005.
162. P. E. Jansson, R. Stenutz and G. Widmalm, *Carbohydr. Res.*, 2006, **341**, 1003.
163. M. A. Rojas-Macias, J. Stahle, T. Lutteke and G. Widmalm, *Glycobiology*, 2015, **25**, 341.

164. Q. Cui, I. A. Lewis, A. D. Hegeman, M. E. Anderson, J. Li, C. F. Schulte, W. M. Westler, H. R. Eghbalnia, M. R. Sussman and J. L. Markley, *Nat. Biotechnol.*, 2008, **26**, 162.
165. D. S. Wishart, T. Jewison, A. C. Guo, M. Wilson, C. Knox, Y. Liu, Y. Djoumbou, R. Mandal, F. Aziat, E. Dong, S. Bouatra, I. Sinelnikov, D. Arndt, J. Xia, P. Liu, F. Yallou, T. Bjorndahl, R. Perez-Pineiro, R. Eisner, F. Allen, V. Neveu, R. Greiner and A. Scalbert, *Nucleic Acids Res.*, 2013, **41**, D801.
166. S. L. Robinette, F. Zhang, L. Bruschweiler-Li and R. Bruschweiler, *Anal. Chem.*, 2008, **80**, 3606.
167. K. Bingol, F. Zhang, L. Bruschweiler-Li and R. Bruschweiler, *Anal. Chem.*, 2012, **84**, 9395.
168. J. Kikuchi, Y. Tsuboi, K. Komatsu, M. Gomi, E. Chikayama and Y. Date, *Anal. Chem.*, 2016, **88**, 659.
169. J. Xia and D. S. Wishart, *Nat. Protoc.*, 2011, **6**, 743.
170. F. Savorani, G. Tomasi and S. B. Engelsen, *J. Magn. Reson.*, 2010, **202**, 190.
171. B. Worley and R. Powers, *ACS Chem. Biol.*, 2014, **9**, 1138.
172. S. Cacciatore, C. Luchinat and L. Tenori, *Proc. Natl. Acad. Sci. U. S. A.*, 2014, **111**, 5117.
173. J. Hao, M. Liebeke, W. Astle, M. De Iorio, J. G. Bundy and T. M. Ebbels, *Nat. Protoc.*, 2014, **9**, 1416.
174. I. A. Lewis, S. C. Schommer and J. L. Markley, *Magn. Reson. Chem.*, 2009, **47**(suppl. 1), S123.
175. J. T. Brindle, H. Antti, E. Holmes, G. Tranter, J. K. Nicholson, H. W. Bethell, S. Clarke, P. M. Schofield, E. McKilligin, D. E. Mosedale and D. J. Grainger, *Nat. Med.*, 2002, **8**, 1439.
176. Y. Wang, E. Holmes, J. K. Nicholson, O. Cloarec, J. Chollet, M. Tanner, B. H. Singer and J. Utzinger, *Proc. Natl. Acad. Sci. U. S. A.*, 2004, **101**, 12676.
177. M. E. Dumas, R. H. Barton, A. Toye, O. Cloarec, C. Blancher, A. Rothwell, J. Fearnside, R. Tatoud, V. Blanc, J. C. Lindon, S. C. Mitchell, E. Holmes, M. I. McCarthy, J. Scott, D. Gauguier and J. K. Nicholson, *Proc. Natl. Acad. Sci. U. S. A.*, 2006, **103**, 12511.
178. M. E. Dumas, S. P. Wilder, M. T. Bihoreau, R. H. Barton, J. F. Fearnside, K. Argoud, L. D'Amato, R. H. Wallis, C. Blancher, H. C. Keun, D. Baunsgaard, J. Scott, U. G. Sidelmann, J. K. Nicholson and D. Gauguier, *Nat. Genet.*, 2007, **39**, 666.
179. B. J. Blaise, J. Giacomotto, B. Elena, M. E. Dumas, P. Toulhoat, L. Segalat and L. Emsley, *Proc. Natl. Acad. Sci. U. S. A.*, 2007, **104**, 19808.
180. Y. Wang, J. Utzinger, J. Saric, J. V. Li, J. Burckhardt, S. Dirnhofer, J. K. Nicholson, B. H. Singer, R. Brun and E. Holmes, *Proc. Natl. Acad. Sci. U. S. A.*, 2008, **105**, 6127.
181. T. A. Clayton, D. Baker, J. C. Lindon, J. R. Everett and J. K. Nicholson, *Proc. Natl. Acad. Sci. U. S. A.*, 2009, **106**, 14728.
182. C. Pungaliya, J. Srinivasan, B. W. Fox, R. U. Malik, A. H. Ludewig, P. W. Sternberg and F. C. Schroeder, *Proc. Natl. Acad. Sci. U. S. A.*, 2009, **106**, 7708.

183. I. A. Lewis, M. E. Campanella, J. L. Markley and P. S. Low, *Proc. Natl. Acad. Sci. U. S. A.*, 2009, **106**, 18515.
184. O. Beckonert, M. Coen, H. C. Keun, Y. Wang, T. M. Ebbels, E. Holmes, J. C. Lindon and J. K. Nicholson, *Nat. Protoc.*, 2010, **5**, 1019.
185. K. Suhre, H. Wallaschofski, J. Raffler, N. Friedrich, R. Haring, K. Michael, C. Wasner, A. Krebs, F. Kronenberg, D. Chang, C. Meisinger, H. E. Wichmann, W. Hoffmann, H. Volzke, U. Volker, A. Teumer, R. Biffar, T. Kocher, S. B. Felix, T. Illig, H. K. Kroemer, C. Gieger, W. Romisch-Margl and M. Nauck, *Nat. Genet.*, 2011, **43**, 565.
186. S. Tiziani, Y. Kang, J. S. Choi, W. Roberts and G. Paternostro, *Nat. Commun.*, 2011, **2**, 545.
187. J. Kettunen, T. Tukiainen, A. P. Sarin, A. Ortega-Alonso, E. Tikkanen, L. P. Lyytikainen, A. J. Kangas, P. Soininen, P. Wurtz, K. Silander, D. M. Dick, R. J. Rose, M. J. Savolainen, J. Viikari, M. Kahonen, T. Lehtimaki, K. H. Pietilainen, M. Inouye, M. I. McCarthy, A. Jula, J. Eriksson, O. T. Raitakari, V. Salomaa, J. Kaprio, M. R. Jarvelin, L. Peltonen, M. Perola, N. B. Freimer, M. Ala-Korpela, A. Palotie and S. Ripatti, *Nat. Genet.*, 2012, **44**, 269.
188. M. J. Claesson, I. B. Jeffery, S. Conde, S. E. Power, E. M. O'Connor, S. Cusack, H. M. B. Harris, M. Coakley, B. Lakshminarayanan, O. O'Sullivan, G. F. Fitzgerald, J. Deane, M. O'Connor, N. Harnedy, K. O'Connor, D. O'Mahony, D. van Sinderen, M. Wallace, L. Brennan, C. Stanton, J. R. Marchesi, A. P. Fitzgerald, F. Shanahan, C. Hill, R. P. Ross and P. W. O'Toole, *Nature*, 2012, **488**, 178.
189. M. I. Smith, T. Yatsunenko, M. J. Manary, I. Trehan, R. Mkakosya, J. Cheng, A. L. Kau, S. S. Rich, P. Concannon, J. C. Mychaleckyj, J. Liu, E. Houpt, J. V. Li, E. Holmes, J. Nicholson, D. Knights, L. K. Ursell, R. Knight and J. I. Gordon, *Science*, 2013, **339**, 548.
190. K. Auro, A. Joensuu, K. Fischer, J. Kettunen, P. Salo, H. Mattsson, M. Niironen, J. Kaprio, J. G. Eriksson, T. Lehtimaki, O. Raitakari, A. Jula, A. Tiitinen, M. Jauhiainen, P. Soininen, A. J. Kangas, M. Kahonen, A. S. Havulinna, M. Ala-Korpela, V. Salomaa, A. Metspalu and M. Perola, *Nat. Commun.*, 2014, **5**, 4708.
191. G. Frost, M. L. Sleeth, M. Sahuri-Arisoylu, B. Lizarbe, S. Cerdan, L. Brody, J. Anastasovska, S. Ghourab, M. Hankir, S. Zhang, D. Carling, J. R. Swann, G. Gibson, A. Viardot, D. Morrison, E. Louise Thomas and J. D. Bell, *Nat. Commun.*, 2014, **5**, 3611.
192. D. Kentner, G. Martano, M. Callon, P. Chiquet, M. Brodmann, O. Burton, A. Wahlander, P. Nanni, N. Delmotte, J. Grossmann, J. Limenitakis, R. Schlapbach, P. Kiefer, J. A. Vorholt, S. Hiller and D. Bumann, *Proc. Natl. Acad. Sci. U. S. A.*, 2014, **111**, 9929.
193. H. Wen, Y. J. An, W. J. Xu, K. W. Kang and S. Park, *Angew. Chem., Int. Ed.*, 2015, **54**, 5374.

Subject Index